Petrology and Sedimentology: Key Concepts and Applications

Petrology and Sedimentology: Key Concepts and Applications

Editor: Seymor White

Callisto Reference
www.callistoreference.com

Callisto Reference,
118-35 Queens Blvd., Suite 400,
Forest Hills, NY 11375, USA

Visit us on the World Wide Web at:
www.callistoreference.com

ISBN: 978-1-63239-954-0 (Hardback)

The publisher's policy is to use permanent paper from mills that operate a sustainable forestry policy. Furthermore, the publisher ensures that the text paper and cover boards used have met acceptable environmental accreditation standards.

Printed in China.

Cataloging-in-Publication Data

Petrology and sedimentology : key concepts and applications / edited by Seymor White.
p. cm.
Includes bibliographical references and index.
ISBN 978-1-63239-954-0
1. Petrology. 2. Sedimentology. 3. Physical geology. I. White, Seymor.
QE431.2 .P48 2018
552--dc23

Table of Contents

Preface

Over the recent decade, advancements and applications have progressed exponentially. This has led to the increased interest in this field and projects are being conducted to enhance knowledge. The main objective of this book is to present some of the critical challenges and provide insights into possible solutions. This book will answer the varied questions that arise in the field and also provide an increased scope for furthering studies.

Petrology studies rocks and the conditions leading to their formation. Together with sedimentology, which studies layers of compressed materials such as clay, sand and silt; these disciplines help in the interpretation of the composition of the Earth's surface. Research in this field helps to identify and extract various minerals found in the Earth's crust and mantle. Tectonic and seismic activity occurring inside the Earth can also be studied by researchers in these fields. This book attempts to understand the multiple branches that fall under the discipline of petrology and sedimentology and how such concepts have practical applications. Students, researchers, experts and all associated with these fields will benefit alike from this book.

I hope that this book, with its visionary approach, will be a valuable addition and will promote interest among readers. Each of the authors has provided their extraordinary competence in their specific fields by providing different perspectives as they come from diverse nations and regions. I thank them for their contributions.

Editor

1

Trace Metal Inventories and Lead Isotopic Composition Chronicle a Forest Fire's Remobilization of Industrial Contaminants Deposited in the Angeles National Forest

Kingsley O. Odigie*, A. Russell Flegal

WIGS Laboratory, Environmental Toxicology, University of California Santa Cruz, Santa Cruz, California, United States of America

Abstract

The amounts of labile trace metals: [Co] (3 to 11 $\mu g\ g^{-1}$), [Cu] (15 to 69 $\mu g\ g^{-1}$), [Ni] (6 to 15 $\mu g\ g^{-1}$), [Pb] (7 to 42 $\mu g\ g^{-1}$), and [Zn] (65 to 500 $\mu g\ g^{-1}$) in ash collected from the 2012 Williams Fire in Los Angeles, California attest to the role of fires in remobilizing industrial metals deposited in forests. These remobilized trace metals may be dispersed by winds, increasing human exposures, and they may be deposited in water bodies, increasing exposures in aquatic ecosystems. Correlations between the concentrations of these trace metals, normalized to Fe, in ash from the fire suggest that Co, Cu, and Ni in most of those samples were predominantly from natural sources, whereas Pb and Zn were enriched in some ash samples. The predominantly anthropogenic source of excess Pb in the ash was further demonstrated by its isotopic ratios ($^{208}Pb/^{207}Pb$: $^{206}Pb/^{207}Pb$) that fell between those of natural Pb and leaded gasoline sold in California during the previous century. These analyses substantiate current human and environmental health concerns with the pyrogenic remobilization of toxic metals, which are compounded by projections of increases in the intensity and frequency of wildfires associated with climate change.

Editor: Ben Bond-Lamberty, DOE Pacific Northwest National Laboratory, United States of America

Funding: This work was supported by Flegal Research Fund and the Department of Energy (DOE) Office of Science Graduate Fellowship Program. All opinions expressed in this paper are the authors' and do not necessarily reflect the policies and views of DOE Office of Science. The funders had no role in study design, data collection and analysis, decision to publish, or preparation of the manuscript.

Competing Interests: The authors have declared that no competing interests exist.

* Email: kodigie@ucsc.edu

Introduction

Historic Emissions of Trace Metals

Extensive anthropogenic emissions of environmentally persistent contaminants have substantially altered the natural biogeochemical cycles of some trace metals in the biosphere over the past century [1–3]. For example, ~90% and 75% of atmospheric lead (Pb) in California and the world, respectively, during the second half of the previous century were attributed to anthropogenic sources, predominantly the combustion of leaded gasoline [1].

These industrial emissions extensively contaminated aerosols and sediments in Southern California, where enormous amounts of leaded gasoline were combusted during the previous century. For example, ~8.9% (412,300 metric tons) of the estimated 4,639,000 metric tons of Pb additives used in gasoline in the US between 1950 and 1982 was in California [4]. Subsequently, Pb concentrations in surface soil in Los Angeles Metropolitan area increased from 16±0.5 µg/g (mean ± SD) between 1919 and 1933 to 79±23 µg/g between 1967 and 1970 [5]; and the flux of Pb to sediments in the coastal San Pedro Basin, which abuts the Los Angeles Metropolitan area, increased by ~7-fold, compared to the natural rate, following the introduction of alkyl-lead gasoline additives in Southern California [6]. Similar temporal increases in fluxes of Pb in age-dated sediments that were attributed to the use of leaded gasoline have been documented in adjacent areas both north and south of Southern California, albeit at lower levels [7,8]. Substantial temporal increases in the fluxes of other trace metals, e.g., zinc (Zn), associated with anthropogenic inputs have also been measured in Southern California [9].

However, efforts to reduce industrial metal contamination, especially Pb contamination, over the past half century have been quite successful in much of the world, notably the US and specifically Los Angeles. For example, the burden of emissions from the past combustion of leaded gasoline on humans is shown by the relatively high average blood Pb levels (PbB) of children (ages 1–5 years) in the US, which declined from 15 µg/dL in 1976–1980 to 3.6 µg/dL in 1988–1991, during the phase-out and subsequent elimination of leaded gasoline in the US, and then to the current level of 1.3 µg/dL [10,11]. Despite these marked improvements, the current average PbB level in the US is still about 80-fold greater than the estimated natural PbB (0.016 µg/dL) level in humans [12] and health impairments are now being reported at Pb levels that were previously considered harmless [11,13]. Similarly, sublethal toxicities of other trace metals (e.g., Cu and Zn) continue to be documented at increasingly lower levels [14].

Moreover, the legacy of historic anthropogenic Pb emissions is evident in terrestrial and aquatic sediments in California and elsewhere [5,7]. Ten years after Pb was phased out from gasoline in California, concentrations of Pb (77±19 µg/g) in soil in Pasadena, California were still 6-fold greater than the natural (baseline) Pb level (~12.5 µg/g) of crustal rock [5]. This persistence of industrial Pb deposits was summarized by Steding

Figure 1. Photograph of the 2012 Williams Fire in Angeles National Forest, California showing the relatively remote and mountainous terrain where the fire occurred. Photo by Freddie W. Duncan, US Forest Service.

et al. [15] and Dunlap et al. [1], who predicted that it could take decades to centuries to flush leaded gasoline depositions from surface sediments in California's Central Valley to San Francisco Bay. Similarly, Semlali et al. [16], based on a study in France, estimated that it would take at least 700 years to reduce the amount of anthropogenic Pb accumulated in soil by 50%.

Remobilization of Trace Metals by Forest Fires

Wildfires remobilize legacy trace metal contaminants deposited in grasslands and forests. Metals are sequestered in sediment, soil organic matter, and vegetation in forests, where they are relatively immobile [17–19]. However, the combustion of vegetation and soil organic matter releases these trace metals in more labile forms [17,20]. In addition, forest fires increase erosion rates by as much as 100-fold and facilitate the rapid transport of remobilized trace metals to draining water bodies, raising contamination levels in aquatic ecosystems [21,22].

While there have been several investigations of the remobilization of mercury by forest fires [23,24], there have been few similar studies on the remobilization of Pb and other trace metals (e.g., Co and Zn). The latter include a recent study in the relatively unpopulated and relatively pristine Santa Barbara area north of Los Angeles [25] and a similar study in Australia [26]. In contrast to those studies, this study was conducted in the Angeles National Forest, which is adjacent to the Los Angeles Metropolitan area, which has a population of ~17 million [27], has many large sources of industrial contaminants [28–30], and ranks as one of the nation's most contaminated metropolitan areas [31]. Moreover, "car is king" in Los Angeles, where leaded gasoline emissions were the dominant source of atmospheric Pb during the previous century and resuspension of that legacy pollution continues to be the principal component of atmospheric Pb today, as documented in mass balance study of atmospheric Pb contamination in the Los Angeles Air Basin [5]. Therefore, we hypothesized that forest fires within that basin would remobilize relatively large amounts of industrial Pb and other industrial metals, and that remobilization would add to the already large amounts of legacy gasoline Pb deposits that now constitute the major source of atmospheric Pb in Los Angeles.

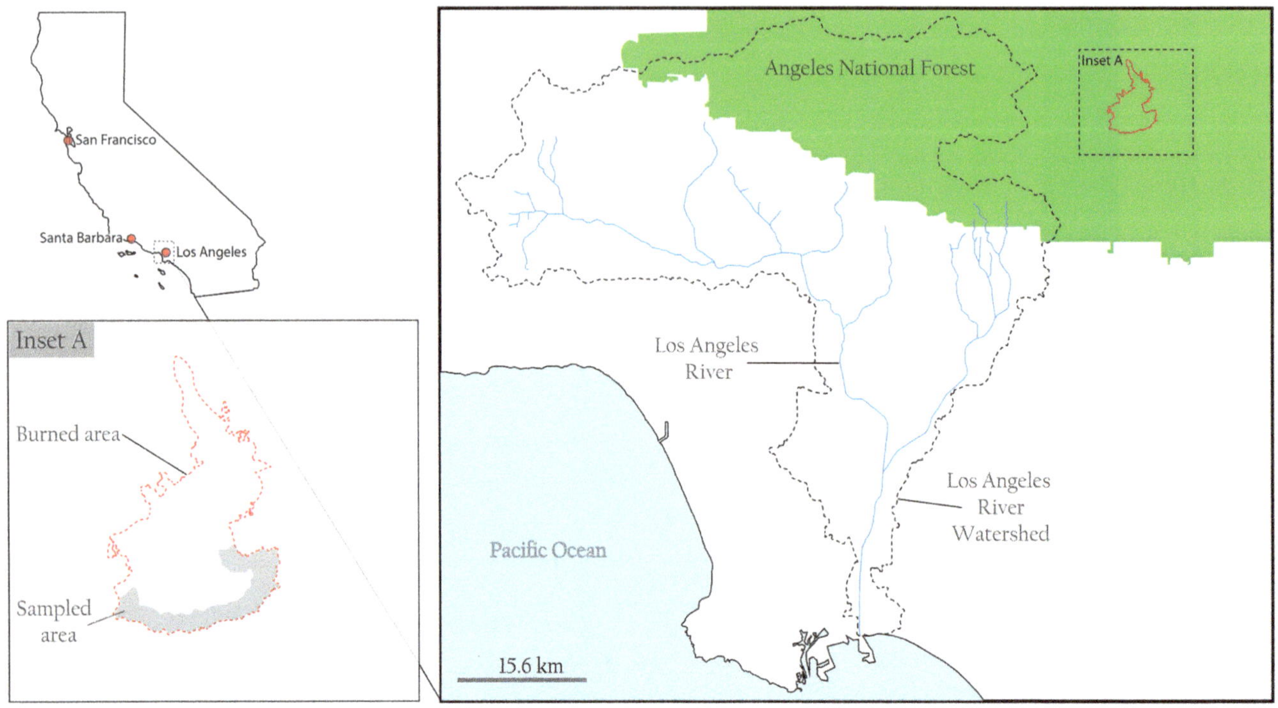

Figure 2. Map showing approximate samples collection sites within the boundary of the 2012 Williams Fire in Angeles National Forest, California.

Table 1. Detection limits of analytes.

	Detection Limit (3 x SD, Standard Deviation of 3 blanks)						
	(ng/g)					(μg/g)	
Element	[Co]	[Cu]	[Ni]	[Pb]	[Zn]	[Al]	[Fe]
Detection Limit	0.002	0.010	0.002	0.008	0.148	0.015	0.035

Table 2. Digestion recoveries (concentrations) for elements in standard reference materials (SRMs).

	% Recovery (Mean ± SD, n = 3)						
SRM	Co	Cu	Ni	Pb	Zn	Al	Fe
NIST SRM 1547 (Peach Leaves)	82.6±3.0	89.3±0.7	74.5±2.8	104±0.8	98.7±1.2	65.7±5.2	86.7±4.7
NIST SRM 2709 (San Joaquin Soil)	52.0±0.4	49.1±0.6	49.1±0.4	77.5±0.7	53.2±0.7	55.6±1.1	93.7±0.5
NIST SRM 1633b (Coal Fly Ash)	43.7±0.7	38.6±0.5	42.6±0.6	62.5±0.3	41.2±0.0	25.6±0.8	76.7±1.9

Table 3. Analytical (instrumental) precision.

SRM	% RSD (n = 5)					% RSD (n = 4)	
	[Co]	[Cu]	[Ni]	[Pb]	[Zn]	[Al]	[Fe]
SRM 1640a	0.9	1.1	0.9	0.5	1.6	–	–
Consistency Standard A	2.3	1.6	2.3	0.2	0.9	0.3	0.5
Consistency Standard B	–	–	–	–	–	0.3	0.6

Materials and Methods

Samples Collection Sites

The Angeles National Forest, which covers ~ 283,280 hectares (700,000 acres), abuts the Los Angeles Metropolitan Area [32]. The September 2012 Williams Fire scorched about 1,696 hectares (4,192 acres) of land in Angeles National Forest from approximately September 2nd to 13th (Fig. 1). The fire occurred in a very steep (~ 30–80% slope) area of the forest that is covered with medium to heavy brush. The fuels, mostly chaparral, brush, and mixed conifers, were approximately 15 to 20 years old [33]. Fourteen ash and 6 soil (from unburned patches) samples were collected from different accessible locations within the burned perimeter of the fire site (Figs. 1 and 2) using established trace metal clean techniques [25]. The samples were then transported in plastic containers to the University of California, Santa Cruz for processing. The collection location (GPS coordinates) for each sample is presented in Table S1 in File S1. No specific permissions were required to access and collect samples from the fire site, and the sampling did not involve interactions with endangered or protected species.

Analytical Methods

Trace metal or ultra-pure (2x sub-boiling quartz distilled) grade reagents and high purity (18.2 MΩ cm) water (Milli-Q) were used for cleaning labware and processing samples. Fourteen ash and 6 soil samples, 3 procedural blanks, and triplicates of 3 National Institute of Standards and Technology (NIST) Standard Reference Materials (SRMs), SRM 1547 (peach leaves), SRM 2709 (San Joaquin soil), and SRM 1633b (coal fly ash), were processed concurrently in a Class-100 HEPA-filtered trace metal clean laboratory. Three ash samples (CWA-1, CWA-8, and CWA-14) and 2 soil samples (CWS-1 and CWS-6) were processed in triplicate.

A concentrated aqua regia (HNO_3: HCl, 1:3 v/v) digestion was used to extract the acid-leachable (labile) trace (Co, Cu, Ni, Pb, and Zn) and major (Al and Fe) metals in the samples and reference materials [34,35] using established methods [8]. These metals were selected to represent a suite of inorganic contaminants emitted from anthropogenic sources at different – from relatively low (e.g., Ni) to high (e.g., Pb) – levels in Los Angeles [29,36,37]. Briefly, approximately 0.3 to 0.5 g of ash, soil or reference material was transferred to a clean Teflon digestion vial, dried overnight at 65°C, refluxed in hot (~130°C) aqua regia for about 6 hours, dried, weighed and redissolved in 10 mL of 1M ultrapure HNO_3.

The digests were analyzed for trace metal (Co, Cu, Ni, Pb, and Zn) concentrations with a Finnigan ELEMENT 2 high resolution inductively coupled plasma – mass spectrometer (HR ICP-MS), and for Al and Fe concentrations with a PerkinElmer Optima 4300 DV inductively coupled plasma – optical emission spectrometer (ICP-OES). The concentrations of Fe were used to compute the enrichment factors of trace metals in the samples. The samples were then analyzed for stable lead isotopic composition after optimizing their concentrations with a Finnigan ELEMENT 2 HR ICP-MS. Internal standards were used to correct for instrumental drifts during analysis, and NIST SRM 981 (common lead) was used as standard to correct for instrumental fractionation of lead isotopic composition. The detection limit (3 x standard deviation of procedural blanks), reproducibility, and recoveries of each analyte are presented in Tables 1–4. The recoveries of metals in the NIST SRM 1547 (peach leaves) are assumed to be most representative of the metals

Table 4. Analytical (instrumental) precision for lead isotopic compositions.

	% RSD (n = 5)		
Sample	$^{208}Pb/^{207}Pb$	$^{206}Pb/^{207}Pb$	$^{207}Pb/^{204}Pb$
CWA-2	0.07	0.04	0.09

in the ash samples since they were derived predominantly from vegetation and soil organic matter.

Results and Discussion

Metal Concentrations

The "acid-leachable" concentrations of Al, Co, Cu, Fe, Ni, Pb, and Zn, and lead isotopic ratios ($^{208}Pb/^{207}Pb$, $^{206}Pb/^{207}Pb$, and $^{207}Pb/^{204}Pb$) of the ash and soil samples are listed in Table 5. The enrichment factors of the trace metals (Co, Cu, Ni, Pb, and Zn) normalized to Fe are listed in Table S2 in File S1. These concentrations and enrichment factors are considered conservative because they may not include metals volatilized during the fire and those contained in refractory aluminosilicate lattices, which require dissolution with concentrated HF [38,39].

Concentrations of Co (3.1 to 11 $\mu g\ g^{-1}$), Cu (15 to 69 $\mu g\ g^{-1}$), Ni (5.9 to 15 $\mu g\ g^{-1}$), Pb (7.4 to 42 $\mu g\ g^{-1}$), and Zn (65 to 500 $\mu g\ g^{-1}$) in the ash varied substantially. The ranges of these concentrations are consistent with those measured in ash collected from the 2009 Jesusita Forest Fire site in Santa Barbara, California (Table 5). The concentrations are also comparable to concentrations of these metals in ash sourced primarily from the experimental combustion of ponderosa pines in Montana [40]. The relatively high Zn levels potentially originated from anthropogenic sources, which have typically included fossil fuel combustion exhaust, galvanized parts and railings, break lining wear, and tire wear [41,42].

Variations in the amounts of these remobilized metals presumably reflect variations in baseline concentrations of the metals, anthropogenic input, and levels of volatilization of organic matter during the fire [43,44]. The concentrations of Co (6.3 to 13 $\mu g\ g^{-1}$), Cu (17 to 32 $\mu g\ g^{-1}$), Ni (6.0 to 21 $\mu g\ g^{-1}$), Pb (8.6 to 140 $\mu g\ g^{-1}$), and Zn (32 to 61 $\mu g\ g^{-1}$) in the soil samples are also comparable to previously reported values of soils in Southern California [45–47]. The concentrations of some metals (e.g., Co and Ni) in the soils approximate their baseline concentrations in Southern California, but the >2-fold increase in Pb concentrations (e.g., 140 $\mu g\ g^{-1}$) indicated that the Angeles National Forest site had received substantial Pb inputs from anthropogenic sources.

Enrichment Factors

The amounts of anthropogenic trace metals in the ash are indicated by the enrichment factors of the metals normalized to Fe: Co (0.9 to 2.5), Cu (0.9 to 6.2), Ni (0.3 to 1.0), Pb (1.9 to 29), and Zn (0.9 to 19) (Table S2 in File S1).

$$EF = \frac{\frac{[X]Sample}{[Fe]Sample}}{\frac{[X]Baseline}{[Fe]Baseline}}$$

We calculated the enrichment factor of each metal by normalizing the concentrations of the metal ($[X]_{Sample}$) and iron ($[Fe]_{Sample}$) in the sample to the baseline concentrations of each metal ($[X]_{Baseline}$) and iron ($[Fe]_{Baseline}$) of ^{210}Pb-dated preindustrial sediment deposited in San Pedro Basin, which drains parts of the Los Angeles Metropolitan Area. Normalization of trace metals to Fe is considered more appropriate than normalization to Al for samples that were not treated with HF because of the recalcitrance of aluminosilicates as indicated by the difference in the recoveries of these metals in the NIST SRMs (Table 2). Using the enrichment classification proposed by Sutherland [48], the ratios indicate that: Co was not measurably enriched (EF <2) in 10 samples and moderately enriched (EF = 2–5) in 4 samples; Cu was not measurably enriched in 5 samples, moderately enriched in 7 samples, and significantly enriched (EF = 5–20) in 2 samples; Ni was not measurably enriched in all 14 ash samples; Pb was not measurably enriched in 1 sample, moderately enriched in 7 samples, significantly enriched in 5 samples, and very highly enriched (EF = 20–40) in 1 sample; and Zn was not measurably enriched in 4 samples, moderately enriched in 5 samples, and significantly enriched in 5 samples. These industrial enrichments were further indicated by the simple linear correlations between the concentrations of the metals and the concentrations of Fe: [Co] ($r = 0.576$, $p = 0.031$, $n = 14$), [Cu] ($r = 0.058$, $p = 0.844$, $n = 14$), [Ni] ($r = 0.612$, $p = 0.020$, $n = 14$), [Pb] ($r = -0.158$, $p = 0.590$, $n = 14$), and [Zn] ($r = -0.221$, $p = 0.448$, $n = 14$) in the ash. Metals that are primarily from natural origin, e.g., [Al] and [Fe] ($r = 0.909$, $p = 0.000007$, $n = 14$), which are relatively abundant and not easy to contaminate, correlate better than those enriched by anthropogenic inputs [49]. The addition of anthropogenic metal, e.g., Pb, to a system often changes its relative abundance to normalizing major elements as indicated by the correlation of [Pb] and [Fe] above. These varying enrichments are also generally consistent with those of these metals in ash collected from the 2009 Jesusita Fire in Santa Barbara, California [e.g., 25].

Lead Isotopic Compositions

The apparent anthropogenic enrichment of Pb in some of the samples is further demonstrated in Figure 3, which is a plot of the lead isotopic ratios ($^{208}Pb/^{207}Pb$: $^{206}Pb/^{207}Pb$) of the ash and soil samples. The plot also includes the lead isotopic compositions of pre-industrial and contaminated sediment deposited in the San Pedro Basin, which drains parts of the Los Angeles Metropolitan area as previously mentioned, and those of leaded gasoline sold in California during the second half of the last century.

The plot indicates that the Pb in the ash is mainly from anthropogenic sources, predominantly the past combustion of leaded gasoline. The plot also indicates that the portion of anthropogenic Pb from the leaded gasoline used between 1964 to 1979 was substantially larger than that from the 1980s leaded gasoline in the ash, which contrasts the results from the Santa Barbara study that indicate that the predominant source of anthropogenic Pb was the combustion of the 1980s leaded gasoline [25]. The predominantly alkyl-lead in the samples is further corroborated by the isotopic compositions of Pb in most (4) of the soil samples (Fig. 3). Two soil samples (CWS-4 and CWS-5) have distinct Pb isotopic compositions compared to the other samples and are assumed to be representative of mixtures between baseline and anthropogenic Pb or variants of baseline Pb isotopic compositions.

Table 5. Concentrations (µg g^{-1}) of trace metals: Co, Cu, Ni, Pb, and Zn; concentrations (mg g^{-1}) of Al and Fe; and lead isotopic composition of ash (CWA) and soil (CWS) samples collected from the 2012 Williams Fire in Angeles National Forest in Southern California.

Sample ID	Concentrations [mean ± *(standard deviation, SD)*]							Lead Isotopic Ratios [mean ± *(SD)*]		
	[Co] (µg g^{-1})	[Cu] (µg g^{-1})	[Ni] (µg g^{-1})	[Pb] (µg g^{-1})	[Zn] (µg g^{-1})	[Al] (mg g^{-1})	[Fe] (mg g^{-1})	$^{208}Pb/^{207}Pb$	$^{206}Pb/^{207}Pb$	$^{207}Pb/^{204}Pb$
CWA-1*	3.90	15.4	6.77	16.0	216	13.8	10.9	2.454	1.189	15.673
	(0.05)	*(0.24)*	*(0.25)*	*(0.04)*	*(2.92)*	*(0.55)*	*(0.29)*	*(0.001)*	*(0.001)*	*(0.004)*
CWA-2	7.54	68.5	10.6	24.0	128	18.4	25.0	2.461	1.185	15.635
CWA-3	3.59	41.3	9.21	42.2	83.0	5.79	6.94	2.438	1.178	15.656
CWA-4	6.76	30.8	10.6	14.8	85.0	18.7	24.4	2.457	1.190	15.633
CWA-5	3.81	38.6	9.72	20.8	65.3	7.86	10.8	2.454	1.192	15.659
CWA-6	8.75	56.0	12.5	27.2	414	13.6	16.2	2.463	1.213	15.684
CWA-7	5.17	65.5	14.3	38.5	203	14.3	13.8	2.451	1.195	15.711
CWA-8*	10.9	58.2	13.3	8.95	374	13.0	17.0	2.457	1.193	15.660
	(0.74)	*(3.24)*	*(0.52)*	*(1.02)*	*(15.7)*	*(0.36)*	*(0.55)*	*(0.001)*	*(0.002)*	*(0.011)*
CWA-9	4.32	17.3	10.5	9.68	96.5	14.3	16.2	2.471	1.186	15.662
CWA-10	7.47	26.3	14.6	20.0	73.6	24.4	31.5	2.458	1.187	15.672
CWA-11	3.13	17.0	5.86	8.23	74.2	7.42	8.71	2.450	1.189	15.733
CWA-12	6.05	20.5	11.2	17.3	65.5	18.6	18.1	2.462	1.183	15.669
CWA-13	6.30	56.8	9.17	11.5	503	7.85	10.7	2.472	1.188	15.664
CWA-14*	6.66	43.3	12.9	7.38	66.6	20.7	19.2	2.450	1.189	15.686
	(0.27)	*(0.23)*	*(0.15)*	*(0.25)*	*(0.68)*	*(0.53)*	*(0.39)*	*(0.001)*	*(0.001)*	*(0.017)*
CWS-1*	7.31	20.0	6.04	20.7	44.5	29.2	52.4	2.459	1.181	15.635
	(0.10)	*(3.89)*	*(0.14)*	*(0.76)*	*(0.46)*	*(2.11)*	*(1.84)*	*(0.002)*	*(0.000)*	*(0.011)*
CWS-2	8.65	19.6	21.1	143	61.2	35.1	46.8	2.461	1.204	15.647
CWS-3	12.5	32.1	16.8	20.3	41.5	32.1	43.5	2.455	1.178	15.614
CWS-4	6.26	25.9	11.8	8.62	32.0	20.7	28.1	2.506	1.196	15.684
CWS-5	8.82	17.0	15.0	10.9	37.3	34.9	40.1	2.491	1.185	15.643
CWS-6*	9.98	20.0	13.1	16.4	44.2	33.5	46.4	2.443	1.160	15.581
	(0.67)	*(0.89)*	*(1.21)*	*(1.81)*	*(0.65)*	*(5.00)*	*(1.37)*	*(0.001)*	*(0.006)*	*(0.024)*
Jesusita Fire ash, Santa Barbara, California (Data Summary, n = 30)										
Mean, [x̄]	4.46	–	22.7	18.8‡	151	16.7‡	13.9‡	2.466‡	1.205‡	15.635‡
Range (Min	1.95	–	8.88	4.29‡	41.6	5.70‡	5.00‡	2.436‡	1.161‡	15.563‡
to	–	–	–	–	–	–	–	–	–	–
Max)	7.23	–	49.1	51.0‡	433	36.0‡	31.1‡	2.481‡	1.218‡	15.716‡

Samples marked with (*) were processed in triplicates and are reported as mean ± (standard deviation, SD). Also listed in the table are the ranges and mean concentrations of metals and lead isotopic compositions of ash collected from the 2009 Jesusita Fire in Santa Barbara, California. Data marked with (‡) were summarized from Odigie and Flegal [25].

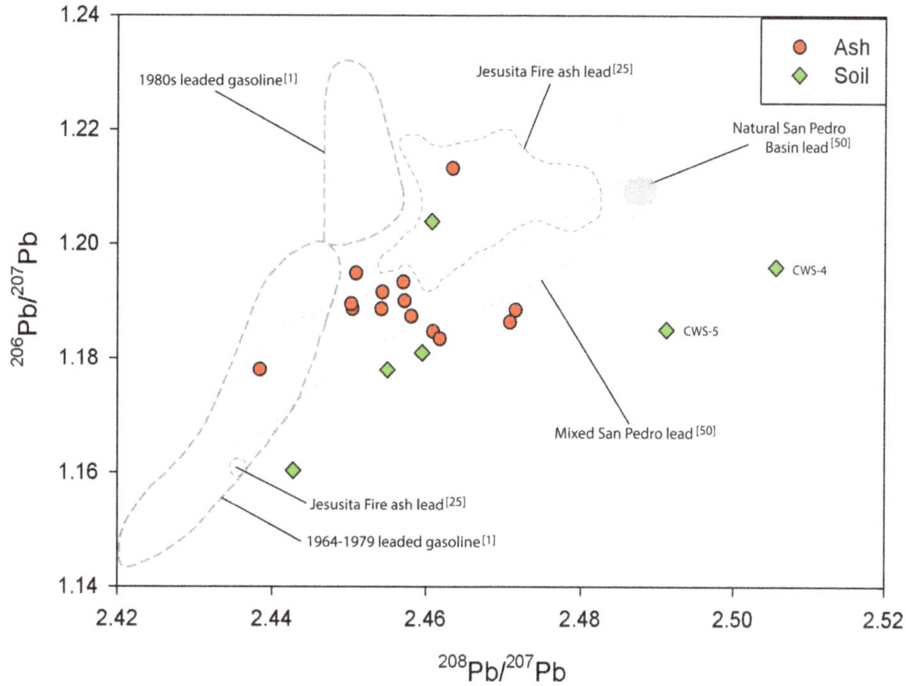

Figure 3. Lead isotopic composition ($^{208}Pb/^{207}Pb$: $^{206}Pb/^{207}Pb$) of ash and soil samples collected from the 2012 Williams Fire site in Southern California. The plot also includes the ranges of isotopic composition of alkyl-lead used in California and those of ^{210}Pb dated sediment cores in the San Pedro Basin. Also included in the plot are the ranges of Pb isotopic composition of ash collected from the 2009 Jesusita Fire in Santa Barbara, California.

The apparent leaded gasoline signature of Pb in the samples attests to the persistence of industrial Pb deposits in the Los Angeles Basin that is consistent with the results of the mass balance study by Harris and Davidson [5]. It is also consistent with previous reports on the persistence of industrial Pb deposits in California's Central Valley, which is north of Los Angeles, based on both mass balance and lead isotopic composition analyses [1,15]. The latter has been principally traced to atmospheric depositions from the combustion of leaded gasoline in California [1].

This attribution is based on the temporal changes in the isotopic compositions of leaded gasoline in California in the previous century. Before 1960, alkyl-lead in the US was primarily manufactured with ores from Australia, Canada, Mexico, and Peru, which had $^{206}Pb/^{207}Pb$ ratios of 1.037, 1.064, 1.192, and 1.200, respectively. Then, the proportion of Pb sourced from the Missouri deposits ($^{206}Pb/^{207}Pb = 1.28$ to 1.33) increased from 9% in 1962 to 82% in 1976 [50]. This change in the sources of Pb in alkyl-lead was subsequently reflected in the Pb isotopic compositions of aerosols in Southern California. There, the $^{206}Pb/^{207}Pb$ ratio of aerosols, which was ~1.15 before 1967, increased to 1.20 by 1974, and then to 1.23 by 1977 [51].

Accelerated Transport of Contaminants to Water Bodies and Atmosphere

In addition to atmospheric contamination, wildfires increase the load of contaminants in aquatic ecosystems. This was recently documented in a study [22] that examined the impact of wildfires on contaminant loadings in Southern California – the same area where the Williams Fire occurred. The results from that study show that the mean fluxes of Cu, Pb, Ni, and Zn from burned sites were 110-fold, 740-fold, 82-fold, and 110-fold greater compared to their respective fluxes at similar unburned (control) sites. That study also found that the mean post-fire concentrations of Cu, Pb, Ni, and Zn in the first stormwater collected from an area that was ~20 km from the closest fire site were each 3 times their pre-fire levels, which they attributed to ash fallout. Similarly, Sabin et al. [52] reported 4-fold, 6-fold, 8-fold, and 13-fold post-fire increases in fluxes of Cu, Zn, Pb, and Ni, respectively, over their mean fluxes in an unburned site in Southern California, which were likewise attributed to forest fires in nearby mountains. Those attributions are consistent with results from a study by Young and Jan [53], who showed that trace metals (Cu, Ni, Pb, and Zn) fallout increased by up to 3-fold in a 10,000 km^2 unburned area that was impacted by smoke plumes and ash from wildfires in the Angeles National Forest in Southern California.

Potential Impact of Climate Change

The mobilization of contaminants by wildfires is a concern, especially in the western US where climate change is expected to dramatically expand fire regimes [54]. Recent studies have linked increases in fire intensity, frequency, and burned area to climate change [55–57]. For example, wildfire frequency has been positively linked ($r = 0.76$, $p < 0.001$, $n = 34$) to regional spring and summer temperatures in western US [54]. The frequency of wildfires and total area burned in western US in the mid-1980s were ~4 times and ~7 times, respectively, higher than their mean values from 1970 to 1986, and this interval was marked by a rise in temperature of <0.9°C [54]. Furthermore, the Intergovernmental Panel on Climate Change has projected that global mean surface temperature could increase by up to 4.8°C by the end of this century (2081–2100) relative to 1986–2005, and researchers have predicted a corresponding increase in the frequency and intensity of wildfires [58–61]. Consequently, it is hypothesized that the pyrogenic remobilization and volatilization of contaminants from forests and grasslands will increase in response to projected climate change. That expected increase is a concern, especially in forests contaminated with radioactive pollutants and at sites used for nuclear waste storage, as recently highlighted for Chernobyl, Ukraine and Los Alamos National Laboratory, US, respectively [62–64]. The results from this and complementary studies provide empirical evidence that wildfires are already remobilizing and volatilizing other contaminants – albeit more pedestrian ones – that constitute a greater global health problem (e.g., Pb).

Supporting Information

File S1 Tables S1–S2. Table S1: Collection locations (coordinates) for ash (CWA) and soil (CWS) samples collected from the 2012 Williams Fire site in the Angeles National Forest in California. Table S2: Enrichment factors (*f*-Fe, normalized to Fe) of trace metals in ash (CWA) samples collected from the 2012 Williams Fire site in the Angeles National Forest in California.

Acknowledgments

Special thanks to Aradhna Tripati and John Mering of UCLA, Martha Thomas, and Priya Ganguli for help with sample collection; Rob Franks of UCSC for analytical support; and all members of the WIGS laboratory, especially Sharon Hibdon, for their support with this project.

Author Contributions

Conceived and designed the experiments: KO ARF. Performed the experiments: KO. Analyzed the data: KO. Contributed reagents/materials/analysis tools: KO ARF. Wrote the paper: KO ARF.

References

1. Dunlap CE, Alpers CN, Bouse R, Taylor HE, Unruh DM, et al. (2008) The persistence of lead from past gasoline emissions and mining drainage in a large riparian system: Evidence from lead isotopes in the Sacramento River, California. Geochim Cosmochim Acta 72: 5935–5948.
2. Nriagu JO (1996) A history of global metal pollution. Science 272: 223–224.
3. Flegal AR, Gallon C, Ganguli PM, Conaway CH (2013) All the Lead in China. Crit Rev Environ Sci Technol 43: 1869–1944.
4. Mielke HW, Laidlaw MAS, Gonzales C (2010) Lead (Pb) legacy from vehicle traffic in eight California urbanized areas: Continuing influence of lead dust on children's health. Sci Total Environ 408: 3965–3975.
5. Harris AR, Davidson CI (2005) The role of resuspended soil in lead flows in the California South Coast Air Basin. Environ Sci Technol 39: 7410–7415.
6. Chow TJ, Bruland KW, Bertine K, Soutar A, Koide M, et al. (1973) Lead pollution: Records in Southern California coastal sediments. Science 181: 551–552.
7. Ritson PI, Bouse RM, Flegal AR, Luoma SN (1999) Stable lead isotopic analyses of historic and contemporary lead contamination of San Francisco Bay estuary. Mar Chem 64: 71–83.
8. Soto-Jimenez MF, Hibdon SA, Rankin CW, Aggarawl J, Ruiz-Fernandez AC, et al. (2006) Chronicling a century of lead pollution in Mexico: Stable lead isotopic composition analyses of dated sediment cores. Environ Sci Technol 40: 764–770.
9. Finney B, Huh C-A (1989) High resolution sedimentary records of heavy metals from the Santa Monica and San Pedro Basins, California. Mar Pollut Bull 20: 181–187.
10. Pirkle JL, Brody DJ, Gunter EW, Kramer RA, Paschal DC, et al. (1994) The decline in blood lead levels in the United States-The National Health and Nutrition Examination Surveys (NHANES). Jama-J Am Med Assoc 272: 284–291.
11. Wheeler W, Brown MJ (2013) Blood lead levels in children aged 1–5 years – United States, 1999–2010. CDC MMWR 62: 245–248.
12. Flegal AR, Smith DR (1992) Lead levels in preindustrial humans. N Engl J Med 326: 1293–1294.
13. WHO (2010) Childhood lead poisoning. Geneva, Switzerland, 74: WHO Document Production Services. 74 p.
14. McGeer JC, Szebedinszky C, McDonald DG, Wood CM (2000) Effects of chronic sublethal exposure to waterborne Cu, Cd or Zn in rainbow trout. 1: Iono-regulatory disturbance and metabolic costs. Aquat Toxicol 50: 231–243.
15. Steding DJ, Dunlap CE, Flegal AR (2000) New isotopic evidence for chronic lead contamination in the San Francisco Bay estuary system: Implications for the persistence of past industrial lead emissions in the biosphere. Proc Natl Acad Sci USA 97: 11181–11186.
16. Semlali RM, Dessogne JB, Monna F, Bolte J, Azimi S, et al. (2004) Modeling lead input and output in soils using lead isotopic geochemistry. Environ Sci Technol 38: 1513–1521.
17. Biswas A, Blum JD, Klaue B, Keeler GJ (2007) Release of mercury from Rocky Mountain forest fires. Global Biogeochem Cycles 21: DOI: 10.1029/2006GB002696.
18. Obrist D, Moosmuller H, Schurmann R, Chen LWA, Kreidenweis SM (2008) Particulate-phase and gaseous elemental mercury emissions during biomass combustion: Controlling factors and correlation with particulate matter emissions. Environ Sci Technol 42: 721–727.
19. Friedli HR, Arellano AF, Cinnirella S, Pirrone N (2009) Initial estimates of mercury emissions to the atmosphere from global biomass burning. Environ Sci Technol 43: 3507–3513.
20. Wiedinmyer C, Friedli H (2007) Mercury emission estimates from fires: An initial inventory for the United States. Environ Sci Technol 41: 8092–8098.
21. Cohen AS, Palacios-Fest MR, McGill J, Swarzenski PW, Verschuren D, et al. (2005) Paleolimnological investigations of anthropogenic environmental change in Lake Tanganyika: I. An introduction to the project. J Paleolimnol 34: 1–18.
22. Stein ED, Brown JS, Hogue TS, Burke MP, Kinoshita A (2012) Stormwater contaminant loading following southern California wildfires. Environ Toxicol Chem 31: 2625–2638.
23. Witt EL, Kolka RK, Nater EA, Wickman TR (2009) Forest fire effects on mercury deposition in the boreal forest. Environ Sci Technol 43: 1776–1782.
24. Biswas A, Blum JD, Keeler GJ (2008) Mercury storage in surface soils in a central Washington forest and estimated release during the 2001 Rex Creek Fire. Sci Tot Environ 404: 129–138.
25. Odigie KO, Flegal AR (2011) Pyrogenic remobilization of historic industrial lead depositions. Environ Sci Technol 45: 6290–6295.
26. Kristensen LJ, Taylor MP, Odigie KO, Hibdon SA, Flegal AR (2014) Lead isotopic compositions of ash sourced from Australian bushfires. Environ Pollut 190: 159–165.
27. US Census (2013) US Census Bureau 2012 population estimate. Available: http://quickfacts.census.gov/qfd/states/06/06037.html. Accessed 2013 May 22.
28. Smail EA, Webb EA, Franks RP, Bruland KW, Sanudo-Wilhelmy SA (2012) Status of metal contamination in surface waters of the coastal ocean off Los Angeles, California since the implementation of the Clean Water Act. Environ Sci Technol 46: 4304–4311.
29. Singh M, Jaques PA, Sioutas C (2002) Size distribution and diurnal characteristics of particle-bound metals in source and receptor sites of the Los Angeles Basin. Atmos Environ 36: 1675–1689.
30. Sardar SB, Fine PM, Sioutas C (2005) Seasonal and spatial variability of the size-resolved chemical composition of particulate matter (PM10) in the Los Angeles Basin. J Geophys Res-Atmos 110: D07S08.
31. Su JG, Jerrett M, Beckerman B, Wilhelm M, Ghosh JK, et al. (2009) Predicting traffic-related air pollution in Los Angeles using a distance decay regression selection strategy. Environ Res 109: 657–670.
32. USDA (2013) Angeles National Forest. Available: http://www.fs.usda.gov/main/angeles/about-forest. Accessed 2013 Feb 19.
33. InciWeb (2012) Camp Williams Fire. Available: http://www.inciweb.org/incident/3230/. Accessed 2013 Feb 19.
34. Kober B, Wessels M, Bollhofer A, Mangini A (1999) Pb isotopes in sediments of Lake Constance, Central Europe constrain the heavy metal pathways and the pollution history of the catchment, the lake and the regional atmosphere. Geochim Cosmochim Acta 63: 1293–1303.
35. Emmanuel S, Erel Y (2002) Implications from concentrations and isotopic data for Pb partitioning processes in soils. Geochim Cosmochim Acta 66: 2517–2527.
36. Bruland KW, Bertine K, Koide M, Goldberg ED (1974) History of metal pollution in Southern California Coastal Zone. Environ Sci Technol 8: 425–432.
37. Hu S, Herner JD, Shafer M, Robertson W, Schauer JJ, et al. (2009) Metals emitted from heavy-duty diesel vehicles equipped with advanced PM and NO_X emission controls. Atmos Environ 43: 2950–2959.
38. Hornberger MI, Luoma SN, van Geen A, Fuller C, Anima R (1999) Historical trends of metals in the sediments of San Francisco Bay, California. Mar Chem 64: 39–55.
39. Schiff KC, Weisberg SB (1999) Iron as a reference element for determining trace metal enrichment in Southern California coastal shelf sediments. Mar Environ Res 48: 161–176.
40. Gabet EJ, Bookter A (2011) Physical, chemical and hydrological properties of Ponderosa pine ash. Int J Wildland Fire 20: 443–452.
41. Councell TB, Duckenfield KU, Landa ER, Callender E (2004) Tire-wear particles as a source of zinc to the environment. Environ Sci Technol 38: 4206–4214.
42. Pakbin P, Ning Z, Shafer MM, Schauer JJ, Sioutas C (2011) Seasonal and Spatial Coarse Particle Elemental Concentrations in the Los Angeles Area. Aerosol Sci Technol 45: 949–U156.

43. Ranali AJ (2004) A summary of the scientific literature on the effects of fire on the concentration of nutrients in surface waters. USGS Open-File Report 2004–1296.
44. Simpson IJ, Akagi SK, Barletta B, Blake NJ, Choi Y, et al. (2011) Boreal forest fire emissions in fresh Canadian smoke plumes: C-1-C-10 volatile organic compounds (VOCs), CO_2, CO, NO_2, NO, HCN and CH_3CN. Atmos Chem Phys 11: 6445–6463.
45. Wu J, Edwards R, He XQ, Liu Z, Kleinman M (2010) Spatial analysis of bioavailable soil lead concentrations in Los Angeles, California. Environ Res 110: 309–317.
46. Holmgren GGS, Meyer MW, Chaney RL, Daniels RB (1993) Cadmium, lead, zinc, copper, and nickel in agricultural soils of the United States of America. J Environ Qual 22: 335–348.
47. Shacklette HT, Boerngen JG (1984) Element concentrations in soils and other surficial materials of the conterminous United States. U.S. Geological Survey Professional Paper 1270.
48. Sutherland RA (2000) Bed sediment-associated trace metals in an urban stream, Oahu, Hawaii. Environ Geol 39: 611–627.
49. Daskalakis KD, O'Connor TP (1995) Normalization and elemental sediment contamination in the Coastal United States. Environ Sci Technol 29: 470–477.
50. Ng A, Patterson CC (1982) Changes of lead and barium with time in California off-shore basin sediments. Geochim Cosmochim Acta 46: 2307–2321.
51. Shirahata H, Elias RW, Patterson CC, Koide M (1980) Chronological variations in concentrations and isotopic compositions of anthropogenic atmospheric lead in sediments of a remote subalpine pond. Geochim Cosmochim Acta 44: 149–162.
52. Sabin LD, Lim JH, Stolzenbach KD, Schiff KC (2005) Contribution of trace metals from atmospheric deposition to stormwater runoff in a small impervious urban catchment. Water Res 39: 3929–3937.
53. Young DR, Jan TK (1977) Fire fallout of metals off California. Mar Pollut Bull 8: 109–112.
54. Westerling AL, Hidalgo HG, Cayan DR, Swetnam TW (2006) Warming and earlier spring increase western US forest wildfire activity. Science 313: 940–943.
55. Gillett NP, Weaver AJ, Zwiers FW, Flannigan MD (2004) Detecting the effect of climate change on Canadian forest fires. Geophys Res Lett 31: DOI: 10.1029/2004GL020876.
56. Westerling AL, Turner MG, Smithwick EAH, Romme WH, Ryan MG (2011) Continued warming could transform Greater Yellowstone fire regimes by mid-21st century. Proc Natl Acad Sci USA 108: 13165–13170.
57. Wang YH, Flannigan M, Anderson K (2010) Correlations between forest fires in British Columbia, Canada, and sea surface temperature of the Pacific Ocean. Ecol Model 221: 122–129.
58. IPCC (2013) Climate Change 2013: The Physical Science Basis: Cambridge Univ. Press, New York.
59. McKenzie D, Gedalof Z, Peterson DL, Mote P (2004) Climatic change, wildfire, and conservation. Conserv Biol 18: 890–902.
60. Schuur EAG, Bockheim J, Canadell JG, Euskirchen E, Field CB, et al. (2008) Vulnerability of permafrost carbon to climate change: Implications for the global carbon cycle. Bioscience 58: 701–714.
61. Flannigan MD, Logan KA, Amiro BD, Skinner WR, Stocks BJ (2005) Future area burned in Canada. Clim Change 72: 1–16.
62. Charles D (2010) Russia's forest fires ignite concerns about Chornobyl's legacy. Science 329: 738–739.
63. Gilbert N (2010) Russia counts environmental cost of wildfires. Nature DOI: 10.1038/news.2010.404.
64. Weinhold B (2011) Fields and forests in flames: Vegetation, smoke & human health. Environ Health Perspect 119: A386–A393.

2

Distance-Decay and Taxa-Area Relationships for Bacteria, Archaea and Methanogenic Archaea in a Tropical Lake Sediment

Davi Pedroni Barreto[1], Ralf Conrad[3], Melanie Klose[3], Peter Claus[3], Alex Enrich-Prast[2,4]*

1 Instituto de Microbiologia Prof. Paulo de Góes, Universidade Federal do Rio de Janeiro, Rio de Janeiro, Brazil, **2** Instituto de Biologia, Universidade Federal do Rio de Janeiro, Rio de Janeiro, Brazil, **3** Max-Planck Institute for Terrestrial Microbiology, Marburg, Hessen, Germany, **4** Department of Water and Environmental Studies, Linköping University, Linköping, Sweden

Abstract

The study of of the distribution of microorganisms through space (and time) allows evaluation of biogeographic patterns, like the species-area index (z). Due to their high dispersal ability, high reproduction rates and low rates of extinction microorganisms tend to be widely distributed, and they are thought to be virtually cosmopolitan and selected primarily by environmental factors. Recent studies have shown that, despite these characteristics, microorganisms may behave like larger organisms and exhibit geographical distribution. In this study, we searched patterns of spatial diversity distribution of bacteria and archaea in a contiguous environment. We collected 26 samples of a lake sediment, distributed in a nested grid, with distances between samples ranging from 0.01 m to 1000 m. The samples were analyzed using T-RFLP (Terminal restriction fragment length polymorphism) targeting *mcrA* (coding for a subunit of methyl-coenzyme M reductase) and the genes of Archaeal and Bacterial 16S rRNA. From the qualitative and quantitative results (relative abundance of operational taxonomic units) we calculated the similarity index for each pair to evaluate the taxa-area and distance decay relationship slopes by linear regression. All results were significant, with *mcrA* genes showing the highest slope, followed by Archaeal and Bacterial 16S rRNA genes. We showed that the microorganisms of a methanogenic community, that is active in a contiguous environment, display spatial distribution and a taxa-area relationship.

Editor: Jonas Waldenström, Linneaus University, Sweden

Funding: This study was funded by the National Council of Scientific and Technological Development (CNPq), trough the project 477260/2011-0. (Brazil - www.cnpq.br) and by the Max Planck Society (Germany - www.mpg.de). The funders had no role in study design, data collection and analysis, decision to publish, or preparation of the manuscript.

Competing Interests: The authors have declared that no competing interests exist.

* Email: aeprast@biologia.ufrj.br

Introduction

The biogeography concept is defined as the study of the distribution and the range of living organisms across space and time. Most studies in this field were traditionally performed targeting macro-organisms such as plants and animals [1]. Since the development of molecular tools, the concepts of biogeography started to be also studied in microorganisms [2–5].

A long-held concept in microbial ecology is that microorganisms are ubiquitously distributed and can be found in any habitat with favorable environmental conditions. This concept was introduced by Martinus Willem Beijerinck and concisely summarized by Lourens Gerhard Marinus Baas Becking in the quote, "Everything is everywhere, the environment selects" [6]. This statement is based on some traits of the microorganisms, such as the small size of individuals and the consequent ease of their dispersal across long distances, high rates of reproduction, short generation times, and large population sizes, leading to a small chance of local extinction.

Free-living eukaryotic microorganisms are often described as occurring ubiquitously. When they are not dominant in some specific environment, it is possible to reanimate the cryptic diversity by changing the environmental conditions *in vitro*[3,7]. A study showed that it is possible to find nearly 80% of all known species of the flagellate genus *Paraphysomonas* in just a small sample of sediment [8], meaning that the global diversity of this genus is well represented by its local diversity. This observation is mostly explained by the high dispersal rate of the flagellates (due to their low size), their extremely short generation times (leading to a low rate of extinction) and also their capacity to generate resistant forms when the environmental conditions are unfavorable [8]. The authors suggested that if eukaryotic microorganisms were ubiquitous, then prokaryotic microorganisms should be ubiquitous as well, since they have an even smaller size and larger populations. Indeed, some studies on prokaryotes suggested global distribution, for example, psychrophilic polar bacteria were found at both the South and the North poles [9].

Recent studies, however, showed that the distribution of microorganisms is not random, and that biogeography patterns of distribution are established [10,11]. For example, the genetic distances between populations of microorganisms were shown to increase with geographic distance, which might represent a speciation process driven by the geographic isolation of the microorganisms [12]. Other studies were able to identify endemic microorganisms, and true geographic isolation in extreme environments like hot springs, pristine soils, salt lakes, and hot

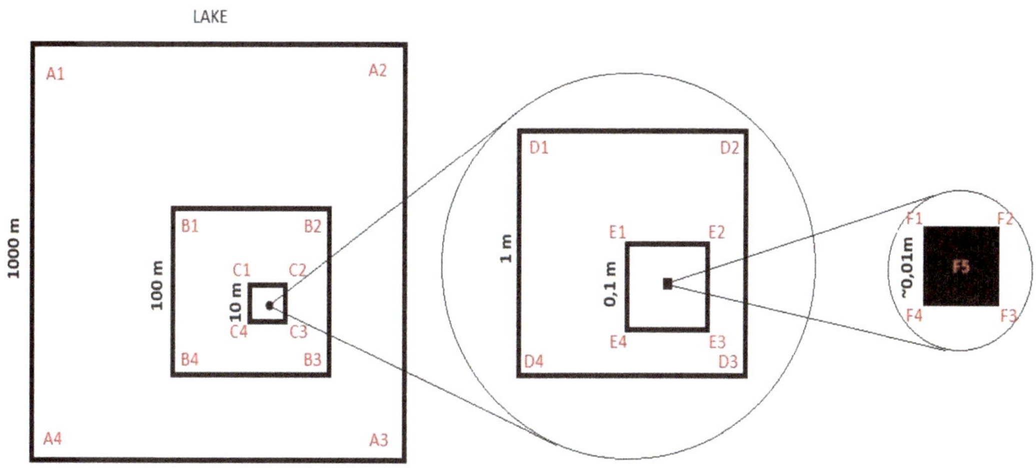

Figure 1. Sampling grid. Designed inside the Lago Negro area, A1-F5 represents the sampling points. (Adapted from (Horner-Devine 2004)).

and cold deserts around the world, all being strong evidence for non-cosmopolitan distribution [10,13–17]. The main difficulty is to evaluate the factors determining the geographical distribution, whether historic evolutionary events (geographic barriers for example) or contemporary ecological environmental factors. [5].

The taxa-area relationship is one of the most consistent general patterns in ecology and well described for macro-organisms [1]. It is represented by the equation $\mathbf{S = c\,A^z}$, where **S** is the number of species, **A** is the area sampled, **c** is a constant that is empirically derived from the taxon and the specific location studied, and the exponent **z**, the power law index (i.e. **z**-value), represents the rate of the increasing number of species along the increasing sampling area (graphical slope). When significant, the z-value may be strong evidence for biographical distribution of the species.

Values for the z exponent have already been described for microorganisms. Interestingly, z-values for microorganisms were often smaller than for macro-organisms. This result may be attributed: to the larger capacity of dispersion; to the lack of a clear "species" resolution; and to the use of molecular fingerprint or sequencing techniques [18–20]. Molecular fingerprint methods, such as T-RFLP (Terminal restriction fragment length polymorphism) and DGGE (Denaturing gradient gel electrophoresis), proved to be an important tool for accessing the diversity of microorganisms in different environments with relatively low costs and little time consumed [21]. However, fingerprinting methods usually are limited as they detect the most common species and thus underestimate the total diversity in a sample [22]. That is mostly because fingerprinting techniques lump different closely related "species" into a single taxonomic unit (often called Operational Taxonomical Unity – OTU), and usually ignore rare species. Nevertheless, fingerprinting techniques are still extremely valuable for rapidly comparing the microbial community composition in different environments [2].

Another parameter of species distribution through space is the distance-decay relationship, which consists of the decay of similarity between different communities as a function of the distance separating them [23], and can also be seen as evidence of a biogeographical pattern. The main difference between the distance-decay approach and the species-area relationship is based on the consideration of the relative abundance of the species in addition of their presence or absence. Bell et al. showed that bacteria in water-filled tree holes, found at the same place, displayed a significant distance-decay relationship [24].

Little is known about the geographical distribution of methanogenic archaea. So far, hot desert soil methanogenic Archaea were shown to be widely spread between different parts of the globe, and this could be found mostly by reactivation of the cryptic methanogenic process *in vitro* [25]. To our best knowledge, there is no description of distance-decay or species-area relationship for non-extremophilic archaea in a contiguous environment. Given the high ecological stability of methanogenic sediments and soils, with a continuous anaerobic environment and a regular input of organic matter, the microbial communities related to the methanogenesis processes tend to be stable through time [26].

Among the processes commonly underlying the biogeographical patterns of distribution of organisms and communities - selection, drift, dispersion and diversification - selection and dispersion are closely related with geographical distances, given the increase of habitat heterogeneity and dispersion limitations with increasing area [27]. Thus, we hypothesized that if there is a change of diversity patterns among different spots of a contiguous lake sediment, it should have a strong correlation with the geographical distance between them, and this change should also present itself differently depending on which part of the microbial community is considered. For this purpose we were targeting three different genes. The 16S rRNA genes of Bacteria and Archaea that are transcribed to generate the structural RNA of the small-subunit ribosomes are universal and strongly conserved genes and are therefore widely used as taxonomic markers [28]. The *mcrA* is a functional gene coding for the alpha subunit of the methyl-coenzyme M reductase, an enzyme being essential and characteristic for the methanogenesis biochemical pathway in Archaea [29]. Thus, we were targeting different groups of prokaryotes. By that we expected to see if there is a differential influence of geographic distance as a factor driving distribution of these three different groups of microorganisms.

We were able to show a geographical distribution pattern of the composition of Bacteria, Archaea and methanogenic Archaea in a contiguous tropical lake sediment and the presence of a significant **z-value** for all three groups. We hypothesized that intrinsic ecological differences between the communities, general diversity profiles, and technical particularities and limitations could be defining the differences in the distribution of these taxonomical groups or at least how we perceived it.

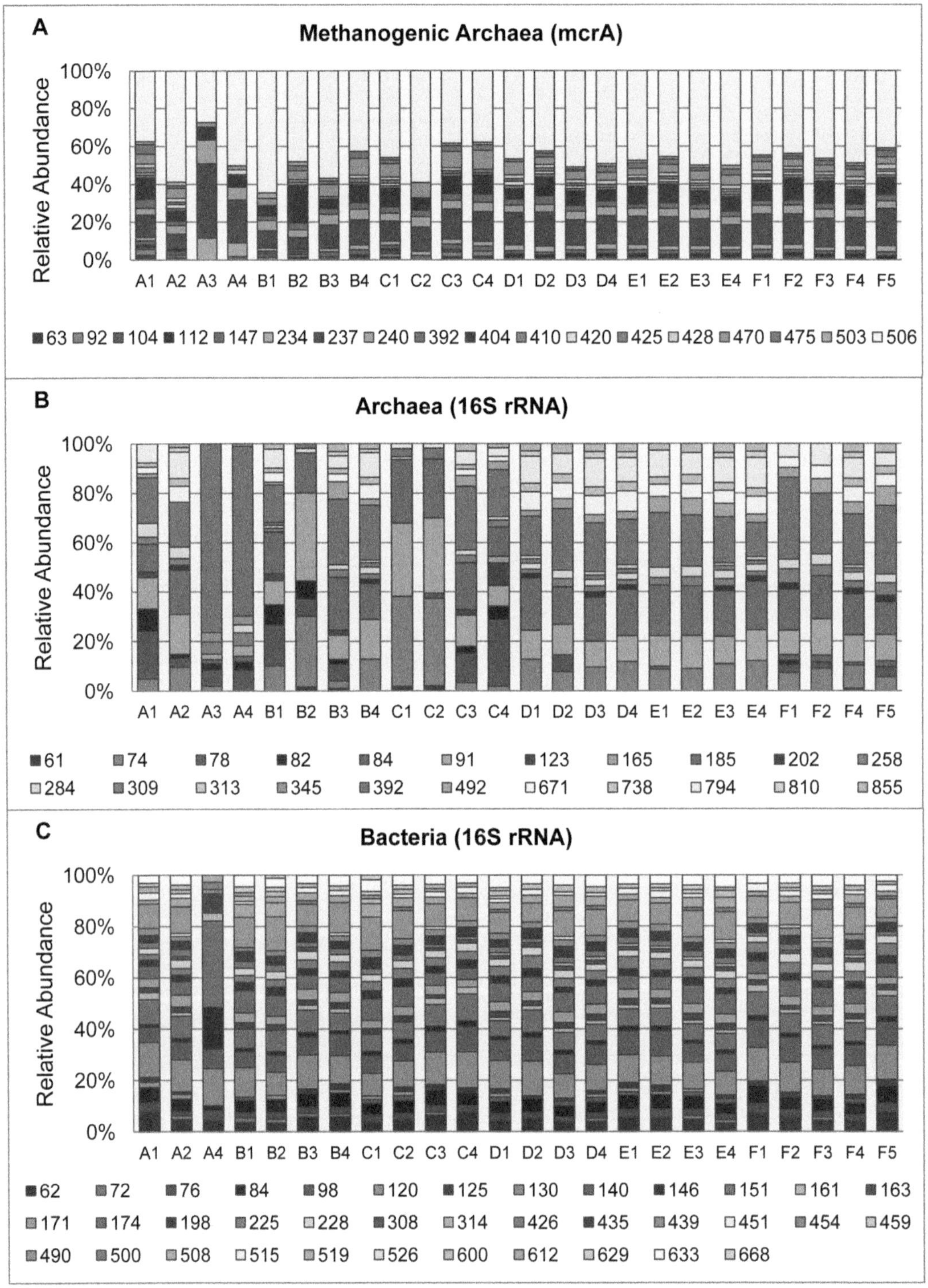

Figure 2. T-RFLP profiles of the three different genes studied in sediment inside the Lago Negro area. A1-F5 represent the sampling points (see Figure 1 for more details), the bar size represents the relative abundance of each OTU, which are defined by their size in base pair (bp). (A) *mcrA*, (B) Archaeal 16S rRNA gene, and (C) Bacterial 16S rRNA gene.

Materials and Methods

Sampling and Study Area

The Pantanal consists of the largest floodplain in South America and is periodically flooded by the Paraguay River and its tributaries. Altitude above sea level varies from 80 to 120 m and the total estimated area is around 138.123 km^2. The water flow continuously carries organic material, and during the flood period the sediment spreads all over the plain constituting the most important source of carbon and nutrients for the methanogenic archaea [30]. The climate is hot and wet in the summer, and cold and dry in the winter. The maximum temperature often surpasses 40°C. Between the months of May and July, the average

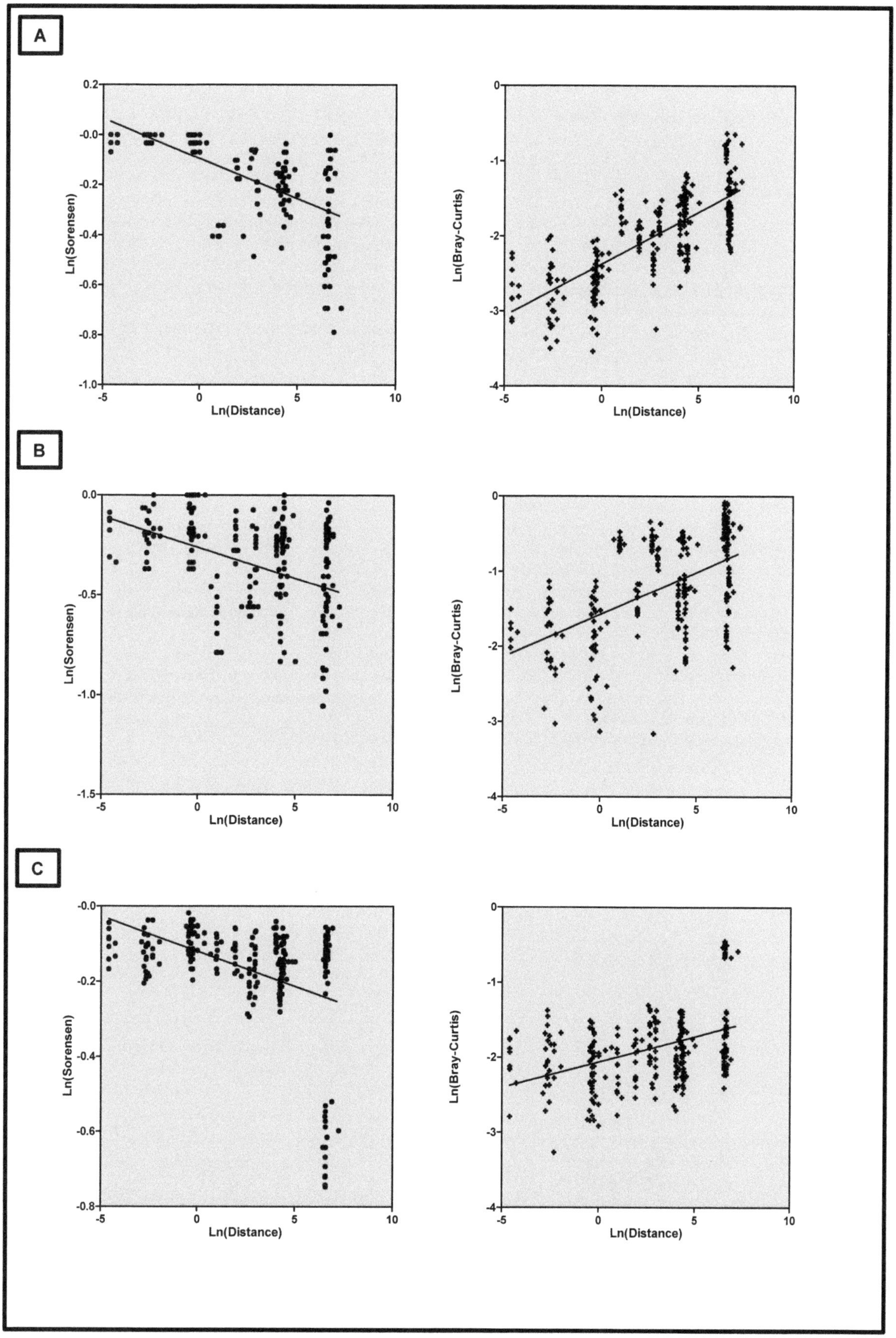
A
B
C
Ln(Sorensen)
Ln(Bray-Curtis)
Ln(Distance)

Figure 3. Distance decay and taxa-area relationships. Log transformed Bray-Curtis and Sørensen indices plotted against the distance separating the points, for the three different genes, lines represent a simple linear regression. (A) *mcrA*; (B) Archaeal 16S rRNA gene and (C) Bacterial 16S rRNA gene.

temperature drops below 20°C, and the minimum temperature may reach 0°C [31].

The study was conducted in a single lake (19°02.651′S 57°30.254′W) called *Lagoa Negra*. No special permission is needed for sediment sampling in non-protected areas in Brazil, so no specific permissions were required for these locations. This study did not involve endangered or protected species.

The lake is located at the west margin of the Paraguay River and a few kilometers east from the cities of Corumbá and Ladário, close to the Bolivia-Brazil border. *Lagoa Negra* is a perennial shallow freshwater lake, with water depth varying between 2 m and 3 m approximately, getting lower or higher following the flood regime. The total area of the lake is 10.8 km^2 (maximum length of 4.1 km and maximum width of 3.5 km, but it can also vary following the Paraguay River flood regime), water pH ranging from 6.9 to 8.3 and conductivity ranging from 140 to 240 $\mu S\ cm^{-1}$ at 25°C. The top layer of sediment is mostly composed of a thin clayey substrate with approximately 2% of sand (>63 µm) and total carbon and organic matter concentrations of 1.8% and 2% respectively [32]. Twenty one sediment cores (7 cm diameter) were collected in the lake in July 2011. The cores were distributed on a nested square grid with distance between points varying from 0.01 m to approximately 1,400 m (Figure 1). The 10 cm top layer of the sediment was homogenized and then 1 ml was sampled and frozen for molecular biology analyses. The central core E was sub sampled within distances of 10 cm and 1 cm between points inside the core area (also 10-cm deep with polyethylene straws). The sediment samples were immediately frozen in liquid nitrogen and then air-shipped to Germany for later molecular biology analyses at the Max-Planck Institute for Terrestrial Microbiology.

T-RFLP

The frozen sediment samples were thawed, and the total DNA was extracted using the FastDNA SPIN Kit for Soil (MP Biomedicals) following the manufacturer's protocol and a three times treatment with 5.5 M guanidine thyocianate as an additional cleaning step during the matrix binding step. The total DNA was quantified using spectophotometry. Preparations were considered to be of good quality when resulting in more than 30 ng/µl of DNA.

Bacterial and Archaeal 16S rRNA gene fragments were PCR-amplified using the primer pairs Eub 9/27f (5′–GAG TTT GAT CMT GGC TCA G–3′) and Eub 907/926r (5′–CCG TCA ATT CMT TTR AGT TT–3′), as described by Lane and Weisburg [33] for Bacteria, and the primer pairs A109f (5′–ACK GCT CAG TAA CAC GT–3′) and A934b (5′–GTG CTC CCC CGC CAA TTC CT–3′), as described by Großkopf et al. [34] for Archaea. For the T-RFLP analyses the forward Bacteria and the backward Archaea primers were labeled with FAM (5-carboxy-fluorescein). Each 50-µl volume of PCR reaction contained 1× GoTaq Flexi Green Buffer (Promega); 1.5 mM $MgCl_2$ (Promega); 200 µM of dNTP's Mix (Fermentas); 0.33 µM of each primer described (Sigma); 1 U GoTaq Flexi DNA Polymerase (Promega) and 10 µg of Bovine Serum Albumin BSA (Roche). Diluted total DNA extract was added as template (1 µl). The reaction was started with a initial denaturation step (94°C for 3 min), followed by 24–28 cycles of denaturation (94°C for 45 sec), annealing (52°C for 45 sec) and extension (72°C for 80 sec), and a final extension step (72°C for 5 min).

For amplification of *mcrA* the primer pair MCRf (5′–TAY GAY CAR ATH TGG YT–3′) and MCRb (5′–ACR TTC ATN GCR TAR TT–3′) was used as described by Springer et al. [29], the forward primer being labeled with FAM for T-RFLP analyses. Each 50-µl of PCR reaction contained 1× MasterAmp PCR PreMix B (Biozym); 0.33 µM of each primer described; 1 U GoTaq Flexi DNA Polymerase (Promega); 10 µg BSA (Roche) and 1 µl of DNA template. The reaction started with an initial denaturation step (94°C for 3 min), followed by 32 cycles of denaturation (94°C for 45 sec), annealing (50°C for 45 sec) and extension (72°C for 90 sec), finished by a final extension step (72°C for 5 min).

The quality of the PCR products was controlled using agarose gel (1.5%) electrophoresis. The DNA was then purified using the GenElute PCR Clean-Up Kit (Sigma) following the manufacturer's instructions, and stored at −20°C.

For T-RFLP, the PCR product was digested using the following restriction enzymes: MspI incubated overnight at 37°C for Bacterial 16S rRNA genes, TaqI incubated for 3 h at 65°C for Archaeal 16S rRNA genes, and Sau96 incubated for 3 h at 37°C for *mcrA*. After the incubation, the digested products were once again cleaned with the SigmaSpin Sequencing Reaction Clean-Up, Post-Reaction Purification Columns (Sigma). The samples were denatured at 94°C for 2 min and loaded into an ABI 3100 automated gene sequencer (Applied Biosystems) for separation of the TRFs. T-RFLP data were retrieved by comparison with an internal standard using GeneScan 3.71 software (Applied Biosystems).

Taxa-area and distance-decay relationships

The T-RFLP profiles were analyzed and standardized as described in Dunbar et. al [35] resulting in T-RFs of 60 to 855

Table 1. Distance-decay regression coefficients based on the Bray-Curtis dissimilarity index **(Y)**; Regression coefficient based on the Sørensen Similarity index; and the exponent **z** calculated by the distance-decay approach values for each one of the target genes calculated by -(regression coefficient)/2.

Gene	*Distance-decay (Y)*	*Sørensen Regr. Coefficient*	*(z)*
mcrA	0.14	−0.032	0.016
Archaea 16S rRNA	0.11	−0.031	0.0155
Bacteria 16S rRNA	0.067	−0.018	0.009

Table 2. Z-values previously described for microorganisms, 99% or 95% represent the taxonomic resolution applied for definition of the OTUs (sequence similarity).

Microbial Community	z-value	Diversity Access	Habitat Org.	Reference
Desert Soil Fungi	0.074	fingerprint (ARISA)	Contiguous	[19]
Salt-Marsh Bacteria (99% OTU)	0.04	sequencing	Contiguous	[18]
Salt-Marsh ß-proteobacteria (99% OTU)	0.019	sequencing	Contiguous	[18]
Salt-marsh Bacteria (95% OTU)	0.019	sequencing	Contiguous	[18]
Salt-Marsh ß-proteobacteria (95% OTU)	0.008*	sequencing	Contiguous	[18]
Water-filled tree holes Bacteria	0.26	fingerprint (DGGE)	Island	[20]
Metal-cutting fluid sump Bacteria	0.26-0.29	fingerprint (DGGE)	Island	[38]
Soil Bacteria	0.03	Fingerprint (T-RFLP)	Noncontiguous	[39]
Tropical Forest Soil Bacteria	0.42 and 0.47	Fingerprint (T-RFLP)	Contiguous	[40]
Freshwater lake sediment Bacteria	0.009	fingerprint (T-RFLP)	Contiguous	this study
Freshwater lake sediment Archaea	0.0155	Fingerprint (T-RFLP)	Contiguous	this study
Freshwater lake sediment methanogenic Archaea	0.016	fingerprint (T-RFLP)	Contiguous	this study

*not significant different from zero.

base pairs (bp) size, each representing more than 1% of the total fluorescence of that sample.

We used the 25 resulting profiles for pair wise calculation of the Bray-Curtis dissimilarity indices [36] and the Sørensen similarity indices resulting in 300 different pairs for each targeted gene. A simple linear regression of the log-transformed data of the Bray-Curtis indices plotted against the distances between the sediment samples from which the pairs originated was used to estimate the slope of the distance-decay relationship [23], in this work we are using the resulting regression coefficients **(Y)** as a parameter for discussion.

The same log-transformed linear regression was used to calculate the slope of the Sørensen indices plotted against the distance between points. The resultant similarity slope was used to calculate the **z-value** of the taxa-area relationship with the formula $\mathbf{\log(S_S) = constant\ -2z\ \log(D)}$, where S_S is the pair-wise similarity between communities and D is the distance between two samples used in the distance-decay approach, where the **z-value** is determined by **−(regression coefficient)/2** as described by Harte et al. [37].

By using both, the taxa-area relationship based on the Sørensen index and the distance-decay based on the Bray-Curtis dissimilarity index we addressed the subject not only by "presence/absence" of a taxon (Sørensen index), but also the relative abundance of a taxon (Bray-Curtis index) provided by the T-RFLP.

In order to avoid randomization patterns that could influence the results, the same calculations were performed utilizing distances smaller than 200 m and 20 m and they showed similar slopes as the complete data set (data not shown).

Results

The analysis of the *mcrA* T-RFLP profile resulted in a total of 18 different OTUs of methanogens throughout the lake area (Figure 2A) of which 5 OTUs were found at all sampling points (237 bp, 240 bp, 404 bp, 470 bp and 506 bp) and one OTU (63 bp), which was the least frequently retrieved OTU, was found at only 6 sampling points. The OTU with 506 bp showed the highest relative abundance in all the samples. The Archaeal 16S rRNA T-RFLP profile showed 22 different OTUs (Figure 2B) with none of them showing dominance over the others, only 2 OTUs were found at all sampling points (91 bp and 392 bp) and the OTUs with 165 bp and 345 bp were found at only 1 and 2 sampling points, respectively. The Bacterial 16S rRNA gene T-RFLP profiles showed the largest number of OTUs among the different genes targeted with a total of 37 different OTUs found at all the sampling points across the lake area, showing a high diversity of dominant groups (Figure 2C). Eight OTUs were recovered from all the sampling points (62 bp, 76 bp, 84 bp, 130 bp, 140 bp, 146 bp, 151 bp and 439 bp); in contrast, 3 other OTUs (459 bp, 526 bp and 612 bp) were recovered from only 3 sampling points. In a literature based affiliation it was possible to relate some of the observed OTUs with previously described ones, keeping in mind that such affiliation can only be tentative (Table S1).

Linear regressions were performed to evaluate the slope coefficient of the correlations between the geographical distance and the communities' similarities, and Mantel tests (with 9999 permutations) were performed to test the significance of these correlations.

All the three different T-RFLP profiles showed a significant distance-decay relationship based on the Bray-Curtis dissimilarity **(Y)** (Figure 3). The *mcrA* gene showed the largest slope of 0.14 (Mantel test $P<0.001$, $R=0.69$), followed by the archaeal 16S rRNA gene with a slope of 0.11 (Mantel test $P=0.005$, $R=0.30$), while the slope for the bacterial 16S rRNA gene was only 0.067 (Mantel test $P=0.006$, $R=0.34$).

The regression coefficient based on the Sørensen similarity indices calculated for all the three genes also showed significant values (Figure 3) of −0.032 ($P<0.001$, $R=-0.60$) for *mcrA*, −0.031 (Mantel test $P=0.03$, $R=-0.45$) for archaeal and −0.018 (Mantel test $P=0.006$, $R=-0.34$) for bacterial 16S rRNA genes. However the calculated z-values were small, thus indicating relatively flat taxa-area relationships (Table 1).

Discussion

Our study indicates a small but significant biogeographical distribution of the microbial community composition in the lake sediment for all the three different genes targeted, which represent the Bacteria, the Archaea, and the methanogens. The conclusion is based on the distance decay relationship of operational taxonomic units of three different microbial genes. All dominant operational taxa (more than 10% relative abundance) from the three genes analyzed in this study were observed at all the sampling points within the lake. Nevertheless, a significant taxa-area relationship could be observed. These values were small if compared to others described for macro-organisms, which was expected given the high capacity of dispersion of the targeted microorganisms. However, the similarity distance-decay, which takes into consideration also the relative abundance of the groups, shows that they were not homogeneously distributed within the entire lake area. We therefore conclude that the most distant samples were the ones, which were most different from each other.

The **z-values** and the distance decay regression coefficients **(Y)** observed in this study were lower than others previously observed for microorganisms in the literature [17–19][36–38], but are the first ones described for Bacteria, Archaea and methanogenic archaea in a single contiguous environment. The existence of significant **z-values** shows that all these microbial groups have a biogeographical distribution. The low values that we found were expected given that we used a fingerprint technique for accessing the microbial diversity of our target environment, and not a high resolution ribotyping technique. At an OTU resolution of 95% sequence similarity of the 16S rRNA gene, Horner-Devine et al. described similar z-values for Bacteria ($z = 0.019$) and Beta-proteobacteria ($z = 0.008$) in salt-marshes [18]. At higher resolutions the z presented higher values (0.04 for Bacteria and 0.019 for Beta-proteobacteria at 99% sequences similarity). It is interesting to note that the lower values described by the mentioned study were presented by the lower taxonomical levels reached, while in this study we observed higher values for *mcrA* and the Archaea than for the Bacteria. Other **z-values** found for other groups of microorganisms are displayed in the Table 2.

The differences between **z-values** and distance decay values **(Y)** described for the three different genes, representing three different groups of microorganisms, may be explained by two hypotheses. The first one is that Bacteria (16S rRNA) have a larger capacity for dispersal than the Archaea (16S rRNA) in general and the methanogenic archaea (*mcrA*) in particular. By sampling the first 10 centimeters of the sediment we were able to access different communities living at different depths at the same time. We assume that the methanogens (all Archaea) were preferable located in the deeper sediment layers, since their activity is inhibited by the presence of other electron acceptors like oxygen, nitrate, iron and sulfate, potentially present in the surface layers of the sediment. On the other hand, Bacteria were not restricted to deeper sediment layers [41–43]. The Bacteria domain showed weaker species-area and distance-decay relationships than the methanogens (*mcrA*) and the Archaea (16S rRNA). The profile of Archaea and Bacteria OTUs was well distributed throughout the lake area, showing the large diversity of micro-habitats that can be exploited by these groups, and that the ability of dispersion seems not to be compromised at smaller scales.

The second hypothesis is based on the taxonomic sensitivity of the T-RFLP method. OTUs can represent a large variety of different taxonomic groups. The universal primers targeting 16S rRNA genes do not represent the entire set of conceivable species, which are many more than the OTUs derived from T-RFLP. Hence, the total diversity of the community is underestimated by T-RFLP [44]. A study targeting 16S rRNA genes of the bacterioplankton of temperate lakes showed that the sequence similarity within a single OTU varied from 73 to 100%, however, sequence homology of 97% is generally used to define bacterial species [5,45]. When utilizing a functional gene such as the *mcrA*, the probability of reaching lower taxonomic levels is higher, because translated genes show diminished conservation and can present a higher variability of codons, and thus a higher chance for being differentiated by terminal fragment size methods [46].

This may explain why the **z-values** were higher for T-RFLP of *mcrA* than of 16S rRNA genes, but it does not explain why the Archaea distance decay and **z-values** were higher than those of the Bacterial since both of them targeted 16S rRNA genes. Several authors described that Bacteria diversity tends to be higher than Archaea diversity at different environments [47–49], thus the T-RFLP fingerprinting technique could be more efficient in accessing a larger part of the total diversity of Archaea 16S rRNA and consequentially being closer to the actual taxa distribution of the environment, thus showing higher values of **z** and **Y**.

Our work shows some important results that contribute to a better understanding of similar ones developed within different environments. Studies of bacterioplankton distribution in lakes showed similar patterns between different and not connected lakes in distant geographic regions, but with variations in their relative abundance [50]. On the other hand bacterioplankton communities distance dissimilarity inside one single lake were described as being weaker than between different lakes in North America, and mostly influenced by different water regimes and partial geographic isolation [51]. Some studies in saline lakes in China, Mongolia and Argentina showed that Bacterial biogeography in these environments was based on contemporary environmental factors (Na^+, CO_3^{2-}, and HCO_3^- ion concentrations, pH and temperature) and geographic distance, while Archaeal biogeography was influenced only by environmental factors [17].

Geographical distances have been previously described as an important factor related with microbial spatial distribution corroborating our findings [52–54]. Rosselló-Mora et al. used metabolic compounds as a comparison parameter between different populations of the extremophilic bacterium *Salinibacter rubium*. They discovered that the divergence among different phenotypes was related to different geographical locations, and geographical distance between the sites[55]. Environmental and geographical factors also influenced magnetotactic bacteria biogeographical distribution [56].

Our distance-decay and taxa-area relationships can possibly be a result of some degree of dispersal limitation coupled with ecological drift, thus maintaining the taxa-area relationship and distance-decay pattern significant. It is believed that drift plays a important role for the distance-decay patterns found among microorganisms living in some degree of dispersal limitation (i.e. in subsurface habitats)[5]. A high dispersal rate is expected inside this contiguous environment with no significant geographical barriers imposing any kind of mobility limitation to microorganisms, and that could be a factor flattening the curves. But as we already stated, a group preferentially found in deeper layers of sediment (as the methanogenic archaea) could be more easily restricted.

This study did not focus on scanning environmental factors that could be driving the observed biogeographical pattern, so selection as a possible driving factor cannot be excluded, and selection processes are also related with geographical distances, as the diversity of habitats tends to increase with an increasing area [5,27]. However, given the small scale of the study and the probable absence of a clear environmental gradient within the Lagoa Negra we don't think it as a major factor. We believe that the correlation that we described between geographical distance and communities' structures are representative of the biogeographical pattern of distribution of the lake microbial community.

In conclusion, we showed that there was a significant geographic distribution of methanogenic archaea, Archaea and even Bacteria, related with geographic distance in a contiguous environment, i.e. the sediment of a tropical lake.

Supporting Information

Table S1 Tentative genetic affiliation of the OTUs. For the three studied genes there are in the literature some genetic affiliation with different phylogenetic groups, we show some of these in this table. * The affiliation with the phylogenetic groups was based on literature data that used the same primers and restriction enzymes for TRFLP. All the affiliations are only tentative, and can only be interpreted as the probable main groups of the lake sediment community.

Acknowledgments

We thanks, the colleagues Roberta Peixoto, Juliana Valle, Ruan Andrade and the Federal University of Mato Grosso do Sul - Brazil (UFMS) for the support during the sampling.

Author Contributions

Conceived and designed the experiments: DPB RC AEP. Performed the experiments: DPB PC MK. Analyzed the data: DPB PC MK. Contributed reagents/materials/analysis tools: RC PC MK AEP. Wrote the paper: DPB RC AEP.

References

1. Begon M, Townsend CR, Harper JL (2006) Ecology: from individuals to ecosystems. 4th ed. Blackwell Publishing Ltd.
2. Martiny JBH, Bohannan BJM, Brown JH, Colwell RK, Fuhrman JA, et al. (2006) Microbial biogeography: putting microorganisms on the map. Nat Rev 4: 102–112.
3. Fenchel T, Esteban GF, Finlay BJ (1997) Local versus Global Diversity of Microorganisms: Cryptic Diversity of Ciliated Protozoa. Oikos 80: 220–225.
4. Fenchel T (2003) Biogeography for bacteria. Science 301: 925.
5. Hanson Ca, Fuhrman Ja, Horner-Devine MC, Martiny JBH (2012) Beyond biogeographic patterns: processes shaping the microbial landscape. Nat Rev Microbiol 10: 497–506.
6. Baas-Becking L, Becking LGMB (1934) Geobiologie of inleiding tot de milieukunde. 18.
7. Finlay BJ (2002) Global dispersal of free-living microbial eukaryote species. Science 296: 1061.
8. Finlay BJ, Clarke KJ (1999) Ubiquitous dispersal of microbial species. Nature 400: 1999.
9. Staley JT, Gosink JJ (1999) Poles apart: biodiversity and biogeography of sea ice bacteria. Annu Rev Microbiol 53: 189–215.
10. Papke RT, Ramsing NB, Bateson MM, Ward DM (2003) Geographical isolation in hot spring cyanobacteria. Environ Microbiol 5: 650–659.
11. Papke RT, Ward DM (2004) The importance of physical isolation to microbial diversification. FEMS Microbiol Ecol 48: 293–303.
12. Diniz-Filho JAF, Telles MPDC (2000) Spatial pattern and genetic diversity estimates are linked in stochastic models of population differentiation. Genet Mol Biol 23: 541–544.
13. Cho J-CC, Tiedje JM (2000) Biogeography and Degree of Endemicity of Fluorescent Pseudomonas Strains in Soil. Appl Environ Microbiol 66: 5448–5456.
14. Oda Y, Star B, Huisman LA (2003) Biogeography of the purple nonsulfur bacterium Rhodopseudomonas palustris. Appl Environ Microbiol 69: 5186–5191.
15. Takacs-Vesbach C, Mitchell K, Jackson-Weaver O, Reysenbach A-L (2008) Volcanic calderas delineate biogeographic provinces among Yellowstone thermophiles. Environ Microbiol 10: 1681–1689.
16. Bahl J, Lau MCY, Smith GJD, Vijaykrishna D, Cary SC, et al. (2011) Ancient origins determine global biogeography of hot and cold desert cyanobacteria. Nat Commun 2: 163.
17. Pagaling E, Wang H, Venables M, Wallace A, Grant WD, et al. (2009) Microbial biogeography of six salt lakes in Inner Mongolia, China, and a salt lake in Argentina. Appl Environ Microbiol 75: 5750–5760.
18. Horner-Devine MC, Lage M, Hughes JB, Bohannan BJM (2004) A taxa-area relationship for bacteria. Nature 432: 750–753.
19. Green JL, Holmes AJ, Westoby M, Oliver I, Briscoe D, et al. (2004) Spatial scaling of microbial eukaryote diversity. Nature 432: 747–750.
20. Bell T, Ager D, Song J-I, Newman Ja, Thompson IP, et al. (2005) Larger islands house more bacterial taxa. Science 308: 1884.
21. Head IM, Saunders JR, Pickup RW (1998) Microbial evolution, diversity, and ecology: a decade of ribosomal RNA analysis of uncultivated microorganisms. Microb Ecol 35: 1–21.
22. Woodcock S, Curtis TP, Head IM, Lunn M (2006) Taxa-area relationships for microbes: the unsampled and the unseen. Ecology 9: 805–812.
23. Nekola JC, White PS (2004) The distance decay of similarity in biogeography and ecology. J Biogeogr 26: 867–878.
24. Bell T (2010) Experimental tests of the bacterial distance-decay relationship. ISME J 4: 1357–1365.
25. Angel R, Claus P, Conrad R (2012) Methanogenic archaea are globally ubiquitous in aerated soils and become active under wet anoxic conditions. ISME J 6: 847–862.
26. Conrad R (2007) Microbial ecology of methanogens and methanotrophs. Adv Agron 96: 1–63.
27. Nemergut DR, Schmidt SK, Fukami T, O'Neill SP, Bilinski TM, et al. (2013) Patterns and processes of microbial community assembly. Microbiol Mol Biol Rev 77: 342–356.
28. Stackebrandt E, Goebel BM (1994) Taxonomic Note: A Place for DNA-DNA Reassociation and 16S rRNA Sequence Analysis in the Present Species Definition in Bacteriology. Int J Syst Bacteriol 44: 846–849.
29. Springer E, Sachs MS, Woese CR, Boone DR (1995) Partial gene sequences for the A subunit of methyl-coenzyme M reductase (mcrI) as a phylogenetic tool for the family Methanosarcinaceae. Int J Syst Bacteriol 45: 554–559.
30. Marani L, Alvalá PCC (2007) Methane emissions from lakes and floodplains in Pantanal, Brazil. Atmos Environ 41: 1627–1633.
31. Guerrini V (1978) Bacia do alto rio Paraguai: estudo climatológico. Brasília: EDIBAP/SAS.
32. Bezerra M de O, Mozeto A (2008) Deposição de carbono orgânico na planície de inundação do rio Paraguai durante o Holoceno Médio. Oecol Bras 12: 155–171.
33. Weisburg WG, Barns SM, Pelletier Da, Lane DJ (1991) 16S ribosomal DNA amplification for phylogenetic study. J Bacteriol 173: 697–703.
34. Großkopf R, Janssen PH, Liesack W (1998) Diversity and structure of the methanogenic community in anoxic rice paddy soil microcosms as examined by cultivation and direct 16S rRNA gene sequence retrieval.
35. Dunbar J, Ticknor LO, Kuske CR (2001) Phylogenetic specificity and reproducibility and new method for analysis of terminal restriction fragment profiles of 16S rRNA genes from bacterial communities. Appl Environ Microbiol 67: 190–197.
36. Roger Bray J, Curtis JTT, Bray JRR (1957) An ordination of the upland forest communities of southern Wisconsin. Ecol Monogr 27: 325–349.
37. Harte J, McCarthy S, Taylor K, Kinzig A, Fischer ML (1999) Estimating Species-Area Relationships from Plot to Landscape Scale Using Species Spatial-Turnover Data. Oikos 86: 45.
38. Van der Gast CJ, Lilley AK, Ager D, Thompson IP (2005) Island size and bacterial diversity in an archipelago of engineering machines. Environ Microbiol 7: 1220–1226.
39. Fierer N, Jackson RB (2006) The diversity and biogeography of soil bacterial communities. Proc Natl Acad Sci U S A 103: 626.
40. Noguez AMM, Arita HTT, Escalante AEE, Forney LJJ, Garcia-Oliva F, et al. (2005) Microbial macroecology: highly structured prokaryotic soil assemblages in a tropical deciduous forest. Glob Ecol Biogeogr 14: 241–248.
41. Falz KZ, Holliger C, Großkopf R, Liesack W, Nozhevnikova AN, et al. (1999) Vertical distribution of methanogens in the anoxic sediment of Rotsee (Switzerland). Appl Environ Microbiol 65: 2402–2408.
42. Lovley DR, Klug MJ (1986) Model for the distribution of sulfate reduction and methanogenesis in freshwater sediments. Geochim Cosmochim Acta 50: 11–18.
43. Chan OC, Claus P, Casper P, Ulrich A, Lueders T, et al. (2005) Vertical distribution of structure and function of the methanogenic archaeal community in Lake Dagow sediment. Environ Microbiol 7: 1139–1149.
44. Liu WT, Marsh TL, Cheng H, Forney LJ (1997) Characterization of microbial diversity by determining terminal restriction fragment length polymorphisms of genes encoding 16S rRNA. Appl Environ Microbiol 63: 4516–4522.
45. Eiler A, Bertilsson S (2004) Composition of freshwater bacterial communities associated with cyanobacterial blooms in four Swedish lakes. Environ Microbiol 6: 1228–1243.
46. Marsh TL (1999) Terminal restriction fragment length polymorphism (T-RFLP): an emerging method for characterizing diversity among homologous populations of amplification products. Curr Opin Microbiol 2: 323–327.
47. Inagaki F, Nunoura T, Nakagawa S, Teske A, Lever M, et al. (2006) Biogeographical distribution and diversity of microbes in methane hydrate-bearing deep marine sediments on the Pacific Ocean Margin. Proc Natl Acad Sci U S A 103: 2815–2820.
48. Ochsenreiter T, Selezi D, Quaiser A, Bonch-Osmolovskaya L, Schleper C (2003) Diversity and abundance of Crenarchaeota in terrestrial habitats studied by 16S RNA surveys and real time PCR. Environ Microbiol 5: 787–797.
49. Bowman JP, McCuaig RD (2003) Biodiversity, community structural shifts, and biogeography of prokaryotes within Antarctic continental shelf sediment. Appl Environ Microbiol 69: 2463.

50. Lindström ES, Leskinen E (2002) Do neighboring lakes share common taxa of bacterioplankton? Comparison of 16S rDNA fingerprints and sequences from three geographic regions. Microb Ecol 44: 1–9.
51. Yannarell AC, Triplett EW (2004) Within and between Lake Variability in the Composition of Bacterioplankton Communities: Investigations Using Multiple Spatial Scales. Appl Environ Microbiol 70: 214.
52. Schauer R, Bienhold C, Ramette A, Harder J (2010) Bacterial diversity and biogeography in deep-sea surface sediments of the South Atlantic Ocean. ISME J 4: 159–170.
53. Galand PE, Potvin M, Casamayor EO, Lovejoy C (2010) Hydrography shapes bacterial biogeography of the deep Arctic Ocean. ISME J 4: 564–576.
54. Xiong J, Liu Y, Lin X, Zhang H, Zeng J, et al. (2012) Geographic distance and pH drive bacterial distribution in alkaline lake sediments across Tibetan Plateau. Environ Microbiol 14: 2457–2466.
55. Rosselló-Mora R, Lucio M (2008) Metabolic evidence for biogeographic isolation of the extremophilic bacterium Salinibacter ruber. ISME J 2: 242–253.
56. Lin W, Wang Y, Gorby Y, Nealson K, Pan Y (2013) Integrating niche-based process and spatial process in biogeography of magnetotactic bacteria. Sci Rep 3: 1643.

3

Invasibility of Mediterranean-Climate Rivers by Non-Native Fish: The Importance of Environmental Drivers and Human Pressures

Maria Ilhéu[1,2]*, Paula Matono[1,2], João Manuel Bernardo[1]

1 Departamento de Paisagem Ambiente e Ordenamento, Escola de Ciências e Tecnologia, Universidade de Évora, Évora, Portugal, 2 Instituto de Ciências Agrárias e Ambientais Mediterrânicas, Universidade de Évora, Évora, Portugal

Abstract

Invasive species are regarded as a biological pressure to natural aquatic communities. Understanding the factors promoting successful invasions is of great conceptual and practical importance. From a practical point of view, it should help to prevent future invasions and to mitigate the effects of recent invaders through early detection and prioritization of management measures. This study aims to identify the environmental determinants of fish invasions in Mediterranean-climate rivers and evaluate the relative importance of natural and human drivers. Fish communities were sampled in 182 undisturbed and 198 disturbed sites by human activities, belonging to 12 river types defined for continental Portugal within the implementation of the European Union's Water Framework Directive. Pumpkinseed sunfish, *Lepomis gibbosus* (L.), and mosquitofish, *Gambusia holbrooki* (Girard), were the most abundant non-native species (NNS) in the southern river types whereas the Iberian gudgeon, *Gobio lozanoi* Doadrio and Madeira, was the dominant NNS in the north/centre. Small northern mountain streams showed null or low frequency of occurrence and abundance of NNS, while southern lowland river types with medium and large drainage areas presented the highest values. The occurrence of NNS was significantly lower in undisturbed sites and the highest density of NNS was associated with high human pressure. Results from variance partitioning showed that natural environmental factors determine the distribution of the most abundant NNS while the increase in their abundance and success is explained mainly by human-induced disturbance factors. This study stresses the high vulnerability of the warm water lowland river types to non-native fish invasions, which is amplified by human-induced degradation.

Editor: Maura (Gee) Geraldine Chapman, University of Sydney, Australia

Funding: This study was partially funded by INAG (Instituto da Água) as part of a larger programme on the implementation of the Water Framework Directive and by the Foundation for Science and Technology (FCT) through the research project (PTDC/AAC-AMB/102541/2008). Additional funding was provided by FEDER under the programme COMPETE and by National Funds through FCT under the Strategic Project PEst-OE/AGR/UI0115/2014. P. Matono was supported by a PhD grant from FCT (SFRH/BD/23435/2005). The funders had no role in study design, data collection and analysis, decision to publish, or preparation of the manuscript.

Competing Interests: The authors have declared that no competing interests exist.

* Email: milheu@uevora.pt

Introduction

The rate and extent of invasions in freshwater ecosystems are particularly alarming in the Mediterranean region, which is among the most heavily invaded ecosystems in the world [1–3]. The Iberian freshwater habitats are one of the most paradigmatic examples where there are constant reports of new invading fish species and colonization of new areas (see [4–10] as examples).

The success of biological invasions can be explained by several factors, with environmental drivers being among the most important ones [11–12]. One of the most frequently stated hypotheses in the biological invasion literature is that species should have a greater chance of success if they are introduced to an area with a climate that closely matches that of their original range [13–15]. Other environmental drivers, such as spatial heterogeneity and environmental variability, may also be important [16–18]. The success of an invading fish may be predicted with reference to environmental conditions at different scales, ranging from habitat to river types or ecoregions. Some of the features of the recipient ecosystem and the success of an introduced species may be determined by human-induced pressure. The "human activity" hypothesis argues that human activities facilitate the establishment of non-native species (NNS) by disturbing natural landscapes and by increasing propagule pressure [2,8]. In Mediterranean-climate rivers, both landscape and human disturbance factors are expected to play a major role in the biological invasions, as these systems are largely governed by stochastic processes and have suffered a long history of human-induced pressure [19].

Although non-native species are not mentioned specifically in the European Union's Water Framework Directive (WFD) [20], in the context of the Directive's objectives NNS represent an important pressure (listed in the WFD annexes), since they can modify the structure of native biota and the ecological functioning of aquatic systems. NNS affect the ecological quality of natural environments in multiple ways and may represent a serious threat to native communities. Potential effects include genetic alterations within populations, spreading of pathogens and parasites, competition with and replacement of native species, and habitat deterioration or modification (see [21] for Iberian freshwater ichthyofauna). All the combined effects may result in changes in ecosystem function and global homogenization [1,3,22–23], with profound impact at ecological, evolutionary, genetic, and economic levels [24–25].

The apparently strong relationship between introduced fish species and degraded river conditions and their potential impact on native species and ecosystem suggest that NNS may be useful indicators of biological integrity and river health [26–28] and they are therefore incorporated in river monitoring schemes under the WFD and elsewhere. The incorporation of NNS into the ecological assessment will require knowledge about the density, distribution and potential risk of each species for each water body. According to the WFD, with the exception of particular cases, the ecological assessment involves the previous derivation of typologies for the aquatic systems. The ecological assessment is performed within groups of similar ecosystems and is expressed as a deviation from the reference conditions, i.e. conditions found in sites without human pressures. In fact, differences in climate, hydrology, geomorphology, geology, soil and vegetation make comparison of river communities difficult if not impossible. The use of river typology, by stratifying the spatial variability, makes the ecological assessment more practical, transparent and understandable by decision-takers, managers, technicians and public, and for these reasons it was the chosen approach on the WFD implementation. Moreover, approaching invasibility at a regional framework is useful for management purposes because it allows the identification of water bodies particularly vulnerable to NNS invasion and helps to regionalize monitoring schemes [29–31].

Understanding the factors promoting successful invasions is of great conceptual and practical importance. From a practical point of view, it should help to prevent future invasions and to mitigate the effects of recent invaders through early detection and prioritization of management measures. Thus, the identification of the environmental determinants of fish invasions is relevant in forecasting the overall impact of invasions on a global scale and prerequisite management authorities to adopt sound conservation policies.

The objectives of the present study were to determine: i) the patterns of non-native fish richness and abundance in Mediterranean-climate river types of Portugal; ii) the environmental drivers that favour the invasiveness of non-native fish within the morpho-climatic gradients; iii) the relationship between human-induced disturbance and non-native fish abundance and iv) the relative importance of environmental variables and human pressures for the occurrence and abundance of non-native fish species.

Methods

Study Area

The study considered 12 river types out of 15 defined for continental Portugal under the implementation of the WFD [32] (Fig. 1), as the 3 very large and highly modified river types were not considered.

The climate of Portugal presents high intra- and inter-annual precipitation and discharge variation, with severe and unpredictable floods between autumn and spring and persistent summer droughts [33], although the influence of factors such as topography and proximity to the Atlantic Ocean causes significant climatic contrasts. The general conditions of the atmospheric circulation cause a decrease in precipitation from north to south and from west (coast) to east (inland), enhanced by orographic asymmetry. Indeed, the mountain barrier in the north and center causes less rainfall in the interior regions. The temperature shows an opposite pattern, increasing from north to south. In a general way, the altitude causes a decrease in temperature and an increase in rainfall.

Portuguese river types reflect two main geographical and climatic gradients: north–south and west–east. The north–south gradient is associated with a decrease in altitude, precipitation, annual discharge, and increasing temperature. The west–east gradient is related to a continental effect, with a precipitation decrease and an increase in temperature extremes. Owing to these features, most rivers are permanent in the north and intermittent in the south. The most relevant climatic and morphological characteristics of the basins of the river-types are presented in Table 1 (see also [32]). The river types may be described as:

M - Mountain rivers of the North: trout rivers with small drainage area located in the North-West and North-Center regions with low mean temperature and high precipitation; N1< 100 km^2 - North rivers of small drainage area: low mean temperature, altitude mainly in the 200–600 m range; N1> 100 km^2 - North rivers of medium and large drainage area: characteristics similar to the previous type; N2 - Medium-large Alto Douro rivers: drainage area larger than 100 km^2, high temperature (similar to the south types), lower rainfall; N3 - Small Alto Douro rivers: drainage area smaller than 100 km^2 and characteristics similar to the previous type; N4 - North-South transition type: located at the center region with intermediate characteristics; L - Littoral type: coastal streams of small-medium dimensions located at the littoral center region, low altitude; S1< 100 km^2 - Small South rivers: a large group of low altitude temporary rivers, in a region of high temperature and low precipitation; S1>100 km^2 - South rivers of medium-large dimension: larger drainage temporary rivers with similar characteristics to the previous type; S2 - South mountain rivers: small streams in the mountain regions of the south with lower temperature and higher precipitation than in the other south river types; S3 - Rivers of Tagus and Sado sedimentary basins: small and medium dimension, low altitude and high mineralization, smaller rivers with temporary regime; S4 - Small chalk rivers of Algarve: low altitude, low precipitation and high temperature.

Native freshwater fish fauna of Portuguese rivers presents relatively low species richness per site and most fish species are endemic cyprinids with high conservation value, particularly in the south. Many of those species are threatened with extinction [34–35].

Sampling and data collection

Sampling was carried out between 2004 and 2006 during spring at 380 sites in the main Portuguese river basins (Fig. 1). The number of sampled sites in each river type was proportional to its basin area: M = 28 sites; N1<100 km^2 = 60 sites; N1> 100 km^2 = 68 sites; N2 = 20 sites; N3 = 28 sites; N4 = 16 sites; L = 33 sites; S1<100 km^2 = 26 sites; S1>100 km^2 = 33 sites; S2 = 22 sites; S3 = 37 sites; S4 = 9 sites. In each river type, sampling sites included undisturbed or least disturbed sites and disturbed ones. For the site selection, a preliminary human-induced pressure screening using GIS and information on pollution loads was followed. The final selection was based on the human disturbance level regarding ten variables [36–37]: land use, urban area, riparian zone, connectivity of the river segment, sediment load, flow regime, morphological condition, presence of artificial lentic water bodies, toxic and acidification levels, and nutrient/organic load. The spatial scale of the assessment depended on the character of each pressure; some were made at local scale, others at the fluvial segment, and some others at the basin scale. Each variable was scored from 1 (minimum

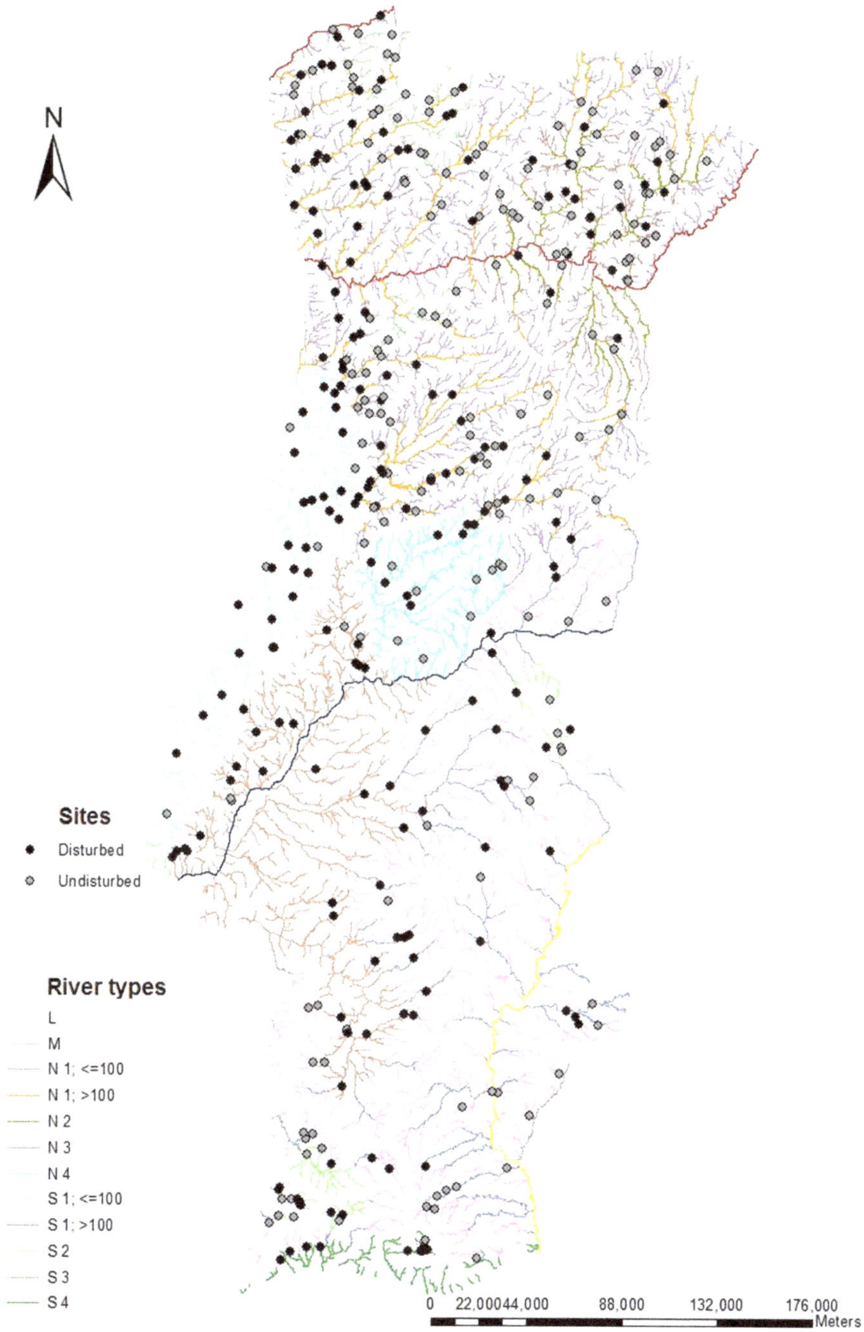

Figure 1. Map of the river types defined to continental Portugal showing undisturbed (grey dots) and disturbed (black dots) sampling sites.

disturbance) to 5 (maximum disturbance) (Tables 2 and 3) and only sites with scores of 1 and/or 2 and only one variable scored with a 3 were considered as undisturbed or least disturbed (i.e., reference sites). The sum of these scores represents the total human pressure. Several water quality variables complemented the evaluation of human pressure in each site, following Standard Methods for the Examination of Water and Wastewater [38]: Biochemical Oxygen Demand (BOD_5, mg L^{-1}), Chemical Oxygen Demand (COD, mg L^{-1}), Total Suspended Solids (TSS, mg L^{-1}), Soluble Reactive Phosphorous (SR-P, mg L^{-1}), nitrite (NO_2^-, mg L^{-1}), nitrate (NO_3^-, mg L^{-1}), ammonium (NH_4^+, mg L^{-1}), and total inorganic dissolved nitrogen (N, mg L^{-1}).

Environmental characterization of sites was based on regional and local variables. Regional variables were obtained from digital cartography with free Internet access and included the drainage area of the basin upstream the site (km^2), distance from source (km), altitude (m), slope (%), mean annual discharge (mm), mean annual air temperature (°C), and mean annual rainfall (mm). Rainfall, temperature, and flow variables were described from 30-year data series. Topographical variables were derived from a Digital Elevation Model (DEM), with a 90-m grid cell resolution

Table 1. Main morpho-climatic characteristics of river types in continental Portugal (mean and SD).

River-Types	Mean Annual Temperature (°C)	Mean Annual Precipitation (mm)	Altitude (m)	Drainage Area (km^2)
M	11.0	1944	506	24.8
northern mountain streams	(1.5)	(379)	(300)	(17)
N1<100 km^2	12.4	1190	413	33
north streams with small drainage area	(1.3)	(358)	(242)	(23)
N1>100 km^2	12.6	1196	274	549
north streams with large drainage area	(1.2)	(374)	(205)	(65)
N2	13.1	596	300	960
streams from Alto Douro with large drainage area	(1.0)	(81)	(141)	(1115)
N3	13.0	671	432	32
streams from Alto Douro with small drainage area	(0.8)	(134)	(160)	(23)
N4	14.1	1065	280	151
transition streams between north and south	(0.7)	(168)	(122)	(361)
L	14.8	941	44	180
littoral streams from west/centre region	(0.3)	(118)	(44)	(671)
S1<100 km^2	15.7	628	183	30
south streams with small drainage area	(0.9)	(86)	(75)	(21)
S1>100 km^2	15.8	587	137	439
south streams with large drainage area	(0.9)	(84)	(68)	(579)
S2	15.4	743	175	60
southern mountain streams	(0.3)	(85)	(147)	(87)
S3	15.6	730	54	388
streams of sedimentary deposits in Tagus and Sado basins	(0.4)	(118)	(46)	(1081)
S4	16.9	632	54	67
southern carsick streams of Algarve	(0.5)	(60)	(57)	(89)

(CGIAR-CSI 2005), using ArcMap 9.1. Local variables were assessed *in situ* at each sampling occasion: water temperature (°C), conductivity (µS cm^{-1}), pH, dissolved oxygen (mg L^{-1}), mean stream wetted width (m), maximum and mean water depth (m), mean current velocity (m s^{-1}), dominant substrate class, adapted from the Wentworth scale [39]: 1 – mud and sand; 2 – gravel; 3 – pebble; 4 – cobble; 5 – boulders; 6 – boulders larger than 0.50 m, riparian vegetation (%), shadow (%), and proportion of different habitat types (pool, run, riffle).

Human disturbance level was evaluated *in situ* regarding the same ten variables described above for site selection. A total of 182 undisturbed and 198 disturbed sites were sampled covering the full human pressure gradient in each river type. The proportion of sites along the degradation gradient was sampled according to their availability within each river type.

Fish were collected by electrofishing following the WFD-compliant and CEN sampling protocol [40–41]. A backpack battery-powered electrofishing equipment was used, wading in shallow reaches (<1.2 m) or from a boat in deeper areas. The power and pulse frequency of the equipment were adjusted to ensure capture efficiency but also that fish were only stunned for a short period of time. Captured specimens were carefully handled, ensuring the minimum possible stress, and were quickly returned to the river. No specimens of endangered or protected species were injured or sacrificed.

The present study was part of a large programme coordinated by the National Water Institute (INAG, Instituto da Água, now part of APA, Agência Portuguesa do Ambiente). The National Institute for the Nature Conservation and Forestry (ICNF, Instituto de Conservação da Natureza e das Florestas (http://www.proforbiomed.eu/project/partners/institute-conservation-nature-and-forest-icnf) provided the necessary fishing permits. All sampling took place in national public rivers, which were under the jurisdiction of the National Water Institute and ICNF. As all the fish manipulation was restricted to the sampling procedures and animal care was taken into account, it was not necessary to obtain permission from the National *Animal Care* and Use *Committee* (Comissão de Ética, Bioética e Bem-estar Animal).

Captures were quantified as density (individuals/100 m^2) and biomass (grams/100 m^2). The degree of invasibility was measured using NNS richness and abundance.

A nonparametric ANCOVA performed using the rank transformation analysis of covariance evaluated the existence of significant differences in the density and number of NNS between river types [42,43]. The total human pressure was included as a covariable, in order to exclude its effects from the analysis.

Mann-Whitney test was used to identify significant differences in the number and density of NNS between undisturbed and disturbed sites. The Z test of proportions [44] evaluated the existence of significant differences in the frequency of occurrence of NNS between undisturbed and disturbed sites.

Table 2. Description, assessment scale and methods, and scoring criteria of the variables; land use, urban area, riparian vegetation, morphological condition and sediment load - used to evaluate the level of anthropogenic disturbance in sampled sites.

Variables	Description	Assessment scale	Score	Criteria	Methods
Land use	Impact of farming/ forestry practices	River segment	5	>40% Agricultural use (intensive agriculture), very severe impact (rice field)	Local expert assessment complemented with Corine Land Cover (2000, 2006)*
			4	>40% Strong impact (area with strong forestry, including clearcuts)	
			3	<40% Moderate impact (subsistence gardens, pastures)	
			2	<40% Small impact (cork and holm oaks, high-growth forest)	
			1	<10% No significant impacts (natural forest and bush)	
	Land cover and bankface characterization	Local	5	Irrigated crops and/or high stocking	Local expert assessment complemented with Corine Land Cover (2000, 2006)*
			4	Horticultural crops, semi-intensive grazing	
			3	Extensive cultures (e.g. pastures, cereal crops, pine, eucalyptus), extensive grazing	
			2	Cork and holm oaks	
			1	Natural	
Urban area	Impact of urban areas	River segment	5	Very severe (location near a city with basic sanitation needs)	Local expert assessment complemented with Corine Land Cover (2000, 2006) *
			4	Town	
			3	Village	
			2	Hamlet	
			1	Negligible (isolated dwellings)	
Riparian vegetation	Deviation from the natural state of the riparian zone	River segment	5	Lack of riparian shrubs and trees (only the presence of annual plants)	Local expert assessment
			4	Fragmented vegetation with bushes and/or the presence of reed	
			3	Second replacement step (dominance of dense brushwood)	
			2	First replacement step (presence of shrub or tree strata with some level of preservation).	
			1	Potential vegetation (presence of shrub and tree strata according to the geo-series)	
Morphological condition	Deviation from the natural state of the stream bed and banks	Local	5	Transverse and longitudinal profile of the channel completely changed, with very few habitats	Local expert assessment
			4	Channelized sector, missing most of the natural habitats	
			3	Channelized sector, missing some types of natural habitats, but maintaining much of the shape of the natural channel	
			2	Poorly changed sector, close to the natural mosaic of habitats.	
			1	Morphological changes absent or negligible	

Table 2. Cont.

Variables	Description	Assessment scale	Score	Criteria	Methods
Sediment load	Deviation from the natural sediment load (both carried in the water column and deposited on the riverbed)	River segment and local	5	>75% of coarse particles of the stream bed are covered with fine sediments (sand, silt, clay)	Local expert assessment
			4	50–75% of coarse particles of the stream bed are covered with fine sediments (sand, silt, clay)	
			3	25–50% of coarse particles of the stream bed are covered with fine sediments (sand, silt, clay)	
			2	5–25% of coarse particles of the bed are covered with fine sediments (sand, silt, clay)	
			1	<5% of coarse particles of the stream bed are covered with fine sediments (sand, silt, clay)	

*information available from http://sniamb.apambiente.pt/clc/frm/.

The response pattern of the most abundant NNS to the human-induced pressure gradient was tested with a nonparametric ANCOVA (detailed above) [42,43] of NNS density between five human pressure classes. The establishment of these five quality classes (High, Good, Moderate, Poor and Bad) for the stressor gradient (total human pressure) to which sites were assigned, followed the approach used in the REFCOND Guidance Document [37]. To account for the possible influence of the environmental variables, a Principal Components Analysis (outputs not shown) was performed in order to extract the most relevant environmental gradient (PCA1), which was then included as a covariable in this analysis. For each species, only data from river types with occurrences were considered.

Univariate Generalized Linear Models (GLM) were used for partitioning the variance [45] among three sets of explanatory variables - environmental, human-induced pressure and spatial trends on the occurrence and density of the most abundant NNS. For each species, only data from river types with NNS occurrences were considered. The set of environmental variables included altitude, mean annual discharge, drainage area of the basin, slope, mean annual temperature, mean water depth, mean stream width, mean current velocity, dominant substrate class, proportion of each habitat type, pH, dissolved oxygen and conductivity. The set of human pressure variables included BOD_5, COD, TSS, N, SRP, and the 10 human pressure variables evaluated at each site (Table 2). The spatial structure of data was incorporated into the analyses to prevent misinterpretation of relations between and within the spatially arranged datasets [46]. Owing to spatial autocorrelation, values of particular variables in neighbouring sites are more or less similar than they would be in a random set of observations [47]. Autocorrelation is a frequently observed feature in spatially sampled biological data that may make the identification of plausible relationships between biota and the environment difficult [48]. The spatial structure of data was explored, including geographical coordinates of sites and their higher and cross product terms, in the modelling procedure (x, y, xy, x^2, y^2, x^2y, xy^2, x^3 and y^3) [47]. The x and y coordinates were centred to zero mean before computing the matrix of spatial descriptors, to reduce collinearity between successive terms when fitting the polynomial. GLMs were performed using a forward selection procedure of the explanatory variables. The best models (minimal adequate) were estimated according to the lowest Akaike Information Criterion (AIC). Plots of residuals (not shown) were examined to complement AIC values and to confirm goodness-of-fit (see [49]). For the occurrence of species (presence–absence data), GLMs were performed using the binomial distribution and logit link function. In order to overcome the problems of distribution fitting resulting from the high number of absences in the density matrix, this continuous response variable was standardized to the maximum value for the species and converted into four classes: 0 (0% of individuals), 1 (0% - 10% of individuals), 2 (10% - 50%) and 3 (>50%). For this transformed density response variable (count data), GLMs were performed using the Poisson distribution and log link function. The existence of over-dispersion in data (variance higher than the mean; dispersion parameter>1) was checked during the analysis. If the value observed was higher than the threshold limit then quasi-binomial and quasi-Poisson distributions should be used, respectively. The total variation within the response variables was decomposed among the three sets of explanatory variables and the percentage of explained deviance calculated for eight different components [50]: (i) pure effect of environmental drivers, (ii) pure effect of anthropogenic disturbance, (iii) pure effect of spatial trends, (iv) combined variation due to the joint effect of environmental and anthropogenic components, (v) combined variation due to the joint effect of environmental and spatial components, (vi) combined variation due to the joint effect of anthropogenic and spatial components, (vii) combined variation due to the joint effect of the three components and (viii) variation not explained by the independent variables included in the analysis.

Multicollinearity among explanatory variables may result in the exclusion of ecologically meaningful variables from the models if another intercorrelated variable or variables happen to explain the variation better in statistical terms. Therefore, some of the most clearly intercorrelated variables (Spearman's rank correlation $|r|>0.75$; $P<0.05$) were initially excluded. The exclusion decision took into account the potential relevance of the variable in the occurrence and distribution of NNS. Furthermore variables were maintained in the models only if their addition did not cause any Variation Inflation Factor (VIF) to exceed 3.0, therefore ensuring that no covariation exists between the selected variables in the models.

Table 3. Description, assessment scale and methods, and scoring criteria of the variables: hydrological regime, toxic and acidification levels, organic and nutrient loads, artificial lentic waters - used to evaluate the level of anthropogenic disturbance in sampled sites.

Variables	Description	Assessment scale	Score	Criteria	Methods
Hydrological regime	Deviation from the natural hydrological regime (flow pattern and/or quantity). Includes all sources of hydrologic alteration, such as significant water abstraction	Local	5	<50% and strong deviation from the natural variability of the flow regime	Local expert assessment complemented with SNIRH
			4	<50% and moderate deviation from the natural variability of the flow regime	
			3	>50% and duration of flood periods close to the natural	
			2	>75% and duration of flood periods close to the natural	
			1	>90% and normal duration of natural flood periods	
		Local	5	<10% of mean annual discharge	Local expert assessment complemented with SNIRH
			4	<15% of mean annual discharge	
			3	>15% of mean annual discharge	
			2	>30% of mean annual discharge	
			1	>90% of mean annual discharge	
Toxic and acidification levels	Deviation from the natural state of toxicity conditions, including acidification and oxygen levels	Local	5	Constant for long periods (months) or frequent occurrence of strong deviations from natural conditions (e.g. pH <5.0, DO <30%)	Local expert assessment complemented with SNIRH
			4	Constant for long periods (months) or frequent occurrence of strong deviations from natural conditions (e.g. pH <5.5, DO <30–50%)	
			3	Occasional deviations (single measurements or episodic) in relation to natural conditions (e.g. pH <5.5, DO <30–50%)	
			2	Occasional deviations (single measurements or episodic) in relation to natural conditions (e.g. pH <6.0)	
			1	Conditions within the normal range of variation	
Organic and nutrient loads	Deviation from the normal values of BOD, COD, ammonium, nitrate and phosphate concentrations	Local	5	>20% of values in classes D or E	SNIRH (classification of water quality for multiple uses, according to the guidelines from the National Water Institute*), complemented with local expert assessment
			4	>10% of values in classes D or E	
			3	>10% of values in class C	
			2	No obvious or too small signs of eutrophication and organic loading	
			1	No signs of eutrophication and organic loading	
Artificial lentic water bodies	Impact related to the presence of artificial lentic water bodies upstream and/or downstream of the site (upstream change in thermal and flow regimes; downstream invasion by exotic species of lentic character)	Local	5	Local immediately downstream of a large reservoir or within the influence area of its backwater	SNIRH and available cartography

Table 3. Cont.

Variables	Description	Assessment scale	Score	Criteria	Methods
			4	Local immediately downstream of a mini-hydro or within the influence area of its backwater	
			3	Local downstream of a reservoir or within the influence area of the reservoir	
			2	Local downstream of a mini-hydro or within the influence area of its backwater	
			1	No influence of reservoirs	
Connectivity	Impact of artificial barriers to fish migration	River basin and segment	5	Permanent artificial barrier	SNIRH, available cartography, documental data and local expert assessment
			4	Occasional passage of some species	
			3	Passage of certain species or only in certain years	
			2	Passage of most species in most years	
			1	No barriers or existence of an effective pass-through device	

*information available from http://snirh.pt/snirh/_dadossintese/qualidadeanuario/boletim/tabela_classes.php.

Prior to the analyses, all data were either log (x+1) (linear measurements) or arcsin [sqrt(x)] (percentages) transformed to improve normality. Statistical analyses were performed using CANOCO 4.5, STATISTICA 6, PRIMER 6, SPSS 21 and BRODGAR 2.6 software applications.

Results

Patterns of richness, distribution and abundance of non-native species

A total of 41 fish species were captured, including 6 diadromous species and 10 NNS: pumpkinseed sunfish *Lepomis gibbosus* (L.), Iberian gudgeon *Gobio lozanoi* Doadrio and Madeira, mosquitofish *Gambusia holbrooki* (Girard), carp *Cyprinus carpio* L, goldfish *Carassius auratus* (L.), largemouth bass *Micropterus salmoides* (Lacépède), chameleon cichlid *Herichthys facetum* (Jenyns), black bullhead *Ameiurus melas* (Rafinesque), pike-perch *Sander lucioperca* (L.), and bleak *Alburnus alburnus* (L.) (Table 4).

NNS occurred in 33% of the sites, representing nearly 11% of both the total mean density (3.58 ind./100 m^2, SD = 12.54) and number of species (0.54, SD = 0.94) per site. Although the absolute values may seem low, they represent an important percentage of the total density and number of species, as Mediterranean streams usually show low species richness and density per site. Biomass values were not shown or included in the analysis, as they followed the density pattern and were therefore redundant.

There were significant differences in the density ($F_{(11,380)} = 8.04$; $P<0.001$) and proportion of non-native species ($F_{(11,380)} = 9.20$; $P<0.001$) among river types considering the total human pressure as a covariable in the analysis. NNS did not occur in the mountain river type or in the southern chalk river type. The small northern stream types, N1<100 km^2, N3 and N4, registered very low densities, frequency of occurrence, and percentage of NNS (Table 4). N2 and S2 also showed low densities of NNS, but these species represented more than 15% of the fish assemblages in each site and registered high values of frequency of occurrence (f. oc.). N1>100 km^2 and S1<100 km^2 presented high values of occurrence and density of NNS, also representing more than 10% of total density and species richness. Littoral and southern river types S1> 100 km^2 and S3 showed the highest occurrence and abundance of NNS; in Littoral type, NNS represented almost 12% of the total density and 15% of the total species richness, S1>100 km^2 and S3 registered percentages between 20% and 30%.

The most frequent and abundant NNS were *L. gibbosus* (mean density = 0.89 ind./100 m^2, SD = 3.3; f. oc. = 0.21), *G. lozanoi* (mean density = 1.79 ind./100 m^2, SD = 10.9; f. oc. = 0.14), and *G. holbrooki* (mean density = 0.54 ind./100 m^2, SD = 3.1; f. oc. = 0.11) (Table 4). The remaining species registered low occurrences and abundances. *L. gibbosus* showed a wide distribution in both northern and southern river types - N1> 100 km^2, N2, L, and mostly southern river types (Table 4), which are associated with high annual temperature and conductivity, large drainage area, and low altitude (Table 1). *G. holbrooki* were the most abundant species in the southern river types, showing a distribution almost limited to these river types, while *G. lozanoi* was the dominant species in north/central river types, exhibiting high occurrence in river types with large drainage area and low altitude, especially in the north/centre region - N1>100 km^2, N2, and L (Table 4). *C. carpio* and *M. salmoides* occurred in several river types, showing higher values in the south. *H. facetum*, *A. melas*, and *A. alburnus*, only occurred in the southern river types S1<100 km^2, S1>100 km^2, and S3 and *S. lucioperca* only occurred in N1>100 km^2, all presenting very low frequencies of occurrence and abundance. *C. auratus* registered nearly vestigial occurrence in N1>100 km^2, N2, and S3 (Table 4).

Overall, disturbed sites showed significantly higher occurrence (f. oc. = 0.46, $P<0.001$) and abundance (mean = 6.41, SD = 16.8, $U_{380} = 12155.00$, $Z = -6.56$, $P<0.001$) of NNS than undisturbed ones (f. oc. = 0.18; mean = 0.5, SD = 1.7). *M. salmoides*, *A. melas*, *S. lucioperca*, and *A. alburnus* only occurred in disturbed sites. All NNS but *H. facetum*, exhibited significantly higher occurrence ($P<0.01$) in disturbed sites. Fish density was significantly higher in

Table 4. Frequency of occurrence (f.oc.), number of species (%) and density (mean and SD) (ind/100 m^2) of non-native fish species (NNS) for total data (a), undisturbed (b) and disturbed (c) sites in each river-type (see Table 1) with NNS occurrence.

		NNS f.oc.	Total	N1<100 km^2	N1>100 km^2	N2	N3	N4	L	S1<100 km^2	S1>100 km^2	S2	S3
NNS f.oc.	a)	0.30		0.1	0.4	0.7	0.07	0.2	0.5	0.4	0.7	0.3	0.5
	b)	0.18		0	0.23	0.67	0.12	0.11	0	0.13	0.50	0.25	0.17
	c)	0.46		0.25	0.58	0.80	0	0.29	0.58	0.70	0.79	0.40	0.52
Nbr. NNS	a)		12.0	3.0	13.1	19.3	2.9	4.6	14.8	13.1	31.1	16.3	19.6
(%)			(21.0)	(9.2)	(19.3)	(14.2)	(11.2)	(10.0)	(19.6)	(23.2)	(30.1)	(31.1)	(25.9)
	b)		5.3	0	5.6	19.0	4.9	2.2	0	2.6	17.3	7.4	4.2
			(12.0)		(10.9)	(15.2)	(14.2)	(6.7)		(7.2)	(19.5)	(15.6)	(10.2)
	c)		18.2	6.4	19.0	20.0	0	7.7	18.8	29.8	41.2	27.0	22.6
			(25.2)	(12.7)	(22.4)	(12.6)		(13.1)	(20.3)	(30.1)	(32.8)	(41.7)	(27.0)
NNS Mean density	a)		3.6	2.3	4.8	1.2	0.07	0.3	9.3	3.6	6.4	1.1	6.3
(ind/100 m^2)			(12.5)	(9.1)	(12.8)	(1.6)	(0.3)	(0.8)	(30.1)	(8.6)	(11.5)	(2.1)	(12.5)
	b)		0.5	0	0.4	1.2	0.1	0.3	0	1.6	1.4	0.7	0.2
			(1.7)		(1.2)	(1.8)	(0.5)	(1.0)		(4.4)	(1.8)	(1.6)	(0.4)
	c)		6.4	4.8	8.2	1.2	0	0.2	11.8	6.9	10.1	1.5	7.5
			(16.8)	(13.0)	(16.3)	(0.9)		(0.4)	(33.6)	(12.4)	(14.1)	(2.7)	(13.3)
Lepomis gibbosus		0.21	0.9	0.1	1.1	0.6	0.01	0.3	0.8	2.3	2.1	0.06	2.3
			(3.3)	(0.9)	(3.7)	(0.9)	(0.02)	(0.8)	(2.1)	(6.7)	(4.4)	(0.3)	(5.4)
Cyprinus carpio		0.03	0.1	0	0	0.01	0	0	0.09	0.01	0.9	0	0.08
			(1.6)			(0.01)			(0.4)	(0.06)	(5.5)		(0.4)
Carassius auratus		0.008	0.01	0	0.01	0.01	0	0	0	0	0	0	0.01
			(0.03)		(0.06)	(0.01)							(0.02)
Gobio lozanoi		0.14	1.8	2.2	3.4	0.6	0.07	0	7.5	0	0	0	1.6
			(10.9)	(9.1)	(10.6)	(1.2)	(0.3)		(29.9)				(6.7)
Micropterus salmoides		0.02	0.07	0	0.2	0	0	0	0.2	0	0.1	0.07	0.08
			(0.7)		(1.5)				(0.6)		(0.6)	(0.3)	(0.5)
Herichtys facetum		0.01	0.03	0	0	0	0	0	0	0.3	0.1	0	0
			(0.4)							(1.3)	(0.3)		
Ameiurus melas		0.003	0.1	0	0	0	0	0	0	0	0	0	1.2
			(2.2)										(7.1)
Gambusia holbrooki		0.11	0.5	0	0.03	0	0	0.01	0.7	1.1	2.7	0.9	1.1
			(3.1)		(0.2)			(0.06)	(2.0)	(2.5)	(9.0)	(2.1)	(3.2)
Sander lucioperca		0.003	0.01	0	0.04	0	0	0	0	0	0	0	0
			(0.1)		(0.3)								
Alburnus alburnus		0.005	0.03	0	0	0	0	0	0	0	0.4	0	0
			(0.5)								(1.6)		

disturbed sites only for the most abundant species, *L. gibbosus* ($U_{380} = 15029.50$, $Z = 3.94$, $P<0.001$), *G. lozanoi* ($U_{380} = 15594.50$, $Z = 3.73$, $P<0.001$), *G. holbrooki* ($U_{380} = 14676.50$, $Z = 5.68$, $P<0.001$), and *M. salmoides* ($U_{380} = 17199.00$, $Z = 2.91$, $P<0.01$). In all river types but N3, the occurrence of NNS was significantly higher in disturbed sites ($P<0.01$). The abundance and percentage of NNS were higher in human-disturbed sites in L, N1<100 km^2, N1>100 km^2, S1<100 km^2, and S1>100 km^2 (Table 4). The majority of river types that did not revealed significant differences between undisturbed and disturbed sites presented null or very low abundance of non-native fishes.

No significant correlations (Spearman's rank correlations IrI < 0.5) were observed between the density of the most abundant NNS and total human pressure, as this relationship was not linear. A marked increase in fish density along the human disturbance gradient considering the environmental gradient (PCA1) as a covariable was observed, but there was a decrease in fish abundance in the extreme of the gradient, when the pressure was maximum (Fig. 2). This was particularly evident for *G. lozanoi* ($F_{(4,246)} = 5.34$, $P<0.001$) and *G. holbrooki* ($F_{(4,235)} = 7.07$, $P<0.001$), as mean fish densities reached the highest values in the poor quality class. The increase in fish density from moderate to poor quality classes was particularly marked for *G. holbrooki*. This species showed a comparatively higher mean density in the bad quality class and a smooth decrease between the poor and bad quality classes. *L. gibbosus* densities showed a considerable dispersion in each quality class. Nevertheless, the highest mean density was observed in moderate human pressure conditions, decreasing in the poor quality class and presenting very low values in the extreme of the pressure gradient (Fig. 2). Overall, the response of this species along the disturbance gradient was less marked ($F_{(4,343)} = 0.30$, $P>0.05$) than that observed for the other species.

Relative importance of pure environmental and human pressure variables

The results obtained above showed that both environmental and human pressure variables play a determinant role in the occurrence and abundance of non-native species. Moreover, the performed analyses revealed the existence of significant covariation ($P<0.001$) between these two groups of variables. Therefore, the analysis of the relationship between the occurrence and abundance of the most abundant NNS and the environmental and human pressure gradients was complemented with a GLM approach in order to simultaneously evaluate the relative influence of environmental variables, human-induced pressure, and spatial trends. All GLMs showed over-dispersion parameters under the threshold limit of 1, and were therefore performed using binomial and Poisson distributions as previously described. For all the response variables, most of the explained variability was accounted for by unique, or "pure", components. Shared effects of explanatory variables were responsible for a comparatively small proportion of data variability (Fig. 3). Together, environmental variables and human pressure accounted for the majority of the explained variation in the occurrence and abundance of the three considered NNS (31.7% to 40.2%) (Fig. 3).

The partition of the variance showed some differences in the relative influence of each variable set among the three NNS. The occurrence of NNS was mainly determined by environmental variables, whereas abundance showed a proportionally higher influence of human pressure variables. *L. gibbosus* occurrence and abundance revealed a high dependence on environmental variables. *G. lozanoi* presented a balanced influence of environmental and pressure variables. *G. holbrooki* showed a strong influence of human pressure on both occurrence and abundance (Fig. 3). The results suggest that these three species have different levels of tolerance to human pressure. In all models, spatial variables (autocorrelation) represented the smaller fraction of the explained variation in both occurrence (0.8% to 4.8%) and abundance of the three NNS (1.4% to 2.5%) (Fig. 3). A higher influence of these variables on the occurrence of *G. lozanoi* and *G. holbrooki* was observed, which is in accordance with a more restricted and concentrated distribution of these two species.

Overall, both regional and local explanatory variables were present in the best ecological models in environmental and human pressure GLMs (Table 5). Amongst environmental GLMs, drainage area, in particular, showed a significant reduction in deviance, and positively shaped all the response variables. Slope (positive coefficient) and percentage of riffles (negative coefficient) were ranked as the second most important predictors respectively in G. lozanoi and L. gibbosus occurrence and abundance. For G. holbrooki only mean annual temperature (positive coefficient) revealed a higher positive influence than drainage area. Concerning human pressure variables, toxicity and acidification levels were the most important predictors and were negatively related to the occurrence and abundance of the three NNS. For L. gibbosus,

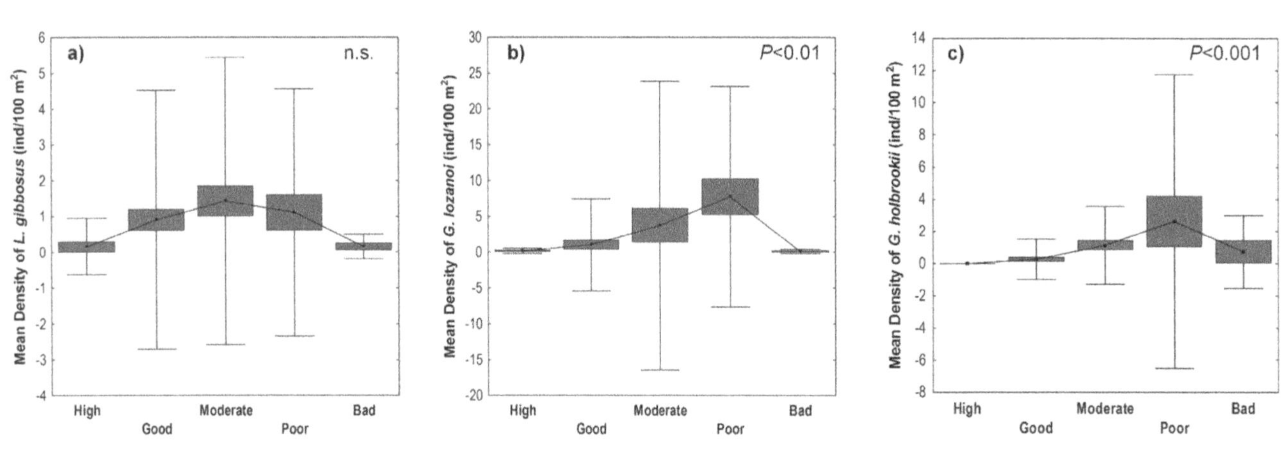

Figure 2. Response pattern of the most abundant NNS to human-induced pressure gradient, established according to 5 environmental quality classes: a) Mean density of *L. gibbosus*; b) Mean density of *G. lozanoi*; c) Mean density of *G. holbrooki*. (■): Mean; box:±SE; whisker:±SD. Significance of ANCOVA results with total human pressure as a covariable is shown.

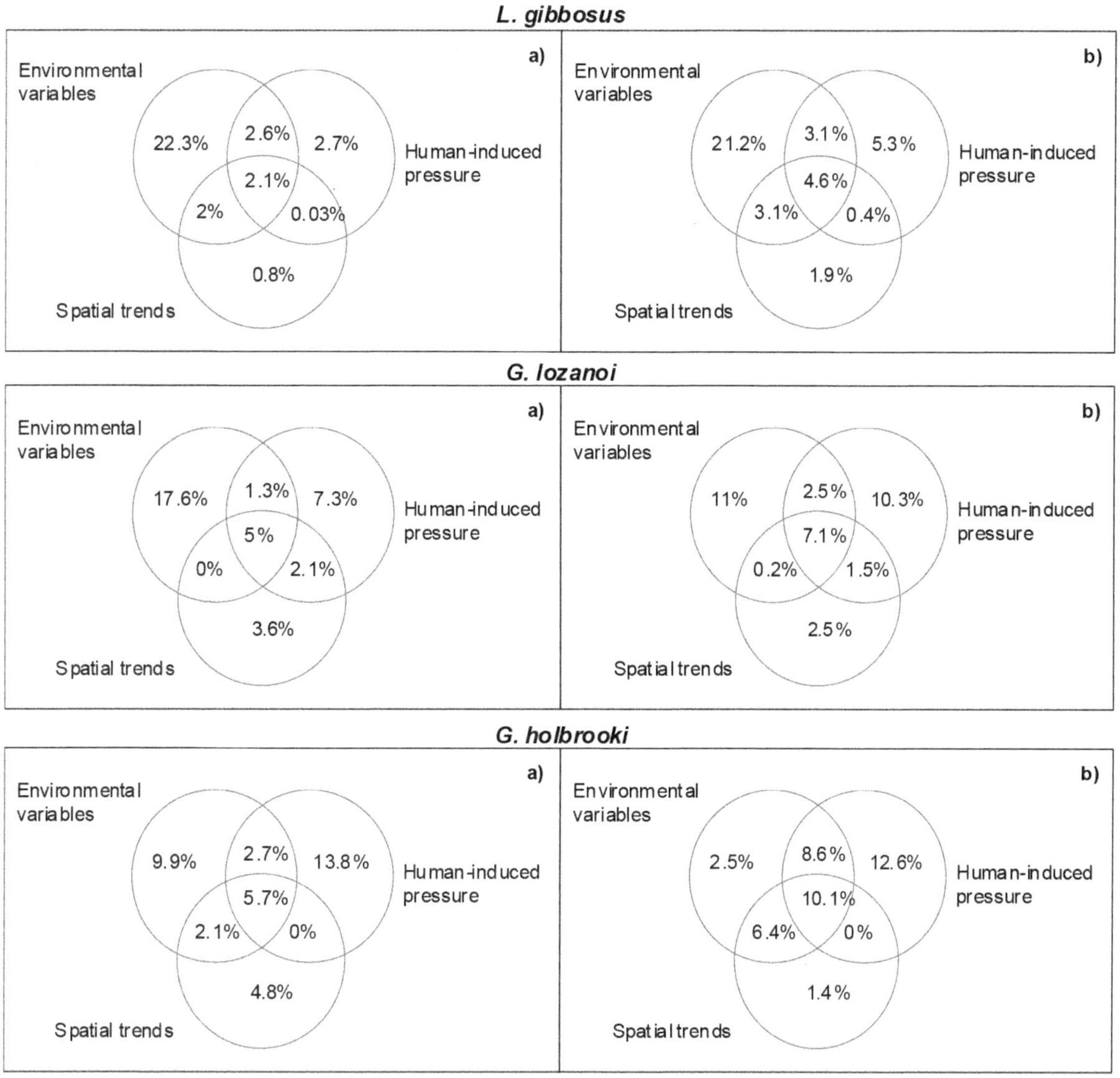

Figure 3. Veen diagrams showing the partition of the variance in the occurrence (a) and mean density (b) of the most abundant NNS explained by environmental variables, human-induced pressure and spatial trends.

TSS and COD also assumed a significant positive influence. Although nutrient/organic loads were retained in both L. gibbosus human pressure GLMs, these variables did not present a relevant decrease of deviance. Total dissolved nitrogen (N) and urban area showed an important relation with G. lozanoi occurrence and density. Sediment load revealed a particular influence on G. holbrooki occurrence, presenting the highest deviance reduction in this model.

For all the models produced, values of VIF were very low (Table 5), indicating the lack of significant multicollinearity between predictors, ensuring that no covariation exists among the variables included in the models, and therefore ensuring the reliability of the conclusions arising from the results.

Discussion

Non-native fish present in Portuguese rivers have different histories. Although rapid spread is possible in a short period, the timing and location of NNS introductions is relevant to the expansion of species distribution and abundance, which explains the relatively narrow distribution and abundance of *A. melas*, *A. alburnus*, and *S. lucioperca*, all recently introduced [9–10,51] at the date of this study, compared with other NNS introduced earlier. The most frequent and abundant species were *L. gibbosus*, *G. holbrooki and G. lozanoi*. In southern rivers the most common and abundant NNS were the former ones while *G. lozanoi* dominated the northern rivers. As such, the discussion will mainly focus on these NNS, which presented the higher invasive potential and are totally established in Portuguese continental waters for more than forty years.

A striking fact regarding the distribution of NNS in this study was their absence from mountain and southern chalk river types. Mountain rivers and small highland streams, with high slope and energy and low productivity (such as M, N4, S4 and N3), seem to be much less vulnerable to invasion by the set of introduced species. The upstream reaches of most Mediterranean basins experience strong seasonal patterns in their environmental conditions. In winter and spring, these streams present high flows whereas in summer they are frequently reduced to small isolated pools. This hydrological regime may prevent the invasion of NNS

Table 5. Summary of the environmental and human pressure GLMs for *L. gibbosus*, *G. lozanoi* and *G. holbrooki* (response variables occurrence and density).

Species	Response variables	Models features	Models AIC	Explanatory variables	Type	Deviance reduction	Coef. sign	VIF	P
L. gibbosus	Occurrence	Binomial distribution		Drainage area	Env	76.31	+	2.14	***
		"logit" link function	274.7	% Riffles	Env	21.95	-	1.5	*
				TSS	Pres	21.93	+	1.09	*
			356.5	Nutrient/organic loads	Pres	3.06	+	2.09	*
	Density	Poisson distribution		Drainage area	Env	69.53	+	2.36	***
		"log" link function		% Riffles	Env	28.44	-	1.63	**
			430.9	Mean annual runoff	Env	7.66	-	1.78	*
				Toxic/acid levels	Pres	18.99	-	1.44	*
				COD	Pres	13.08	+	1.31	**
				TSS	Pres	5.66	+	1.12	*
			496.1	Nutrient/organic loads	Pres	2.68	+	2.19	*
G. lozanoi	Occurrence	Binomial distribution		Drainage area	Env	38.40	+	2.21	**
		"logit" link function	216.3	Slope	Env	11.22	+	1.61	*
				Dissolved N	Pres	17.62	+	1.42	**
				Toxic/acid levels	Pres	13.53	-	1.92	*
				COD	Pres	5.39	-	2.27	*
			236.1	TSS	Pres	4.31	+	1.29	**
	Density	Poisson distribution		Drainage area	Env	26.26	+	2.73	**
		"log" link function	286.4	Slope	Env	13.73	+	1.69	*
				COD	Pres	15.06	-	2.17	**
				Urban area	Pres	11.41	+	2.62	*
				TSS	Pres	4.94	+	1.22	*
				Toxic/acid levels	Pres	6.16	-	1.69	*
			293.38	Land use	Pres	6.13	-	2.43	*
G. holbrooki	Occurrence	Binomial distribution		Mean annual temperature	Env	22.88	+	2.12	*
		"logit" link function		Drainage area	Env	14.87	+	2.38	**
			186.1	% Riffles	Env	7.86	-	1.44	*
				Sediment load	Pres	22.06	+	2.31	***
			186.7	Toxic/acid levels	Pres	13.21	-	1.98	***
	Density	Poisson distribution		Mean annual temperature	Env	26.06	+	2.25	*
		"log" link function	247.3	Drainage area	Env	9.48	+	2.29	*
				Toxic/acid levels	Pres	42.29	-	1.77	**
				Sediment load	Pres	9.70	+	1.36	*
			240.8	Artif. lentic water bodies	Pres	6.40	+	2.75	*

Explanatory variables type: environmental (Env) and human-induced pressure (Pres). Significance levels (*P*) at
*$P<0.05$.
**$P<0.01$.
***$P<0.001$.

which are poorly adapted to high discharge events, including flash flows (e.g. [52–53]). Conversely, NNS occurrence and abundance were particularly meaningful in river types with medium and large drainage areas, both in the northern "cool–warm water" and in the south. The high proportion of non-native fishes in the middle and lower reaches of Mediterranean-climate rivers was also reported in other studies (e.g. [54–56]). In general, the pattern of NNS distribution and abundance among river types followed the pattern of human-induced disturbance. Although human-induced disturbance occurred among all river types, some areas exhibited higher degradation. Overall, the sites located in littoral regions, both in the centre-north and in the southern region, which are associated with higher human density, presented higher degraded conditions and higher occurrence and abundance of non-native fishes. The least disturbed river types were located in higher altitude regions and with small drainage areas, where NNS were absent or present in very low densities. Most of those streams are located in isolated areas with difficult human access, far from the main human pathways. Remote areas with low human disturbance receive much lower propagule pressure than areas hosting immense human settlement [57].

Sites with higher human pressure exhibited significantly higher occurrence, abundance, and percentage of NNS. This pattern was observed in all river types and for all species, excepting those with very low abundance of NNS and *H. facetum*. Several species (*A. alburnus, A. melas, S. lucioperca*) occurred exclusively in impaired sites. The response of the most abundant NNS to the human pressure gradient was shaped by a marked increase in their abundance along the quality classes. Though the same general pattern was observed for the three species, *G. lozanoi* and *G. holbrooki* were particularly responsive to degradation, reaching the highest abundances in the poor quality class.

Non-native freshwater fish have commonly been documented to succeed in degraded aquatic habitats in many areas of the world [28,58–61]. Disturbed systems and communities may attract biological invasions more than pristine ecosystems due to the redistribution of space and energy resources and may promote new vacant niches for the most adaptable and tolerant invaders [11,62]. One of the most important human pressures in Mediterranean-climate rivers is related to nutrient and organic loads, which under the highly favourable climatic conditions increase the aquatic productivity and food availability. The resource availability theory argues that invasibility is directly influenced by nutrient enrichment and resource availability [11,63], which may facilitate biological contamination by reducing resource limitation and therefore competition. The high proportion of NNS in medium and large lowland rivers where productivity is naturally higher corroborates this hypothesis. Moreover, these river types (Littoral, N1>100 km^2 and S1> 100 km^2) are also the most affected by hydrological disturbances, flow regulation, connectivity loss, and dam construction. Dams typically create lentic conditions that favour non-native fishes, thus presenting higher abundance in regulated streams than in unregulated ones [64–65]. These river types are particularly vulnerable to non-native fish invasions, as some species, namely *L. gibbosus* and *G. holbrooki*, prefer enriched lentic habitats (e.g. [53,66]).

This study reveals that natural landscape/environmental drivers and human-induced disturbance are both important factors determining the distribution and invasiveness of NNS, and their relative importance vary from species to species.

The occurrence of *G. lozanoi* was mostly explained by environmental variables, while its abundance was equally determined by landscape/habitat and human-disturbance factors. In the Portuguese continental waters *G. lozanoi* populations were particularly abundant in N1>100 km^2 and littoral river types which is consistent with previous work associating this species to lowland rivers with slow flow [67]. Being absent from high altitude rivers and warm waters, this species occurrence was positively related to slope and drainage area (larger areas providing water flow all year long). For this reason, *G. lozanoi* does not occur in the warmer southern and central rivers, which present low or no flow during summer. The species abundance was positively correlated with human-induced hydrological disturbance, habitat modification, water quality degradation (TSS, COD, acidification) and agricultural areas. Other studies also evidenced this species response to anthropogenic hydrological disturbance [64,68].

G. holbrooki is mainly present in southern lowland river types (S1>100 km^2, S1<100 km^2 and S3) with the highest water temperature and conductivity. This species occurrence was positively correlated with temperature and drainage area. Although G. *holbrooki* is able to withstand wide temperature ranges, it prefers warm water temperatures [69–70]. Conversely, the lower temperatures of the northern rivers could limit the species proliferation, due to the effect of latitude on life-history traits, namely the reproductive potential [71]. At the habitat scale, *G. holbrooki* was negatively associated with riffles, that is, it displays a preference for standing or slow flowing waters rich in organic detritus and muddy sediments. High river discharges tend to displace individuals and eliminate populations [52,58,72], as this species has no swimming abilities and behavioural response to fast flowing waters [73]. The invasiveness of *G. holbrooki* seems to be mostly determined by human-induced disturbance factors. Its abundance was strongly related to indicators of disturbance describing local in-stream habitat modifications and organic and sediment loads. The response to degradation is partially related to the species tolerance to a wide range of environmental conditions, including pH (from 3.9 to 8.8, [74]) and dissolved oxygen (0.28 mg/L, [69–70]. The ability to tolerate low dissolved oxygen enables it to survive in anoxic eutrophic waters with high organic and nutrient loads, and it is often the only fish species present in these water bodies (Ilhéu pers. obs.). Moreover, this species is also tolerant to a wide range of pollutants, including organic wastes, phenols, pesticides, and heavy metals, due to the species' phenotypic plasticity [75–76].

L. gibbosus is one of the most widespread NNS in Portugal and a very successful invader. This species was particularly frequent and abundant in warm water rivers of southern and central Portugal – littoral and southern river types, with the exception of southern mountain rivers – while in the north it occurred mainly in larger rivers (N1>100 km^2 and N2). *L. gibbosus* responded positively to a certain degree of anthropogenic disturbance, namely hydrological and habitat modification, and nutrient and organic loads, probably because these conditions lead to an increase in suitable habitats and also offer feeding advantages [77]. The intermediate tolerance of *L. gibbosus* to degradation, also reported by other authors [78], supports the idea that human-induced disturbance is not a requisite for successful invasion by all introduced species [79–80], and that a particular site may contain introduced organisms simply because they were introduced there and natural environmental conditions were favourable to their establishment. Indeed, this species abundance and distribution was mostly explained by pure environmental variables; its density was positively related to drainage area and negatively related with runoff and percentage of riffles, which is consistent with its higher occurrences and abundance in the warm water lowland rivers. Sunfishes exhibit low ability to withstand high current velocities and are often reduced in number by flood events [73,81].

Differences in swimming ability or behavioural response to high current velocities between native and non-native species may be the mechanism responsible for the observed differential removals [65–66]. Furthermore, *L. gibbosus* is known to be invasive in some southern and central European locations but not in England or Norway [82–84], being the climatic conditions the limiting factor in the species invasiveness [85]. This illustrates the high vulnerability of the warm water rivers of southern Europe, where environmental conditions tend to favour the proliferation of this type of species, being high flows the major constraint.

The invasive potential of NNS seems to be determined by different environmental factors that may interact direct and indirectly. However, other factors may also explain the success of many NNS invasions, namely the spatial structure and the biota features, including species life-histories and interspecific interactions [86–88]. Thus further studies integrating all these factors should be performed in order to enhance the ability to predict the vulnerability of ecosystems to invasive species.

Conclusions

This study reveals that natural landscape/environmental drivers are major factors determining the distribution of non-indigenous species in Portuguese rivers, while their abundance seems to be largely promoted by human-induced disturbance. The importance and role of different invasion factors are context dependent, because of the interaction between species traits and environment, regarding habitat, climate match, and human disturbance, and thus the river type approach seems to be useful to predict NNS invasiveness. This study also stresses the high vulnerability of the warm water lowland river types to invasion by NNS where conditions are amplified by human-induced degradation. Although some warm water species may be present in most basins (e.g. *L. gibbosus*), they become successful invaders mainly in large lowland rivers, particularly in the south. In contrast, high energy and low productivity rivers, typically at higher altitudes, seem to resist better to the invasion of NNS even when subjected to anthropogenic disturbance. These are relevant issues for assessment of the ecological status of aquatic systems, river basin management and river rehabilitation programmes.

Acknowledgments

We are grateful to several persons and their teams, who have taken part in data collection and entry: field colleagues from Universidade de Évora, M. T. Ferreira from Instituto Superior de Agronomia/Universidade de Lisboa, P. Raposo de Almeida from Instituto de Oceanografia/ Universidade de Évora, N. Formigo from Universidade do Porto and R. Cortes from Universidade de Trás-os-Montes e Alto Douro. We are also grateful to Dárcio Sousa for GIS help and A. Márcia Barbosa for statistical advice in GLM procedure. Valuable comments and critical review of an earlier draft of this manuscript by Thierry Oberdorff are deeply acknowledged, as well as the insightful comments of three anonymous referees, which have improved the manuscript.

Author Contributions

Conceived and designed the experiments: MI PM JMB. Performed the experiments: MI PM JMB. Analyzed the data: PM MI. Contributed reagents/materials/analysis tools: MI JMB. Wrote the paper: MI PM JMB.

References

1. Clavero M, Garcia-Berthou E (2006) Homogenization dynamics and introduction routes of invasive freshwater fish in the Iberian Peninsula. Ecol Appl 16: 2313–2324.
2. Leprieur F, Beauchard O, Blanchet S, Oberdorff T, Brosse S (2008) Fish invasions in the world's river systems: when natural processes are blurred by human activities. PloS Biol 6: 404–410.
3. Marr SM, Olden JD, Leprieur F, Arismendi I, Ćaleta M, et al. (2013) A global assessment of freshwater fish introductions in Mediterranean-climate regions. Hydrobiologia 719: 317–329.
4. Benejam L, Carol J, Benito J, Garcia-Berthou E (2007) On the spread of the European catfish (*Silurus glanis*) in the Iberian Peninsula: first record in the Llobregat river basin. Limnetica 26: 169–171.
5. Caiola N, de Sostoa A (2002) First record of the Asiatic cyprinid *Pseudorasbora parva* in the Iberian Peninsula. J Fish Biol 61: 1058–1060.
6. Comesaña J, Ayres C (2009) New data on the distribution of pumpkinseed *Lepomis gibbosus* and largemouth bass *Micropterus salmoides*, and of non-endemic Iberian gudgeon *Gobio lozanoi* in the Galicia region (NW Spain). Aquat Invasions 4: 425–427.
7. Miranda R, Leunda PM, Oscoz J, Vilches A, Tobes I, et al. (2010) Additional records of non-native freshwater fishes for the Ebro River basin (Northeast Spain). Aquat Invasions 5: 291–296.
8. Pérez-Bote JP, Romero RR (2009) First record of *Sander lucioperca* (Perciformes, Percidae) in the Alqueva reservoir, Guadiana basin (SW Iberian Peninsula). Limnetica 28: 225–228.
9. Ribeiro F, Chaves ML, Marques TA, Moreira da Costa L (2006) First records of *Ameiurus melas* (Siluriformes, Ictaluridae) in the Alqueva Reservoir, Portugal. Cybium 30: 283–284.
10. Ribeiro F, Gante HF, Sousa G, Filipe AF, Alves MJ, et al. (2009) New records, distribution and dispersal pathways of *Sander lucioperca* (Linnaeus, 1758) in Iberian freshwaters. Cybium 33: 255–256.
11. Davis M, Grime J, Thompson K (2000) Fluctuating resources in plant communities: a general theory of invasibility. J Ecol 88: 528–534.
12. Ruesink JL (2005) Global analysis of factors affecting the outcome of freshwater fish introductions. Conserv Biol 19: 1883–1893.
13. Bomford M, Barry SC, Lawrence E (2009) Predicting establishment success of introduced freshwater fishes: a role for climate matching. Biol Invasions 12: 2559–2571.
14. Duncan RP, Bomford M, Forsyth DM, Conibear L (2001) High predictability in introduction outcomes and the geograpical range size of introduced Australian birds: a role for climate. J Anim Ecol 70: 621–32.
15. Williamson M (1996) Biological Invasions. London: Chapman and Hall. 244 p.
16. Fausch KD, Taniguchi Y, Nakano S, Grossman G, Townsend C (2001) Flood disturbance regimes influence rainbow trout invasion success among five holarctic regions. Ecol Appl 11: 1438–1455.
17. Davies KF, Chesson P, Hamson S, Inouye BD, Melbourne BA, et al. (2005) Spatial heterogeneity explains the scale dependence of the native-exotic diversity relationship. Ecology 86: 1602–1610.
18. Blanchet S, Leprieur F, Beauchard O, Staes J, Oberdorff T, et al. (2009) Broad-scale determinants of non-native fish species richness are context-dependent. P Roy Soc Lond B Bio 276: 2385–2394.
19. Gasith A, Resh VH (1999) Streams in Mediterranean-climate regions: abiotic influences and biotic responses to predictable seasonal events. Annu Rev Ecol Syst 31: 51–58.
20. European Commission (2000) Directive 2000/60/EC of the European Parliament and of the Council of 23 October 2000. Establishing a framework for Community action in the field of water policy. Official Journal of the European Communities L327.
21. Leunda PM (2010) Impacts of non-native fishes on Iberian freshwater ichthyofauna: current knowledge and gaps. Aquat Invasions 5: 239–262.
22. Olden JD, Poff NL (2003) Towards a mechanistic understanding and prediction of biotic homogenization. Am Nat 162: 442–460.
23. Rahel FJ (2000) Homogenization of fish faunas across the United States. Science 288: 854–856.
24. Leprieur F, Olden J, Lek S, Brosse S (2009) Contrasting patterns and mechanisms of spatial turnover for native and exotic freshwater fish in Europe. J Biogeogr 36: 1899–1912.
25. Gozlan RE (2010) The cost of non-native aquatic species introductions in Spain: fact or fiction? Aquat Invasions 5: 231–238.
26. Harris JH, Silveira R (1999) Large-scale assessments of river health using an Index of Biotic Integrity with low-diversity fish communities. Freshwater Biol 41: 235–252.
27. Joy MK, Death RG (2004) Application of the index of biotic integrity methodology to New Zealand freshwater fish communities. Environ Manage 34: 415–428.
28. Kennard MJ, Arthington AH, Pusey BJ, Harch BD (2005) Are alien fish a reliable indicator of river health? Freshwater Biol 50: 174–193.
29. Angermeier PL, Smogor RA, Stauffer JR (2000) Regional frameworks and candidate metrics for assessing biotic integrity in Mid-Atlantic highland streams. Trans Am Fish Soc 129: 962–981.
30. Hughes RM, Larsen DP (1988) Ecoregions: an approach to surface water protection. J Water Pollut Control Fed 60: 486–493.
31. Schmutz S, Melcher A, Frangez C, Haidvogl G, Beier U, et al. (2007) Spatially based methods to assess the ecological status of riverine fish assemblages in European ecoregions. Fisheries Manag Ecol 14: 441–452.

32. INAG IP (2008) River Typology in Continental Portugal under the implementation of Water Framework Directive – Abiotic characterization. Lisboa: Instituto da Água, IP (in Portuguese).
33. Miranda R, Oscoz J, Leunda PM, García-Fresca C, Escala MC (2005) Effects of weir construction on fish population structure in the River Erro (North of Spain). Ann Limnol-Int J Lim 41: 7–13.
34. Cabral MJ, Almeida J, Almeida PR, Dellinger TR, Ferrand de Almeida N, et al. (2005) Livro Vermelho dos Vertebrados de Portugal. Lisboa: Instituto da Conservação da Natureza. 659 p.
35. Smith KG, Darwall WRT, editors (2006) The status and distribution of freshwater fish endemic to the Mediterranean basin. Gland, Switzerland, and Cambridge: IUCN. 34 p.
36. FAME (2004) Development, Evaluation and Implementation of a Standardised Fish-based Assessment Method for the Ecological Status of European Rivers-A Contribution to the Water Framework Directive. Final Report (Co-ordinator: Schmutz S). Institute for Hydrobiology and Aquatic Ecosystem Management, University of Natural Resources and Applied Life Sciences, Vienna. Available: http://fame.boku.ac.at. Accessed: 2007 Jul 23.
37. CIS-WFD (2003) Guidance on Establishing Reference Conditions and Ecological Status Class Boundaries for Inland Surface Waters. Final Version. Brussels: EU Common Implementation Strategy for the Water Framework Directive.
38. Clesceri LS, Greenberg AE, Eaton AD (1998) Standard methods for the examination of water and wastewater, 20th Edition. Washington: American Public Health Association, American Water Works Association, Water Environmental Federation.
39. Giller PS, Malmqvist B (1998) The biology of streams and rivers. New York: Oxford University Press. 296 p.
40. INAG IP (2008) Manual for biological assessment of water quality in rivers according to Water Framework Directive. Sampling and analysis protocol for fish fauna. Lisboa: Instituto da Água, I.P. (in Portuguese).
41. CEN (2003) Water Quality–Sampling of Fish with Electricity. European standard–EN 14011:2003. Brussels: European Committee for Standardization.
42. Quade D (1967) Rank analysis of covariance. J Am Stat Assoc 62: 1187–1200.
43. Conover WJ, Iman RL (1982) Analysis of covariance using the rank transformation. Biometrics 38: 715–724.
44. Daniel WW (1987) Biostatistics: A foundation for analysis in the health sciences. New York: Wiley. 734 p.
45. Borcard D, Legendre P, Drapeau P (1992) Partialling out the spatial component of ecological variation. Ecology 73: 1045–1055.
46. Hinch SG, Sommers KM, Collins NC (1994) Spatial autocorrelation and assessment of habitat-abundance relationships in littoral zone fish. Can J Fish Aquat Sci 51: 701–712.
47. Legendre P (1993) Spatial autocorrelation: Trouble or new paradigm. Ecology 74: 1659–1673.
48. Legendre P, Fortin MJ (1989) Spatial pattern and ecological analysis. Vegetation 80: 107–138.
49. Zuur AF, Ieno EN, Smith GM (2007) Analyzing Ecological Data. New York: Springer. 672 p.
50. Andersen MJ, Cribble NA (1998) Partitioning the variation among spatial, temporal and environmental components in a multivariate dataset. Aust J Ecol 23: 158–167.
51. Ribeiro F, Collares-Pereira MJ, Moyle PB (2009) Non-native fish in the fresh waters of Portugal, Azores and Madeira Islands: a growing threat to aquatic biodiversity. Fisheries Manag Ecol 16: 255–264.
52. Meffe GK (1984) Effects of abiotic disturbance on coexistence of predator-prey fish species. Ecology 65: 1525–1534.
53. Bernardo JM, Ilhéu M, Matono P, Costa AM (2003) Interannual variation of fish assemblage structure in a Mediterranean river: implications of streamflow on the dominance of native exotic species. River Res Appl 19: 1–12.
54. Ilhéu M (2004) Patterns of habitat use by freshwater fish in Mediterranean streams. PhD Thesis, University of Évora.
55. Vila-Gispert A, Alcaraz C, García-Berthou E (2005) Life-history traits of invasive fish in small Mediterranean streams. Biol Invasions 7: 107–116.
56. Hermoso V, Clavero M, Blanco-Garrido F, Prenda J (2011) Invasive species and habitat degradation in Iberian streams: an analysis of their role in freshwater fish diversity loss. Ecol. Appl 21: 175–188.
57. Drake JA, Mooney HA, DiCastri F, Grove RH, Kruger FJ, et al. (1989) Biological Invasions: a Global Perspective. New York: Wiley. 525 p.
58. Arthington AH, Hamlet S, Bluhdorn DR (1990) The role of habitat disturbance in the establishment of introduced warm-water fishes in Australia. In: Pollard DA, editor.Introduced and translocated fishes and their ecological effect.Can-Canberra: Australian Government Publishing Service. pp. 61–66.
59. Gido KB, Brown JH (1999) Invasion of North American drainages by alien fish species. Freshwater Biol 42: 387–399.
60. Meador MR, Brown LR, Short T (2003) Relations between introduced fish and environmental Conditions at large geographic scales. Ecol Indic 3: 81–92.
61. Ross R, Lellis WA, Bennett RM, Johnson CS (2001) Landscape determinants of non indigenous fish invasions. Biol Invasions 3: 347–361.
62. Belote RT, Jones RH, Hood SM, Wender B (2008) Diversity-invasibility across an experimental disturbance gradient in Appalachian forests. Ecology 89: 183–192.
63. González AL, Kominoski JS, Danger M, Ishida S, Iwai N et al. (2010) Can ecological stoichiometry help explain patterns of biological invasions? Oikos 119: 779–790.
64. Alexandre CM, Almeida PR (2010) The impact of small physical obstacles on the structure of freshwater fish assemblages. River Res Appl 26: 977–994.
65. Power ME, Dietrich WE, Finlay JC (1996) Dams and downstream aquatic biodiversity: Potential food web consequences of hydrologic and geomorphic change. Environ Manage 20: 887–895.
66. Godinho FN, Ferreira MT (2000) Composition of endemic fish assemblages in relation to exotic species and river regulation in a temperate stream. Biol Invasions 2: 231–244.
67. Doadrio I, Madeira MJ (2004) A new species of the genus Gobio Cuvier, 1816 (Actynopterigii, Cyprinidae) from the Iberian Peninsula and Southwestern France. Graellsia 60: 107–116.
68. Lobón-Cervia J, Montanes C, de Sostoa A (1991) Influence of environment upon the life history of gudgeon, *Gobio gobio* (L.): a recent and successful colonizer of the Iberian Peninsula. J Fish Biol 39: 285–300.
69. Lloyd LN (1984) Exotic Fish: Useful Additions or "Animal Weeds"? J Austr New Guinea Fishes Assoc 1: 31–42.
70. Pyke GH (2005) A review of the biology of *Gambusia affinis* and *G. holbrooki*. Rev Fish Biol Fisher 15: 339–365.
71. Benejam L, Alcaraz C, Sasal P, Simon-Levert G, Garcıa-Berthou E (2009) Life history and parasites of the invasive mosquitofish (*Gambusia holbrooki*) along a latitudinal gradient. Biol Invasions 1: 2265–2277.
72. Galat DL, Robertson B (1992) Response of endangered *Poeciliopsis occidentalis sonoriensis* in the Rio Yaqui drainage, Arizona, to introduced *Gambusia affinis*. Environ Biol Fish 33: 249–264.
73. Minckley WL, Meffe GK (1987) Differential selection by flooding in stream fish communities of the arid American Southwest. In: Matthews WJ, Heins DC, editors.Community and evolutionary ecology of North American stream fishes.Norman: University of Oklahoma Press. pp. 96–104.
74. Brown-Peterson N, Peterson MS (1990) Comparative life history of female mosquitofish, *Gambusia affinis*, in tidal freshwater and oligohaline habitats. Environ Biol Fish 27: 33–41.
75. Andreasen JK (1985) Insecticide resistance in mosquitofish of the lower Rio Grande Valley of Texas - an ecological hazard? Arch Environ Con Tox 14: 573–577.
76. Saiki MK, Martin BA, May TW (2004) Reproductive status of western mosquitofish inhabiting selenium-contaminated waters in the Grassland Water District, Merced County, California. Arch Environ Con Tox 47: 363–369.
77. Almeida D, Almodóvar A, Nicola GG, Elvira B (2009) Feeding tactics and body condition of two introduced populations of pumpkinseed *Lepomis gibbosus*: taking advantages of human disturbances? Ecol Freshw Fish 18: 15–23.
78. Halliwell DB, Langdon RW, Daniels RA, Kurtenbach JP, Jacobson RA (1999) Classification of freshwater fish species of the northeastern United States for use in the development of indices of biological integrity, with regional applications. In: Simon TP, editor.Assessing the Sustainability and Biological Integrity of Water Resource Quality Using Fish Communities.Boca Raton: CRC Press. pp. 301–333.
79. Niemela J, Spence JR (1991) Distribution and abundance of an exotic groundbeetle (Carabidae): a test of community impact. Oikos 62: 351–335.
80. Townsend CR (1996) Invasion biology and ecological impacts of brown trout *Salmo trutta* in New Zealand. Biol Conserv 78: 13–22.
81. Schultz AA, Maughan OE, Bonar SA, Matter WJ (2003) Effects of flooding on abundance of native and nonnative fishes downstream from a small impoundment. N Am J Fish Manage 23: 503–511.
82. Copp GH, Fox MG, Przybylski M, Godinho FN, Vila-Gispert A (2004) Life-time growth patterns of pumpkinseed *Lepomis gibbosus* introduced to Europe, relative to native North American populations. Folia Zool 53: 237–254.
83. García-Berthou E, Alcaraz C, Pou-Rovira Q, Zamora L, Coenders G, et al. (2005) Introduction pathways and establishment rates of invasive aquatic species in Europe. Can J Fish Aquat Sci 62: 453–463.
84. Sterud E, Jorgensen A (2006) Pumpkinseed *Lepomis gibbosus* (Linnaeus, 1758) (Centrarchidae) and associated parasites introduced to Norway. Aquat Invasions 1: 278–280.
85. Villeneuve F, Copp GH, Fox MG, Stakėnas S (2005) Interpopulation variation in growth and life-history traits of the introduced sunfish, pumpkinseed *Lepomis gibbosus*, in southern England. J Appl Ichthyol 21: 275–281.
86. Allen CR, Nemec KT, Wardwell DA, Hoffman JD, Brust M, et al. (2013) Predictors of regional establishment success and spread of introduced non-indigenous vertebrates. Glob Ecol Biogeogr 22: 889–899.
87. Filipe AF, Magalhães F, Collares-Pereira MJ (2010) Native and introduced fish species richness in Mediterranean streams: the role of multiple landscape influences. Divers Distrib 16: 773–785.
88. Olden JD, Poff LN, Bestgen KR (2006) Life-history strategies predict fish invasions and extirpations in the Colorado River Basin. Ecol Monograph 76: 25–40.

4

Vertical Profiles of Soil Water Content as Influenced by Environmental Factors in a Small Catchment on the Hilly-Gully Loess Plateau

Bing Wang[1], Fenxiang Wen[1], Jiangtao Wu[1], Xiaojun Wang[1]*, Yani Hu[2]

1 College of Environmental Science and Resources, Shanxi University, Taiyuan, China, **2** Library, Hebei University of Science and Technology, Shijiazhuang, China

Abstract

Characterization of soil water content (SWC) profiles at catchment scale has profound implications for understanding hydrological processes of the terrestrial water cycle, thereby contributing to sustainable water management and ecological restoration in arid and semi-arid regions. This study described the vertical profiles of SWC at the small catchment scale on the hilly and gully Loess Plateau in Northeast China, and evaluated the influences of selected environmental factors (land-use type, topography and landform) on average SWC within 300 cm depth. Soils were sampled from 101 points across a small catchment before and after the rainy season. Cluster analysis showed that soil profiles with high-level SWC in a stable trend (from top to bottom) were most commonly present in the catchment, especially in the gully related to terrace. Woodland soil profiles had low-level SWC with vertical variations in a descending or stable trend. Most abandoned farmland and grassland soil profiles had medium-level SWC with vertical variations in varying trends. No soil profiles had low-level SWC with vertical variations in an ascending trend. Multi-regression analysis showed that average SWC was significantly affected by land-use type in different soil layers (0–20, 20–160, and 160–300 cm), generally in descending order of terrace, abandoned farmland, grassland, and woodland. There was a significant negative correlation between average SWC and gradient along the whole profile ($P<0.05$). Landform significantly affected SWC in the surface soil layer (0–20 cm) before the rainy season but throughout the whole profile after the rainy season, with lower levels on the ridge than in the gully. Altitude only strongly affected SWC after the rainy season. The results indicated that land-use type, gradient, landform, and altitude should be considered in spatial SWC estimation and sustainable water management in these small catchments on the Loess Plateau as well as in other complex terrains with similar settings.

Editor: Andrew C. Singer, NERC Centre for Ecology & Hydrology, United Kingdom

Funding: This work was supported by the Natural Science Foundation of China (No. 41201277 and 41101025) and the Natural Science Foundation of Shanxi Province of China (No. 2014011034-2). The funders had no role in study design, data collection and analysis, decision to publish, or preparation of the manuscript.

Competing Interests: The authors have declared that no competing interests exist.

* Email: xjwang@sxu.edu.cn

Introduction

Soil water content (SWC) is a critical factor for plant growth and a determinant of plant distribution in arid and semiarid areas such as China's Loess Plateau [1, 2]. Vertical distribution of SWC can greatly affect soil water movement [3], thereby greatly affecting the biomass production and water use efficiency of plants (e.g., switchgrass) under water stress [4]. Plant-available water stored in the soil profile has a buffering capacity, which, in deep layers, prolongs or alleviates the effects of seasonal or inter-annual drought on plant growth and soil water flux to the atmosphere [5–7]. Research has provided strong evidence that deep soil water depletion plays a key role in sustainable agriculture, ecological restoration, and terrestrial water cycling on the Loess Plateau [8–10]. However, measurement of SWC profiles has been frequently conducted at different spatial scales. The results thus need to be converted before comparison analysis or practical uses. The SWC profile in small catchment is considered to be at a moderate scale for data exchanging. In particular, small catchment is thought to be the basic unit for integrated soil and water loss management in complicated terrain of the Loess Plateau [11, 12]. Characterization of SWC profiles and evaluation of relevant influencing factors at the small catchment scale have implications for hydrological modeling of soil water dynamics and sustainable management of soil water resources in similar areas.

Classical statistics is frequently used to analyze the variability of SWC profiles at the small catchment scale, which involves the estimation of descriptive parameters such as average (mean), variance, standard deviation (STD), and coefficient of variation (CV). Average SWC at individual soil depth intervals or across the whole soil profile is extensively determined. The CV of SWC is also routinely calculated as the temporal variable in a certain period of time or the spatial variable across a specific area. The SWC profile can be divided into distinct intervals by considering its average and CV which exhibit complex spatial-temporal relationships in several plots or watersheds [13–16]. Additionally, ranking method, clustering method, and semivariogram model have been applied for the division of SWC profile [3, 17–20]. However, the above-mentioned methods cannot clearly reflect the variation trend in SWC profiles. Thus, great effort has been made

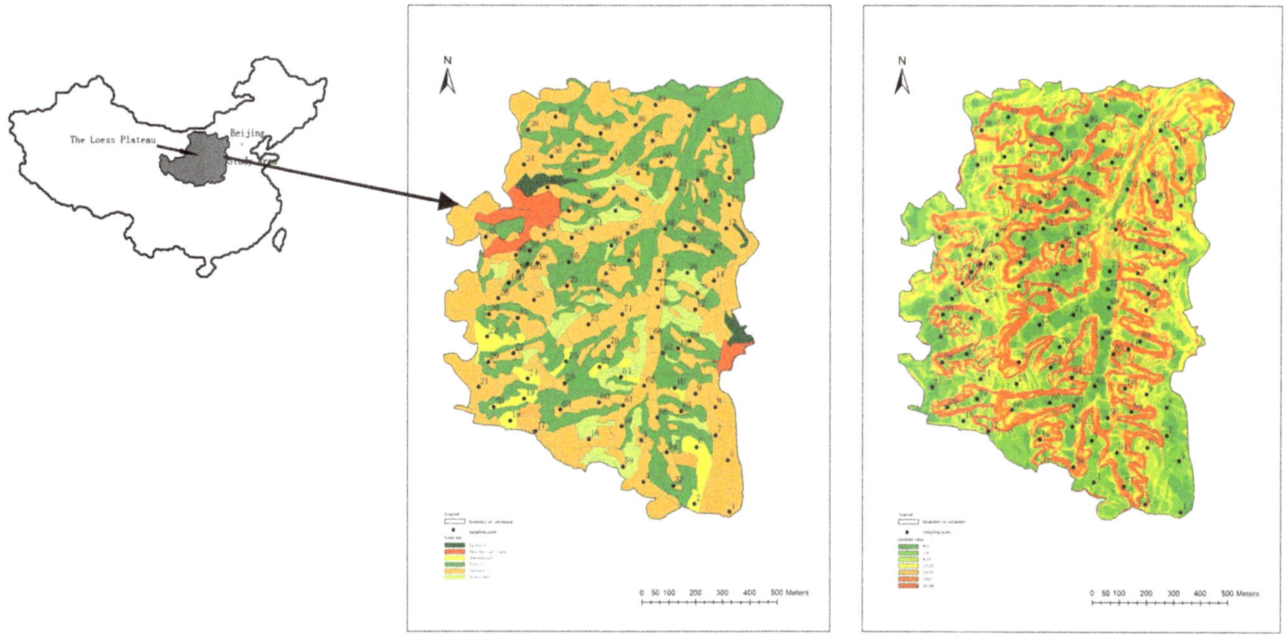

Figure 1. Location of Sanyanjing catchment and distribution of 101 sampling points in the Sanyanjing catchment.

to describe the vertical profiles of SWC through comparing variation curves or variation ranges, in small watersheds related to different land-use types, vegetation species, and/or terrain factors [3, 17, 18, 20–22]. If a massive sample size is involved, however, it becomes difficult to distinguish the vertical profiles and major influencing factors of SWC by direct comparisons.

In recent decades, a great number of studies have been conducted on the spatiotemporal variability of SWC and related influencing factors worldwide. Canton [23] pointed out that wasteland-scale spatial variability of SWC is mainly controlled by surface cover and soil properties in a semi-arid region of Spain, where surface cover counteracts the influence of terrain factors (including gradient, aspect, topographic wetness index, and distance from the river) on SWC distribution. Burnt [24] reported a topographic index which can simulate changes in high-level SWC in a humid climate zone of Devin County, UK. O'loughlin [25] estimated the spatial pattern of SWC distribution in a small catchment using humidity index model based on digital terrain dataset. Hawley [15] discovered that topography is the major factor responsible for the spatial distribution of SWC in an agricultural region, where resultant SWC variation is diminished by vegetation in a moist climate zone in Chickasha, USA. In arid and semi-arid areas, catchment-scale distribution of SWC is strongly affected by land-use/vegetation and topographic indices, e.g., land-use type, soil organic matter content, tillage, soil physical properties, gradient, and aspects [6, 17–19, 21, 26].

Many researchers have focused on the quantification of environmental parameters such as topographic factor, vegetation type, soil texture, and land-use type, in attempt to evaluate their impacts on the variability of SWC. At the small catchment scale, little information is available on the major factors affecting vertical profiles of SWC in cinnamon soil (Haplic Lixisols, FAO) zone on the hilly and gully Loess Plateau [3, 19]. As a regional water reservoir experiencing depletion, the plateau region requires measurements and characterization of deep SWC profiles for the thick soil layer. However, soil sampling for SWC profile analysis at the catchment scale has been commonly conducted at <200 cm depth [18–20, 23]. Wang et al. [21] exceptionally examined SWC along the 0–21 m soil profile on the Loess Plateau, but the reliability of their tests might be affected by a small sample size (11 sites). Deep soil sampling at a larger number of sites and statistical analysis of parameters involving soil depth information will contribute to better understanding of the vertical profiles and influencing factors of SWC.

In the present study, we characterized the vertical profiles of SWC in a small catchment on the Loess Plateau by cluster analysis of two descriptive parameters (mean and regression gradient). Sampling was carried out in the 0–300 cm profile at 101 points throughout the catchment before and after the rainy season, to meet the demand for deep depth, spatial representativeness and temporal comparability. The influences of selected environmental factors (land-use type, topographic factors, and landform) on average SWC were examined by multi-linear regression [6, 26–32]. The results were discussed in order to provide new insights to the vertical profiles and influencing factors of SWC on the Loess Plateau, further providing reference data for sustainable management of water resource in small catchment areas in the semi-arid region with complex terrain.

Material and Methods

1 Site description

This study was conducted in Sanyanjing catchment (112°2′13″, 37°46′23″), which is located on the east margin of the Loess Plateau in Shouyang county, mid-east Shanxi province, China (Figure 1). The catchment has a total area of 1.32 km^2 and the elevation ranges from 1001 to 1160 m. It is a hilly and gully area with mostly deep gully erosion slopes. The landform consists of ridge and gully.

The catchment area has a semi-arid continental climate (Cwa by Koppen Climate Classification) with an average annual precipitation of 474.2 mm (1967–1999). Snow in the winter accounts for ~8% and rainfall in July to September for ~73% of annual precipitation. Monthly average precipitation, potential evapotranspiration and precipitation in 2013 are shown in Figure 2. Annual mean temperature in this area is 8.1°C, with a maximum of 34.7°C and a minimum −20.6°C. The soil type is cinnamon soil (Haplic lixisols, FAO), which consists of 54–62% silt and 10.95–30.15% sand with the bulk density of 1.3–1.4 g/cm^3. Soil texture was measured using a particle size analyzer (SEDIMAT 4–12, UGT, Germany). Soil bulk density was determined through sampling with cutting rings (inner diameter 5.0 cm, volume 100 cm^3) and drying in an oven (105°C, 24 h). The profile of soil texture related to different landforms is listed in Table 1. Maximum soil depth is mostly down to 300 cm on the ridge and bare rock could be rarely seen only on the northern margin of the gully area.

The distribution of land-use types across the catchment is shown in Figure 1. Terrace is the dominant land-use type, accounting for about 60% of the study area. Few terraces had been abandoned for natural restoration of vegetation because of the Grain for Green project since 2000. Grassland is mainly covered with herbs and semi-shrubs, which had never been reclaimed for several decades. About 80% of the woodland is covered with semi-shrubs at steep slopes unsuitable for sampling.

2 Soil sampling

Ethics statement. Sampling activities at the farmland were allowed by the owners. No specific permissions were required at other locations because they were not privately-owned or protected in any way and the field activities did not involve any endangered or protected species.

A total of 101 sampling points were designed in a 150 m×150 m grid throughout the catchment area by considering major land-use types, including terrace (83), abandoned farmland (9), grassland (3), and woodland (6). Soil sampling was carried out during two periods in 2013, from April 29 to May 4 (before the rainy season) and from October 28 to November 1 (after the rainy season). No precipitation occurred during the two sampling periods or a week before sampling. Each sample was taken at 20 cm intervals along the 0–300 cm soil profile using an auger (inner diameter 5.0 cm). The samples were kept in capped aluminum boxes for transportation. Measurement of SWC was conducted using an oven-drying method (105°C, 24 h). At the majority of the sampling points, soils were collected along a vertical profile over 300 cm. A few exceptions were in the north of the gully at the lowest altitude where weathered rock was occasionally encountered. Background information of the 101 sampling points is summarized in Table 2.

3 Data analysis

To identify the variability of SWC profiles at the catchment scale, descriptive parameters were calculated for each profile. Further, we calculated the linear regression coefficient (K value) between SWC and soil depth to represent the variation trend of SWC vertical profiles and the mean value to describe the average level of SWC along the 0–300 cm profile.

The SWC profiles were classified using a combined cluster analysis of the K and mean values. Cluster analysis is the process of grouping a set to data objects into multiple groups (or clusters), so that objects within a cluster share high similarity but are dissimilar to those in other clusters. In this approach, dissimilarities and similarities are assessed based on the attribute values describing the objects and often involve distance measures. Cluster analysis is a statistical classification method for discovering whether the individuals of a population gall into different groups by making quantitative comparisons of multiple characteristics [34]. Here a combined cluster analysis was conducted in three steps: 1) cluster of the mean to three groups, which present the average level of SWC along the vertical soil profile, 2) cluster of K to three groups, which reflect the variation trend of SWC profiles (top to bottom), and 3) combination of the two sets of groups into nine new groups using the between-groups linkage method with squared Euclidean distance criteria [34].

For group comparisons, SWC profiles (0–300 cm) of the same group were averaged and re-plotted. The average curves of SWC were compared between groups to identify the major factors influencing SWC in individual soil layers (0–20, 20–160, and 160–300 cm). On the basis of cluster analysis, the influences of land-use type, topography and landform on average SWC in individual soil layers were examined by multi-regression analysis. The independent variables were land-use type, landform type, Sin(gradient), Sin(aspect), flow accumulation (calculated cell numbers to a grid cell from surrounding cells with the ArcGIS hydrology analysis

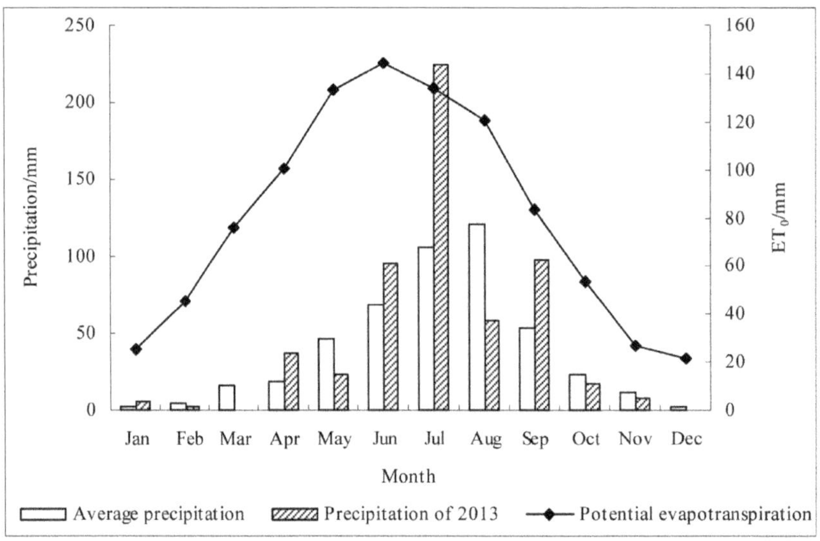

Figure 2. Average annual precipitation and potential evapotranspiration of 1967–1999 and precipitation in 2013 in the Sanyanjing catchment, Shanxi province, China.

Table 1. Soil texture in vertical profiles related to different landforms in the Sanyanjing catchment in Shanxi province, China.

Point description	Soil depth (cm)	Sand (%)		Silt (%)		Clay (%)
		>0.05 mm (%)	0.05–0.02 mm (%)	0.02–0.0063 mm (%)	0.0063–0.002 mm (%)	<0.002 mm (%)
	0–20	26.05	30.40	20.60	7.00	15.95
	20–40	21.35	34.80	20.80	6.90	16.15
	40–60	30.25	25.90	21.60	4.50	17.75
	60–80	28.25	28.80	21.70	2.80	18.45
	80–100	18.45	38.00	20.80	6.70	16.05
	100–120	26.45	30.10	20.40	5.20	17.85
Terrace at	120–140	26.95	27.90	22.00	4.40	18.75
ridge	140–160	16.35	32.60	24.20	8.10	18.75
(Point 33)	160–180	27.75	22.40	22.70	7.90	19.25
	180–200	36.55	17.80	20.60	8.00	17.05
	200–220	18.45	29.00	24.30	9.00	19.25
	220–240	20.45	28.50	23.20	8.60	19.25
	240–260	22.05	27.30	23.90	8.80	17.95
	260–280	31.35	20.30	22.20	9.40	16.75
	280–300	24.65	29.80	20.70	5.90	18.95
	0–20	30.15	20.00	23.20	6.40	20.25
	20–40	27.75	21.80	21.90	6.10	22.45
	40–60	25.95	26.80	20.50	4.50	22.25
	60–80	23.85	27.09	21.20	1.20	26.65
	80–100	0.95	31.00	20.10	20.10	27.85
	100–120	10.95	23.40	18.40	18.40	28.85
Terrace at	120–140	7.45	27.30	19.20	19.20	26.85
ridge	140–160	17.45	19.50	18.90	18.90	25.25
(Point 39)	160–180	10.45	25.80	18.20	18.20	27.35
	180–200	15.75	24.70	17.00	17.00	25.55
	200–220	13.35	21.40	18.60	18.60	28.05
	220–240	20.85	8.80	22.00	22.00	26.35
	240–260	20.25	12.40	20.00	20.00	27.35
	260–280	12.15	23.70	19.10	19.10	25.95
	280–300	24.95	29.00	20.50	0.50	25.05

module), and elevation. The former two factors were categorical variables converted into dummy variables before introduced into the regression analysis; and the latter four factors were continuous variables produced using digital elevation model at 1-m resolution.

SWC data were statistically analyzed in SPSS13.0 (SPSS Inc., Chicago, IL, USA), and topographic features were analyzed in ArcGIS 10.0 (ESRI, Redlands, CA, USA). SWC profiles were drawn in Microsoft Excel 2010 (Microsoft Corp., Redmond, WA, USA) and then clustered in SPSS 13.0 by considering descriptive parameters (maximum, minimum, mean, CV, STD, and K). Multi-regression analysis was performed in SPSS 13.0, with a *P*-value less than 0.05 considered statistically significant.

Results

1 Vertical profiles and descriptive parameters of SWC

The vertical profiles (0–300 cm) of SWC at 101 sampling points before the rainy season were drawn (Figure 3). These SWC profiles showed dynamic variations across the catchment study area, with substantial differences in the soil layers below 100 cm. At a few sampling points, there were obvious soil water depletion (e.g., 67, 75, 85, and 94) and an increasing trend (top to bottom) of SWC (e.g., 12 and 36). High degrees of soil desiccation were rarely detected in the lower soil layers, and low SWC was mainly found in the lower soil layers of woodland.

Descriptive parameters such as maximum, minimum, mean, and CV, STD are commonly used to reveal the spatial-temporal variability of SWC. However, these parameters cannot reflect the variation trend of SWC vertical profiles. To this end, the K value of SWC to profile depth was introduced for quantification of variation trend of SWC vertical profiles (Figure 4). Results showed that before the rainy season, SWC substantially varied between 5.87% and 34.72%, whereas the mean, STD, CV, and K values respectively ranged from 10.57% to 21.76%, 0.47 to 4.53, 3% to 24%, and −0.0405 to 0.0274 along the vertical soil profile (0–

Table 2. Background information of 101 soil sampling points in the Sanyanjing catchment study area in Shanxi province, China.

Land-use type	Vegetation	Landform type	Soil profile/cm	Sampling points
Terrace (n = 83)	Maize	Ridge	300	1, 3–5, 7–9, 12–18, 21–22, 25–26, 28, 32–35, 37, 51, 53, 91, 95, 96, 98
			260	99
		Gully	300	29, 30, 38–46, 48–50, 54–60, 62, 63, 65, 66, 68–74, 77–84, 87–90
			280	47
			260	31, 61
			220	52
			160	86
	Millet	Ridge	300	11
	Maize +five-year-walnut	Gully	300	27
Abandoned farmland	Subshrubs + herbs	Ridge	300	6, 10, 23
(n = 9)			240	2
	Subshrubs + herbs + few ulmus pumila	Gully	220	19
	Robinia peseudoacacia + subshrubs + herbs	Ridge	300	24
	Robinia peseudoacacia+ subshrubs + herbs	Gully	300	64
	Herbs + few almond-apricot	Gully	300	20
	Poplar + subshrubs +herbs	Gully	300	75
Grassland	Subshrubs +herbs	Ridge	300	92, 93
(n = 3)		Gully	300	76
Woodland	Poplar	Gully	300	97
(n = 6)			280	85
	Poplar + herbs	Gully	140	101
	Poplar + subshrubs	Gully	280	100
	Poplar + subshrubs + herbs	Gully	300	67, 94

300 cm). The ranges of the parameters after the rainy season were generally similar with those before the rainy season.

According to the division criteria of Nielson [24], CV in the range of 10–100% indicates moderate variability. Thus, the vertical variability of SWC at all sampling points in Sanyanjing catchment (Figure 4) can be classified to the medium degree. K is the linear regression coefficient between SWC and soil depth. A positive value of K indicates that SWC increases with increasing soil depth. Inversely, a negative value of K indicates that SWC decreases with increasing soil depth. The positive and negative K values of SWC data (Figure 4) are indicative of different variation trends of SWC vertical profiles in the catchment.

2 Clustering of SWC profiles

The vertical profile of SWC across the catchment can be described more clearly using cluster analysis. The 101 SWC profiles before the rainy season were classified into the first three groups by considering the mean value of SWC (Figure 5a), and the second three groups by considering the K value of SWC to soil depth (Figure 5b). The mean value of SWC ranged from 10.57% to 13.13% (low level), 14.15% to 16.86% (medium level), and 17.13% to 21.76% (high level) in the first three groups, whereas the K value of SWC to soil depth ranged from −0.0405 to 0.0106 (decreasing trend), 0.0144 to 0.0163 (stable trend), and 0.0194 to 0.0274 (increasing trend) in the second three groups. By combining the two cluster series, we obtained nine groups of SWC profiles (Table 3).

Before the rainy season, vertical profiles of SWC in groups 1–3 featured low-level SWC (Table 3). In group 1, SWC decreased along the vertical profile (0–300 cm) in woodland (2) and abandoned farmland (1) located in the gully area. In group 2, SWC remained stable along the vertical profile in terrace (2) and grassland (1) located on the ridge as well as woodland (1) located in the gully (Table 4). No sampling points were classified into group 3 with increasing SWC along the vertical profile.

Vertical profiles of SWC in groups 4–6 featured medium-level SWC (Table 3). In group 4, SWC decreased along the vertical profile on the ridge related to terrace (4) as well as in the gully related to abandoned farmland (1) and woodland (1) located in the gully. In group 5, SWC remained stable along the vertical profile in terrace (7) mostly located in the gully, few terrace (3) and abandoned farmland (1) located on the ridge, and grassland (1) and woodland (1) located in the gully. In group 6, SWC increased along the vertical profile in terrace (2) located on the ridge (Table 4).

Vertical profiles of SWC in groups 7–9 featured high-level SWC (Table 3). In group 7, SWC decreased along the vertical profile in terrace (5), grassland (1), and abandoned farmland (1) located on the ridge, and terrace (2), abandoned farmland (1) and woodland (1) located in the gully. In group 8, SWC remained stable along

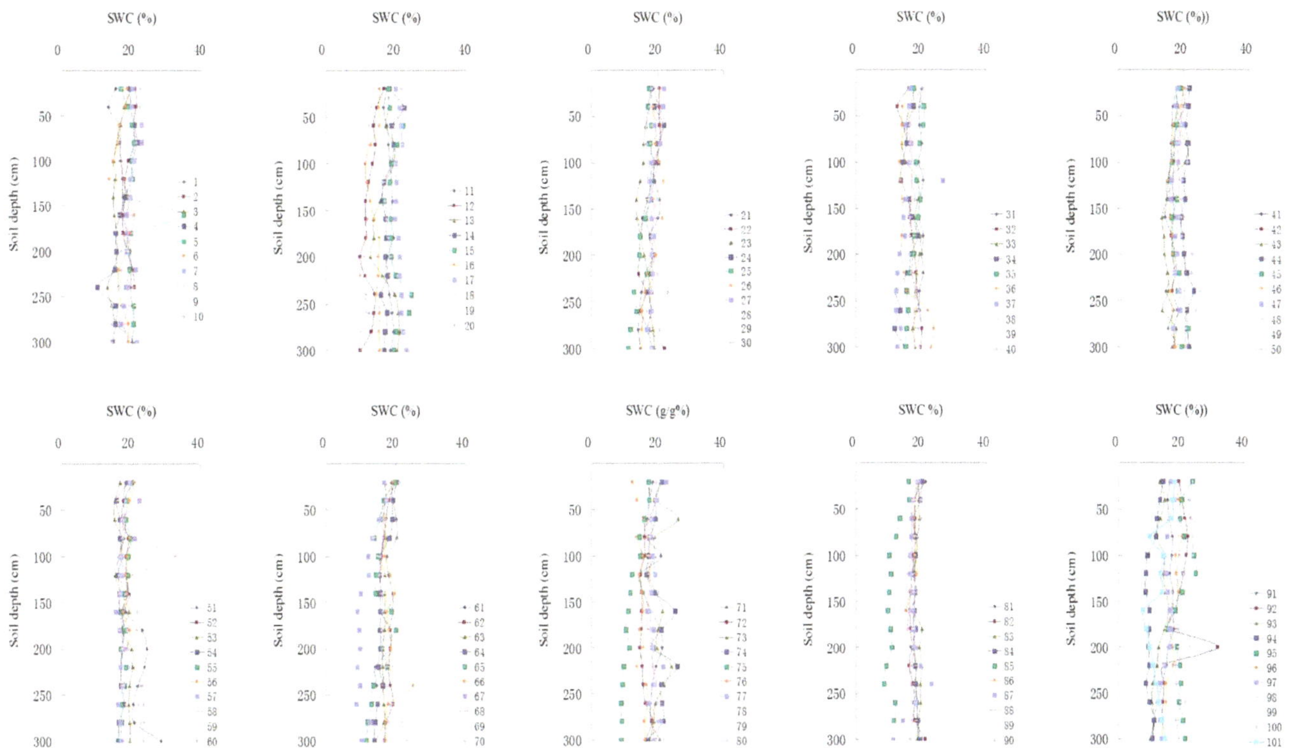

Figure 3. Vertical profiles of soil water at 101 sampling points in the Sanyanjing catchment.

the vertical profile at up to 58 of 101 sampling points, far more than other groups. Most sampling points of group 2 were located in the gully related to terrace (41), and few were on the ridge related to terrace (14) and abandoned farmland (3). In group 9, SWC increased along the vertical profile in terrace located on the ridge (3) and in the gully (1) (Table 4).

Similar grouping of SWC profiles was obtained with data collected after the rainy season. Overall, soil profiles of group 8 with high-level SWC in a stable trend were most commonly present in the catchment, more after the rainy season than before the rainy season. Group 3 of SWC profiles with low level and increasing trend was absent in the study area.

3 The relationships between average SWC and selected environmental factors

According to cluster analysis, there were nine combinations of SWC profiles in terms of average level and variation trend. We averaged SWC profiles of the same group and plotted the average curves (Figure 6), to examine differences of SWC profiles among various types. From Figure 4, we divided the whole soil profile (0–30 cm) into three layers (0–20, 20–160, and 160–300 cm) for multiple linear regression analysis. The results showed that selected environmental factors had significant linear correlations with average SWC at individual layers of 0–20 cm ($P<0.001$, $R^2=0.30$; $P<0.001$, $R^2=0.37$), 20–160 cm ($P=0.01$, $R^2=0.19$; $P<0.001$, $R^2=0.39$), and 160–300 cm ($P<0.001$, $R^2=0.32$; $P<0.001$, $R^2=0.43$; Table 5).

Before the rainy season, average SWC in the lower soil layer (10–20 cm) was significantly lower in grassland and woodland than in terrace, with no significant difference between abandoned farmland and terrace ($P_{D51}=0.109$, $P_{D52}=0.003$, $P_{D53}=0.047$, $P_{X1}=0.013$, and $P_{D61}=0.005$; Table 5). Additionally, average SWC decreased with increasing gradient, with higher levels on the ridge than in the gully.

In the lower soil layer (20–160 cm), average SWC decreased significantly with increasing gradient ($P=0.026$; Table 5), and was significantly lower in woodland than in the other three types of land-use types, with no significant differences among the latter three types. Other environmental factors had no significant linear correlation with average SWC ($P>0.05$).

In the deeper soil layer (160–300 cm), there also existed a significantly negative correlation between Sin(gradient) and average SWC ($P=0.001$; Table 5). Average SWC obviously increased with increasing gradient and was significantly higher in terrace than in abandoned farmland, grassland, and woodland (in descending order).

Similar results can be seen in the data collected after the rainy season. That is, land-use type was the major environmental factor affecting average SWC, whereas landform and altitude strongly affected average SWC only in specific periods and soil layers (Table 5). In the whole vertical profile (0–300 cm), SWC occurred at high levels from upper to deeper layers in terrace, with the lowest level in woodland. Compared with data of terrace, average SWC was relatively low in grassland and woodland in the top (0–20 cm) and deeper soil layers (160–300 cm), with significantly low levels in abandoned farmland soils in the deeper layer only. In the deeper soil layer (160–300 cm), average SWC varied with different land-use types in descending order of terrace > abandoned farmland > grassland > woodland.

Discussion

1 Vertical profiles of SWC at the catchment scale

According to Wang [21], the variability of SWC (as indicated by the CV) varies notably across the whole Loess Plateau, i.e., 15% in

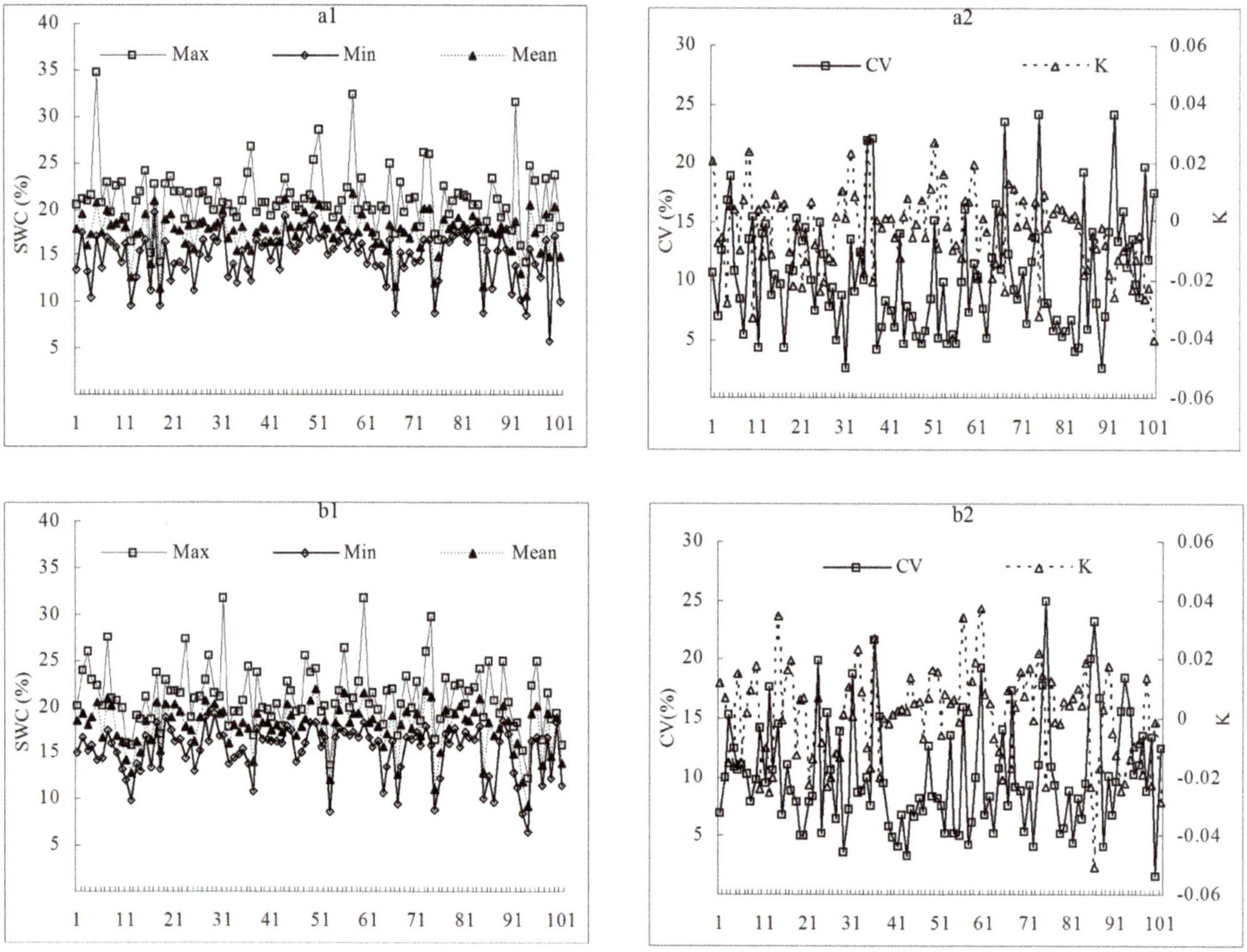

Figure 4. Statistical parameters of soil water content at 101 sampling points across the Sanyanjing catchment. (a. before the rainy season; and b. after the rainy season.)

Changwu and 55% in Shenmu. In the small catchment of Sanyanjing, SWC profiles exhibited weak and medium degrees of variability at 0–300 cm depth [33], with CV in the range of 3–24% (Figure 2). The lower variability of SWC profiles in our study area may be related to the higher SWC levels across the catchment (Pearson correlation coefficient between average SWC and CV,

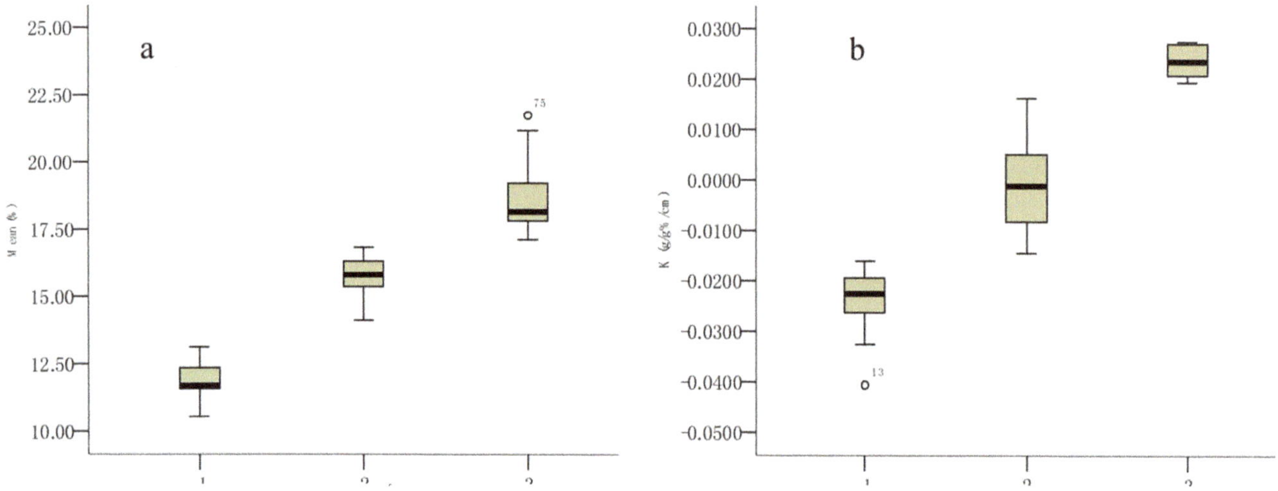

Figure 5. Grouping of 101 vertical soil water profiles in the Sanyanjing catchment before the rainy season by cluster analysis of the mean value (a) and regression gradient (K, b).

Table 3. Combined grouping of 101 vertical profiles of soil water content (0–300 cm) in Sanyanjing catchment by cluster analysis of the mean value and regression gradient.

Cluster by mean	Cluster by K	Combined grouping	Quantity of points		Point Nos.	
			Before the rainy season	After the rainy season	Before the rainy season	After the rainy season
1	1	1	3	0	65,67, 85	-
	2	2	4	1	12,18,93,94	94
	3	3	0	0	-	-
2	1	4	6	15	25,37,64,91,99,101	11,12,18,37,53,64,67,75,76,85,91,93,97,99,101
	2	5	13	0	3,16,23,34,41,43,56,63,65,70,72,76,97	3,16,23,34,41, 43,56,63,65,70,72,76,97
	3	6	2	1	32,36	14
3	1	7	11	1	4,10,19,21,26,35,61,86,92,96,100	86
	2	8	58	80	2,5–8,11,13–15,17, 20,22,24,27–31,33, 38–40,42,44–50,52, 53,55,56–59,62,66, 68,69,71,73,74, 77–84,87–90,95,98	2–10,13,15,16, 17,19–36,38–52,54,55,57–59, 61–63,65,66,68,69–74, 77–84, 87–90,92,95,96,98,100
	3	9	4	3	1,9,51,60	1, 51,60

Table 4. Grouping richness of 101 vertical profiles of SWC (0–300 cm) in relation to different land-use types in Sanyanjing catchment before the rainy season.

Land-use type	Grouping								
	1	2	3	4	5	6	7	8	9
Terrace	0	2	0	4	10	2	7	54	4
Abandoned	1	0	0	1	1	0	2	0	0
Grassland	0	1	0	0	1	0	1	0	0
Woodland	2	1	0	1	1	0	1	0	0

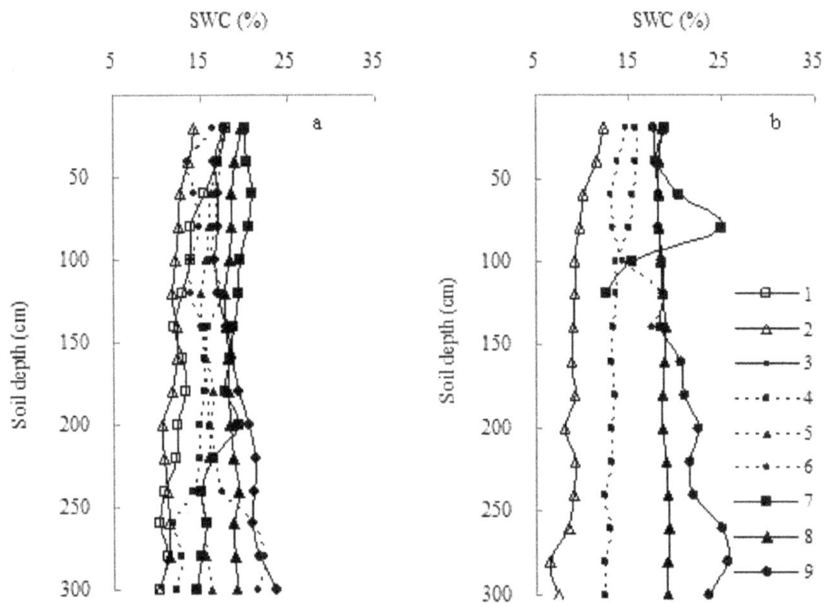

Figure 6. Vertical soil water profiles in relation to different groups in the Sanyanjing catchment study area before and after the rainy season (a. before the rainy season; and b. after the rainy season).

−0.40; $P<0.01$). Qiu [17] found that wetter soil with greater vertical variations in an increasing trend along the SWC profile (mean 13.03%; and STD, 2.3%) is representative in a dry year in Danangou catchment on the Loess Plateau, where the land-use pattern (including slope farmland, terrace, and orchard) differs from that in our study area.

Cluster analysis of the mean and K values provides a clear description for the overall variability of SWC in the vertical profile. Based on combined grouping, the 101 vertical SWC profiles were classified into nine groups with high, medium, and low levels associated with increasing, stable, and decreasing trends (Table 3). More than half of the SWC profiles were obtained from terrace soils in the gully and classified into group 8 (58/101 before the rainy season and 80/101 after the rainy season) with high-level SWC in a stable trend (Tables 3, 4). Despite that all sampling points of woodland were also located in the gully, their average SWC remained the lowest among different land-use types and mostly descended along the vertical profile (Table 4). The above differences can be attributed to the lower soil water consumption by maize crop in the terrace, which generally has shallower root distribution and less above-ground biomass than trees in the woodland. Our observations coincide with previous findings on the Loess Plateau that soil water conditions of terrace, gully farmland, and dam land are better than that of artificial woodland. The latter land-use type is associated with soil desiccation, especially in deep soil layers [2, 22, 35, 36].

Although the cluster analysis divided vertical SWC profiles into nine groups, only eight types were present in the Sanyanjing catchment and no sampling points were classified into group 3 (i.e., low-level SWC with an increasing trend from top to bottom). According to previous research in semi-arid regions, if SWC occurs at low level in the upper soil layer, deep-root crops, shrubs, and trees will consume more soil water in the deeper soil layers through root extraction [37–40]. Additionally, it is hard to achieve soil water recharge in the deeper soil layers by precipitation infiltration because the depth of soil water infiltration is shallow. Therefore, soil desiccation exists in the lower soil layer in case of no groundwater recharge [8–10]. These mechanisms explain the absence of high-level SWC with an increasing trend along the vertical profile in the small catchment of Sanyanjing (Table 3).

2 Effects of environmental factors on average SWC at the catchment scale

Consistent with cluster analysis (Table 3), multiple regression analysis showed that land-use type had a significant effect on soil water status in the small catchment of Sanyanjing (Table 4, 5). This result coincides with the data previously reported in small catchments on the Loess Plateau [17, 18, 22, 23, 29]. For example, Zhang [23] concluded that average SWC (20–200 cm) descends with different land-use types (farmland > grassland > shrub land > and woodland, n = 80) in the small catchment of Zhifanggou. Bai [42] found that average SWC (0–500 cm) ranges from of 9% to 16% in orchard, gradient farmland, terrace, and grassland, but remains less than 10% in shrub land and most woodland (n = 91) in Nangou catchment in the central area of Loess Plateau, Ansai, Shaanxi. The consistency of the data demonstrates that cluster analysis is a reliable method for characterization of SWC profiles.

The effect of land-use type on SWC can be related to the differences existing in anthropogenic activity and vegetation type [22]. Average SWC was found significantly higher in terrace and abandoned farmland than in grassland and woodland along the 0–300 cm profile (Table 5). Abandoned farmland and terrace are associated with artificial tillage in the surface soil layer, which improves soil porosity and loosens soil structure, further enhancing soil water infiltration [22, 43]. Additionally, soil water consumption by crops is less than that in grassland and woodland due to lower leaf area index [21], contributing to the accumulation of SWC. The above mechanisms account for the greater average of SWC profiles with a stable trend to soil depth in terrace and abandoned farmland.

Difference in root distribution is another factor contributing the effect of land-use type on SWC [40]. In the Sanyanjing catchment, average SWC of woodland was higher in the 0–20 cm soil layer but lower in the 20–160 and 160–300 cm soil

Table 5. Multi-linear regression analysis of soil water content and selected environmental factors in three layers (0–20, 20–160, and 160–300 cm) of the vertical soil profile in the Sanyanjing catchment study area.

				Before the	rainy season				
Model		Y_1			Y_2			Y_3	
	Unstandardized coefficients B	Standardized coefficients (Beta)	Sig.	Unstandardized coefficients B	Standardized coefficients (Beta)	Sig.	Unstandardized coefficients B	Standardized coefficients (Beta)	Sig.
Constant	14.714		0.066	6.031		0.498	18.514		0.069
X_1	−2.743	−0.23	0.013	−2.751	−0.22	0.026	−4.471	−0.296	0.001
X_2	−0.112	−0.037	0.68	0.042	0.013	0.889	−0.261	−0.068	0.449
X_3	4.86E−06	0.025	0.783	−7.80E−06	−0.038	0.694	1.20E−05	−0.05	0.578
X_4	0.005	0.081	0.507	0.011	0.177	0.175	0	0.005	0.97
D_{51}	1.09	0.146	0.109	0.019	0.002	0.98	−0.788	−0.084	0.354
D_{52}	−3.472	−0.278	0.003	−0.845	−0.064	0.503	−2.553	−0.162	0.073
D_{53}	−1.715	−0.191	0.047	−3.216	−0.342	0.001	−4.745	−0.384	0
D_{61}	−1.512	−0.348	0.005	−0.607	−0.134	0.307	−0.833	−0.151	0.212
	$R^2 = 0.30$	($P<0.001$)		$R^2 = 0.19$	($P = 0.001$)		$R^2 = 0.32$	($P<0.001$)	
				After the	**rainy season**				
Model		Y_1			Y_2			Y_3	
	Unstandardized coefficients B	Standardized coefficients (Beta)	Sig.	Unstandardized coefficients B	Standardized coefficients (Beta)	Sig.	Unstandardized coefficients B	Standardized coefficients (Beta)	Sig.
Constant	11.426		0.111	−5.811		0.452	−9.045		0.41
X_1	−2.01	−0.178	0.041	−2.763	−0.221	0.01	−3.688	−0.206	0.015
X_2	0.313	0.11	0.201	−0.132	−0.042	0.618	0.294	0.063	0.44
X_3	6.30E−06	−0.034	0.692	1.80E−05	−0.089	0.285	1.60E−05	−0.054	0.508
X_4	0.007	0.125	0.278	0.023	0.359	0.002	0.027	0.291	0.009
D_{51}	−0.622	−0.088	0.307	−1.161	−0.149	0.08	−1.81	0.162	0.053
D_{52}	−2.374	−0.2	0.021	−2.372	−0.181	0.032	−4.166	−0.222	0.008
D_{53}	−2.599	−0.305	0.001	−4.807	−0.51	0	−6.862	−0.467	0
D_{61}	−2.082	−0.506	0	−1.512	−0.332	0.004	−3.14	−0.476	0
	$R^2 = 0.37$	($P<0.001$)		$R^2 = 0.39$	($P = 0.001$)		$R^2 = 0.43$	($P<0.001$)	

Dependent Variable: Y_1 (soil water content of 0–20 cm layer).
Y_2 (average soil water content of 20–160 cm layer).
Y_3 (average soil water content of 160–300 cm layer).
Independent Variables: X_1 = Sin(gradient), X_2 = Sin(aspect), X_3 = flowaccu, X_4 = elevation.
Dummy Variables: X_5 = terrace, (D_{51}, D_{52}, D_{53}) = (0,0,0); X_5 = abandoned farmland, (D_{51}, D_{52}, D_{53}) = (1,0,0).
X_5 = grassland, (D_{51}, D_{52}, D_{53}) = (0,1,0); X_5 = woodland, (D_{51}, D_{52}, D_{53}) = (0,0,1).
X_6 = ridge, (D_{61}) = 1; X_6 = gully, (D_{61}) = 0.
D represents sub-variable; binary variables 0 and 1 for the absence and presence of some land-use type or landform, respectively.

layers than data of grassland (Table 5). The varying trends of SWC profiles between grassland and woodland can be related to different distribution of root system in individual soil layers and stratified root extraction of soil water. The rooting depth of maize crop is reported to be approximately 100 cm and most maize roots are distributed in the soil layer of 0–20 cm, shallower than average rooting depths in grassland (20–60 cm) and woodland (20–100 cm) [40]. Diverse root distribution patterns can lead to different levels of soil water consumption by plants, contributing to great variability of SWC level.

In addition to land-use type, topographic factors strongly affected SWC in the study area (Table 4). This is because the distribution of wind and solar radiation varies with different topographic conditions, leading to different levels of soil evaporation, runoff on gradient, and soil water infiltration [41]. Gradient negatively affected SWC in the soil layers of 0–20, 20–160, and 160–300 cm (Table 5), possibly due to the increased runoff with increasing gradient and resultant reduction of precipitation infiltration [17, 27, 29, 41]. Other topographic factors including aspect and flow accumulation had no significant effects on average SWC in the three soil layers (Table 5). Similarly, Gómez [29] referred that aspect has no obvious influence on SWC in burned and unburned areas. Shi [19] suggested that aspect and catchment area significantly affect SWC during the wet period only, whereas elevation has a significant effect on SWC in arid and humid periods but not in semi-arid and semi-humid periods. In the present study, we found the effect of elevation on SWC of the three layers varying with the period of time and being significant after the rainy season only.

As for the landform type, location of sampling points significantly affected SWC only in the surface layer (0–20 cm) before the rainy season and throughout all the three layers (0–20, 20–160, and 160–300 cm) after the rainy season, with greater values in the gully than on the ridge (Table 5). The effect of landform type on SWC can be related to different levels of soil evaporation as affected by wind strength and solar radiation and soil physical properties. Similarly, Zhang [22] suggested that average SWC descends with different landforms as gully > terrace > slop land > hill top.

Overall, land-use type is the most significant factor affecting SWC while topographic factors and landform type are interacting jointly at the catchment-scale. Because the impact of environmental factors on SWC varies in different periods, it is necessary to increase the observation frequency, in order to better understand the spatiotemporal distribution and influencing factors of SWC in the small catchment. Such work will provide reference data for selecting reasonable environmental parameters in catchment scale SWC simulation over different periods of time.

Conclusions

In this study, cluster analysis enables catchment-scale characterization of soil water profiles in terms of average level and variation trend along the vertical profile, allowing for simple and clear interpretation of the results. A total of nine groups of soil water profiles are recognized but those with low-level soil water content and a decreasing trend are not present in the Sanyanjing catchment. Land-use type, gradient, landform type, and altitude are the major environmental factors significantly influencing average soil water content in the hilly and gully catchment with complex terrain. The former two factors strongly affect soil water content along the 0–300 cm soil profile, whereas effects exerted by the latter two factors vary by soil layer and season.

Understanding the vertical profile of soil water content and evaluation of related major influencing factors in individual soil layers can help with sustainable land use and water management in catchment areas on the hilly and gully Loess Plateau as well as in arid and semi-arid areas with complex terrain. For better estimation of soil water profiles in small catchments, other factors such as fertilization, coverage, and soil physical properties may be considered with respect to specific soil layers.

Acknowledgments

We give our thanks to Prof./Dr. Wenzhao Liu (Institute of Soil and Water Conservation, Northwest A & F University, Yangling) for suggestions on experimental arrangement.

Author Contributions

Conceived and designed the experiments: BW FW JW XW YH. Performed the experiments: BW FW JW XW YH. Analyzed the data: BW FW JW XW YH. Contributed reagents/materials/analysis tools: BW FW JW XW YH. Wrote the paper: BW FW JW XW YH.

References

1. Engelbrecht BMJ, Comita LS, Condit R, Kursar TA, Tyree MT, et al. (2007) Drought sensitivity shapes species distribution patterns in tropical forests. Nature 447: 80–82.
2. Yang L, Wei W, Mo BR, Chen LD (2011) Soil water under different artificial vegetation restoration in the semi-hilly region of the Loess Plateau. Acta Ecologica Sinica 31: 3060–3068 (in Chinese with English abstract).
3. Xing G, Zhang XM, Fei XL, Wu YX (2012) Study on soil moisture content under different land use types in Sunjiacha basin. Agricultural Research in the Arid Areas 30: 225–229 (in Chinese with English abstract).
4. Li JW, Zuo HT, Li QF, Fan XF, Hou XC (2011) Effect of soil water spatial distribution pattern on switchgrass during first growing season. Acta Agrista Sinica 19: 43–50 (in Chinese with English abstract).
5. Jipp PH, Nepstad DC, Cassel DK, Carvalho C (1998) Deep soil moisture storage and transpiration in forests and pastures of seasonally-dry Amazonia. Climatic Change 39: 395–412.
6. Grassini P, You JS, Hubbard KG, Cassman KG (2010) Soil water recharge in a semi-arid temperate climate in the central US Great Plains. Agricultural Water Management 97: 1063–1069.
7. Markewitz D, Devine S, Davidson EA, Brando P, Nepstad DC (2010) Soil moisture depletion under simulated drought in the Amazon: impacts on deep root uptake. New Phytologist 187: 592–607.
8. Li YS (2001) Fluctuation of yield on high-yield field and desiccation of the soil on dryland. Acta Pedologica Sinica 38: 353–356 (in Chinese with English abstract).
9. Huang MB, Dang TH, Gallichand J, Goulet M (2003) Effect of increased fertilizer applications to wheat crop on soil-water depletion in the Loess Plateau, China. Agricultural Water Management 58: 267–278.
10. Liu WZ, Zhang XC, Dang TH, Zhu OY, Li Z, et al. (2010) Soil water dynamics and deep soil recharge in a record wet year in the southern Loess Plateau of China. Agricultural Water Management 97: 1133–1138.
11. Beldring S, Gottschalk L, Seibert J, Tallaksen LM (1999) Distribution of soil moisture and groundwater levels at patch and catchment scales. Agricultural and Forest Meteorology 98–99: 305–324.
12. Li B, Rodell M (2013) Spatial variability and its scale dependency of observed and modeled soil moisture over different climate regions. Hydrology and Earth System Sciences 17: 1177–1188.
13. Henninger DL, Petersen GW, Engman ET (1976) Surface soil moisture within a watershed: Variations, factors influencing, and relationship to surface runoff. Soil Science Society of American Journal 40: 773–776.
14. Jones EB, Owe M, Schmugge TJ (1982) Soil moisture variation patterns observed in Hand county, South Dakota. Water Recources Bulletin 18: 949–954.
15. Hawley ME, Jackson TJ, Mccuen RH (1983) Surface soil moisture variation on small agricultural watersheds. Journal of Hydrology 62: 179–200.
16. Robinson M, Dean TJ (1993) Measurement of near surface soil water content using a capacitance probe. Hydrological Processes 7: 77–86.
17. Qiu Y, Fu BJ, Wang J, Chen LD (2000) Quantitative analysis of relationships between spatial and temporal variation of soil moisture content and

environmental factors at a gully catchment. Acta Ecologica Sinica 20: 741–747 (in Chinese with English abstract).
18. Zeng C, Shao MA, Wang QJ, Zhang J (2011) Effects of land use on temporal-spatial variability of soil water and soil-water conservation. Acta Agriculturae Scandinavica Section B-Soil and Plant Science 61: 1–13.
19. Shi ZH, Zhu HD, Chen J, Fang NF, Ai L (2012) Spatial heterogeneity of soil moisture and its relationships with environmental factors at small catchment level. Chinese Journal of Applied Ecology 23: 889–895 (in Chinese with English abstract).
20. Chen LD, Huang ZL, Gong J, Fu BJ, Huang YL (2007) The effect of land cover/vegetation on soil water dynamic in the hilly area of the loess plateau, China. Catena 70: 200–208.
21. Wang YQ, Shao MA, Liu ZP, Orton R (2013) Regional-scale variation and distribution patterns of soil saturated hydraulic conductivities in surface and subsurface layers in the loessial soils of China. Journal of Hydrology 487: 13–23.
22. Zhang R, Cao H, Wang YQ, Huang CQ, Tan WF (2012) spatial variability of soil moisture and its influence factors in watershed of gully region on the loess plateau. Research of Soil and Water Conservation 19: 52–58 (in Chinese with English abstract).
23. Canton Y, Sole-benet A, Domingo F (2004) Temporal and spatial patterns of soil moisture in semiarid badlands of SE Spain. Journal of Hydrology 285: 199–214.
24. Burnt TP, Butcher DP (1985) Topographic controls of soil moisture distributions. Journal of Soil Science 36: 469–486.
25. O'loughlin EM (1981) Saturation regions in catchments and their relations to soil and topographic properties. Journal of Hydrology 53: 229–246.
26. Huang J, WU P, Zhao XN (2012) Effects of rainfall intensity, underlying surface and slope gradient on soil infiltration under simulated rainfall experiments. Catena 104: 93–102
27. Qiu Y, FU BJ, Wang J, Chen LD (2003) Spatiotemporal prediction of soil moisture content using multiple-linear regression in a small catchment of the Loess Plateau, China. Catena 54: 173–195.
28. Qiu Y, Fu B, Wang J, Chen L, Meng Q, Zhang Y (2010) Spatial prediction of soil moisture content using multiple-linear regressions in a gully catchment of the Loess Plateau, China. Journal of Arid Environments 74: 208–220.
29. Gómez-Plaza A, Martínez-Mena M, Albaladejo J, Castillo VM (2001) Factors regulating spatial distribution of soil water content in small semiarid catchments. Journal of Hydrology 253: 211–226.
30. Dripps WR, Bradbury KR (2007) A simple daily soil-water balance model for estimating the spatial and temporal distribution of groundwater recharge in temperate humid areas. Hydrogeology Journal 15: 433–444.
31. Yao XL, Fu BJ, Lu YH, Sun FX, Wang S, et al. (2013) Comparison of four spatial interpolation methods for estimating soil moisture in a complex terrain catchment. PLoS One 8(1): e54660.
32. Wang MB, Li HJ (1995) Quantitative study on the soil water dynamics of various forest plantations in the loess plateau region in northwestern Shanxi. Acta Ecologica Sinica 15: 172–184 (in Chinese with English abstract).
33. Wang YQ, Zhang XC, Han FP (2008) Profile variability of soil properties in check dam on the Loess Plateau and its functions. Environmental Science 29: 1020–1026 (in Chinese with English abstract).
34. Jain AK (2010) Data clustering: 50 years beyond K-means. Pattern Recognition Letters 31: 651–666.
35. Huang YL, Chen LD, Fu BJ, Wang YL (2005) Spatial pattern of soil water and its influencing factors in gully catchment of the Loess Plateau. Journal of Natural Resources 20: 483–492 (in Chinese with English abstract).
36. Zou JL, Shao MA, Gong SH (2011) Effects of different vegetation and soil types on profile variability of soil moisture. Research of Soil and Water Conservation 18: 12–17 (in Chinese with English abstract).
37. Kizito F, Dragila M,Se'ne M, Lufafa A, Diedhiou I, et al. (2006) Seasonal soil water variation and root patterns between two semi-arid shrubs co-existing with Pearl millet in Senegal, West Africa. Journal of Arid Environments 67: 436–455.
38. Li J, Chen B, Li XF, Zhao YJ, Ciren YJ, et al. (2008) Effects of deep soil desiccation on artificial forestlands in different vegetation zones on the Loess Plateau, China. Acta Ecologica Sinica 28: 1429–1445 (in Chinese with English abstract).
39. Cheng LP, Liu WZ (2013) Long term effects of farming system on soil water content and dry soil layer in deep loess profile of Loess Tableland in China. Journal of Integrative Agriculture. 13(6): 1382–1392.
40. Wang XZ, Jiao F (2011) Partition of soil moisture profiles based on sequential clustering method. Journal of Northwest A &F University 39: 191–201,196 (in Chinese with English abstract).
41. Fu XL, Shao MA, Wei XR, Wang HM, Zeng C (2013) Effects of monovegetation restoration types on soil water distribution and balance on a hillslope in northern Loess Plateau of China. Journal of Hydrologic Engineering 18: 413–421.
42. Bai TL, Yang QK, Shen J (2009) Soil variability of soil moisture vertical distribution and related affecting factors in hilly and gully watershed region of Loess Plateau. Chinese Journal of Ecology 28: 2508–2514 (in Chinese with English abstract).
43. Lian G, Guo XD, Fu BJ, Hu CX (2006) Spatial variability of bulk density and soil water in a small catchment of the Loess Plateau. Acta Ecologica Sinica 26: 647–654 (in Chinese with English abstract).

5

Empirical Evidence Reveals Seasonally Dependent Reduction in Nitrification in Coastal Sediments Subjected to Near Future Ocean Acidification

Ulrike Braeckman[1]*, Carl Van Colen[1], Katja Guilini[1], Dirk Van Gansbeke[1], Karline Soetaert[2], Magda Vincx[1], Jan Vanaverbeke[1]

1 Ghent University, Department of Biology, Marine Biology Research Group, Ghent, Belgium, **2** Netherlands Institute for Sea Research, Department of Ecosystem Studies, Yerseke, The Netherlands

Abstract

Research so far has provided little evidence that benthic biogeochemical cycling is affected by ocean acidification under realistic climate change scenarios. We measured nutrient exchange and sediment community oxygen consumption (SCOC) rates to estimate nitrification in natural coastal permeable and fine sandy sediments under pre-phytoplankton bloom and bloom conditions. Ocean acidification, as mimicked in the laboratory by a realistic pH decrease of 0.3, significantly reduced SCOC on average by 60% and benthic nitrification rates on average by 94% in both sediment types in February (pre-bloom period), but not in April (bloom period). No changes in macrofauna functional community (density, structural and functional diversity) were observed between ambient and acidified conditions, suggesting that changes in benthic biogeochemical cycling were predominantly mediated by changes in the activity of the microbial community during the short-term incubations (14 days), rather than by changes in engineering effects of bioturbating and bio-irrigating macrofauna. As benthic nitrification makes up the gross of ocean nitrification, a slowdown of this nitrogen cycling pathway in both permeable and fine sediments in winter, could therefore have global impacts on coupled nitrification-denitrification and hence eventually on pelagic nutrient availability.

Editor: Kay C. Vopel, Auckland University of Technology, New Zealand

Funding: U.B. was financially supported by FWO project nr G.0033.11 and the Special Research Fund of Ghent University. C.V.C. and K.G. acknowledge a postdoctoral fellowship from the Flemish Fund for Scientific Research (FWO). K.G. also received funding from the European Community's Seventh Framework Programme (FP7/2007-2013) under grant agreement no. 265847. Additional funding was provided by the Special Research Fund of Ghent University (BOF-GOA 01GA1911W). The funders had no role in study design, data collection and analysis, decision to publish, or preparation of the manuscript.

Competing Interests: The authors have declared that no competing interests exist.

* Email: Ulrike.Braeckman@UGent.be

Introduction

Over the past 250 years, the atmospheric CO_2 concentrations have increased by nearly 40% as a consequence of human activities [1]. This increase is partly counteracted by the capacity of the oceans to absorb CO_2, which occurs currently at a rate of about 10^6 metric tons of CO_2 per hour [2]. The latter process led to a decrease in surface-ocean pH by about 0.1 units since the start of the industrial revolution [3]. Climate change models predict a further decrease of 0.35 units by the end of the century for open ocean waters [3,4] and recent measurements for coastal zones even reveal acidification rates that are an order of magnitude higher [5,6].

This decrease in pH is known to have a direct or indirect negative effect on many nektonic, pelagic and benthic organisms [7–10] and thus marine food web structures in general [11]. Although knowledge on the effects of ocean acidification on the level of organisms and populations is increasing fast, the understanding of how biogeochemical processes are affected in acidified seawater is lagging behind [12]. As the ocean carbon cycle is linked with the cycles of major nutrient elements, it is to be expected that ocean acidification will have large consequences for marine ecosystem functioning [13,14]. With respect to the N-cycle, currently available knowledge suggests that ocean acidification will lead to increased N_2 fixation while nitrification may decrease [13]. This would reduce the supply of oxidized nitrogen substrate to denitrifiers and reduce levels of nitrate-supported primary production that may instigate shifts in plankton communities [13]. Most empirical evidence illustrates that ocean acidification decreases nitrification rates in pelagic environments [15–18] but the consequences for N-cycling in soft-sediments underlying an acidified water column are far less understood. As coastal sediments are spatially and temporally heterogeneous, the available studies do not allow a proper generalisation, e.g. due to the absence of seasonal replication and the limited amount of sediment types investigated [19–25]. Part of the benthic variability in coastal soft-sediments is linked to the macrobenthic communities that vary both spatially (e.g. sediment type) and temporally (e.g. seasonal demographics). Furthermore, the impact of these communities and species on biogeochemical cycling might fluctuate seasonally [26–28] and depends on the environmental

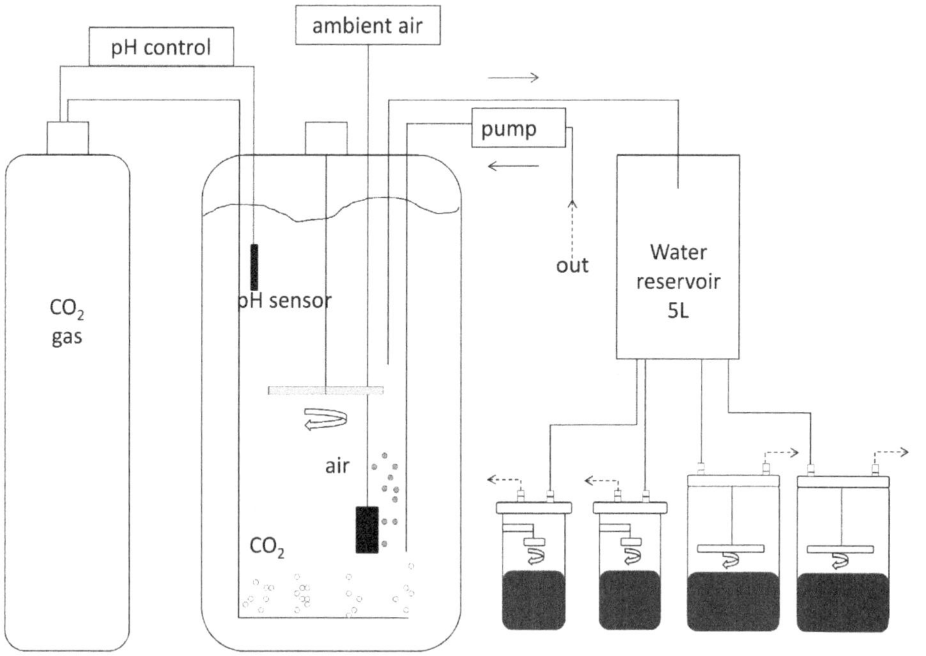

Figure 1. Scheme of experimental set-up. Dashed arrows on core outflow represent tubes that lead to the pump, directing them back into the tank. See also text for explanation. The same set-up without the CO_2-gas was used as a control.

context such as temperature or algal bloom deposition. A decrease in seawater pH has been shown to affect both macrobenthic bioturbation and bio-irrigation activities [18,22,24], which alter redox gradients and microbial communities that regulate the cycling of energy and matter.

To investigate ocean acidification effects on benthic nitrogen cycling and organic matter mineralization, we measured multiple response variables associated with benthic nitrogen cycling in closed core incubations, using two different sediment types in control and manipulated sea water carbonate chemistry. These incubations were executed during two seasons of the year representing the period prior to the annual phytoplankton bloom and the actual bloom period. Nitrification consumes NH_x (NH_4^++ NH_3) and O_2, and produces NO_x (NO_2^-+NO_3^-) thus any effect of acidification on nitrification may be reflected in the fluxes of these substances, as well as alter the oxygen penetration depth in the sediment. We constructed integrated mass budgets of O_2, NO_x and NH_x [29] to estimate nitrification rates from the measured fluxes.

Material and Methods

Sampling and experimental set-up

We sampled a fine sandy sediment station (St. 115bis: 51° 09.2′N; 02° 37.2′E, 13 m depth) and a station with coarse, permeable sediment (St. 330: 51° 26.0′N; 02° 48.5′E, 20 m depth), before (February) and during (April) the annual phytoplankton bloom of 2012. Both stations are located in the subtidal part of the Belgian Part of the North Sea, characterized by turbid waters with a light extinction coefficient of 0.36 m^{-1} [30], excluding light penetration to the sea floor, hence also precluding microphytobenthos growth. Sampling was carried out with the *RV* Simon Stevin. At each station, a CTD cast was performed to record the water temperature.

Sediment was collected by means of a Reineck box-core at both stations. At St. 115bis with fine sandy sediment (median grain size 180 µm with 14% of mud; [31]), four separate Reineck box-corers were subsampled with a single Plexiglass tube each (internal diameter ø: 10 cm; H: 25 cm). At St. 330 with permeable sediment (median grain size 360 µm, devoid of mud; permeability: 5.3 10^{-10} m^2, [31]), we measured fluxes in centrally stirred chambers (Plexiglass, ø: 19 cm; H: 30 cm) to create a pore water flow similar to *in situ* conditions [32]. As these chambers were too large to insert in the Reineck box-core, they were filled with homogenized sediment from the upper 10 cm of the sediment from 7 Reineck box-corers. The experimental collections differed between the two sites rendering them not strictly comparable. No specific permits were required for the described field study: the location is not privately-owned or protected in any way and the field study did not involve endangered or protected species.

Sediment cores were transported within 10 h to a temperature-controlled room at *in situ* temperature (recorded from the CTD cast: February: 5°C, April: 9°C) in the lab and immediately submerged in well-aerated sea water in two set-ups (one control set-up and one set-up to be acidified; Fig. 1). Teflon coated magnets rotated by a central magnet were adjusted 5 cm above the sediment surface of the fine sandy sediment cores to ensure water mixing. The rate of water circulation was kept well below the resuspension threshold. Lids equipped with a flat stirring disc were fixed onto the permeable sediment chambers with the stirring disc rotating at 12 rpm 5.4 cm above the sediment surface. From 200 L tanks, a flow-through of seawater through inlet and outlet ports in the lids was ensured with a peristaltic pump (Watson-Marlow 520S), refreshing the core volume ca. 2× h^{-1}. The outflow of the cores was recycled and aerated in the 200 L tanks. Our set-up encompassed duplicate cores for each pH treatment – sediment type combination in each month.

Manipulation of carbonate chemistry

After 24 h of acclimatization, the acidification process was started. To create a predicted pH likely to establish within the current century in coastal waters (~pH 7.7; [33]), the carbonate chemistry of natural sea water was manipulated through controlled pumping of 100% CO_2 gas at the bottom of a 200 L seawater tank (~pH 8.0), using Dulcometer technology (ProMinent) connected to a pH electrode (Dulcotest PHE-112SE) mounted in the tank [33] (Fig. 1). This method was chosen as it best replicates ocean acidification by altering dissolved inorganic carbon (DIC) while keeping total alkalinity constant [34]. Every other day the pH electrode was calibrated using Hanna Instruments' pH NBS buffers. The control seawater unit had the same set-up except for the CO_2 gas supply and was bubbled with ambient air.

The sediments were incubated under experimental conditions for 14 days in the dark. During this acclimatization period, temperature, salinity (WTW COND 330), pH (Hanna Instruments), oxygen (Pyroscience needle-type optodes) and alkalinity (Gran titration [35], were monitored daily. Carbonate system parameters (pCO_2, dissolved inorganic carbon and calcite and aragonite saturation states) of the overlying water in each tank for the experiment duration were calculated from alkalinity, pH, temperature and salinity in CO2SYS [36] and can be found in Table S1.

Sediment core incubations

We quantified sediment community oxygen consumption (SCOC) and dissolved inorganic nitrogen (DIN) exchange using a direct flux approach by incubating undisturbed sediment cores. From the resulting oxygen and DIN fluxes, we modeled nitrification rates using an integrated mass budget approach [28,29].

One week after acclimatization, depth profiles of oxygen were measured in the sediment with O_2 microsensors (100 µm tip size, Unisense) in vertical increments of 250 µm (3 replicates in each core), up to 10 mm depth. pH profiles were measured in depth intervals of 1000 µm with pH microsensors (tip size 500 µm, Unisense), starting just above the sediment surface downwards, until a constant pH was reached (max. 3.5 cm). A two-point calibration was performed in pH buffers 4.01 and 7.00 before the measurements. Electrodes were connected to a picoammeter (oxygen) or pH-meter and output was displayed on an online PC using SensorTracePro (Unisense) software.

After a 14-day acclimatization period, dark incubations were initiated. Sediment cores were uncoupled from the recirculation system, closed airtight for a period long enough to measure steady changes in oxygen and nutrient concentrations while ensuring that oxygen concentration in the sediment-overlying sea water did not drop below 50% saturation. Oxygen concentration in the sediment-overlying sea water was continuously monitored with Oxygen Spot Sensors (OXSP5, Pyroscience) glued to the inner wall of the tubes. The sensors were operated with an optical oxygen meter (FireStingO2, Pyroscience), connected with a lens-spot adapter (SPADLNS) and Spot Fiber (SPFIB). Incubation time depended on the sediment type and temperature (i.e. longer incubation times for permeable sediment and measurements at low temperatures). Tubes of half the volume of a normal fine sandy sediment tube (i.e. 12.5 cm height) were incubated with tank water to estimate the influence of processes occurring only in the water column. Six (February) or five (April) times, bottom water samples were taken with a glass syringe for the determination of oxygen concentrations (in 12 mL Winkler bottles) and DIN (10 mL), the latter filtered through Whatman GF/F filters. At the same time, tank water was carefully injected into the overlying water of the tubes to compensate for the sampled volume. Tank samples were taken for oxygen and DIN analyses to correct for the dilution in the tubes during this additional sampling. Oxygen samples were stored in the same temperature-controlled room in the dark until further analysis (within 3 days); DIN samples were stored frozen (−20°C).

At the end of the incubations, the tubes were opened and the sediment was subsampled for pigments and organic carbon and nitrogen (each 2 ml from upper 2 cm). The remaining sediment was sieved on a 1 mm mesh to sample the macrofauna, which was preserved in ethanol. Macrobenthos specimens were identified to the lowest possible taxonomic level (typically species level), counted and biomasses were determined (blotted wet weights). Apart from deriving the structural characteristics, density, biomass and species richness, also the functional descriptors Bioturbation Potential index (BPi) and Bioturbation Potential of the Community (BPc) were calculated [37,38].

Laboratory analyses and flux calculation

Pigments were determined by HPLC (Gilson) analysis according to Wright and Jeffrey [39]. Total organic C and N content was analyzed with an Element Analyzer N1500 (Carlo Erba). Oxygen was analyzed by automated Winkler titration [40] and DIN by automated colorimetric techniques (SKALAR). Oxygen and DIN fluxes were calculated from the significant regression slopes of concentration over time compensating for dilution by refill water. Finally, a correction for the processes occurring in the water column was made, by subtracting the rates measured in the water incubations from the total measured rate.

Mass budget modeling

The fluxes of O_2, NO_x, and NH_x across the sediment–water interface were used to estimate rates of nitrification, denitrification and total nitrogen mineralization. This was done by constructing an integrated mass balance of oxygen, nitrate and ammonium over the entire sediment column [29]. See Materials S1 for detailed methodology.

It must be noted that the DIN concentrations of the 'tank water' used in the core incubations (cf. Supra: 2.3 Sediment core incubations) differed from the field DIN concentrations [41]: while NH_x concentrations were rather similar in the field and in the lab, NO_x concentrations were one order of magnitude higher in the tank than in the field. The reported NO_x fluxes should therefore be considered as "potential fluxes".

We chose to make measurements of several aspects of the N cycle rather than analyze a large number of replicates of fewer variables using a variance based statistical approach. The latter strategy allows for detection of significant differences in the measured variables, but we aimed to obtain a more holistic view by performing measurements of different N-cycle related processes covarying with ammonium fluxes (nitrate fluxes, oxygen fluxes) in combination with modeling of individual flux terms subjected to overall mass balance constraints. This allowed us to (1) assess the robustness of single measurements and (2) understand why patterns were observed.

Statistical approach

To test differences in oxygen and pH profiles, a multivariate data matrix was constructed in which each depth horizon is considered a "variable" and each pH measurement a measure of "abundance" [24]. As not all pH profiles were measured to the same depth, including all data would make the design too imbalanced. Therefore, only the major depth horizons were considered (sediment–water interface till 5 mm below the sediment–water interface, in steps of 1 mm). Permutational ANOVAs (Permanova) were carried out to test for differences in

Table 1. Results of Generalized Least Squares for differences in chlorophyll-a content of the sediment, macrobenthic univariate descriptors and measured and estimated sediment processes among the experimental factors pH treatment, Sediment and Month.

	Factor	L-ratio	p-value
Chlorophyll-*a*	pH×Sediment×Month	7.01	0.008
Macrobenthic density	Sediment	13.12	<0.001
	Month	4.63	0.03
Macrobenthic species richness	Sediment	10.28	0.001
Macrobenthic biomass	Sediment	10.33	0.001
Bioturbation potential, BPc	Sediment	12.06	<0.001
	Month	4.09	0.043
SCOC	pH	5.44	0.02
	Sediment×Month	30.48	<0.001
NH_x effluxes	Month	6.42	0.011
	Sediment	29.37	<0.001
NO_x effluxes	Sediment×Month	9.48	0.002
Nitrification	pH×Month	8.07	0.004
	Sediment×Month	18.68	<0.001
Total N mineralization	Sediment×Month	17.10	<0.0001

Only significant results ($p<0.05$) are shown.

these sediment oxygen and pH profiles. A Euclidean distance similarity matrix was built and subsequently fully crossed 3-way Permanova's were run with factors Month, Sediment and pH Treatment. To test the effect of these factors on macrobenthic community structure, a Bray-Curtis similarity matrix was constructed. In case of significant interaction of factors, pair-wise tests were run to further investigate the observed differences. For sediment oxygen and pH profiles, SIMPER analysis was used to determine which depth was responsible for any differences identified, whereas for macrobenthic community analysis, this test indicated the species that characterized the community. Homogeneity of multivariate dispersion ('variance') was tested with PERMDISP for any of the significant terms in Permanova; if significant, this test indicates that observed patterns can be a result of both treatment and dispersion. The difference in pH in the water tanks during the acclimatization period was also tested with Permanova, using a two-factor design with pH treatment and time.

While testing the effect of pH Treatment, Sediment and Month on the univariate variables chlorophyll-*a*, SCOC, DIN fluxes, estimated nitrification, total N mineralization and macrobenthic characteristics, PERMDISP often pointed at heterogeneity of variances that complicated the interpretation of the significant Permanova's. Therefore, we adopted a linear model with a generalized least-squares extension [42–44], which allows unequal variances among treatment combinations to be modeled as a variance covariance matrix [42,43]. Following West et al. [42] and Zuur et al. [44], the most appropriate variance covariate matrix was determined using AIC scores in conjunction with plots of fitted values versus residuals with different variance covariate terms relating to the independent variables, using restricted maximum likelihood (REML, [42]). This procedure resulted in the use of a variance structure that allowed for different variances per stratum for sediment type and pH treatment (varIdent function, R package nlme). The fixed component of the model was then refined by manual backwards stepwise selection using maximum likelihood (ML) to remove insignificant variable terms. Following Underwood [45], the highest order significant interactions in the minimal adequate model were examined, but the nested levels within these were not. The importance of the highest order term was estimated using a likelihood ratio (L-ratio) test to compare the full minimal adequate model with a model in which the relevant variable and all the interaction terms that it was involved in, was omitted. Pair-wise tests were carried out within the R package contrast [46].

All multivariate analyses were carried out within PRIMER v6.0 with Permanova+ add-on software [47,48]. Univariate models were performed in the free statistical environment R (http://cran.r-project.org). Results are shown as mean ± standard deviation.

Results

Evaluation of experimental system

A significant difference in pH of 0.31 ± 0.03 between control and acidified treatments was established in the water tanks in both months (Permanova, pseudo-F >36.8, $p=0.001$) (Table S1). Temperature and salinity remained constant and oxygen concentration stayed saturated over time. All other seawater carbonate variables remained constant over time as well (Table S1).

Pigments

Chlorophyll-*a* concentrations were affected by the interactive effects of pH Treatment, Sediment and Month (Table 1, 2). In the fine sandy sediments, chl-*a* concentrations were low in February. The bloom deposition was evidenced by higher chl-*a* concentrations at the sediment in April. Chlorophyll-*a* concentrations in permeable sediments were an order of magnitude lower in February, but also increased in April (Table 1, 2).

Macrobenthos

Macrobenthic community structure was only dependent on sediment type (Permanova, pseudo-F = 9.29 $p=0.001$) and the two sediment types were strongly dissimilar in terms of macrobenthic community structure (99.91%; SIMPER analysis). The fine sandy sediment community was dominated (94% in terms of

Table 2. Average ± sd of sediment chlorophyll-*a*, measured and estimated sediment processes and macrobenthic parameters for each of the factors (pH treatment, Month, Sediment or their interactions) with a significant effect.

	pH		Month	Month×pH		Sediment	Sediment×Month		Sediment×Month×pH			
Chlorophyll-*a*	-		-			-			**Permeable**	February	Control	0.05±0.07
											Acidified	0.02±0.03
										April	Control	0.13±0.07
											Acidified	0.13±0.04
									Fine sandy	February	Control	0.29±0.04
											Acidified	0.68±0.33
										April	Control	0.81±0.47
											Acidified	11.03±4.48
SCOC ($mmol\ O_2\ m^{-2}\ d^{-1}$)	**Control**	11.73±8.09	-			**Permeable**	February	5.87±4.98	-			
	Acidified	6.63±7.45					April	1.25±2.36				
						Fine sandy	February	10.33±3.57				
							April	19.26±3.21				
NHx efflux ($mmol\ N\ m^{-2}\ d^{-1}$)	-		**February**		1.09±1.55	**Permeable**		0.31±0.39	-			
			April		1.00±1.00	**Fine sandy**		1.79±1.41				
NOx efflux ($mmol\ N\ m^{-2}\ d^{-1}$)	-		-			**Permeable**	February	8.66±1.22	-			
							April	0.25±0.47				
						Fine sandy		3.04±2.53				
Nitrification ($mmol\ N\ m^{-2}\ d^{-1}$)	-		**February**	Control	3.74±2.21	**Permeable**	February	3.15±2.36	-			
				Acidified	0.22±1.14		April	0.02±0.87				
			April		0.87±1.24	**Fine sandy**		1.17±1.72				
N mineralization ($mmol\ N\ m^{-2}\ d^{-1}$)	-		-			**Permeable**	February	4.88±1.23	-			
							April	0.42±0.64				
						Fine sandy		2.97±0.84				
Density (ind. m^{-2})	-		**February**		725±879	**Permeable**		44±41	-			
			April		370±412	**Fine sandy**		1050±659				
Biomass ($g\ m^{-2}$)	-		-			**Permeable**		3±5	-			
						Fine sandy		189±137				
Species richness ($core^{-1}$)*	-		-			**Permeable**	*per 283.53 cm^2	1.13±0.99	-			
						Fine sandy	*per 78.54 cm^2	3.13±1.25				
BPc (m^{-2})	-		**February**		898±1102	**Permeable**		57±51	-			
			April		601±671	**Fine sandy**		1442±785				

numbers) by the small surface-modifying *Magelona johnstoni* and the biodiffusing *Scoloplos armiger* polychaete species and the surface-modifying bivalve *Macoma balthica* (SIMPER analysis; for functional group classification see [38]). The permeable sediment community consisted 94% of the biodiffusing polychaetes *Nephtys cirrosa* and *Ophelia limacina* and the surface-modifying amphipod *Urothoe brevicornis* (SIMPER analysis). Neither structural (density, biomass, species richness) nor functional (BPc) univariate macrobenthic characteristics were affected by pH treatment. Density, biomass, species richness and BPc only differed among sediments (Table 1). The fine sandy sediment clearly displayed higher macrobenthic density, species richness, biomass and BPc values than the permeable sediment (Table 2). Density and BPc differed also marginally among months (Table 1): in February, density and BPc were higher than in April (Table 2).

pH and oxygen sediment profiles

pH profiles in the upper 5 mm of the sediment (Fig. 2) were significantly affected by the interactive effects of Sediment and pH (Permanova: pseudo-F = 17.31, p = 0.001) and Month and pH (Permanova: pseudo-F = 4.23, p = 0.001). However, for both tests, some heterogeneity of variances was detected (Permdisp: p = 0.001), which indicates that there is also a significant dispersion effect. Pair-wise tests revealed significant differences between pH profiles of acidified and control treatments in permeable sediments ($p<0.05$), and a marginally significant difference ($p = 0.054$) between pH profiles of acidified and control treatments in April. The surface and the first two millimeters made up 60% of the difference between pH profiles of acidified and control permeable sediments (Simper). In these layers, the pH was on average 0.24±0.03 units lower than in the respective control surface sediment layers. In April, the difference between acidified and control sediments was mainly (54%) found in the deeper 3–5 mm, where pH values were 0.03±0.03 units lower in the acidified than in the control sediment.

Figure 3 suggests a deeper sediment oxygen penetration in acidified fine sediments in February than their respective controls, but these effects were not significant. Only the interactive effects of Month and Sediment influenced sediment oxygen penetration marginally (Permanova: pseudo-F = 2.90, p = 0.042), where the oxygen content of fine sandy sediments was higher in February than in April (pair-wise tests: $p<0.05$) and that of permeable sediments stayed equal (pair-wise tests: $p>0.05$).

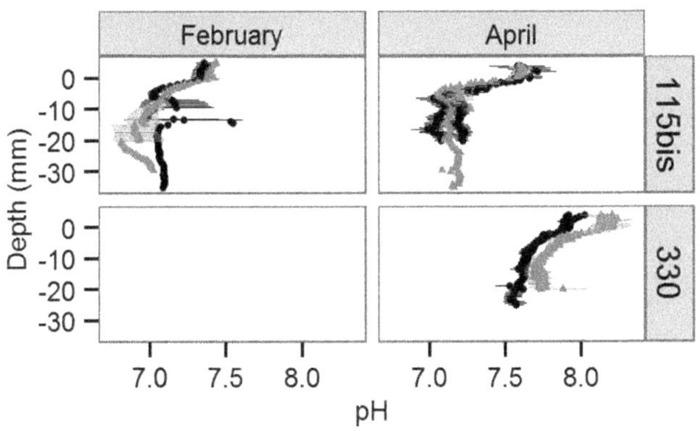

Figure 2. Sediment pH profiles. Mean ± sd pH in control (▲) and acidified (●) treatments of the two sediment types (115bis: fine sandy; 330: permeable) and two seasons (February and April). No data for permeable sediment in February.

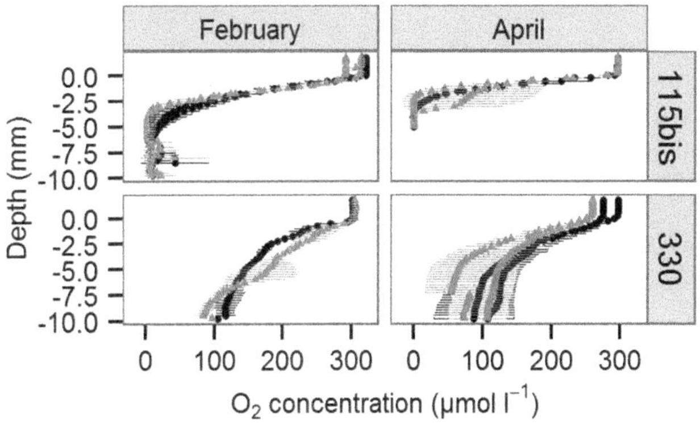

Figure 3. Sediment O_2 profiles. Mean ± sd O_2 concentration in control (▲) and acidified (●) treatments of the two sediment types (115bis: fine sandy; 330: permeable) and two seasons (February and April).

Fluxes at the sediment-water interface

SCOC differed among pH treatment (Table 1, 2; Fig. 4), and was significantly lower in acidified treatments (59±35%) than in the respective controls. SCOC also differed among sediments and months (Table 1, 2). Significantly higher SCOC was measured in February compared to April in the permeable sediments (pair-wise tests: $p<0.002$). In the fine sandy sediments, SCOC was higher in April than in February (pair-wise tests: $p<0.02$).

All NH_x fluxes were directed towards the water column, and only differed between sediments (Table 1, 2; Fig. 4), with lower effluxes in permeable than in fine sandy sediment, and between months (Table 1, 2), with slightly higher effluxes in February than in April. Also note the high variability among NH_x effluxes, especially in the acidified treatments (Fig. 4).

NO_x fluxes were mainly directed into the sediment and differed only among months and sediments (Table 1, 2; Fig. 4). In the permeable sediments, NO_x influxes were highest in February compared to April (pair-wise tests: $p<0.05$), whereas they were equal in both months in the fine sandy sediments (pair-wise tests: $p>0.05$).

The nitrification rates estimated by the mass budgets were affected by the interactive effects of pH Treatment and Month and Sediment and Month (Table 1, 2; Fig. 4). Only in February, estimated nitrification rates were significantly reduced by 94% in sediments underlying an acidified water column compared to control sediments. Nitrification estimates were significantly higher in permeable sediments in February compared to April.

Total N mineralization estimates were only affected by the interactive effects of Sediment and Month (Table 1, 2; Fig. 4). Total N mineralization estimates were significantly higher in permeable sediments in February compared to April.

Discussion

Our results indicate that ocean acidification in coastal sediments reduces sediment community oxygen consumption and nitrification, but does not significantly affect total N mineralization. To the best of our knowledge, this is the first observation of reduced sediment community oxygen consumption (60% reduction) in sediments underlying an acidified water column. SCOC is the sum of all oxygen consuming processes in the sediment, which include nitrification, oxidation of reduced substances other than ammonia and nitrite, but also the oxygen uptake of the fauna present and bacteria, whether or not stimulated by macrofaunal bio-irrigation.

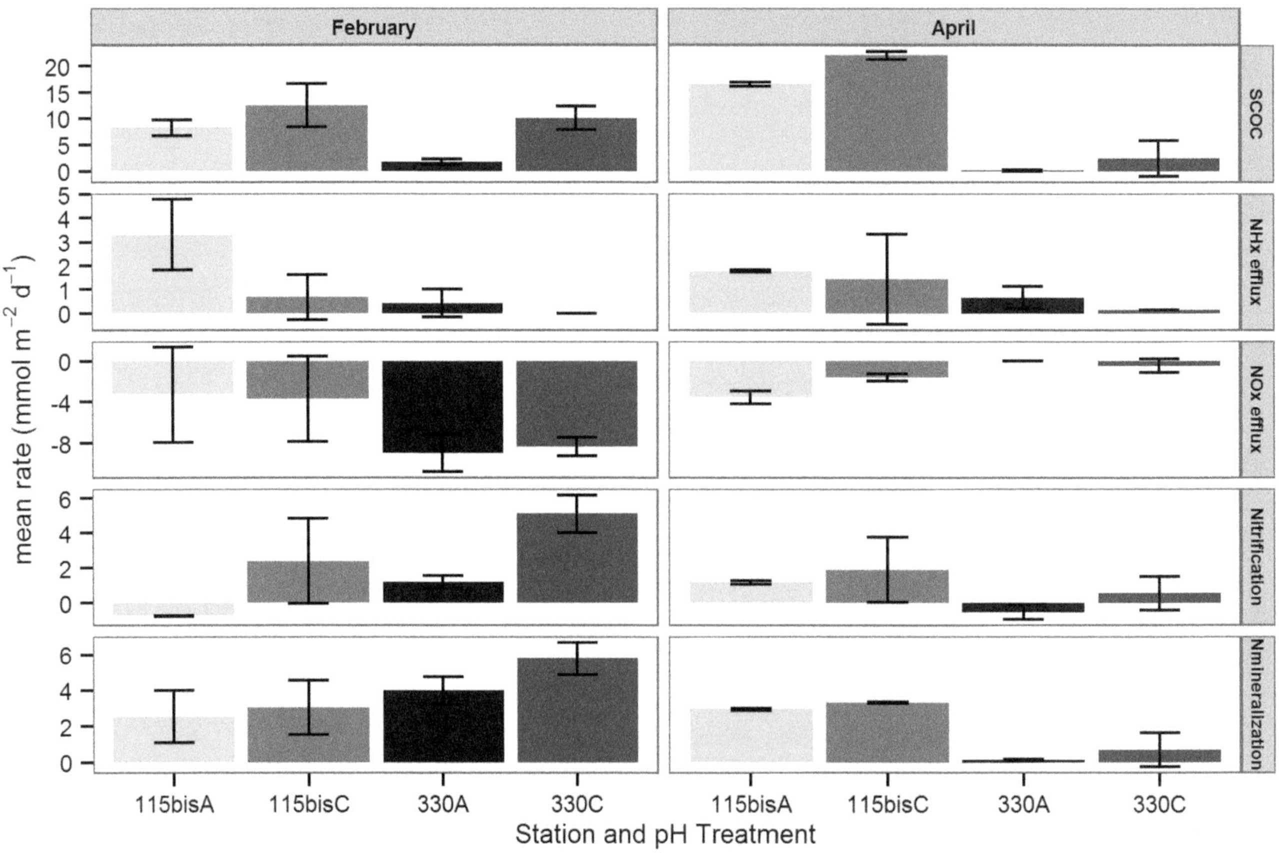

Figure 4. Sediment-water fluxes. Measured Sediment Community Oxygen Consumption (SCOC, mmol O_2 m^{-2} d^{-1}), NH_x and NO_x effluxes and estimated nitrification and total N mineralization (mmol N m^{-2} d^{-1}) (mean ± sd) from control and acidified treatments of the two sediment types (115bis: fine sandy; 330: permeable) and two seasons (February and April).

The effect of ocean acidification on faunal stimulation of SCOC through bio-irrigation is expected to be minor, since no real bio-irrigators were present in the macrofauna communities and such effects are only expected on the longer term (at least several weeks) [19,21,22]. As total N mineralization remained constant, the decrease in SCOC can be mainly attributed to a lower demand by nitrifying organisms. This effect was found in two biologically and biogeochemically contrasting sediments that were incubated at a realistic scenario of reduced seawater pH by 2100. A 94% reduction in nitrification rate was observed before but not during the annual phytoplankton bloom deposition.

Our results are in contrast with the two earlier studies that did not find a significant reduction in ammonia oxidation (which is the first step in nitrification) in the North Sea, in muddy sediments in January (pre-bloom) [17], nor in non-bioturbated muddy surface sediments in September (post-bloom) [18], nor in permeable sediments in August (post-bloom) [17,18]. These different seasonal and sediment settings might complicate the comparison with our study. On the other hand, other North Sea studies interpreted fluxes of reduced and oxidized N-species across the sediment–water interface, and suggested a decreased benthic nitrification mainly at much more extreme pH reductions (pH range: 5.6–7.6) [19,20,23,24] than we applied.

The potential to detect significant trends at realistic acidification scenarios depends on the accuracy of the applied methods. Measuring fluxes in closed tube incubations and correcting for processes occurring in the water column offers the advantage of a holistic approach and balanced interpretation of the on-going processes in each sediment core, but there still exists large variability between replicate experimental units (see relatively large error bars on our figures – but nonetheless are the fluxes in this study in the same range as fluxes earlier measured in the area [41] - and the range of variation in Kitidis et al. [17]). As such, the lower nitrification rates in April, especially in permeable sediments, combined with a relatively high variability among replicates, probably prevent a statistically detectable decrease in April. In some acidified treatments (fine sandy sediment in February, permeable sediments in April), estimated nitrification rates were slightly negative, indicating a violation of the mass budget assumption that all reduced substances other than ammonia are reoxidized within the sediment [41]. This suggests that, in addition to ammonia and nitrite oxidation, other oxidation processes are affected by acidification as well. This could have included NH_4^+ efflux rates, which were also highly variable among replicates. More dedicated experiments, where also the fluxes of other reduced substances are measured and replication is increased, are needed to enhance our understanding of stressor effects (e.g. acidification) on benthic ecosystem functioning.

How can the observed decrease in estimated nitrification rates be explained? The lower pH imposed on the sea water reduced nitrification in both our sediments, by altering the pH within the sediment matrix. In the fine sandy sediments, we observed steep pH profiles in accordance with earlier results [20,24,49]. These profiles are driven by microbially mediated redox reactions linked to mineralization of organic matter and the dissolution and precipitation of minerals such as $CaCO_3$ (reviewed in [10]).

Hence, benthic microbial communities involved in nitrification in fine sandy sediments are likely more adapted to low pH. In addition, the dissolution of minerals acts as a buffer towards pH changes in the water column [10,50–52]. The very similar pH profiles in our fine sandy acidified and control sediments suggest that these sediments were buffered against water column acidification, which resulted in relatively small effects. In contrast, the pH in the permeable sediment was consistently lowered over the upper 1 cm of the sediment under the acidification scenario (Fig. 3). These pH changes may affect biogeochemical rates in several ways. First of all, the first step in nitrification (ammonia oxidation) requires NH_3 rather than NH_4^+ [53,54]. Since seawater buffers pH by shifting the NH_3/NH_4^+ balance towards the larger NH_4^+ ion at lower pH, a decrease in substrate for ammonia oxidation takes place during ocean acidification. In the presented experiments, a decrease in pH of 0.3 would imply a substrate decline of about 50% [55]. Secondly, large infaunal bioturbators can stimulate nitrification [56], but when these are negatively affected by pH reductions in the overlying water, the stimulating effect disappears and a negative effect on ammonia oxidising microbial organisms is observed [18]. Bioturbation is an instrumental mechanism for ammonia-oxidizing bacteria (AOB) but not for ammonia-oxidizing archaea (AOA) [57]. Consequently, bioturbation-mediated change in AOB communities has been suggested to explain the outcome of ocean acidification effects on ammonium oxidation in a pH range of 7.90–6.80 [18]. Quantification of faunal activity (bioturbation and bio-irrigation) was not specifically targeted in our experiments but we did not observe effects of acidification on the structural and functional characteristics of the macrofauna communities in our sediments, which were not inhabited by strong bio-irrigators or bioturbators. Moreover, bioturbation is not expected to decrease in short-term experiments at realistic pH declines [19,21,22].

We speculate that the observed reduction in nitrification in the present short-term study could result from rapid acidification-induced changes in microbial activity and composition, particularly in terms of the realized ratio of archaea to bacteria [58,59]. Within buffered cohesive sediments, only minor changes in the active benthic total microbial communities from surface sediments were observed in a 2-week acidification experiment [60]. However, in the same experiment, Tait et al. [61] revealed changes in the relative abundance of benthic AOB and AOA under reduced pH circumstances. In addition, under strong acidification scenarios, sedimentary AOB show drastic reductions in amoA transcription, while sedimentary AOA transcripts increase [61]. This suggests that AOB and AOA in marine sediments have different pH optima, and the impact of elevated CO_2 on N cycling may be dependent on the relative abundances of these two major microbial groups [61].

Finally, we should be aware that short-term effects of acidification do not necessarily reflect the net effect of climate forcing; long-term studies more likely include seasonal and other natural variability that bioturbators and the interaction that species undergo [22,62]. As benthic nitrification makes up most of ocean nitrification, a reduction of this nitrogen cycling in both permeable and fine sediments in winter, when coastal benthic nitrification is generally highest [63,64], could have global-wide impacts on coupled nitrification-denitrification and hence eventually on pelagic nutrient availability. High NH_x effluxes could as such precipitate a positive feedback on acidification, if eutrophication is accelerated by the NH_x flux, via oxidation of senescent phytoplankton biomass [65].

Supporting Information

Table S1 Monitoring data of 14 days prior to the incubations. Minimum, maximum, average and standard error of the monitored variables and calculated variables (pCO2, DIC and Ωaragonite and Ωcalcite) in the tanks of the different experiments.

Acknowledgments

We are grateful to the crew of RV Zeeleeuw for help with sampling at sea, to Yves Israël for the development of the experimental set-up, to Joris Bril, Bart Beuselinck, Niels Viaene and Liesbet Colson for help with sample processing.

Author Contributions

Conceived and designed the experiments: JV UB CVC. Performed the experiments: UB JV CVC KG. Analyzed the data: UB JV KS. Contributed reagents/materials/analysis tools: MV DVG KS. Wrote the paper: UB JV CVC KG KS MV.

References

1. Solomon S, Qin D, Manning M, Chen Z, Marquis M, et al. (2007) The physical science basis. Contribution of working group I to the fourth assessment report of the intergovernemental panel on climate change. Cambridge Univ. Press, Cambridge, UK, and New York.
2. Brewer PG (2009) A changing ocean seen with clarity. Proc Natl Acad Sci 106: 12213–12214.
3. Caldeira K, Wickett ME (2003) Anthropogenic carbon and ocean pH. Nature 425: 365.
4. Orr JC, Fabry VJ, Aumont O, Bopp L, Doney SC, et al. (2005) Anthropogenic ocean acidification over the twenty-first century and its impact on calcifying organisms. Nature 437: 681–686.
5. Wootton JT, Pfister CA, Forester JD (2008) Dynamic patterns and ecological impacts of declining ocean pH in a high-resolution multi-year dataset. Proc Natl Acad Sci 105: 18848–18853.
6. Provoost P, van Heuven S, Soetaert K, Laane R, Middelburg JJ (2010) Long-term record of pH in the Dutch coastal zone: a major role for eutrophication-induced changes. Biogeosciences 7: 3869–3878.
7. Andersson AJ, Mackenzie FT, Gattuso J-P (2011) Effects of ocean acidification on benthic processes, organisms, and ecosystems. Ocean acidification. Oxford University Press, Oxford. pp. 122–153.
8. Pörtner H-O, Gutowska M, Ishimatsu A, Lucassen A, Melzner F, et al. (2011) Effects of ocean acidification on nektonic organisms. Ocean acidification. Oxford University Press, Oxford. pp. 154–175.
9. Riebesell U, Tortell PD (2011) Effects of ocean acidification on pelagic organisms and ecosystems. Ocean acidification. Oxford University Press, Oxford. pp. 99–112.
10. Widdicombe S, Spicer JI, Kitidis V (2011) Effects of ocean acidification on sediment fauna. Ocean acidification. Oxford University Press, Oxford. pp. 113–122.
11. Kroeker KJ, Micheli F, Gambi MC, Martz TR (2011) Divergent ecosystem responses within a benthic marine community to ocean acidification. Proc Natl Acad Sci 108: 14515–14520.
12. Gehlen M, Gruber N, Gangsto R, Bopp L, Oschlies A (2011) Biogeochemical consequences of ocean acidification and feedbacks to the earth system. Ocean acidification. Oxford University Press, Oxford. pp. 230–248.
13. Hutchins DA, Mulholland MR, Fu F (2009) Nutrient cycles and marine microbes in a CO_2-enriched ocean. Oceanography 22.
14. Voss M, Bange HW, Dippner JW, Middelburg JJ, Montoya JP, et al. (2013) The marine nitrogen cycle: recent discoveries, uncertainties and the potential relevance of climate change. Philos Trans R Soc B Biol Sci 368: 20130121.
15. Huesemann MH, Skillman AD, Crecelius EA (2002) The inhibition of marine nitrification by ocean disposal of carbon dioxide. Mar Pollut Bull 44: 142–148.
16. Beman JM, Chow C-E, King AL, Feng Y, Fuhrman JA, et al. (2011) Global declines in oceanic nitrification rates as a consequence of ocean acidification. Proc Natl Acad Sci 108: 208–213.

17. Kitidis V, Laverock B, McNeill LC, Beesley A, Cummings D, et al. (2011) Impact of ocean acidification on benthic and water column ammonia oxidation. Geophys Res Lett 38: L21603.
18. Laverock B, Kitidis V, Tait K, Gilbert JA, Osborn AM, et al. (2013) Bioturbation determines the response of benthic ammonia-oxidizing microorganisms to ocean acidification. Philos Trans R Soc B Biol Sci 368: 20120441.
19. Widdicombe S, Needham HR (2007) Impact of CO_2-induced seawater acidification on the burrowing activity of *Nereis virens* and sediment nutrient flux. Mar Ecol Prog Ser 341: 111–122.
20. Widdicombe S, Dashfield SL, McNeill CL, Needham HR, Beesley A, et al. (2009) Effects of CO_2 induced seawater acidification on infaunal diversity and sediment nutrient fluxes. Mar Ecol Prog Ser 379: 59–75.
21. Wood HL, Widdicombe S, Spicer JI (2009) The influence of hypercapnia and the infaunal brittlestar *Amphiura filiformis* on sediment nutrient flux- will ocean acidification affect nutrient exchange? Biogeosciences 6: 2015–2024.
22. Godbold JA, Solan M (2013) Long-term effects of warming and ocean acidification are modified by seasonal variation in species responses and environmental conditions. Philos Trans R Soc B Biol Sci 368: 20130186.
23. Murray F, Widdicombe S, McNeill CL, Solan M (2013) Consequences of a simulated rapid ocean acidification event for benthic ecosystem processes and functions. Mar Pollut Bull 73: 435–442.
24. Widdicombe S, Beesley A, Berge JA, Dashfield SL, McNeill CL, et al. (2013) Impact of elevated levels of CO_2 on animal mediated ecosystem function: The modification of sediment nutrient fluxes by burrowing urchins. Mar Pollut Bull 73: 416–427.
25. Gazeau F, van Rijswijk P, Pozzato L, Middelburg JJ (2014) Impacts of Ocean Acidification on Sediment Processes in Shallow Waters of the Arctic Ocean. Plos One 9: e94068.
26. Ouellette D, Desrosiers G, Gagne JP, Gilbert F, Poggiale JC, et al. (2004) Effects of temperature on in vitro sediment reworking processes by a gallery biodiffusor, the polychaete *Neanthes virens*. Mar Ecol Prog Ser 266: 185–193.
27. Maire O, Duchene JC, Rosenberg R, de Mendonca J, Gremare A (2006) Effects of food availability on sediment reworking in *Abra ovata* and *A. nitida*. Mar Ecol Prog Ser 319: 135–153.
28. Braeckman U, Provoost P, Gribsholt B, Van Gansbeke D, Middelburg JJ, et al. (2010) Role of macrofauna functional traits and density in biogeochemical fluxes and bioturbation. Mar Ecol Prog Ser 399: 173–186.
29. Soetaert K, Herman PMJ, Middelburg JJ, Heip C, Smith CL, et al. (2001) Numerical modelling of the shelf break ecosystem: reproducing benthic and pelagic measurements. Deep-Sea Res Part II 48: 3141–3177.
30. Lancelot C, Spitz Y, Gypens N, Ruddick K, Becquevort S, et al. (2005) Modelling diatom and *Phaeocystis* blooms and nutrient cycles in the Southern Bight of the North Sea: the MIRO model. Mar Ecol Prog Ser 289: 63–78.
31. Van Oevelen D, Soetaert K, Franco MA, Moodley L, van IJzerloo L, et al. (2009) Organic matter input and processing in two contrasting North Sea sediments: insights from stable isotope and biomass data. Mar Ecol Prog Ser 380: 19–32.
32. Huettel M, Rusch A (2000) Transport and degradation of phytoplankton in permeable sediment. Limnol Oceanogr 45: 534–549.
33. Caldeira K, Wickett ME (2005) Ocean model predictions of chemistry changes from carbon dioxide emissions to the atmosphere and ocean. J Geophys Res 110: C09S04.
34. Riebesell U, Fabry VJ, Hansson L, Gattuso JP (2010) Guide to best practices for ocean acidification research and data reporting. Luxembourg: Publications Office of the European Union. 260 p.
35. Dickson AG, Sabine CL, Christian JR (2007) Guide to best practices for ocean CO_2 measurements. Available: http://aquaticcommons.org/1443/.
36. Pierrot D, Lewis E, Wallace DWR (2006) CO2SYS DOS Program developed for CO_2 system calculations. ORNL/CDIAC-105. Carbon Dioxide Information Analysis Center, Oak Ridge National Laboratory, US Department of Energy, Oak Ridge, TN.
37. Solan M, Cardinale BJ, Downing AL, Engelhardt KAM, Ruesink JL, et al. (2004) Extinction and ecosystem function in the marine benthos. Science 306: 1177–1180.
38. Queirós AM, Birchenough SNR, Bremner J, Godbold JA, Parker RE, et al. (2013) A bioturbation classification of European marine infaunal invertebrates. Ecol Evol 13: 3958–3985.
39. Wright SW, Jeffrey SW (1997) High-resolution HPLC system for chlorophylls and carotenoids of marine phytoplankton. Jeffrey Sw Mantoura Rfc Wright Sw Eds Phytoplankton Pigments Ocean Guid Mod Methods Unesco Paris Pp 327–341.
40. Parsons TR, Maita Y, Lalli CM (1984) A manual of chemical and biological methods for seawater analysis. Pergamon Press New York.
41. Braeckman U, Foshtomi MY, Van Gansbeke D, Meysman F, Soetaert K, et al. (2014) Variable importance of macrofaunal functional biodiversity for biogeochemical cycling in temperate coastal sediments. Ecosystems 17: 720–737.
42. West B, Welch KB, Galecki AT (2006) Linear Mixed Models: A Practical Guide Using Statistical Software. Taylor & Francis. 382 p.
43. Pinheiro JC, Bates D (2009) Mixed-Effects Models in S and S-PLUS. Springer. 538 p.
44. Zuur A, Ieno EN, Walker N (2009) Mixed Effects Models and Extensions in Ecology with R. Springer. 580 p.
45. Underwood AJ (1997) Experiments in ecology: their logical design and interpretation using analysis of variance. Cambridge University Press.
46. Kuhn M, Weston S, Wing J, Forester J, Thaler T (2013) Contrast: A collection of contrast methods. R package version 0.19. Available at http://CRAN.R-project.org/package=contrast.
47. Clarke KR, Gorley RN (2006) Primer v6: User Manual/Tutorial. Primer-E Plymouth.
48. Anderson MJ, Gorley RN, Clarke KR (2008) PERMANOVA+ for PRIMER: guide to software and statistical methods. Primer-E Plymouth.
49. Dashfield SL, Somerfield PJ, Widdicombe S, Austen MC, Nimmo M (2008) Impacts of ocean acidification and burrowing urchins on within-sediment pH profiles and subtidal nematode communities. J Exp Mar Biol Ecol 365: 46–52.
50. Wenzhöfer F, Adler M, Kohls O, Hensen C, Strotmann B, et al. (2001) Calcite dissolution driven by benthic mineralization in the deep-sea: *in situ* measurements of Ca^{2+}, pH, pCO_2 and O_2. Geochim Cosmochim Acta 65: 2677–2690.
51. Jourabchi P, Cappellen PV, Regnier P (2005) Quantitative interpretation of pH distributions in aquatic sediments: A reaction-transport modeling approach. Am J Sci 305: 919–956.
52. Morse JW, Andersson AJ, Mackenzie FT (2006) Initial responses of carbonate-rich shelf sediments to rising atmospheric pCO_2 and "ocean acidification": Role of high Mg-calcites. Geochim Cosmochim Acta 70: 5814–5830.
53. Suzuki I, Dular U, Kwok SC (1974) Ammonia or Ammonium Ion as Substrate for Oxidation by *Nitrosomonas europaea* Cells and Extracts. J Bacteriol 120: 556–558.
54. Ward BB (1987) Kinetic studies on ammonia and methane oxidation by *Nitrosococcus oceanus*. Arch Microbiol 147: 126–133.
55. Zeebe RE, Wolf-Gladrow D (2001) CO_2 in Seawater: Equilibrium, Kinetics, Isotopes: Equilibrium, Kinetics, Isotopes. Elsevier. 362 p.
56. Stief P (2013) Stimulation of microbial nitrogen cycling in aquatic ecosystems by benthic macrofauna: mechanisms and environmental implications. Biogeosciences 10: 7829–7846.
57. Laverock B, Tait K, Gilbert J, Osborn A, Widdicombe S (2013) Impacts of bioturbation on temporal variation in bacterial and archaeal nitrogen-cycling gene abundance in coastal sediments. Environ Microbiol Reports 6: 113–121.
58. Wyatt NJ, Kitidis V, Woodward EMS, Rees AP, Widdicombe S, et al. (2010) Effects of high CO_2 on the fixed nitrogen inventory of the Western English Channel. J Plankton Res 32: 631–641.
59. Gilbertson WW, Solan M, Prosser JI (2012) Differential effects of microorganism–invertebrate interactions on benthic nitrogen cycling. Fems Microbiol Ecol 82: 11–22.
60. Tait K, Laverock B, Shaw J, Somerfield PJ, Widdicombe S (2013) Minor impact of ocean acidification to the composition of the active microbial community in an Arctic sediment. Environ Microbiol Reports 5: 851–860.
61. Tait K, Laverock B, Widdicombe S (2013) Response of an Arctic Sediment Nitrogen Cycling Community to Increased CO_2. Estuaries Coasts: 1–12.
62. Godbold JA, Calosi P (2013) Ocean acidification and climate change: advances in ecology and evolution. Philos Trans R Soc B Biol Sci 368: 20120448.
63. Kemp WM, Sampou P, Caffrey J, Mayer M, Henriksen K, et al. (1990) Ammonium recycling versus denitrification in Chesapeake Bay sediments. Limnol Oceanogr 35: 1545–1563.
64. Lohse L, Malschaert JFP, Slomp CP, Helder W, van Raaphorst W (1993) Nitrogen cycling in North Sea sediments: interaction of denitrification and nitrification in offshore and coastal areas. Mar Ecol Prog Ser 101: 283–283.
65. Sunda WG, Cai W-J (2012) Eutrophication Induced CO_2-Acidification of Subsurface Coastal Waters: Interactive Effects of Temperature, Salinity, and Atmospheric PCO_2. Environ Sci Technol 46: 10651–10659.

6

Geomorphic and Ecological Disturbance and Recovery from Two Small Dams and Their Removal

Desirée D. Tullos[1]*, Debra S. Finn[2], Cara Walter[1]

1 Water Resources Engineering, Oregon State University, Corvallis, Oregon, United States of America, **2** Department of Integrative Biology, Oregon State University, Corvallis, Oregon, United States of America

Abstract

Dams are known to impact river channels and ecosystems, both during their lifetime and in their decommissioning. In this study, we applied a before-after-control-impact design associated with two small dam removals to investigate abiotic and biotic recovery trajectories from both the elimination of the press disturbance associated with the presence of dams and the introduction of a pulse disturbance associated with removal of dams. The two case studies represent different geomorphic and ecological conditions that we expected to represent low and high sensitivities to the pulse disturbance of dam removal: the 4 m tall, gravel-filled Brownsville Dam on the wadeable Calapooia River and the 12.5 m tall, sand and gravel-filled Savage Rapids Dam on the largely non-wadeable Rogue River. We evaluated both geomorphic and ecological responses annually for two years post removal, and asked if functional traits of the macroinvertebrate assemblages provided more persistent signals of ecological disturbance than taxonomically defined assemblages over the period of study. Results indicate that: 1) the presence of the dams constituted a strong ecological press disturbance to the near-downstream reaches on both rivers, despite the fact that both rivers passed unregulated flow and sediment during the high flow season; 2) ecological recovery from this press disturbance occurred within the year following the restoration action of dam removal, whereas signals of geomorphic disturbance from the pulse of released sediment persisted two years post-removal, and 3) the strength of the press disturbance and the rapid ecological recovery were detected regardless of whether recovery was assessed by taxonomic or functional assemblages and for both case studies, in spite of their different geomorphic settings.

Editor: Maharaj K. Pandit, University of Delhi, India

Funding: The authors gratefully acknowledge funding support from the Oregon Watershed Enhancement Board, award number 207-091 and the National Oceanic and Atmospheric Association (NOAA), National Marine Fisheries Service, award number NA07NMF4630201 [DT, CW]. The funders played no role in the study design, data collection and analysis, decision to publish, or preparation of the manuscript.

Competing Interests: The authors have declared that no competing interests exist.

* Email: desiree.tullos@oregonstate.edu

Introduction

Dam removal is increasingly implemented to address aging infrastructure and river restoration [1]. However, most dams have accumulated decades of sediment behind them that can become a concern for natural resources managers due to the potential for downstream deposition following removal. The deposition associated with sediment released during and following dam removal has the potential to generate an ecologically significant disturbance [2], where a disturbance is broadly defined as a discrete event that falls outside a predictable range for an ecosystem [3]. Physical disturbances (e.g. flooding, landslides) are defined as impacts to geomorphic systems that modify bed forms and features as channels react, relax, and respond to the disturbance [4]. Ecologically, physical disturbances result in the death or displacement of resident organisms [5–6]. Recent literature has emphasized the basis of [7], the need for [8], and results from (see [9] for review) studies that investigate interactions between abiotic and biotic responses to physical disturbance in order to describe how the timing and intensity of habitat disturbance, controlled by spatial and temporal variability in geomorphic processes, play an essential role in structuring biological communities.

In the context of dam removal, the sediment pulse released with decommissioning a dam can be considered a discrete event that acts as both a geomorphic and an ecological disturbance [10], and the biotic and abiotic responses to that disturbance can vary depending on the sensitivity of individual reaches to disturbance [11]. From a geomorphic perspective, rivers that are sensitive to disturbance undergo rapid and large changes as a result of a) features that make the channel not resistant to change, b) inadequate complexity and connectivity, c) large magnitudes of change in input conditions, and d) inadequate energy to process the disturbance [12]. In practice, this means that the sensitivity of the physical system will vary with features of the sediment pulse, including the material size, released volume, and timing of sediment release [13–18], relative to the background dynamics and geomorphic processes of the river [16,19]. Ecologically, the sensitivity to disturbance is similarly defined by the prevailing environmental variability and stability [20]. For example, stream-dwelling species that occupy naturally variable habitats are typically adapted to the local disturbance regime [21]. This variability in sensitivities to physical disturbances leads to a range of responses to dam removal, such that concerns regarding some sediment pulses in some systems are not warranted [22], whereas other dam removals clearly generate physically and ecologically significant changes to the system [23]. Despite this knowledge of general processes surrounding the disturbance of dam removal, it is not clear under what conditions the dam removal generates disturbances that are of adequate magnitude and duration to impact aquatic ecosystems, what the spatial and temporal extent of the disturbance from any dam removal will be, or what recovery trajectory a system may take following the disturbance of dam removal.

Both geomorphologic and ecological recovery are often stated as the key objective of river restoration projects (e.g. [24–25]), and yet recovery from dam removal is a complex interaction of impacts associated with the press disturbance (cf. [26]) imposed by the dam over its lifetime and the pulse disturbance of the removal of the dam. Recovery is defined in the field of geomorphology as the return to a prior landform [27], and in the field of ecology as the return to a prior ecological state [28]. Dam removal can facilitate recovery of pre-dam conditions by restoring the natural regime of material fluxes [1], but the disturbance associated with a pulse release of sediment can generate varied effects on habitat complexity [29–31] and direct ecological impacts (e.g. [30,32]). Hence, the trajectory of geomorphic and ecological recovery is likely to be influenced both by removal of the press disturbance represented by the dam itself and by generating a pulse disturbance associated with the removal event.

As in other restoration actions, benthic macroinvertebrate assemblages are shown to be useful indicators of the ecological disturbance and recovery trajectory following dam removal, and there is accumulating evidence that negative effects of dam removal on benthos are transient (<1 year post dam removal [30,33–37]). Existing studies have primarily relied on measures of change in the taxonomic structure of macroinvertebrate assemblages. However, individual taxa possess unique traits that may benefit them in the altered habitats immediately downstream of dams (e.g. mussels [38]) or in the unstable habitats (e.g. drifting and multivoltine invertebrates [39]) that can occur downstream of dam removals. Shifts in the dominance of such traits downstream of dam removals may provide more targeted insight regarding the longer-term effects of both the press and pulse disturbances associated with dams and their removal. Many macroinvertebrate taxa have been well characterized according to suites of functional traits that are expected to respond directly to habitat and disturbance filters (cf. [40]) associated with changing physical conditions [39,41–43]. Disturbance effects on some of these functional groups, especially those with sensitivities to fine sediment deposition (e.g. a clinging habit, respiration with external gills), may persist well beyond the shorter-term effects assessed generally with taxonomic characterization of the community [32,44].

In this study, we applied a before-after-control-impact (BACI) design associated with two small dam removals in an attempt to investigate recovery trajectories from both the press disturbance associated with the presence of dams and the pulse disturbance associated with removal of the dams. The two study sites occupy distinct physiographic settings that we expected to represent low and high sensitivities to the pulse disturbance of dam removal, given differences in background environmental variability and the relative size of the dams. We evaluated both geomorphic and ecological responses annually for one year prior to and two years following dam removal, and asked if functional traits of the macroinvertebrate assemblages could provide more persistent signals of ecological disturbance than taxonomically defined assemblages over the period of study.

Materials and Methods

Study sites

The Calapooia River runs 121 km west from Tidbits Mountain in the Western Cascade Range to Albany, Oregon where it joins the Willamette River (Figure 1). Brownsville Dam (122°56′0.05″W 44°23′15.67″N) was a run of river, low-head dam (Table 1). The dam was originally built as a log crib dam in the late 1800s as a diversion and was rebuilt after failure in the 1964 flood [45–46]. The study area includes two reaches: 1) a 0.67 km control reach (C-US), a length equal to twenty times active channel width, that was located 2 km upstream from the dam; and 2) a 0.67 km (C-DS) impact reach starting immediately downstream of the dam (Figure 1). Permission for access was not necessary since work was conducted within the high-water elevation of the river.

The Rogue River flows 346 km west from springs on Mount McLoughlin in the Cascade Range near Crater Lake to the Pacific Ocean at Gold Beach, Oregon (Figure 1). Savage Rapids Dam (123°13′46.85″W 42°25′14.30″N) was built in 1921 as an irrigation diversion (Table 1) [47] during the summer, passing peaks flows during winter rainfall and spring snowmelt runoff [48] unregulated. Decommissioning of Savage Rapids Dam began with the removal of the right bays of the dam in April–June 2009 during which time there was a small release of sediment associated with drawdown and the partial failure of the coffer dam [49]. The coffer dam remained in place and flow was forced through the left bays and fish ladder until October–November 2009, when the coffer dam and remaining infrastructure was removed, a pilot channel was constructed, and the river was returned to a free flowing river [50]. The study area consisted of three reaches (Figure 1): 1) a 1.5 km control reach (R-US), approximately 40 times summer wetted width, located 4 km upstream from the dam; 2) one 1.8 km impact reach, starting immediately downstream of the dam (R-DS1); and 3) a second 2.4 km impact reach (R-DS2), starting immediately below R-DS1. Permission for access was not necessary since work was conducted within the high-water elevation of the river.

These two sites represent potential differences in the sensitivity to the pulse release of sediment with dam removal. For example, the Calapooia River downstream of Brownsville Dam was expected to be less sensitive to the sediment pulse associated with the dam removal than the Rogue River downstream of Savage Rapids Dam. Reasons due to a lower erosional efficiency (cf. [18]) and coarse material in the reservoir, relative to the bed downstream (Table 1), on the Calapooia River. In contrast, higher erosional efficiency and a relatively small grain size of material stored in the reservoir indicated that the Rogue River may be sensitive to the geomorphic disturbance associated with a released sediment pulse in the sense that it is not resistant to change and has adequate energy to process the disturbance. The higher sensitivity at Savage Rapids Dam should confer shorter response and relaxation times [12]. However, while the two sites likely possess differences in their sensitivity to the sediment pulse and vary in the size of the dams and rivers (Table 1), they both a) stored in reservoir the equivalent volume of approximately one to two years of sediment yield from the basin; and b) passed sediment through or over the dam and thus downstream reaches were not supply limited.

Data collection

Summer field surveys were conducted prior to and following dam removal. On the Calapooia River, the pre-removal survey was conducted in July 2007, prior to the dam removal in September 2007. Post-removal surveys on the Calapooia River

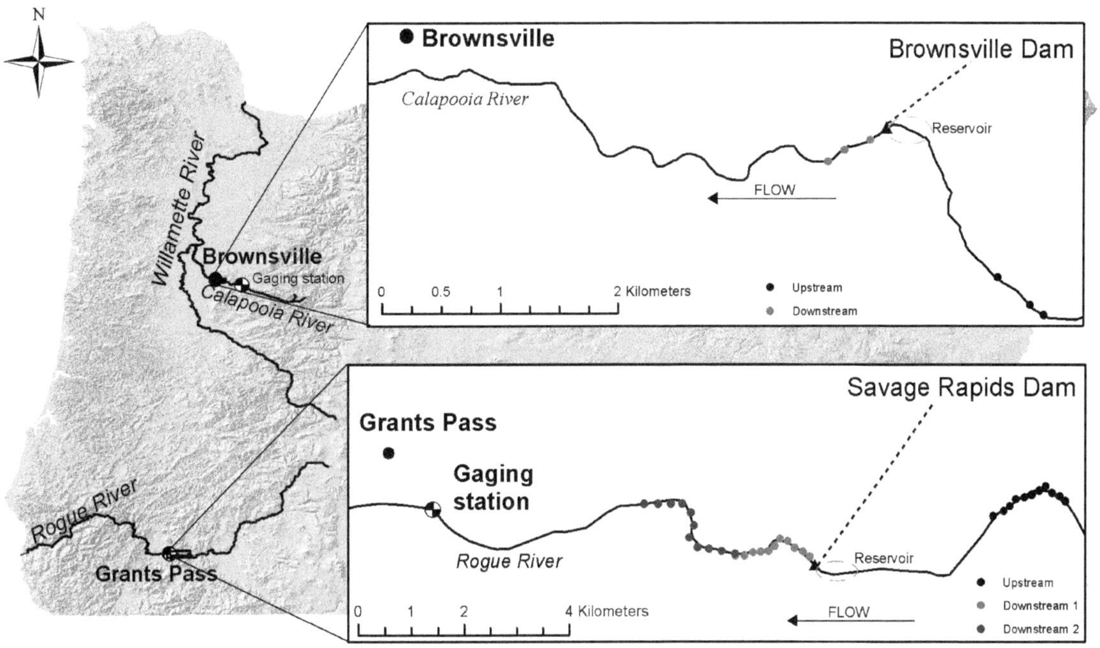

Figure 1. Site locations. Each dot represents a macroinvertebrate sampling location: 3 riffles per reach on the Calapooia River, and 11 transects per reach on the Rogue River.

were conducted in August 2008 and July 2009. On the Rogue River, pre-removal surveys were conducted in September 2008. Surveys were also conducted in July 2009, a time between in-water work periods when some construction had occurred but the channel had not yet been returned to a free-flowing river. Post-removal samples on the Rogue River occurred in August 2010 and July 2011. Surveys included benthic macroinvertebrates sampling, bed material substrate characterization, and topographic surveys. As described below, survey methods varied between the two sites due to the sizes of the rivers at the study sites. The Calapooia River is primarily wadeable while the Rogue River is not wadeable. All data from this study have been published in a data library online at http://rivers.bee.oregonstate.edu/.

Hydrology. Despite growing evidence [18,67] that the sequence of flows post dam removal has limited effect of the rates and styles of response to dam removal, we present the annual peak

Table 1. Characteristics of the study reaches and sediment reservoirs.

Dam name	Brownsville	Savage Rapids
River	Calapooia	Rogue
Drainage area above dam (km^2)	404	6369
Location of dam (River km)	62	173
Dominant catchment land use	private forest and agriculture	public and private forest
Catchment mean annual precipitation (mm)	1730	1400
Dam function	mill and aesthetic diversion	irrigation diversion
Year removed	2007	2009
Reason for removal	fish passage and safety concerns	fish passage concerns
Barrier height (m)	2.4-4	9.1–12.5
Active channel width (m)	35	90
Avg. Width: Depth	34	21
Slope (m/m)	0.002	0.003
Stored sediment volume (m^3)	17,000	543,000
Erosional efficiency	10^{-5}	10^{-4}
$D50_R/D50_D$	2.8	0.18
$D50_R$ (m)	0.06	0.008

Barrier height varies with season due to installation of flashboards (Brownsville Dam) and stop logs (Savage Rapids Dam). Erosional efficiency was calculated as dimensionless ratio of the volume of sediment eroded from the reservoir to the volume of streamflow delivered to the site across the study period. $D50_R$ = median grain size of the reservoir sediments. $D50_D$ = median grain size of the river bed downstream of the dam.

flow at both study sites for context. Hydrologic records were summarized as the annual peak discharge for the historical and study time periods (Figure 2). For the Calapooia River, there has not been a published record of discharge since the U.S. Geological Survey (USGS) stopped gaging the Calapooia River at Holley (#14172000), 16 km upstream of Brownsville Dam, in 1990. Therefore, we used two nearby USGS gages, South Santiam below Cascadia (#14185000) and Mohawk River at Springfield (#14165000), in basins of similar hydrogeology and land uses, to create a log-transformed regression model for annual peak flow for the Calapooia at Holley based upon 35 years of concurrent historical annual peaks [51]. The Rogue River is gauged by the USGS at Grants Pass (#14361500), 8.6 km downstream from the Savage Rapids Dam.

Geomorphology. On the Calapooia River, sediment samples were collected at two riffles per reach as 100-particle counts [52] for post removal years, as bulk samples [53] on two bars per reach in all years, and as bulk samples on two riffles per reach in pre-removal years [22]. Topographic surveys consisted of points taken with a total station (Nikon DTM 352) or real time kinetic (RTK) global positioning system (GPS) (Topcon Hiper Lite +) at slope breaks along four evenly-spaced cross sections per channel unit (riffle, glide, pool, run), along a longitudinal profile, and across bar surfaces [51].

On the Rogue River, water depth was measured and bed material was characterized by type (e.g. sand, gravel, bedrock) at 100–120 points on the thalweg along each study reach as part of Environmental Monitoring and Assessment Program (EMAP) surveys in 2008, 2009, 2010, and 2011. In addition, bankfull height above the low flow water surface was measured at eleven cross sections within each reach as part of the EMAP surveys [54]. More detailed bathymetry and water surface elevation were surveyed using a raft-mounted Acoustic Doppler Current Profiler (Workhorse Rio Grande) paired with a RTK GPS (Topcon GR-3). Surveys were conducted as one to three longitudinal profiles in 2010 and 2011, with variation in the number of profiles due to equipment failures and time constraints on field work. Pre-removal bathymetric surveys in the reservoir and downstream were performed by the U.S. Bureau of Reclamation (USBR) in 1999 and 2002.

Biotic communities. For the Calapooia River site, benthic macroinvertebrates were collected in riffles by disturbing sediment in front of a D-frame net for 60 seconds [55]. Samples were collected in three randomly selected locations per riffle at three riffles per reach on July 18, 2007, August 4, 2008, and July 14,

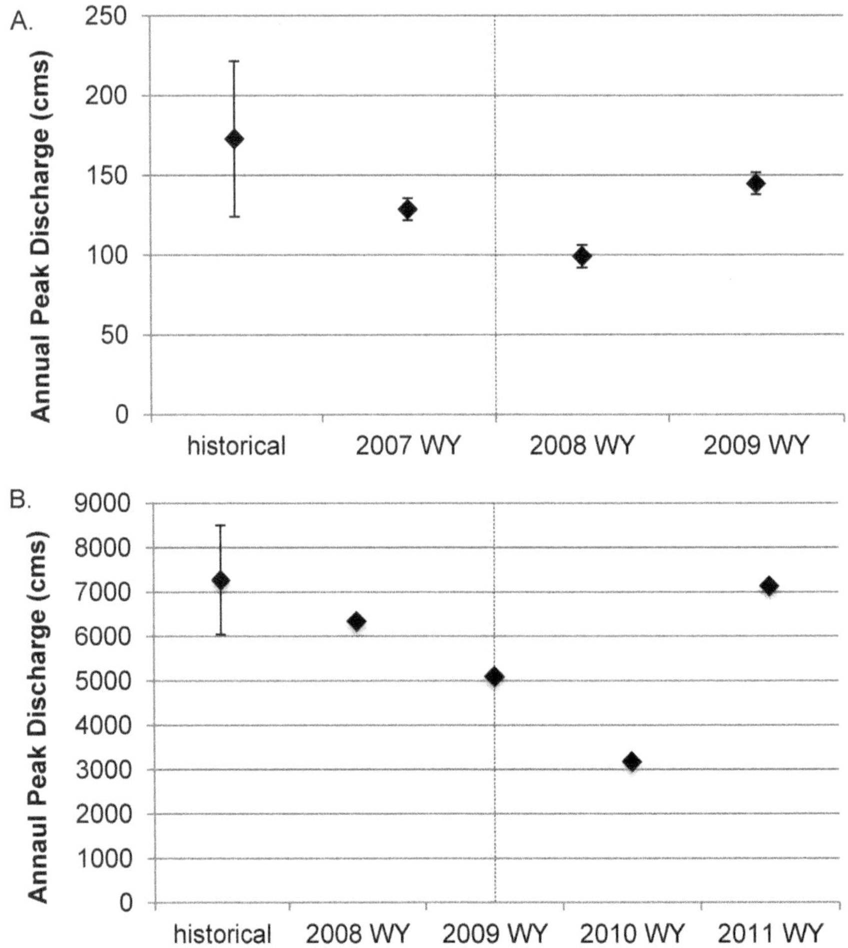

Figure 2. Annual peak discharge for the historical period of record and over the study period for the a) Calapooia at Holley, located 16 km upstream of Brownsville Dam, and b) Rogue River at Grants Pass located 8.6 km downstream of Savage Rapids Dam. Error bars on historical annual peaks represent 2 standard errors. Error bars for the Calapooia River for study years reflect estimation error associated with multiple regression estimate of peak discharge. Vertical dashed lined indicates approximate date of dam removal.

2009. On the Rogue River, samples were collected at the margin of each of eleven transects per reach using a D-frame net according to the EMAP non-wadeable protocol [54] on September 24–26 2008, July 29–30 2009, August 17–19 2010, and July 25–26 2011. All specimens were preserved in 70% ethanol.

Characterization of geomorphic responses

Channel slope. For the Calapooia River, the mean channel slope for each reach was calculated as the average baseflow water surface slope based on edge of water elevations from the topographic surveys of cross sections for each year. For the Rogue River, channel slope was not measured in 2008 and 2009. Therefore, changes in slope on the Rogue River were not considered as part of this analysis.

D_{gm}. For the Calapooia River, the geometric mean diameter (D_{gm}) [56] for each reach was calculated as the average of the D_{gm} for all particle counts and surface bulk samples for that reach. For the Rogue River, type of substrate determined according to EMAP protocols [54] was converted to D_{gm} following the methods of [57] and averaged over each reach.

Variability of thalweg depth. For each of the sites, thalweg variability was calculated from thalweg profiles as a measure of habitat variability [58–59]. The standard deviation of thalweg depth was calculated for each study reach at each study site based on regularly-spaced thalweg depths. On the Calapooia River, thalweg depths were not directly measured. We calculated the thalweg depths as the vertical difference between a Triangulated Irregular Network (TIN) surface derived from the edge of water elevations for cross sections and bars, and the elevations of the longitudinal profile. As the longitudinal profile of the thalweg was measured at slope breaks and not regular intervals, we linearly interpolated the thalweg depth at a uniform spacing of 0.6 m from the thalweg depth derived from the longitudinal profile. On the Rogue River, thalweg depth was measured at constant intervals along each reach following the EMAP protocol [54].

Relative Bed Stability. We analyzed the stability of the riverbed using the modified relative bed stability (RBS*) index (Eqn 1) [60], a ratio between the mean particle diameter, D_{gm}, and the critical diameter at bankfull flow, D^*_{cbf}. The ratio is interpreted in relation to unity such that a ratio larger than 1 indicates the bed is likely stable during bankfull flows, whereas a ratio smaller than 1 indicates the bed is likely mobile during bankfull flows.

$$RBS^* = \frac{D_{gm}}{D^*_{cbf}} = \frac{D_{gm}}{(0.604 R^*_{bf} S/\theta)} = \frac{1.66\theta D_{gm}}{(R_{bf}(C_p/C_t)^{1/3} S)} \quad (1)$$

where: D_{gm} = geometric mean bed surface particle diameter (m); D^*_{cbf} = critical diameter (m) of bed surface particle at bankfull flow, averaged across the reach, and adjusted for shear stress reductions due to wood and depth variation; R_{bf} = bankfull hydraulic radius≈$0.65d_{th-bf}$, where d_{th-bf} = mean thalweg depth + bankfull height above water surface (m); R^*_{bf} = effective bankfull hydraulic radius (m) adjusted for wood and depth; S = energy slope, approximately the reach-scale water surface slope as a dimensionless ratio (m/m); θ = Shields number calculated from particle Reynolds number at bankfull flow $\{R_{ep} = [(gR_{bf}S)0.5D_{gm}]/\nu\}$; C_p = stream reach-scale particle (grain) resistance at bankfull flow, calculated from reach wide mean relative submergence of D_{gm}; and C_t = reach-scale hydraulic resistance at bankfull flow, calculated from bankfull thalweg mean depth and thalweg mean residual depth.

All parameters for calculating RBS* on the Rogue River were estimated based on the EMAP protocol [60], with residual depths calculated following a method consistent with [61]. On the Calapooia River, components for RBS* were calculated from the topographic surveys and analysis of sediment samples. For these assessments, bankfull height was estimated based on bank indicators observed in the field and verified as the highest elevation of depositional features in the Calapooia River. To estimate the residual depth on the Calapooia River, we calculated the elevation difference between each riffle crest and the thalweg in the pool immediately upstream of the crest, then averaged the residual pool depths across each reach.

Characterization of biotic responses

We identified all macroinvertebrates to the family level for insects, amphipods, freshwater clams, and snails. Mites, oligochaetes, leeches, and flatworms were recorded at this coarser level (e.g. mites, oligochaetes, etc) of resolution. Our biological analyses relied on multivariate approaches applied to either taxonomically characterized assemblages for both insects and non-insects or on a functional characterization of only the insect component of the assemblages. We focused functional analysis on the insects because numerous functional traits are better understood for insects than for other benthic taxa. We assigned traits to each insect family following [39]. The suite of 20 traits, with a range of two to six trait modalities each (Table 1 in [39]), represents a variety of life history, mobility, morphological, and ecological functions that are expected to respond to varying aquatic habitat conditions (e.g. [62,41]). Poff et al. [39] assigned trait modalities at the genus level, which we applied at the family level herein. In a few instances in which there was not a clear consensus modality for a particular trait across genera within families, we resolved the issue either by identifying our specimens to the genus level, when feasible, and assigning the appropriate trait modality or by choosing the most commonly occurring trait modality across the genera known to occur regionally within a family.

We used the software package PC-ORD [63] to run all multivariate analyses on the macroinvertebrate assemblages characterized both taxonomically and functionally. The fundamental sample units for these analyses, henceforth referred to as sites by year, are the samples from each year for each riffle (sum of three D-frame samples each) for the Calapooia River and for each EMAP transect (a single D-frame sample each) for the Rogue River (Figure 1). For each of the following multivariate analyses at the taxonomic level, we log(n+1) transformed raw abundance data to create the input matrices of *sites by year × taxon abundance*. We created input matrices for the multivariate functional analyses by multiplying matrices of *sites by year × raw taxon abundance* by a matrix of *taxa × trait modalities* for each of the twenty traits. The resulting matrices of *sites by year × absolute abundance of each trait modality* were then relativized, as the relative abundance of modalities within each trait at each site, to create final input matrices of *sites by year × relative abundance of each trait modality*.

We analyzed relationships of sites by year in multivariate ordination space. We used non-metric multidimensional scaling (NMS) to evaluate the influence of sample year and reach on the multivariate axis values of each assemblage. We used the multi-response permutation procedure (MRPP) in PC-ORD to assess whether there were differences in overall assemblage structure among sample years. MRPP results reported here include both an overall test for differences among groups, based on the significance of the statistic *A*, the chance-corrected within-group agreement, and tests for all possible pairwise comparisons between groups. We also used MRPP on single-year subsets of the two datasets to evaluate whether upstream and downstream reaches were significantly different in assemblage structure, either taxonomically or functionally defined, within any sample year. Significant differences between upstream control sites and downstream impacted sites following, but not prior to, dam removal would support a hypothesis of disturbance effects from the dam removal that persist up to or beyond a one year period, indicative of a year-plus recovery period. For each yearly subset, and for both rivers, we also conducted an indicator species analysis (ISA) [64] on the functionally-defined assemblage data in order to assess whether particular trait modalities were indicative of either upstream control or downstream reaches impacted by the dam and its removal. Under a hypothesis of downstream disturbance effects lasting beyond one year, we expected that trait modalities reflecting disturbance tolerance (e.g. multivoltinism, fast development, heavy body armoring, abundance in the drift) would be indicative of downstream reaches, while trait modalities reflecting intolerance to disturbed conditions (e.g. semivoltinism, slow development, respiration with gills) would be indicative of upstream reference reaches. For both the MRPP and ISA, we used a Bonferroni correction for multiple comparisons with alpha values equal to 0.05 to determine statistical significance.

In addition to the assessments of multivariate assemblage structure at the taxonomic and functional levels, we also quantified single metrics, one taxonomic and one functional, that are thought to be useful indicators of disturbance. We calculated percent of the total abundance of individuals that were Ephemeroptera, Plecoptera, or Trichoptera (%EPT), and percent abundance of the "clinger" modality of the habit trait (%clingers). Percent EPT is thought to respond negatively to disturbance and is a metric used frequently in bioassessment [65–66]. Although thus far not as widely used, %clingers describes the prevalence of a habit that requires abundant interstitial space in the substrate and a dominance of larger particle sizes, and clinger habit is expected to respond negatively to disturbance, particularly fine sediment deposition [44]. For each of these metrics, we used t-tests to assess differences between upstream and downstream reaches in each sample year and at both rivers. For these univariate analyses, we compared the reaches C-US and C-DS in the Calapooia, and the reaches R-US and R-DS1 in the Rogue. We excluded reach R-DS2 from this analysis because none of the prior multivariate analyses suggested any ecological response this far downstream of the former dam.

Results

Hydrologic context

On the Calapooia River, all three years of the study period experienced peak flows that were lower than the historical mean (Figure 2A). The first year post-removal was exceptionally low as it was less than two standard errors below the mean, whereas the pre-removal and second year post-removal experienced an annual peak flow that, while still low, were more typical of the historical peaks. The Rogue River experienced nearly the average annual peak flow in 2008, prior to dam removal, and an average annual peak flow in 2011, two years post-removal. Water year 2009, the year of the removal, experienced a peak flow of average magnitude whereas the lowest annual peak flow occurred in the water year following the dam removal.

Geomorphic responses

Changes in physical habitat on the Calapooia River (Table 2A) indicate channel responses were associated both with the natural dynamism of the river and the removal of Brownsville dam. The downstream (C-DS) channel slope steepened in the first year post removal, without a similar change at the upstream reference site (C-US). In the second year post removal, the channel slope at C-DS flattened while the upstream site steepened. This pattern of steepening and flattening of the channel slope downstream of the dam removal is one indicator of the geomorphic disturbance and recovery associated with the pulse release of sediment.

A second indicator of geomorphic response, bed variability, is represented by the standard deviation (SD) of the thalweg profile and the mean residual depth of the channel. On the Calapooia River, the thalweg SD increased in C-US while decreasing in C-DS between the pre-removal survey and the first year following removal (Table 2A). Also during the first year post-removal, mean residual depth did not change at C-US but decreased in C-DS as a result of sediment deposition in a pool immediately below the dam (Figure 3A). By 2009, the thalweg variability increased at C-US while C-DS underwent a further reduction in the thalweg SD. The downstream decrease in SD for 2009 was accompanied by a net increase in residual depth, associated with scouring of pools and deposition along some planar features in the channel. The anomalous, relative to the upstream control, decrease in variability, and reduction then increase in mean residual depth, downstream of the dam removal illustrates the effect of the sediment pulse on reducing thalweg relief in the Calapooia River.

Finally, based on RBS* values (Table 2A), the bed in both C-US and C-DS was unstable before and after the dam removal and the reaches follow similar patterns of bed stability over time, though magnitudes of changes vary between the reaches. In the upstream reach, RBS* tracks changes in the dominant grain size (Figure 4A), increasing in the first year post-removal, then dropping in the second year post-removal. This trend indicates the basin underwent coarsening of the substrate during the lowest flows of the study in the 2008 water year (Figure 2A), resulting in higher stability of the bed, followed by a reduction in grain size and decreased bed stability during the more moderate flows of the 2009 water year. Similar trends were observed in both the grain size and RBS* for the C-DS, indicating that the sediment pulse had limited effect on the bed stability (Figure 4A) and that changes in bed mobility were more closely associated with the natural dynamism of the river than the dam removal.

On the Rogue River (Table 2B), we found evidence of geomorphic response to disturbance both with respect to bed variability and mobility. Variability of the bed was generally lowest at the upstream reference site for all years, and after a drop in 2009, remained constant over the study period. Likewise, mean residual depth varied little between years, indicative of a relatively stable channel. In R-DS1, thalweg variability exhibited little change during and for the first year following removal, though mean residual depth was reduced between 2009 and 2010. The reduction in residual depth reflects deposition in the deep pool, located ~180–760 m downstream of the former dam. However, since no change in elevation was observed for the other two other pools in R-DS1 (Figure 3B), the range of depths did not appreciably adjust between 2009 and 2010 and thus the thalweg depth SD was steady. By 2011, the thalweg depth variability drops dramatically, accompanied by further reduction in the mean residual depth. These changes in bed variability were associated with deposition in the two pools located farther downstream in R-DS1 and aggradation on riffles located ~1050–2350 m downstream of the dam. In R-DS2, the thalweg depth SD increased incrementally in 2009 and 2010, followed by a drop in variability in 2011. Mean residual depth followed a similar pattern of increase through 2010 followed by reduction in 2011. The increase in variability and residual depth through 2010 appears to be related

Table 2. Geomorphic measures of disturbance from the sediment pulse.

A) Brownsville Dam on the Calapooia River

Upstream (C-US)

Year	Channel			Substrate	
	slope	Mean residual depth (m)	Std. Dev. of thalweg depth (m)	Dgm (mm)	RBS*
2007	0.0018	0.3	0.25	0.02	0.14
2008	0.0017	0.3	0.28	0.04	0.31
2009	0.0020	0.3	0.31	0.02	0.12

Downstream (C-DS)

	slope	Mean residual depth (m)	Std. Dev. of thalweg depth (m)	Dgm (mm)	RBS*
2007	0.0028	0.3	0.32	0.01	0.13
2008	0.0033	0.2	0.27	0.04	0.28
2009	0.0025	0.3	0.25	0.02	0.22

B) Savage Rapids Dam on the Rogue River

Upstream (R-US)

Year	Channel		Substrate	
	Mean residual depth (m)	Std. Dev. of thalweg depth (m)	Dgm (mm)	RBS*
2008	0.9	0.9	0.04	1.75
2009	0.8	0.7	0.05	2.05
2010	0.9	0.7	0.05	1.65
2011	0.9	0.7	0.02	1.38

Downstream (R-DS1)

	Mean residual depth (m)	Std. Dev. of thalweg depth (m)	Dgm (mm)	RBS*
2008	1.3	1.5	0.16	2.98
2009	1.4	1.4	0.05	0.79
2010	0.9	1.5	0.01	0.23
2011	0.4	0.5	0.01	0.18

Downstream (R-DS2)

	Mean residual depth (m)	Std. Dev. of thalweg depth (m)	Dgm (mm)	RBS*
2008	1.3	2.5	0.24	6.16
2009	2.4	2.6	0.25	4.57
2010	2.9	2.8	0.01	0.27
2011	2.5	2.3	0.01	0.21

The year is underlined for sampling pre-removal for both sites, and italicized for sampling during the year of removal for the Rogue River.

to minor adjustments in the longitudinal profile independent of the dam removal whereas the reduction in variability and residual depth between 2010 and 2011 are associated with post-removal deposition in a pool located 1790–2170 m downstream of the dam. The time series of reduction in thalweg depth and variability in R-DS1 reflected the impact of pool filling immediately following removal, while the impact of the sediment pulse on relief was delayed in R-DS2 until two years post-removal.

Finally, RBS* indicates that the bed was generally stable at bankfull flows for all years in R-US and that dam removal reduced bed stability for the downstream reaches (Table 2B). For R-DS1, the bed was stable in 2008 but not in the year of or the years following dam removal. At R-DS2, the bed was stabile prior to and the year of dam removal, but not in the two years following removal. Bed stability in 2008 for R-DS1 and R-DS2, and 2009 for R-DS2 was primarily a function of the prevalence of boulders and bedrock (Figure 4B). However, the stability of the bed drops below one for R-DS1 in 2009, likely associated with the small amount of sediment that was flushed out of the reservoir with the spring 2009 drawdown of the reservoir for construction, which is

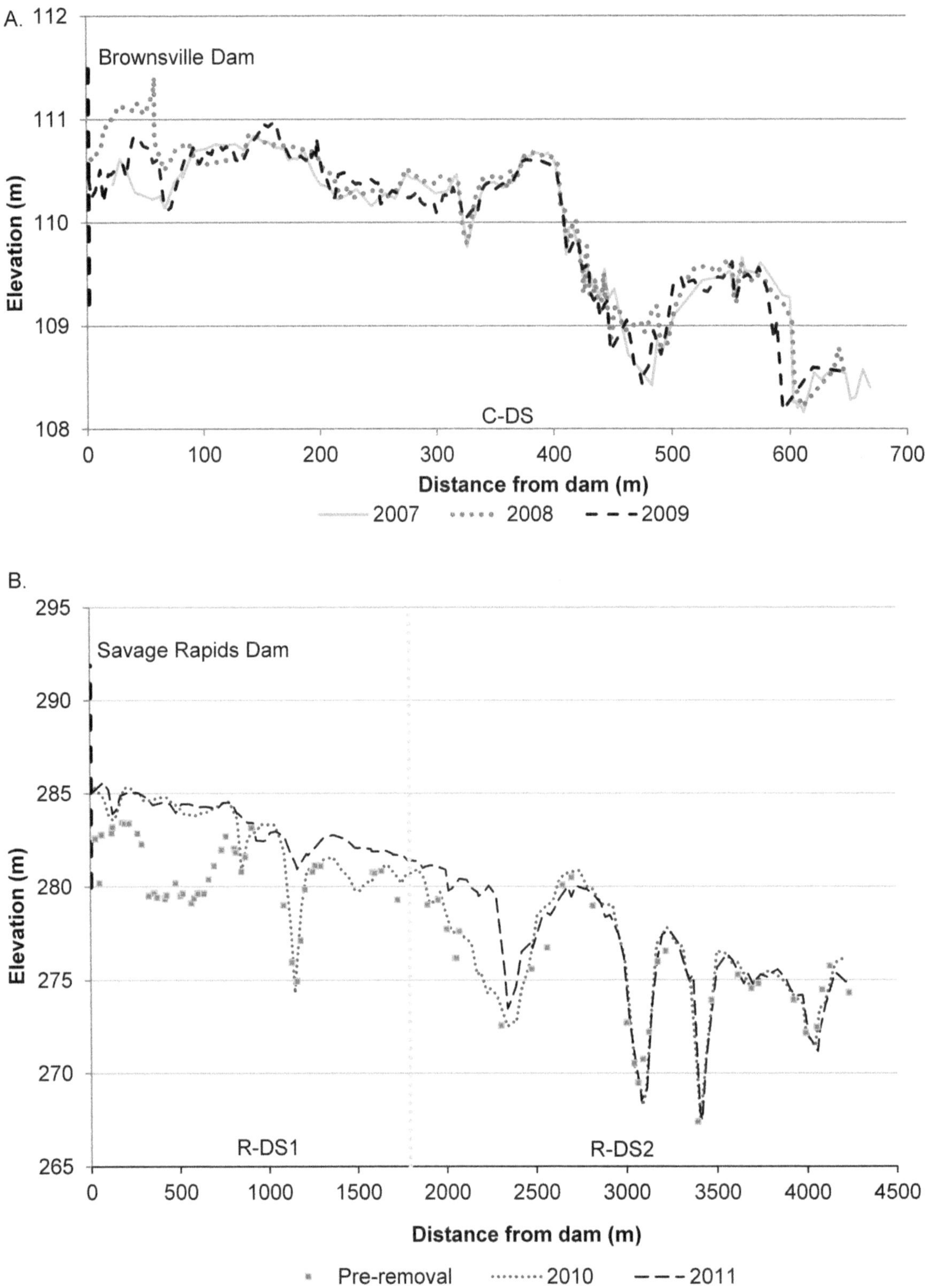

Figure 3. Longitudinal profiles downstream from the dams over time at a) Calapooia River and b) Rogue River. On the Rogue River, no longitudinal profile was collected during 2009, and the pre-removal profile is a compilation of surveys from 1999 in the farthest US and DS reaches and from 2002 for the main reservoir, both conducted by the US Bureau of Reclamation.

evident as fining in R-DS1. In 2010, the first year post-removal, the grain size and RBS* values at R-DS1 were further reduced, and remained low in 2011, the second year post removal. R-DS2 also underwent substantial fining in 2010 and the grain size and bed stability remained low in 2011. Thus, while the grain size at the upstream control site varied little between years, the release of

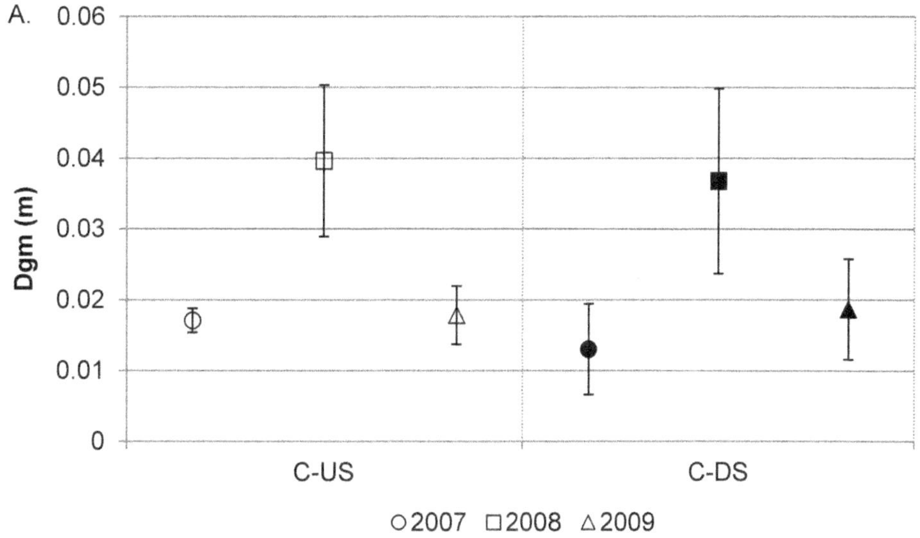

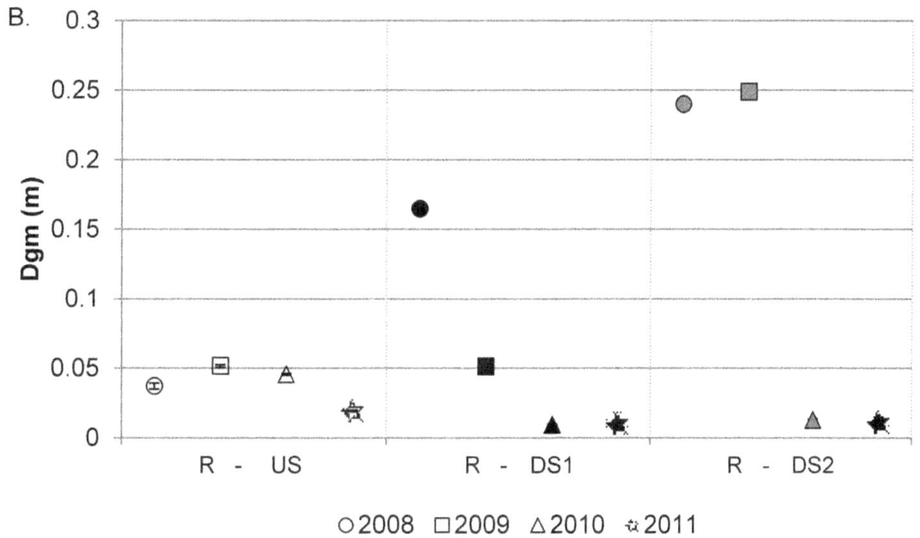

Figure 4. Plot of geometric mean grain size by reach over time at a) Calapooia River and b) Rogue River. Symbols represent the mean value, and lines represent two standard errors from the mean.

predominantly sand from behind the Savage Rapids Dam resulted in a reduction in dominant grain size downstream, with a concurrent reduction in bed stability due to the greater slope in R-DS1 and higher depths in R-DS2 relative to R-US.

Biotic responses

In total, we identified 49 taxa, including 39 insect families, across all sites and years on the Calapooia River. We identified 38 taxa, including 27 insect families, across all sites and years on the Rogue River.

When characterized taxonomically, multivariate analysis of the full datasets revealed similar general patterns on the two rivers, including significant differences in assemblage structure among years, regardless of sample reach (Table 3). Both rivers also exhibited significant differences in taxonomic assemblage structure in upstream vs. downstream reach prior to initiation of dam removal activities, but no evidence of differences between reaches within any year following dam removal. The upstream-downstream differences on the Rogue River, when present, were only apparent in the pairwise comparison of reaches R-US vs. R-DS1. This pair of reaches was significantly different prior to the initiation of dam removal activities in 2008 and approached significance ($p = 0.06$) in 2009, the year during dam removal. Assemblages in reach R-DS2, over 1.7 km downstream of the former dam, were not different from either R-US or R-DS1 in any year.

For the functionally characterized insect assemblages (Table 4), patterns were broadly similar to those revealed for the taxonomically defined full macroinvertebrate assemblages. The functional assemblages in the Calapooia River were significantly different among all sample years in the full dataset, and the only significant differences between reaches C-US and C-DS within years was in 2007, prior to dam removal. On the Rogue River, functional assemblages also differ significantly among years in the full dataset, but the only significant pairwise difference was between 2008 and 2010, the year prior to and the first year following dam removal.

Table 3. Results of MRPP analysis for differences among taxonomically defined assemblages in both rivers.

River	Dataset	Groups	Pairs	*A*	p
Calapooia	full	years		0.2	**<0.0001**
			07/08	0.2	**0.001**
			07/09	0.3	**0.0006**
			08/09	0.1	**0.001**
	2007	reaches		0.2	**0.02**
	2008	reaches		−0.05	0.7
	2009	reaches		0.01	0.4
Rogue	full	years		0.09	**<0.0001**
			08/09	0.1	**<0.0001**
			08/10	0.1	**<0.0001**
			08/11	0.07	**<0.0001**
			09/10	0.02	**0.006**
			09/11	0.03	**<0.0001**
			10/11	0.05	**<0.0001**
	2008	reaches		0.04	0.02
			US/DS1	0.06	**0.002**
			US/DS2	0.02	0.08
			DS1/DS2	0.003	0.5
	2009	reaches		0.02	0.06
			US/DS1	0.02	0.06
			US/DS2	0.02	0.1
			DS1/DS2	0.008	0.2
	2010	reaches		0.004	0.3
			US/DS1	0.002	0.4
			US/DS2	0.007	0.3
			DS1/DS2	0.002	0.4
	2011	reaches		0.02	0.09
			US/DS1	0.02	0.1
			US/DS2	0.02	0.1
			DS1/DS2	0.004	0.4

The first row of statistics for each group structure represents the test across the all groups comprised in that dataset, and the "pairs" column shows statistics for pairwise comparisons within the respective group. *A* is chance-corrected within-group agreement. Bolded p-values indicate significance at alpha = 0.05 after corrections for multiple comparisons.

Unlike the taxonomically defined assemblages, functional assemblages on the Rogue River did not differ significantly between any of the three reaches within any sample year.

To some extent, the differences observed in the pre- and post-removal assemblages were illustrated in the NMS plots. Taxonomic structure clearly delineated C-US and C-DS on the Calapooia River prior to dam removal along Axis 2, whereas the post-removal sites were not separated in ordination space (Figure 5A). Similar results are observed with ordinations of the functionally defined assemblages, with upstream and downstream sites on the Calapooia River clearly separated in ordination space prior to dam removal but not following dam removal (Figure 6A). In contrast, the pre-removal taxonomic distinction between R-US and R-DS1 is not evident on the Rogue River (Figure 5B), although upstream and downstream sites were separated to some extent along Axis 1 for the functionally defined assemblages (Figure 6B) prior to, but not following, dam removal.

Indicator species analysis on spatial distributions of trait modalities confirm the convergence of upstream and downstream sites post-removal, but also revealed somewhat different year-to-year patterns of upstream vs. downstream functional indicators between the two rivers (Table 5). We determined statistical significance at $\alpha = 0.05$ following a Bonferroni correction, but "potential" indicators, those with an uncorrected $p <= 0.05$, were also listed in Table 5 to show trends. On the Calapooia River, the greatest abundance of functional indicators were in the year prior to dam removal (2007), in both reaches C-US and C-DS, followed by minimal, nonsignificant indicators distinguishing the two reaches in years following dam removal (Table 5A). Significant upstream indicators in the Calapooia River in 2007 were swimming habit, poorly synchronized emergence, and multivoltinism. Significant downstream indicators were well-synchronized emergence, univoltinism, and the herbivorous trophic group. In the same year, there were four other potential indicators of reach C-US, and three

Table 4. Results of MRPP analysis for differences among functionally-defined assemblages in both rivers, according to a suite of 20 life-history, mobility, morphological, and ecological traits.

River	Dataset	Groups	Pairs	A	p
Calapooia	full	years		0.5	**<0.0001**
			07/08	0.2	**0.002**
			07/09	0.5	**0.0005**
			08/09	0.5	**0.0007**
	2007	reaches		0.2	**0.03**
	2008	reaches		−0.03	0.8
	2009	reaches		0.03	0.4
Rogue	full	years		0.03	**0.003**
			08/09	0.02	0.02
			08/10	0.06	**0.0005**
			08/11	0.01	0.09
			09/10	0.005	0.2
			09/11	0.005	0.2
			10/11	0.02	0.04
	2008	reaches		0.007	0.3
			US/DS1	0.01	0.2
			US/DS2	0.03	0.1
			DS1/DS2	−0.03	1
	2009	reaches		0.03	0.1
			US/DS1	0.01	0.2
			US/DS2	0.02	0.2
			DS1/DS2	0.03	0.1
	2010	reaches		0.04	0.1
			US/DS1	0.03	0.1
			US/DS2	−0.002	0.4
			DS1/DS2	0.05	0.07
	2011	reaches		0.03	0.2
			US/DS1	0.05	0.09
			US/DS2	−0.03	1
			DS1/DS2	0.04	0.1

This table is organized and includes the same statistics as described for Table 3.

other such indicators at C-DS. At both one and two years post-dam-removal on the Calapooia, there were no significant indicators for either reach, although multivoltinism and a swimming habit scored uncorrected *p*-values of 0.04 and 0.05, respectively, as indicators of C-DS one year after dam removal. There were no significant indicators in either of the two post-removal years on the Calapooia River that clearly distinguished the upstream site.

On the Rogue River, there were no significant indicators distinguishing the reaches for any of the years (Table 5B). However, in total, there were six potential indicator trait modalities distinguishing reaches R-US and R-DS1 in 2008, the year prior to dam removal, and eight potential indicator trait modalities in 2009, the year during removal. R-DS2 had no potential indicators in either of these years. The 2009 analysis suggested that multivoltine insects that do not use gills for respiration, have some degree of armoring, good swimming ability, and poorly synchronized emergence were found in greater abundance in R-DS1 relative to the R-US and R-DS2 reaches in the year that the dam was being removed. In 2010, the year following full dam removal, there were potential indicators of semivoltine insects with non-seasonal development and poor armoring in R-US in contrast to the single potential indicators in R-DS1, multivoltinism, and R-DS2, fast-seasonal development. However, by two years following dam removal, there were no potential indicator trait modalities distinguishing any of the Rogue River sample reaches.

The univariate indicators %clingers (Figure 7) and %EPT (Figure 8) illustrate upstream-downstream trends across years, particularly in the Calapooia River, although only one of the within-year *t*-tests produced significant differences. With no statistical difference in % EPT across the years on the Calapooia River (Figure 8A), the only statistically significant difference on the Calapooia River was greater %clingers in C-DS than C-US in 2007, the year prior to dam removal (Figure 7A). In the two years following dam removal, the two reaches on the Calapooia River became statistically indistinguishable for %clingers. Values of %EPT were greater, though nonsignificant ($p = 0.11$), in C-US relative to C-DS for the Calapooia River prior to dam removal,

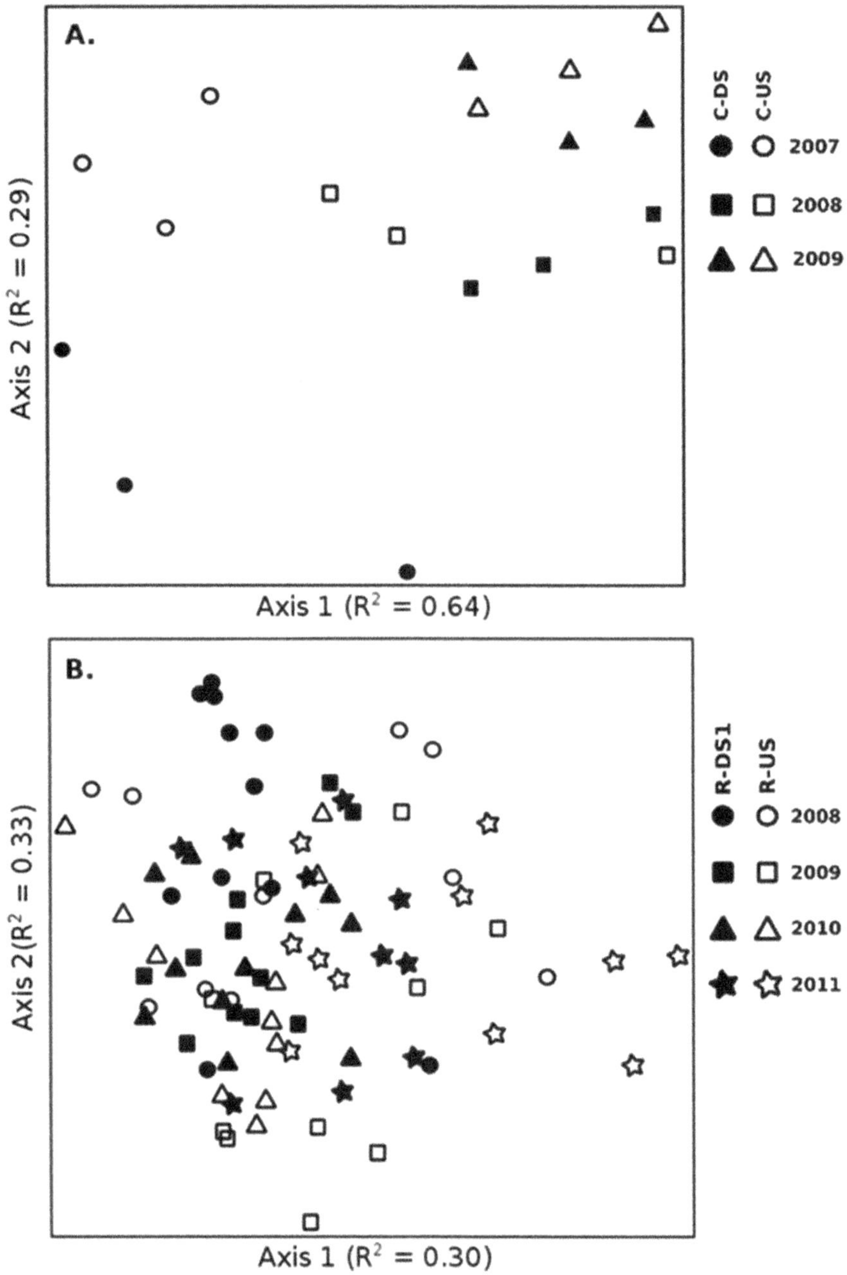

Figure 5. NMS ordination plot according to the taxonomic structure of A) 3 upstream and 3 downstream riffles associated with the Brownsville Dam removal on the Calapooia River over three collection years, and B) 11 upstream and 11 downstream transects associated with the Savage Rapids Dam removal on the Rogue River over four collection years. Note that the 11 transects from reach "DS2" have been removed from the figure to reduce clutter (see Table 3 for MRPP analysis of these data, including the DS2 reach). A) Final stress for the two-dimensional solution = 10.1. B) Final stress for a three-dimensional solution was 15.2, and the unpictured third axis had $R^2 = 0.21$.

followed by convergence of these reaches in the two years following dam removal. Year-to-year trends were difficult to assess and were nonsignificant on the Rogue River due to high reach-scale variability for both %EPT and %clingers (Figures 7B, 8B).

Discussion

Dams are known to generate geomorphic and ecological impacts, considered to be press disturbances [68–69], but it is also possible [2] that the act of decommissioning these dams results in a pulse disturbance, both ecologically [24] and geomorphically [70]. We aimed to investigate recovery from both press and pulse disturbances in two physiographic settings with different dam characteristics, and to assess whether geomorphic and ecological recovery was consistent and concurrent. Overall, our results suggest that 1) the signature of geomorphic disturbance varied between the sites reflective of the physiographic setting and sensitivity of the rivers to the introduced sediment pulse, 2) the geomorphic disturbance from the sediment pulse persisted two years post-removal but ecological recovery of the macroinvertebrate assemblages, whether characterized taxonomically or functionally, occurred within a single year, 3) the signal of

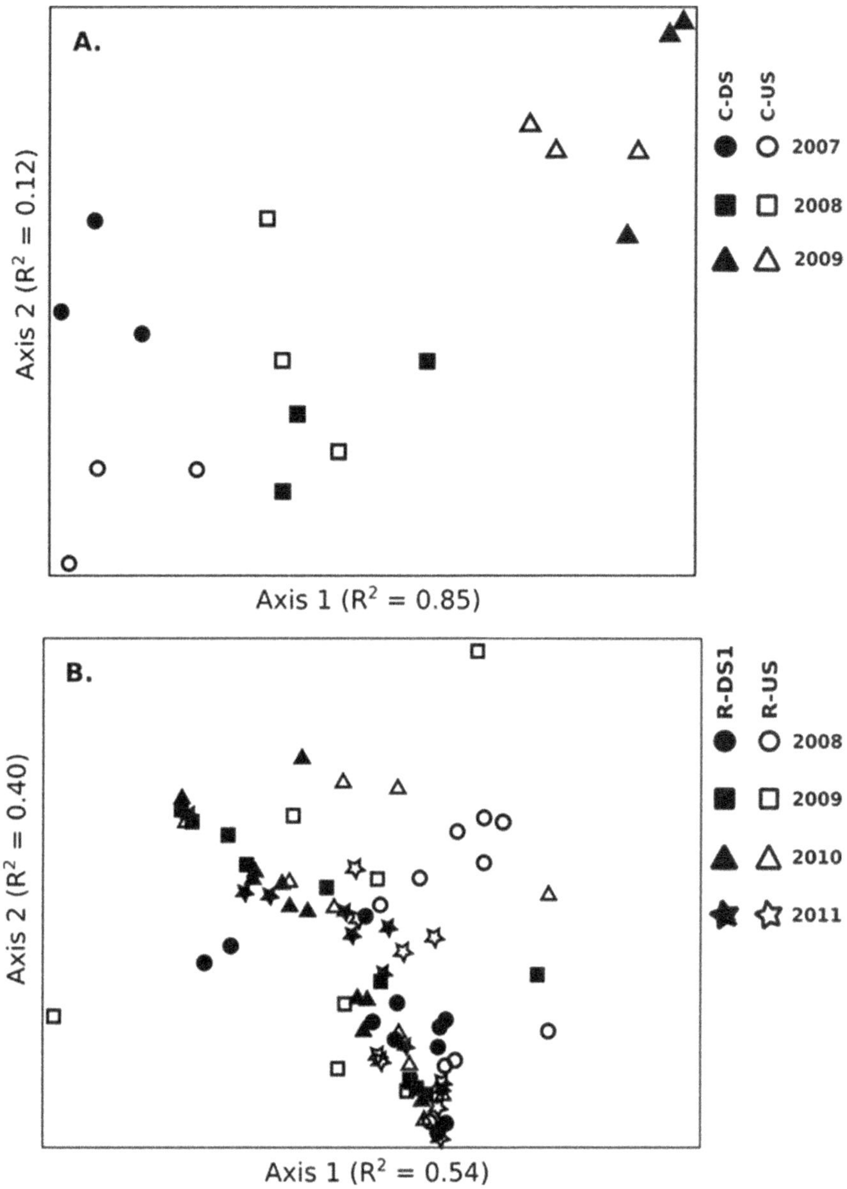

Figure 6. NMS ordination plot according to the functional community structure (based on 20 benthic insect traits) of A) 3 upstream and 3 downstream riffles associated with the Brownsville Dam removal on the Calapooia River over three collection years, and B) 11 upstream and 11 downstream transects associated with the Savage Rapids Dam removal on the Rogue River over four collection years. Note that the 11 transects from reach R-DS2 have been removed from the figure to reduce clutter (see Table 4 for MRPP analysis of these data, including the R-DS2 reach). The symbols follow the same explanation as in Figure 2. A) Final stress for the two-dimensional solution = 6.4. B) Final stress for the two-dimensional solution was 13.2.

ecological recovery from removing the press disturbance of the dams is stronger than the signal of the pulse disturbance introduced by the removal of the dams, and 4) defining assemblages according to traits confirmed the impact of the press and pulse disturbances, but did not provide a stronger indication of lingering ecological disturbance than taxonomically-defined assemblages in our two systems.

From the geomorphic perspective, two physical signatures of the recovery trajectory emerged from the sites that are associated with the local geomorphology and characteristics of the sediment pulse. On the Calapooia River, we found evidence of disturbance and recovery based on 1) an initial increase then subsequent recovery of channel slope which was unmatched in the upstream control reach, and 2) a decrease in bed variability and residual depth that initially reduced bed relief. The release of coarse material to the reach downstream of Brownsville Dam had limited impact on the dominant grain size or bed stability, a predictable result given the small size of the dam and the transport of sediment over the dam for several decades. In contrast to the Calapooia River, the channel response to the sediment pulse on the Rogue River was generally reflected in a reduction in thalweg variability and residual depth, with impacts to the farther downstream reach delayed until two years following removal. In addition, the release of sand to the downstream reaches resulted in a substantial reduction in grain size and bed stability. Thus, the sediment pulse at both sites resulted in reduction in bed relief. However, the pulse

Table 5. Indicator values (IV) and p-values from Indicator Species Analysis (ISA) comparing reaches for each trait modality with a *p*-value $<=0.05$ for A) Calapooia River and B) Rogue River.

A) Calapooia	**C-US**			**C-DS**		
Year	**Trait**	**IV**	**p**	**Trait**	**IV**	**p**
2007	poorly synchronized emergence	67.4	**0.0002**	well synchronized emergence	60	**0.0002**
	multivoltine	89.3	**0.0002**	univoltine	59.5	**0.0002**
	swim (habit)	87.8	**0.0002**	no swim ability	59.8	0.01
	collector-gatherers	59.7	0.004	herbivores	63.1	**0.0004**
	drift absent	61	0.01	clingers	53.6	0.004
	weak swim ability	55.6	0.02	good armoring	74.5	0.03
	high crawl rate	63.4	0.05			
2008	[none]			multivoltine	59.5	0.04
				swim (habit)	58.9	0.05
2009	[none]			high crawl rate	64.8	0.03

B) Rogue	**R-US**			**R-DS1**			**R-DS2**		
Year	**Trait**	**IV**	**p**	**Trait**	**IV**	**p**	**Trait**	**IV**	**p**
2008	cool stenotherm	53.3	0.03	cold/warm eurytherm	34.4	0.002	[none]		
	rare in drift	55.4	0.005						
	high crawl rate	48.4	0.01						
	long-lived	47.3	0.05						
	semivoltine	38.1	0.05						
2009	univoltine	41	0.04	multivoltine	48.5	0.02	[none]		
	respiration with gills	36.2	0.03	poorly synchronized emergence	53.4	0.01			
	no armoring	36	0.05	swim (habit)	49.8	0.01			
	weak flight	35.9	0.02	streamlined shape	48.1	0.04			
2010	semivoltine	46	0.03	multivoltine	45.1	0.05			
	poor armoring	45.9	0.03						
	nonseasonal development	45.9	0.03				fast-seasonal development	36.1	0.008
2011	[none]			[none]			[none]		

Indicator trait modalities are organized into upstream and downstream reach columns by year-defined rows. Significant *p*-values following a correction for multiple comparisons are indicated in bold.

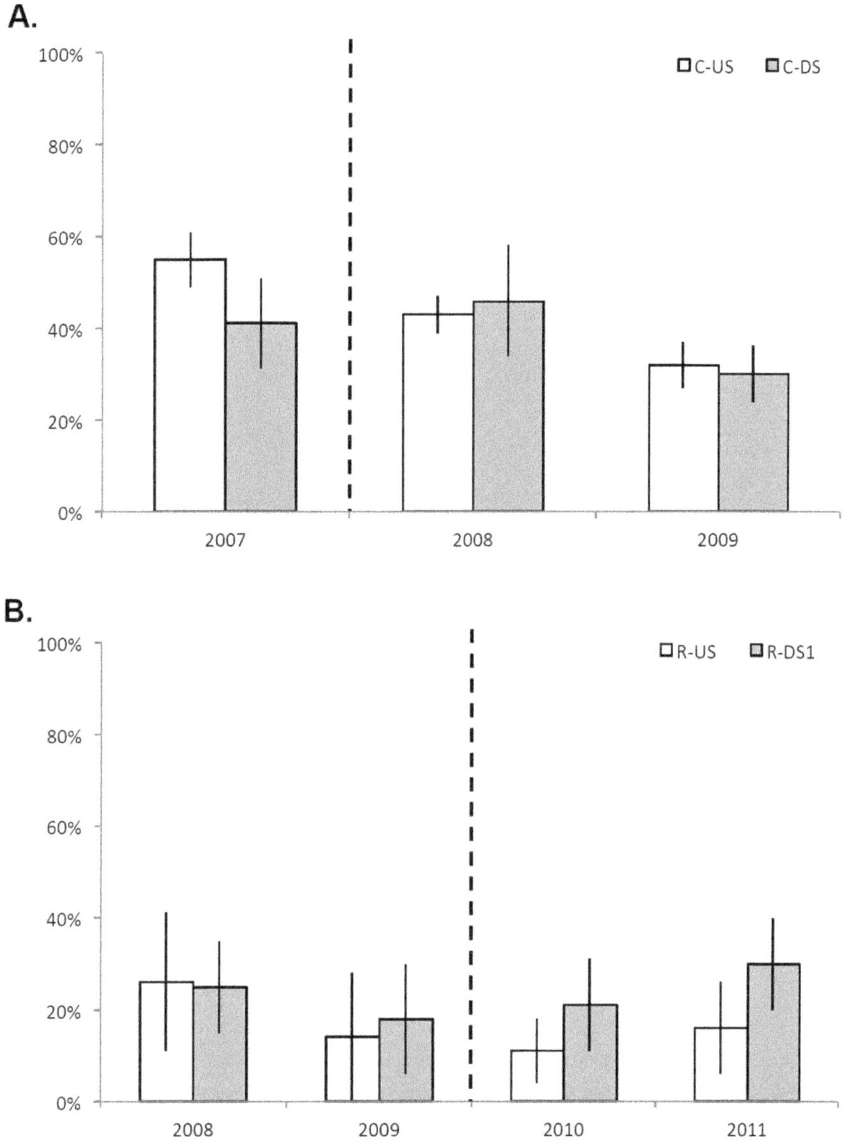

Figure 7. Mean %clingers (expressed as relative abundance, +/−2 standard errors) per reach across A) riffle sample units on the Calapooia, and B) transects sampled on the Rogue. In both panels, year is on the x-axis, and lighter-colored bars represent the upstream reaches. Vertical dashed line shows timing of dam removal. Note that, in the interest of space, the second downstream reach is not plotted for the Rogue due to the lack of change over time.

of coarse sediment released to the Calapooia River was large enough to locally impact the channel slope but not bed material size and mobility, whereas the sand released to the Rogue River reduced bed material size and mobility.

While both sites exhibited some evidence of pulse disturbance for the two years following dam removal, we found limited evidence of the impact of the dam removals on the benthic macroinvertebrate assemblages. Instead, benthic assemblages appeared to rapidly respond to the removal of the press disturbance of the dam, rather than the introduction of a pulse disturbance associated with the release of sediment following removal. We inferred the ecological press disturbance of dams on both rivers through detection of significant differences in assemblage structure between upstream and downstream reaches in the years prior to dam removal. These upstream/downstream differences were detectable both taxonomically and functionally only in the year prior to dam removal. Downstream assemblages moved rapidly towards similarity with the upstream assemblages post-removal, as illustrated by the lack of significant differences between the upstream and downstream sites within a single year post-removal. This pattern suggests that ecological recovery occurred within a single year once dam removal activities had ceased, regardless of geomorphological differences between the two rivers. This rate of recovery is consistent with other studies [30,33–37] that have documented annual-scale recovery from dam removal sediment pulses. In addition to the shorter timescale for ecological recovery, we found some evidence that the spatial extent for ecological disturbance is also smaller than for geomorphic disturbance. Reach R-DS2, the second downstream reach on the Rogue River, did not reflect any effects of press or pulse disturbance ecologically, but did reflect the pulse disturbance geomorphically in 2010 and 2011, suggesting that the spatial scale of the dam-removal disturbance was larger for the abiotic system than our measure of the biotic system.

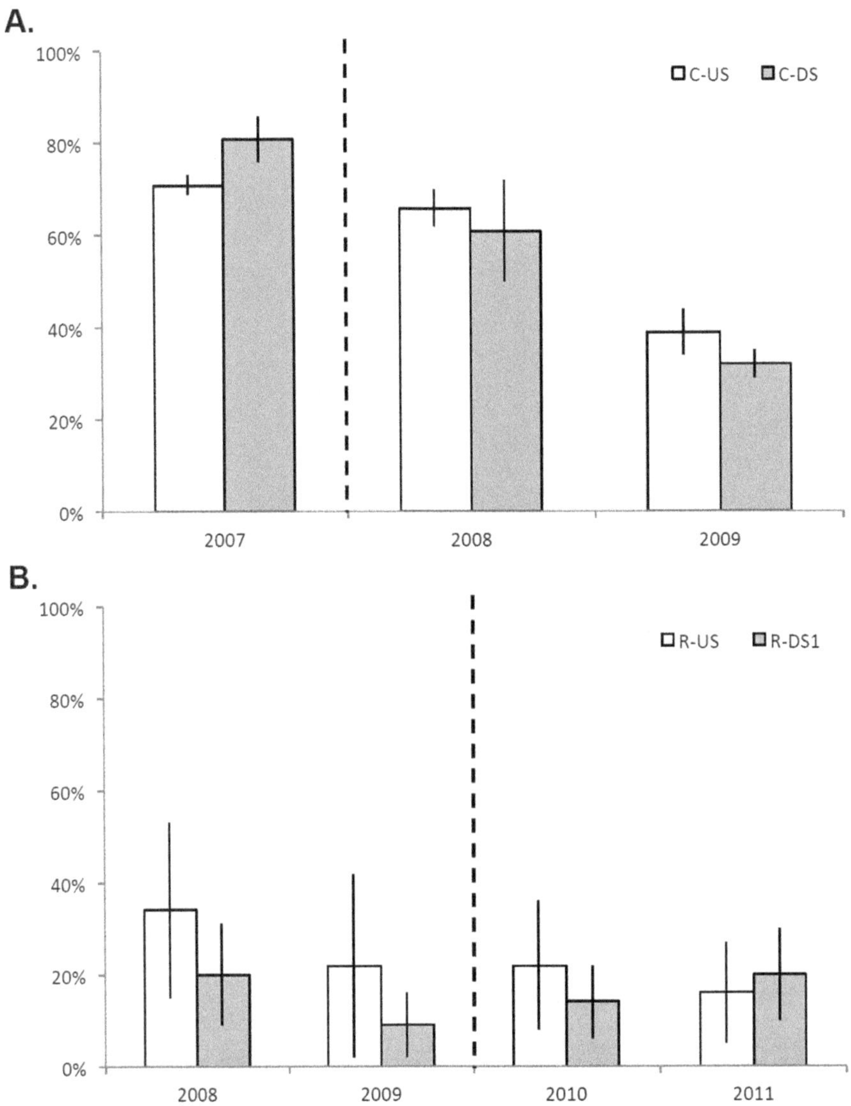

Figure 8. Mean %EPT (expressed as a proportion, +/−2 standard errors) per reach across A) riffle sample units on the Calapooia, and B) transects sampled on the Rogue. In both panels, year is on the x-axis, and lighter-colored bars represent the upstream reaches. Vertical dashed line shows timing of dam removal. Note that the second downstream reach is not plotted for the Rogue River due to the lack of change over time.

We expected that, because individual trait modalities are often linked tightly to disturbance or other environmental filters, functional analysis would provide additional insight into the biotic effects of dam removal [44] over taxonomic approaches. As reflected in the MRPP, ISA, and ordination plots, the results of both the taxonomic and functional trait analyses clearly highlight that the upstream and downstream sites were significantly different prior to dam removal and converged towards similarity post-removal. However, taxonomic differences were stronger than functional differences on the Rogue River, and investigation of the individual indicator traits does not produce a clear interpretation of the disturbance of the dam or its removal. At the Calapooia River, pre-removal trait modalities associated with the upstream reach included both those that tend to be associated with disturbance (e.g. multivoltinism) and those associated with stability (e.g. poorly synchronized emergence, absence from the drift, high crawling rate). Similarly, the trait modalities at the downstream reach on the Calapooia River included a mix of traits that reflect both habitat stability (e.g. univoltinism, well-synchronized emergence, and the clinger habit) and disturbance (e.g. good armoring). Post-removal on the Calapooia River, the lack of distinguishing traits upstream and the decreasing number of disturbance-related traits downstream is indicative of converging functional modalities post removal. Across all years on the Rogue River, the benthic assemblages in the upstream reach were dominated by trait modalities representative of stable habitats, including indicators of long larval life span, high crawl rate, no or poor armoring, and rarity in the drift. In contrast, downstream sites did not strongly associate with any functional modalities prior to dam removal, except eurythermal taxa. During the year of dam removal, the reach immediately below the dam was dominated by traits associated with disturbance. We attribute this shift in the downstream community to dam decommissioning activities, including the potential transport of small volumes of fine sediment from the reservoir during the spring 2009 drawdown and subsequent construction. Only one disturbance-related trait persisted in the two reaches below the dam in the first year post-removal, and the assemblages had no distinguishing trait modalities two years post-removal. Thus, a mix of traits associated with the dams and their removal appear to characterize the communities in the year post-removal, but the dominance of any traits was eliminated at both sites by the second year following removal.

We anticipated that both clinger habit and %EPT would provide a mechanistic link between geomorphic and ecological responses to dam removal. Both respond negatively to disturbance, particularly deposition of sediments [32,44]. However, our results indicate that these univariate measures alone were not strong indicators of the sediment pulse, largely due to the high variability in the assemblages. The only significant difference in %clingers upstream vs downstream of a dam was observed on the Calapooia River prior to dam removal, where the downstream impact reach had higher % clinger taxa than the upstream control site, potentially associated with the clay hardpan substrate downstream prior to removal. There was no difference in percent clingers between upstream and downstream reaches on the Rogue River regardless of year. Clingers also did not emerge as a potential indicator trait on the Rogue River, but did on the Calapooia River and only for the clay hardpan bed in the pre-removal downstream reach C-DS. The lack of differentiation between reaches using the %clinger metric, despite evidence from the physical data that the channels were unstable, suggests that the metric may not always represent a strong link between habitat stability and ecological response. We found a similarly weak link between habitat stability and %EPT, where no clear trend in % EPT is observed across the years at either site. While we did find a trend of %EPT in R-DS1 gradually surpassing %EPT at R-US in the years following the initiation of dam removal, the differences are small relative to within-year variability in the communities.

Conclusions

The increasing frequency of dam removals [71] provides opportunities to study the linkages between abiotic and biotic responses to disturbance. To investigate these linkages, we monitored changes in characteristics of the channels and benthic macroinvertebrate communities in a BACI study design at two sites in different physiographic settings. In particular, we were interested in decoupling the effects of the elimination of the press disturbance of dams and the introduction of a pulse disturbance with dam decommissioning due to sediment releases. Observations of spatial and temporal patterns in channel features and benthic assemblages indicate that: 1) the presence of the dams constituted a stronger ecological disturbance to the near-downstream reaches on both rivers, as predicted by the serial discontinuity concept, than the pulse disturbance of the dam removal, despite the fact that both rivers passed unregulated flow and sediment during the high flow season; 2) ecological recovery from this press disturbance occurred within the year following the restoration action of dam removal, despite signals of lingering geomorphic disturbance from the sediment released with dam removal, and 3) the analysis of functional traits further confirmed our finding that upstream and downstream sites were ecologically distinct prior to dam removal, but did not add any evidence that the ecological disturbance persisted beyond the period reflected in taxonomically-defined assemblages. These results provide insight into the spatial and temporal extent of the geomorphic and ecological disturbances from dam removal, as well as the trajectories of the recovery from the disturbance of dams and their removal. However, the relatively small sizes of the dams and physiographic settings of the Pacific Northwest is clearly limited in scope and thus results should be confirmed under conditions of larger sediment pulses under a range of physiographic settings.

Acknowledgments

We gratefully acknowledge Homero Flores Cervantes, Matt Cox, Bill Gerth, Anna Leitschuh, and Lisa Thompson who helped with data processing and a multitude of undergraduates who supported field data collection. We also thank Dave Lytle and his lab group for support and discussion during data analysis.

Author Contributions

Conceived and designed the experiments: DT DF CW. Performed the experiments: DT CW. Analyzed the data: DT DF CW. Contributed reagents/materials/analysis tools: DT DF CW. Contributed to the writing of the manuscript: DT DF CW.

References

1. Poff L, Hart D (2002) How dams vary and why it matters for the emerging science of dam removal. Bioscience 52: 659–668.
2. Stanley EH, Doyle MW (2003) Trading off: The Ecological Effects of Dam Removal. Frontiers in Ecology and the Environment 1: 15–22.
3. Resh VH, Brown AV, Covich AP, Gurtz ME, Li HW, et al. (1988) The role of disturbance in stream ecology. Journal of the North American Benthological Society 7: 433–455.
4. Graf WL (1977) The Rate Law in fluvial geomorphology. American Journal of Science 277: 178–191.
5. Sousa WP (1984) The role of disturbance in natural communities. Annu Rev Ecol Syst 353–391.
6. Poff NL (1992) Why disturbances can be predictable: a perspective on the definition of disturbance in streams. Journal of the North American Benthological Society 86–92.
7. Montgomery DR (1999) Process Domains and the River Continuum, Journal of the American Water Resources Association 35: 397–410.
8. Reinhardt L, Jerolmack D, Cardinale BJ, Vanacker V, Wright J (2010) Dynamic interactions of life and its landscape: Feedbacks at the interface of geomorphology and ecology. Earth Surface Processes and Landforms 35: 78–101.
9. Wheaton JP, Gibbins C, Wainwright J, Larsen L, McElroy B (2011) Preface: Multiscale Feedbacks in Ecogeomorphology. Geomorphology 126: 265–268.
10. Doyle MW, Stanley EH, Orr CH, Selle AR, Sethi SA, et al. (2005) Stream Ecosystem Response to Small Dam Removal: Lessons from the Heartland. Geomorphology 71: 227–244.
11. Brierley G, Fryirs K (2005) Geomorphology and River Management: Applications of the River Styles Framework. Oxford: Blackwell Publishing. 398 p.
12. Knighton D (1998) Fluvial Forms and Processes: A New Perspective. New York: Oxford University Press. 383 p.
13. Madej MA, Ozaki V (1996) Channel response to sediment wave propagation and movement, Redwood Creek, California, USA. Earth Surface Processes and Landforms 21: 911–927.
14. Lisle TE, Cui Y, Parker G, Pizzuto JE, Dodd AM (2001) The dominance of dispersion in the evolution of bed material waves in gravel-bed rivers. Earth Surface Processes and Landforms 26: 1409–1420.
15. Pizzuto J (2002) Effects of dam removal on river form and process. BioScience 52: 683–691.
16. Doyle MW, Stanley EH, Harbor JM (2002) Geomorphic Analogies for Assessing Probable Channel Response to Dam Removal. Journal of the American Water Resources Association 38: 1567–1579.
17. Doyle MW, Stanley EH, Harbor JM (2003) Channel adjustments following two dam removals in Wisconsin. Water Resources Research 39: 1–15.
18. Sawaske SR, Freyberg DL (2012) A comparison of past small dam removals in highly sediment-impacted systems in the U.S. Geomorphology 151–152: 50–58.
19. Madej MA (2001) Development of channel organization and roughness following sediment pulses in single-thread, gravel bed rivers. Water Resources Research 37: 2259–2272.
20. Poff NL, Ward JV (1990) Physical habitat template of lotic systems: recovery in the context of historical pattern of spatiotemporal heterogeneity. Environmental Management 14: 629–645.

21. Lytle DA, Poff NL (2004) Adaptation to natural flow regimes. Trends in Ecology and Evolution 190: 1–7.
22. Kibler K, Tullos D, Kondolf M (2011) Evolving Expectations of Dam Removal Outcomes: Downstream Geomorphic Effects Following Removal of a Small, Gravel-Filled Dam. Journal of the American Water Resources Association 47: 408–423.
23. East AE, Pess GR, Boundry JA, Magirl CA, Ritchie AC, et al. (2014) Large-scale dam removal on the Elwha River, Washington, USA: River channel and floodplain geomorphic change. Geomorphology. In press.
24. Palmer MA, Bernhardt ES, Allan JD, Lake PS, Alexander G, et al. (2005) Standards for ecologically successful river restoration. Journal of Applied Ecology 42: 208–217.
25. Feld CK, Birk S, Bradley DC, Hering D, Kail J, et al. (2011) Chapter 3 - From Natural to Degraded Rivers and Back Again: A Test of Restoration Ecology Theory and Practice, 1st ed, Advances in Ecological Research. Elsevier Ltd.
26. Bender EA, Case TJ, Gilpin ME (1984) Perturbation Experiments in Community Ecology: Theory and Practice. Ecology 65: 1–13.
27. Wolman MG, Gerson R (1978) Relative scales of time and effectiveness of climate in watershed geomorphology. Earth Surface Processes and Landforms 3: 189–208.
28. Niemi GJ, Devore P, Detenbeck N, Taylor D, Lima A, et al. (1990) Overview of case studies on recovery of aquatic system from disturbance. Environ Manage 14: 571–587.
29. Bushaw-Newton KL, Hart DD, Pizzuto JE, Thomson JR, Egan J, et al. (2003) An Integrative Approach Towards Understanding Ecological Responses To Dam Removal: the Manatawny Creek Study. J Am Water Resour Assoc 38: 1581–1599.
30. Stanley EH, Luebke MA, Doyle MW, Marshall DW (2002) Short-term changes in channel form and macroinvertebrate communities following low-head dam removal. Journal of the North American Benthological Society 21, 172–187.
31. Zunka J, Tullos D, Lancaster S (2014) Effects of sediment pulses on bed relief in bar-pool channels. Earth Surface Processes and Landforms. In press.
32. Zuellig RE, Kondratieff BC, Rhodes HA (2002) Benthos recovery after an episodic sediment release into a Colorado Rocky Mountain River. Western North American Naturalist 62, 59–72.
33. Pollard AI, Reed T (2004) Benthic invertebrate assemblage change following dam removal in a Wisconsin stream. Hydrobiologia 513, 51–58.
34. Thomson JR, Hart DD, Charles DF, Nightengale TL, Winter DM (2005) Effects of removal of a small dam on downstream macroinvertebrate and algal assemblages in a Pennsylvania stream. Journal of the North American Benthological Society 24: 192–207.
35. Orr CH, Kroiss SJ, Rogers KL, Stanley EH (2008) Downstream benthic responses to small dam removal in a coldwater stream. River Res Appl 24: 804–822. doi:10.1002/rra.
36. Chiu MC, Yeh CH, Sun YH, Kuo MH (2013) Short-term effects of dam removal on macroinvertebrates in a Taiwan stream. Aquat Ecol 47: 245–252.36.
37. Renöfält BM, Lejon A, Jonsson M, Nilsson C (2013) Long-term taxon-specific responses of macroinvertebrates to dam removal in a mid-sized Swedish stream. River Res Appl 29: 1082–1089.
38. Sethi SA, Selle AR, Doyle MW, Stanley EH, Kitchel HE (2004) Response of unionid mussels to dam removal in Koshkonong Creek, Wisconsin (USA). Hydrobiologia 525: 157–165.
39. Poff N, Olden J, Vieira N, Finn D, Simmons M, et al. (2006) Functional trait niches of North American lotic insects: traits-based ecological applications in light of phylogenetic relationships. Journal of the North American Benthological Society 25: 730–755.
40. Poff N (1997) Landscape filters and species traits: towards mechanistic understanding and prediction in stream ecology. Journal of the North American Benthological Society 16: 391–409.
41. Tullos D, Penrose D, Jennings G, Cope G (2009) Analysis of Functional Traits in Reconfigured Channels: Implications for the Bioassessment and Disturbance of River Restoration. Journal of the North American Benthological Society 28: 80–92.
42. Dolédec S, Statzner B, Bournard M (1999) Species traits for future bioassessment across ecoregions: patterns along a human-impacted river. Freshwater Biology 42: 737–758.
43. Gayraud S, Statzner B, Bady P, Haybach A, Schöll F, et al. (2003) Invertebrate traits for the biomonitoring of large European rivers: an initial assessment of alternative metrics. Freshwater Biology 48: 2045–2064.
44. Pollard AI, Yuan LL (2009) Assessing the consistency of response metrics of the invertebrate benthos: a comparison of trait- and identity-based measures. Freshwater Biology 55: 1420–1429.
45. McCowan K (2007) Dam's end washes away a bit of history, The Register Guard, Eugene, OR, 28 August: A1.
46. Runyon J, Andrus C, Schwindt R (2004) Calapooia River Watershed Assessment. Calapooia Watershed Council: Brownsville, OR.
47. U.S. Bureau of Reclamation (2001) Savage Rapids Dam, Sediment Evaluation Study. U.S. Department of the Interior. 286 p.
48. Jones KL, O'Connor JE, Keith MK, Mangano JF, Wallick JR (2012) Preliminary assessment of channel stability and bed-material transport in the Rogue River basin, southwestern Oregon: U.S. Geological Survey Open-File Report 2011–1280, 96 p.
49. U.S. Bureau of Reclamation (2009a) Savage Rapids Dam Removal and Replacement Pumping Facilities Status Report 13 (April 1, 2009 through June 30, 2009). Available: http://www.usbr.gov/pn/programs/lcao_misc/savage/reports/SRStatus13.pdf.
50. U.S. Bureau of Reclamation (2009b) Savage Rapids Dam Removal and Replacement Pumping Facilities Status Report 14, Final (July 1, 2009 through November 23, 2009) Available: http://www.usbr.gov/pn/programs/lcao_misc/savage/reports/srstatus14.pdf.
51. Walter C, Tullos DD (2010) Downstream channel changes after a small dam removal: using aerial photos and measurement error for context; Calapooia River, Oregon. River Research and Applications 26: 1220–1245.
52. Wolman MG (1954) A method of sampling coarse river-bed material. Trans Am Geophys Union 35: 951–956.
53. Church MA, McLean DG, Wolcott JF (1987) River Bed Gravels: Sampling and Analysis. In: Thorne CR, Bathurst JC, Hey RD, editors. Sediment Transport in Gravel-bed Rivers. Chichester: John Wiley and Sons Ltd. pp. 43–79.
54. EPA (2000) Environmental Monitoring and Assessment Program - Surface Waters: Field Operations and Methods for Measuring the Ecological Condition of Non-Wadeable Rivers and Streams. Available: http://www.epa.gov/emap/html/pubs/docs/groupdocs/surfwatr/field/R5_remap.pdf.
55. EPA (2001) Environmental Monitoring and Assessment Program - Surface Waters: Western Pilot Study Field Operations Manual For Wadeable Streams. Available: http://www.epa.gov/emap/html/pubs/docs/groupdocs/surfwatr/field/ewwsm01.pdf.
56. Bunte K, Abt SR (2001) Sampling surface and subsurface particle-size distributions in wadable gravel- and cobble-bed streams for analyses in sediment transport, hydraulics, and streambed monitoring. Gen. Tech. Rep. RMRS-GTR-74. Fort Collins, CO: U.S. Department of Agriculture, Forest Service, Rocky Mountain Research Station.
57. Faustini J, Kaufmann PR (2008) Adequacy of visually classified particle count statistics from regional stream habitat surveys. Journal of the American Water Resources Association 43: 1293–1315.
58. Bartley R, Rutherfurd I (2002) Techniques to quantify the variability of thalweg profiles. The Structure, Function and Management Implications of Fluvial Sedimentary Systems 276: 35–44.
59. Mossop B, Bradford MJ (2006) Using thalweg profiling to assess and monitor juvenile salmon (Oncorhynchus spp.) habitat in small streams. Canadian Journal of Fisheries and Aquatic Sciences 63: 1515–1525.
60. Kaufmann PR, Faustini JM, Larsen DP, Shirazi MA (2008) A roughness-corrected index of relative bed stability for regional stream surveys. Geomorphology 99: 150–170.
61. Kaufmann PR, Levine P, Robison EG, Seeliger C, Peck DV (1999) Quantifying Physical Habitat in Wadeable Streams. EPA/620/R-99/003. U.S. Environmental Protection Agency, Washington, DC.
62. Finn D, Poff N (2005) Variability and convergence in benthic communities along the longitudinal gradients of four physically similar Rocky Mountain streams. Freshwater Biology 50: 243–261.
63. McCune B, Mefford M (2006) PC-ORD. Multivariate analysis of ecological data. Version 5.
64. McCune B, Grace J (2002) Analysis of ecological communities. MjM Software Design, Gleneden Beach.
65. Lenat DR (1988) Water quality assessment of streams using a qualitative collection method for benthic macroinvertebrates. Journal of the North American Benthological Society 222–233.
66. Karr JR, Chu EW (1999) Restoring Life in Running Waters: Better Biological Monitoring. Island Press, Washington D.C.
67. Pearson AJ, Snyder NP, Collins MJ (2011) Rates and processes of channel response to dam removal with a sand-filled impoundment. Water Resources Research 47(8): 1–15.
68. Ward JV, Stanford JA (1983) The serial discontinuity concept of lotic ecosystems. In: Fontaine TD, Bartell SM, editors. "Dynamics of Lotic Ecosystems". Ann Arbor: Science Publications. pp. 29–42.
69. Van Looy K, Tormos T, Souchon Y (2014) Disentangling dam impacts in river networks. Ecological Indicators 37: 10–20.
70. Lisle TE (2008) The Evolution of sediment waves influenced by varying transport capacity in heterogeneous rivers. In Habersack, Piegay, Rinaldi, editors. Gravel-Bed Rivers VI: From Process Understanding to River restoration. Elsevier. pp. 443–472.
71. Service RF (2011) Will Busting Dams Boost Salmon? Science 334: 888–892.

7

Abundance and Dynamics of Antibiotic Resistance Genes and Integrons in Lake Sediment Microcosms

Björn Berglund[1]*, Ghazanfar Ali Khan[2], Richard Lindberg[2], Jerker Fick[2], Per-Eric Lindgren[1,3]

1 Department of Clinical and Experimental Medicine, Linköping University, Linköping, Sweden, 2 Department of Chemistry, Umeå University, Umeå, Sweden, 3 Department of Microbiology, Medical Services, County Hospital Ryhov, Jönköping, Sweden

Abstract

Antibiotic resistance in bacteria causing disease is an ever growing threat to the world. Recently, environmental bacteria have become established as important both as sources of antibiotic resistance genes and in disseminating resistance genes. Low levels of antibiotics and other pharmaceuticals are regularly released into water environments via wastewater, and the concern is that such environmental contamination may serve to create hotspots for antibiotic resistance gene selection and dissemination. In this study, microcosms were created from water and sediments gathered from a lake in Sweden only lightly affected by human activities. The microcosms were exposed to a mixture of antibiotics of varying environmentally relevant concentrations (i.e., concentrations commonly encountered in wastewaters) in order to investigate the effect of low levels of antibiotics on antibiotic resistance gene abundances and dynamics in a previously uncontaminated environment. Antibiotic concentrations were measured using liquid chromatography-tandem mass spectrometry. Abundances of seven antibiotic resistance genes and the class 1 integron integrase gene, *intI1*, were quantified using real-time PCR. Resistance genes *sulI* and *ermB* were quantified in the microcosm sediments with mean abundances 5 and 15 gene copies/10^6 16S rRNA gene copies, respectively. Class 1 integrons were determined in the sediments with a mean concentration of 3.8×10^4 copies/10^6 16S rRNA gene copies. The antibiotic treatment had no observable effect on antibiotic resistance gene or integron abundances.

Editor: Dwayne Elias, Oak Ridge National Laboratory, United States of America

Funding: The study was funded by the Swedish Research Council for Environment, Agricultural Sciences and Spatial Planning (contract number: 210-2006-2132). The funders had no role in study design, data collection and analysis, decision to publish, or preparation of the manuscript.

Competing Interests: The authors have declared that no competing interests exist.

* Email: bjorn.berglund@liu.se

Introduction

When antibiotics were introduced for therapeutic use in the 1930s, their success at treating bacterial infections completely changed clinical practice and saved countless lives. However, the use of antibiotics also drives the development and enrichment of antibiotic resistance determinants. Today, antibiotic resistant bacteria are widespread and the rapid dissemination of multi-drug resistance in particular, has brought considerable concerns as to whether humanity is about to return to the pre-antibiotic era [1].

While studies investigating antibiotic resistance have classically been constrained to clinical contexts, it has recently been recognised that the environment is likely to play a major role in the development and dissemination of antibiotic resistance genes (ARGs). Antibiotics and ARGs originate from environmental bacteria, and it is likely that anthropogenic contamination with antibiotics can drive selection for resistance, which in turn may find its way into pathogenic bacteria [2]. Water environments in particular are important in the circulation and accumulation of antibiotic discharge and are therefore likely to serve as hotspots for development and dissemination of ARGs [3]. ARGs in water environments have been reported to correlate with anthropogenic activities [4–6], and extremely high concentrations of antibiotic pollutants have been shown to correlate to high ARG levels [7–8]. Furthermore, it has been shown that antibiotic concentrations below minimum inhibitory concentrations (MICs) can select for resistant bacteria [9]. It is clear that studies are needed to elucidate what effect low-level antibiotic contamination (i.e. concentrations ranging from ng/L to low µg/L) has on the resistance development in the environmental bacterial community.

Mobile genetic elements containing resistance determinants are also considered important for widespread dissemination of ARGs [2]. Integrons are a type of genetic assembly platform often associated with mobile genetic elements. They often harbour ARG cassettes and have the ability to capture new gene cassettes with enzymes known as integrases [10]. Prevalence of integrons has been reported to be elevated in anthropogenically impacted sites, indicating a potential correlation between pollution and the ability of antibiotic resistance to be disseminated [11–13]. The total number and function of different environmental gene cassettes available for integrons are still largely unknown. Previous studies have estimated the total number of different genes cassettes to be over 2,000 in 50 m^2 of soil [14] and about 1,000 in marine sediments [15]. Even though many integron gene cassettes are likely to encode non-resistance functions, the huge diversity of cassettes which environmental integrons are likely to be able to access suggests that these genetic elements play a significant role in bacterial adaptability [14].

Table 1. The experimental set-up consisted of three series of microcosms with varying concentrations of antibiotics added (1x, 10x and 1000x), and additional control microcosms without antibiotic addition.

Antibiotic	Nominal concentration (µg/L)		
	1x	10x	1000x
Azithromycin (AZI)	0.01	0.1	10
Ciprofloxacin (CIP)	0.02	0.2	20
Clarithromycin (CLA)	0.04	0.4	40
Clindamycin (CLI)	0.02	0.2	20
Doxycycline (DOX)	0.01	0.1	10
Erythromycin (ERY)	0.2	2	200
Norfloxacin (NOR)	0.1	1	100
Oxytetracycline (OXY)	0.04	0.4	40
Sulfamethoxazole (SUL)	0.1	1	100
Tetracycline (TET)	0.1	1	100
Trimethoprim (TRI)	0.1	1	100

Microcosms are useful tools in ecological studies involving large and complex microbial systems. The main advantages of microcosms are their high capacity for experimental control and that they enable the study of selective processes in highly complex systems [16]. The effects of antibiotics on a bacterial community and the dissemination of resistance in aquatic environments are typical phenomena which can be studied with microcosms. Related examples include studies which have investigated tetracycline resistance in agricultural soils [17], effects of amoxicillin on bacteria in manured soil [18] and dissemination of a multi-drug resistance plasmid in wastewater sludge [19].

The aim of this study was to investigate the effects of low, environmentally relevant levels of antibiotics (i.e. ng/L to low µg/L) on the abundance of ARGs and class 1 integrons of a bacterial community from a water environment only lightly affected by human activities. Microcosms consisting of water and sediment from Lake Nydala, Sweden, were set up under standardised laboratory conditions. Mixtures of antibiotics of different concentration levels were added to the microcosms, and samples were taken over 100 days. Antibiotic concentrations were determined in the water phase and in the sediment phase using liquid chromatography-mass spectrometry (LC-MS). Seven different ARGs and *intI1*, the integrase gene on class 1 integrons, were determined in the sediment phase using quantitative real-time PCR.

Materials and Methods

Microcosms

Sediment and water were taken from Lake Nydala (63° 49.081N, 20° 20.784E) during ice cover in January 2009. No specific permission was required for the sampling and no endangered or protected species were involved. Lake Nydala does not receive any sewage water discharges. Sediment was taken with a HON-Kajak corer [20] and given that the sampling depth was restricted to 5 cm, the sediment taken was mainly under oxic conditions. The sediment was wet-sieved (4 mm mesh), homogenised and subsamples were taken for measuring dry weight (105° C) and loss on ignition. The experiments were performed similar to OECD Guideline 308 [21] and consisted of 500 mL amber glass flasks (Schott, Mainz, Germany) filled with 200 g of sediment and 300 mL of lake water. Microcosms were equilibrated for seven days prior to spiking with antibiotics. During the equilibration and the study period, the microcosms were slowly shaken and kept at 10° C. At the beginning of the study period, antibiotics were spiked into the water phase in each microcosm, at the nominal concentrations shown in Table 1. The antibiotics used were azithromycin (AZI), ciprofloxacin (CIP), clarithromycin (CLA), clindamycin (CLI), doxycycline (DOX), erythromycin (ERY), norfloxacin (NOR), oxytetracycline (OXY), sulfamethoxazole (SUL), tetracycline (TET) and trimethoprim (TRI), all of which were purchased from Sigma-Aldrich (Steinheim, Germany) and of HPLC grade purity (>98%). Samples were taken of water and sediments immediately prior to spiking, and after 1, 7, 14, 28, 50, 76 and 100 days.

Antibiotic quantification using LC-MS/MS

Water samples were pre-treated as described in [22]. In short, the samples were filtered through a 0.45 µm membrane filter, acidified and isotopically labelled internal standards were added to each sample prior to solid phase extraction. Before extraction of the sediment samples (0.1 g, dry weight) 50 ng of each internal and surrogate standard were added. Extraction was sequentially performed first using 1.5 mL ethylacetate and methanol (1:1 mixture) followed by 1.5 mL methanol and water (7:3 mixture) with 5% triethylamine. Samples were homogenised for four minutes at 42,000 oscillations per minute, using a Mini Beadbeater (Biospec, Bartlesville, OK) with zirconium beads and then centrifuged at 14,000 revolutions per minute for 10 min. The protocol was followed for both eluent mixtures, the supernatants were combined, evaporated to 20 µL and reconstituted in 1 mL water and acetonitrile (95:5 mixture) with 0.1% formic acid.

Antibiotic concentrations in the water and sediment samples were determined by chemical analysis using liquid chromatography-tandem mass spectrometry, as described in [22]. In short, a triple stage quadrupole MS/MS TSQ Quantum Ultra EMR (Thermo Fisher Scientific, San Jose, CA) coupled with an Accela LC pump (Thermo Fisher Scientific) and a PAL HTC autosampler (CTC Analytics AG, Zwingen, Switzerland) were used as analytical system.

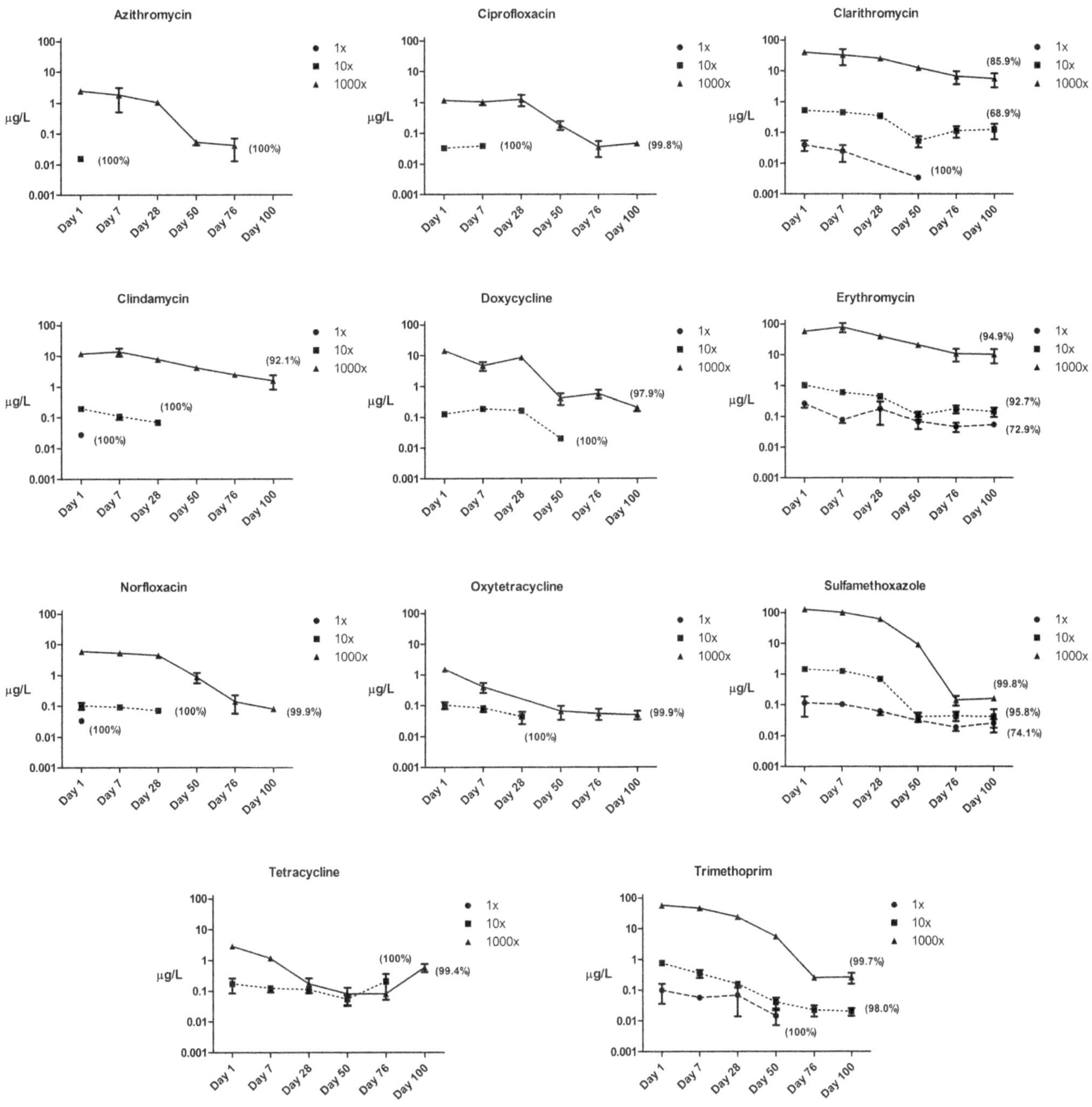

Figure 1. Antibiotics quantified in the water phase over 100 days in microcosms for three microcosms (1x, 10x and 1000x antibiotic concentration). Error bars indicate the standard error of the mean. The reduction (relative decrease of the initial concentration as compared to day '100') is presented in parenthesis after each line. Sampling points without dots indicate that the antibiotic concentration was below the limit of quantification except for the quantification of oxytetracycline at day '28' in the 1000x microcosm which could not be performed due to sample loss.

DNA extraction

200 mg of sediment from each sample was weighed and DNA was extracted using FastDNA SPIN Kit for Soil and the FastPrep Instrument (MP Biomedicals, Santa Ana, CA) according to the manufacturer's protocol. Extracted DNA was stored in −20°C before subsequent analyses. DNA concentrations in extracted DNA from different amounts of starting sediment were measured using Quant-iT PicoGreen dsDNA reagent (Invitrogen, Carlsbad, CA) to ensure a linear yield of extracted DNA.

Gene quantification with real-time PCR

Quantitative real-time PCR was used for determining gene concentrations in the extracted DNA from the microcosm sediments. The analysed genes were tetracycline resistance genes *tetA* and *tetB*, macrolide-lincosamide-streptogramin B resistance gene *ermB*, sulphonamide resistance gene *sulI*, trimethoprim resistance gene *dfrA1*, quinolone resistance gene *qnrS*, glycopeptide resistance gene *vanB* and the integrase gene on class 1 integrons, *intI1*. Gene concentrations were normalised by the measured 16S rRNA gene concentration of each sample. 16S rRNA gene concentrations were normalised against extraction

Table 2. Antibiotics quantified in the sediment phase over 100 days in microcosms of differing initial antibiotic concentrations (1x, 10x and 1000x).

		Concentration by day (ng/g)					
		1	7	28	50	76	100
	AZI	-	-	-	-	-	-
	CIP	-	-	-	-	-	-
	CLA	-	-	-	-	-	-
	CLI	-	-	-	-	-	-
	DOX	-	-	-	-	-	-
1x	ERY	-	-	-	-	-	-
	NOR	-	-	-	-	-	-
	OXY	-	-	-	-	-	-
	SUL	-	-	-	-	-	-
	TET	-	-	-	-	-	-
	TRI	-	-	-	-	-	-
	AZI	-	-	-	-	-	-
	CIP	-	-	-	-	-	-
	CLA	-	-	-	-	-	2099
	CLI	-	-	-	-	-	254
	DOX	-	-	-	-	-	-
10x	ERY	-	-	-	-	-	-
	NOR	-	-	-	-	-	-
	OXY	-	-	-	-	-	-
	SUL	-	-	-	-	-	-
	TET	-	-	-	-	-	-
	TRI	-	-	-	-	-	1647
	AZI	-	-	-	-	N.A.	-
	CIP	-	-	-	-	N.A.	-
	CLA	3161	14454	6937	8376	N.A.	30373
	CLI	1972	7660	3725	8698	N.A.	12013
	DOX	-	-	-	-	N.A.	-
1000x	ERY	-	-	-	-	N.A.	-
	NOR	-	-	-	-	N.A.	-
	OXY	-	-	-	-	N.A.	-
	SUL	1606	1432	-	-	N.A.	-
	TET	-	-	-	-	N.A.	-

Table 2. Cont.

	Concentration by day (ng/g)					
	1	7	28	50	76	100
TRI	3429	17374	5113	3191	N.A.	5286

'-': below limit of quantification.
'N.A.': not analysed due to sample loss.

mass when used separately in statistical analysis. All real-time PCRs were carried out on a CFX96 Real-Time PCR Detection System (Bio-Rad Laboratories, Hercules, CA). Quantification method, primers, primer concentrations and thermal cycling protocols were used as in [22].

Statistical methods

Two-way Repeated Measures ANOVA was used to determine the effects of initial antibiotic level and time spent in the microcosms on gene concentrations. Where time was found as a significant factor, One-way ANOVA with subsequent Bonferroni multiple comparisons test was used to determine the relevant differences. All statistical analyses were carried out using Prism 5 for Windows v.5.00.

Ethics statement

No specific permits were required for the described study. The field study did not involve any endangered or protected species.

Results

Quantification of antibiotics

The antibiotic concentrations in the water phase declined over time in all microcosms (Fig. 1). For the majority of antibiotics, the reduction in concentration from the initial added concentration was close to 100% at day '100'. Only CLA in the 10x and 1000x microcosms (68.9% and 85.9% reduction respectively) and ERY and SUL in the 1x microcosm (72.9% and 74.1% respectively) showed antibiotic reductions below 90%. The ratios of the antibiotic concentrations in the microcosms of different concentrations adhered fairly well to the theoretical values. The means±SEM of the ratios (based on all sampling points where quantification was possible) were 6.4±1.3, 550±103 and 69±8.9 for the 10x:1x, 1000x:1x and 1000x:10x microcosms respectively.

In the sediment phase, no antibiotics were detected in the microcosms with 1x concentration (Table 2). In the microcosms with 10x concentration, only CLA, CLI and TRI could be detected, and only at day '100'. In the 1000x microcosms, CLA, CLI and TRI were detected at all days sampled after antibiotic addition (means 13, 6.8 and 6.9 µg/g respectively). SUL was also detected, but only in days '1' and '7'. AZI, CIP, DOX, ERY, NOR, OXY and TET were not detected at any sampling point in the sediments.

No antibiotics were detected in the day '0' samples (i.e. before antibiotics were added) or in the controls (i.e. the microcosms in which no antibiotics were added) in either the water or the sediment phase (data not shown).

Gene quantification with real-time PCR

Sulphonamide resistance gene *sulI* was detected in 16 out of 75 samples (Table 3). Concentrations were sufficiently high for quantification in only 6 out of these 16 samples (mean value: 5 gene copies/10^6 16S rRNA gene copies). Macrolide-lincosamide-streptogramin B resistance gene *ermB* was detected in 4 out of 75 samples (Table 3), with sufficiently high concentrations to be quantifiable in 3 of these samples (mean value: 15 gene copies/10^6 16S rRNA gene copies). No trends in prevalence or concentration could be observed for either *sulI* or *ermB* in relation to time points or in relation to antibiotic concentration. Trimethoprim resistance gene *dfrA1* could only be detected in 1 sample (at day 1 in a microcosm treated with antibiotics at the 1000x concentration) at a concentration below the limit of quantification (data not shown). Tetracycline resistance genes *tetA* and *tetB*, glycopeptide resistance

Table 3. Quantification and detection of antibiotic resistance genes *sulI* and *ermB* in microcosm sediments with values in units of gene copies/10^6 16S rRNA gene copies.

		Day 1	Day 7	Day 28	Day 50	Day 76	Day 100
sulI	0x	3	+	+	-	+	4
	1x	+	-	7	-	4	+
	10x	+	-	-	-	-	+
	1000x	-	-	+	-	5	-
ermB	0x	-	-	11	-	-	-
	1x	-	-	+	6	27	-
	10x	-	-	-	-	-	-
	1000x	-	-	-	-	-	-

'-': below limit of detection.
'+': detected, but below quantification limit.

gene *vanB* and quinolone resistance gene *qnrS* could not be detected in any sample.

intI1, the integrase gene on class 1 integrons, was quantified in all samples (mean value: 3.8×10^4 gene copies/10^6 16S rRNA gene copies) (Fig. 2). Concentrations of *intI1* did not differ significantly between different time points or between different antibiotic concentration levels. The mean concentration of *intI1* in the day '0' samples ($n = 3$), i.e. before antibiotics were added, was 4.9×10^4 gene copies/10^6 16S rRNA gene copies (data not shown).

Concentrations of 16S rRNA gene copies did not appear to vary with different initial concentrations of input antibiotics, although variation could be significantly attributed to the number of days in the microcosms ($p<0.001$). 16S rRNA gene copy concentrations were lower at day '50' and day '76' than at day '1' ($p<0.05$) (Fig. 2).

Discussion

Initial antibiotic concentrations in the gathered water and sediments were below the limit of quantification. Furthermore, no antibiotics were detectable in the control microcosms during the run of the experiment. In the water phase, the ratio of antibiotic concentrations between the microcosms of different starting concentrations also tended to correspond to the initial ratio. The antibiotic quantification also showed that concentrations were reduced over time (in most microcosms, and for most antibiotics, by almost 100% over 100 days) in the water phase. The same trend was not observable for the antibiotics which were quantifiable in the sediment phase. Instead, antibiotic concentrations tended to either increase (i.e. accumulate) or remain unchanged, likely due to saturation of the sediments. Due to the high limit of quantification in the sediments, it is difficult to speculate how much of the antibiotics remained in the sediment. It is unclear how antibiotics bound to sediments affect bacteria, although it has been reported that even tightly adsorbed to clay particles, antibiotic compounds are biologically active [23].

ARGs could only be quantified and detected in a few of the samples. *sulI* was the most prevalent gene (detected in 16 out of 75 samples) and the mean of the concentration out of the quantifiable concentrations ($n = 6$) was 5 gene copies/10^6 16S rRNA gene copies. This is comparable to levels measured in [4] where *sulI* was measured in river sediments from a pristine location at concentrations around 10^0 gene copies/10^6 rRNA gene copies. *ermB* was also detected in a few samples and *dfrA1* in only one sample. *tetA*, *tetB*, *vanB* and *qnrS* were not detected in any sample. It should be noted that *tetA* and *tetB* have been found to occur stably over time in Swedish wastewaters [24], and *vanB* have been found in Swedish hospitals [25], although these environments can hardly be described as pristine. That these genes are not present in sediments collected from an environment only tangentially perturbed by anthropogenic activities is not surprising, although it should be noted that ARGs have been shown to exist in pristine environments [26–27]. The most prevalent ARG found, *sulI*, did not appear to increase in abundance over time, indicating that the antibiotics had no effect.

The class 1 integron integrase gene *intI1* was quantified in all samples. Concentrations did not significantly vary either when looking at different antibiotic concentrations or over time. This indicates that the antibiotics did not affect the concentration of *intI1*. This is noteworthy since several earlier studies have shown that integron abundances increase in anthropogenically impacted environments [6,11,13]. Notably, the mean concentration in the day '0' samples was in the same range as the mean for the concentrations of later dates (i.e. after antibiotics were added).

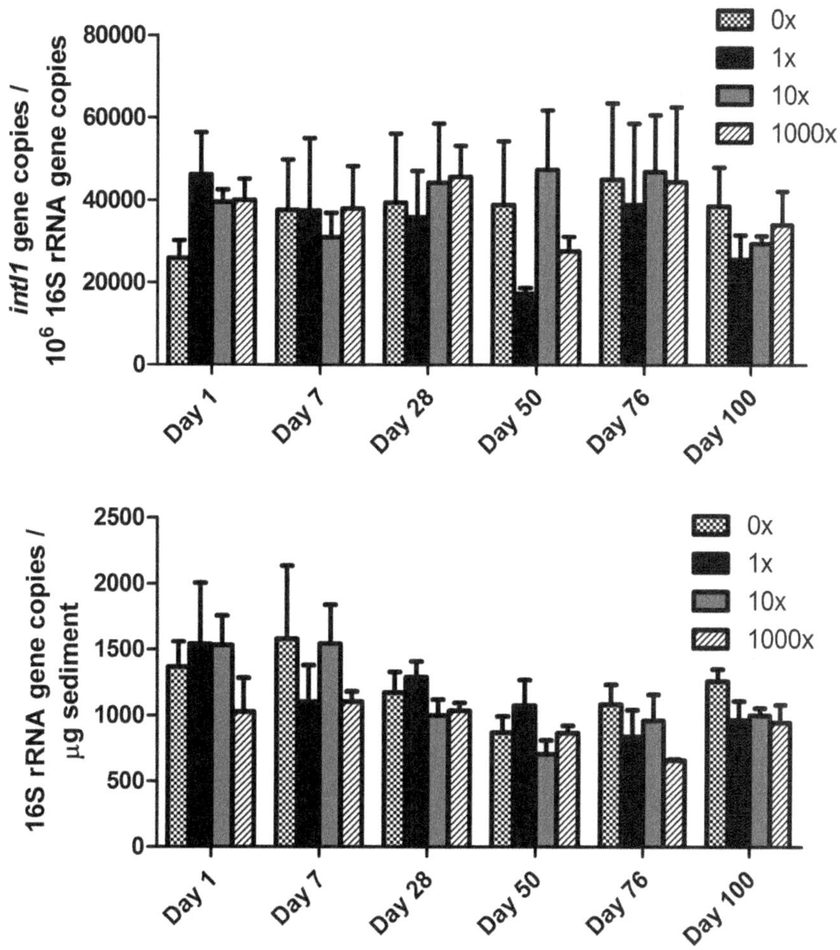

Figure 2. Mean by day of *intI1* and 16S rRNA gene concentrations measured in the microcosm sediments. Different coloured bars denote the microcosm which had different starting concentrations of antibiotics added. Error bars denote the standard error of the mean.

This indicates that the abundance of the *intI1* gene was not particularly affected during the run-time of the experiment. Care should be taken when making this comparison however, since only three samples were used to calculate the day '0' mean.

Time was determined to be a statistically relevant parameter affecting the concentration of the 16S rRNA gene in the microcosms ($p<0.001$). The trend was towards lower concentrations at the later days (i.e. day '50' and day '76') than day '1', at the start of the experiment ($p<0.05$). The statistical tests applied indicated that this decline was independent of antibiotic treatment. It is likely that the declining concentration of 16S rRNA genes is indicative of a reduction in bacteria due to low nutritional content in the sediments, low oxygen availability or other environmental factors. In an environment where the bacterial community is dying due to a lack of vital nutrients, maintaining antibiotic resistance functionality may be too costly to be beneficial. In such a case, the decline in 16S rRNA genes may explain the observation that the ARG concentrations remain the same throughout the experiment. Another explanation could be that the composition of the bacterial community in the microcosms changed as bacteria with a lower 16S rRNA gene copy number became dominant at the expense of bacteria with a higher copy number. Although, even if this was the case, antibiotics could still be ruled out as having a significant effect on this development, as the same trends were seen in the control microcosms and the spiked microcosms, irrespective of antibiotic concentrations. If the decline in 16S rRNA gene copies is an indication that the bacteria in the microcosms did not grow, it may also explain why the antibiotics did not have a significant effect, as some antibiotics exhibit significantly less killing power on slow and non-growing bacteria [28].

The initial concentrations of antibiotics (defined as 1x concentration) added to the microcosms were chosen to mimic the concentrations commonly encountered in effluent-dominated surface waters and similarly perturbed aquatic environments. An increase in antibiotic resistance gene or integron abundance could not be observed in any of the concentration ranges. In [22] it was shown that antibiotic concentrations equal to the 10x level, did not affect resistance gene abundances in a series of constructed wetlands. In [9] it was shown that antibiotic resistant bacteria could be selected for at antibiotic concentrations below MIC-levels. This was performed for concentrations of TET at 15 µg/L and for CIP as low as 0.1 µg/L. These are concentrations lower by about one order of magnitude compared to the initial concentrations used in the 1000x microcosms. The lack of effect seen in this experiment at these concentrations could be explained by the possible lack of bacterial growth and that the bacterial community used in this experiment was considerably more complex.

The experiment parameters in this study were set to simulate a bacterial community in a previously unperturbed environment exposed to a mixture of antibiotics of concentrations likely to occur in an anthropogenic contamination event. The addition of antibiotics had no discernible effect on the abundance of resistance genes or integrons measured in this experiment (including the 16S rRNA gene) at any of the concentrations. That no effect could be seen in the microcosm which was exposed to a thousand times the environmentally relevant concentrations is notable, as the levels have been shown to select for resistant bacteria in less complex systems. Further studies are needed to establish what, if any, effect these higher antibiotic levels have on bacterial communities in pristine and lightly affected environments.

Author Contributions

Conceived and designed the experiments: BB GAK JF PEL. Performed the experiments: BB GAK RL JF. Analyzed the data: BB GAK RL JF. Contributed reagents/materials/analysis tools: JF PEL. Wrote the paper: BB JF PEL.

References

1. Davies J, Davies D (2010) Origins and evolution of antibiotic resistance. Microbiol Mol Biol Rev 74(3): 417–433.
2. Gillings MR (2013) Evolutionary consequences of antibiotic use for the resistome, mobilome, and microbial pangenome. Front Microbiol 4: 4.
3. Baquero F, Martínez JL, Cantón R (2008) Antibiotics and antibiotic resistance in water environments. Curr Opin Biotechnol 19(3): 260–265.
4. Pei R, Kim SC, Carlson KH, Pruden A (2006) Effect of river landscape on the sediment concentrations of antibiotics and corresponding antibiotic resistance genes (ARG). Water Res 40(12): 2427–2435.
5. Pruden A, Arabi M, Storteboom HN (2012) Correlation between upstream human activities and riverine antibiotic resistance genes. Environ Sci Technol 46(21): 11541–11549.
6. Khan GA, Berglund B, Khan MK, Lindgren PE, Fick J (2013) Occurrence and abundance of antibiotics and resistance genes in rivers, canal and near drug formulation facilities – a study in Pakistan. PLoS One 8(6): e62712.
7. Li D, Yu T, Zhang Y, Yang M, Li Z, et al. (2010) Antibiotic resistance characteristics of environmental bacteria from an oxytetracycline production wastewater treatment plant and the receiving river. Appl Environ Microbiol 76(11): 3444–3451.
8. Kristiansson E, Fick J, Janzon A, Grabic R, Rutgersson C, et al. (2011) Pyrosequencing of antibiotic-contaminated river sediments reveals high levels of resistance and gene transfer elements. PLoS One 6(2): e17038.
9. Gullberg E, Cao S, Berg OG, Ilbäck C, Sandegren L, et al. (2011) Selection of resistant bacteria at very low antibiotic concentrations. PLoS Pathog 7(7): e1002158.
10. Mazel D (2006) Integrons: agents of bacterial evolution. Nat Rev Microbiol 4(8): 608–620.
11. Wright MS, Baker-Austin C, Lindell AH, Stepanauskas R, Stokes HW, et al. (2008) Influence of industrial contamination on mobile genetic elements: class 1 integron abundance and gene cassette structure in aquatic bacterial communities. ISME J 2(4): 417–428.
12. Rosewarne CP, Pettigrove V, Stokes HW, Parsons YM (2010) Class 1 integrons in benthic communities: abundance, association with Tn402-like transposition modules and evidence for coselection with heavy-metal resistance. FEMS Microbiol Ecol 72(1): 35–46.
13. Gaze WH, Zhang L, Abdouslam NA, Hawkey PM, Calvo-Bado L, et al. (2011) Impacts of anthropogenic activity on the ecology of class 1 integrons and integron-associated genes in the environment. ISME J 5(8): 1253–1261.
14. Michael CA, Gillings MR, Holmes AJ, Hughes L, Andrew NR, et al. (2004) Mobile gene cassettes: a fundamental resource for bacterial evolution. Am Nat 164(1): 1–12.
15. Koenig JE, Boucher Y, Charlebois RL, Nesbø C, Zhaxybayeva O, et al. (2008) Integron-associated gene cassettes in Halifax Harbour: assessment of a mobile gene pool in marine sediments. Environ Microbiol 10(4): 1024–1038.
16. Jessup CM, Kassen R, Forde SE, Kerr B, Buckling A, et al. (2004) Big questions, small worlds: microbial models systems in ecology. Trends Ecol Evol 19(4): 189–197.
17. Schmitt H, Stoob K, Hamscher G, Smit E, Seinen W (2006) Tetracyclines and tetracycline resistance in agricultural soils: microcosm and field studies. Microb Ecol 51(3): 267–276.
18. Bihn CT, Heuer H, Gomes NC, Kotzerke A, Fulle M, et al. (2007) Short-term effects of amoxicillin on bacterial communities in manured soil. FEMS Microbiol Ecol 62(3): 290–302.
19. Merlin C, Bonot S, Courtois S, Block JC (2011) Persistence and dissemination of the multiple-antibiotic-resistance plasmid pB10 in the microbial communities of wastewater sludge microcosms. Water Res 45(9): 2897–2905.
20. Renberg I (1991) The HON-Kajak sediment corer. J Paleolimnol 6(2): 167–170.
21. OECD (2002) Aerobic and anaerobic transformation in aquatic sediment systems; Guideline for Testing of Chemicals No. 308. OECD: Paris.
22. Berglund B, Khan GA, Weisner SEB, Ehde PM, Fick J, et al. (2014) Efficient removal of antibiotics in surface-flow constructed wetlands, with no observed impact on antibiotic resistance genes. Sci Tot Environ 476–477: 29–37.
23. Chander Y, Kumar K, Goyal SM, Gupta SC (2005) Antibacterial activity of soil-bound antibiotics. J Environ Qual 34(6): 1952–1957.
24. Börjesson S, Mattsson A, Lindgren PE (2010) Genes encoding tetracycline resistance in a full-scale municipal wastewater treatment plant investigated during one year. J Water Health 8(2): 247–256.
25. Söderblom T, Aspevall O, Erntell M, Hedin G, Heimer D, et al. (2010) Alarming spread of vancomycin resistant enterococci in Sweden since 2007. Euro Surveill 15(29). pii: 19620.
26. D'Costa VM, King CE, Kalan L, Morar M, Sung WW, et al. (2011) Antibiotic resistance is ancient. Nature 477(7365): 457–461.
27. Bhullar K, Waglechner N, Pawlowski A, Koteva K, Banks ED, et al. (2012) Antibiotic resistance is prevalent in an isolated cave microbiome. PLoS One 7(4): e34953.
28. Eng RH, Padberg FT, Smith SM, Tan EN, Cherubin CE (1991) Bactericidal effects of antibiotics on slowly growing and nongrowing bacteria. Antimicrob Agents Chemother 35(9): 1824–1828.

8

Factors Influencing Bank Geomorphology and Erosion of the Haw River, a High Order River in North Carolina, since European Settlement

Janet Macfall*, Paul Robinette, David Welch

Center for Environmental Studies, Elon University, Elon, North Carolina, United States of America

Abstract

The Haw River, a high order river in the southeastern United States, is characterized by severe bank erosion and geomorphic change from historical conditions of clear waters and connected floodplains. In 2014 it was named one of the 10 most threatened rivers in the United States by American Rivers. Like many developed areas, the region has a history of disturbance including extensive upland soil loss from agriculture, dams, and upstream urbanization. The primary objective of this study was to identify the mechanisms controlling channel form and erosion of the Haw River. Field measurements including bank height, bankfull height, bank angle, root depth and density, riparian land cover and slope, surface protection, river width, and bank retreat were collected at 87 sites along 43.5 km of river. A Bank Erosion Hazard Index (BEHI) was calculated for each study site. Mean bank height was 11.8 m, mean width was 84.3 m, and bank retreat for 2005/2007-2011/2013 was 2.3 m. The greatest bank heights, BEHI values, and bank retreat were adjacent to riparian areas with low slope (<2). This is in contrast to previous studies which identify high slope as a risk factor for erosion. Most of the soils in low slope riparian areas were alluvial, suggesting sediment deposition from upland row crop agriculture and/or flooding. Bank retreat was not correlated to bank heights or BEHI values. Historical dams (1.2–3 m height) were not a significant factor. Erosion of the Haw River in the study section of the river (25% of the river length) contributed 205,320 m^3 of sediment and 3759 kg of P annually. Concentration of suspended solids in the river increased with discharge. In conclusion, the Haw River is an unstable system, with river bank erosion and geomodification potential influenced by riparian slope and varied flows.

Editor: Xiaoyan Yang, Chinese Academy of Sciences, China

Funding: This work was supported by the North Carolina Clean Water Management Trust Fund (JM). Haw River Assembly was the grant recipient, with funds passed to JM through Elon University to financially support all of this research. The funders had no role in study design, data collection and analysis, decision to publish, or preparation of the manuscript.

Competing Interests: The authors have declared that no competing interests exist.

* Email: macfallj@elon.edu

Introduction

Streams and rivers are considered to be in a state of dynamic equilibrium when the sediment delivered to the channel is in balance with the capacity of the stream to transport and discharge that sediment [1]. Stream channels alternatively experience periods of alluvial deposition, followed by erosional downcutting of the alluvium, followed by periods of additional deposition. These cycles have created a landscape of terraces and floodplains, sculpted by the streams and rivers flowing through them [2]. Globally, changes in land use, climate and other factors have altered the historic patterns of transport and discharge, with significant changes to river shape, processes, sediment dynamics and water quality [2]. With these changes, soil erosion has been identified as a significant challenge in both developing and developed countries.

A landscape perspective of rivers and their watersheds demonstrates the influence of land use and disturbance on river structure and ecology at multiple scales [5]. European settlement in the southeastern United States began a period of forest clearing in the 1700's, followed by row crop agriculture [3]. These practices had deleterious ecological consequences to surface waters in the form of increased sediment loads and habitat degradation [3,4,6–9].

Before urbanization and agricultural clearing, streams in the Piedmont region of the southeastern United States, had low levels of suspended solids and high connectivity between the stream and surrounding floodplains [3,4]. It has been estimated that 25 km^3 of soil have eroded from agricultural lands in the Piedmont region between the coastal plain and the Appalachian Mountains, with an average of 14 cm of topsoil lost from North Carolina since the early 1700's [3]. This erosion from agriculture has left a legacy of upland gullies and sediment deposition near and in streams and rivers [3].

Another anthropogenic disturbance to streams and rivers is the proliferation of dams and artificial water bodies. There are over 2 million artificial surface water impoundments including 82,000 dams in the continental United States, with some dating from the 17th century [10,11]. Both natural and man-made dams can have profound impacts on the ecology and geomorphology of rivers, altering patterns of sediment transport and deposition, water and energy flow, and aquatic habitat [10,12,13]. Downstream channel degradation due to dams has been documented for more than 85 years and in many cases has been extreme [14–17]. Downstream changes often include channel incision, channel pattern change (braided to single-thread or vice-versa), loss or encroachment of vegetation, and bank collapse [13,18,19]. Upstream of the dam there may be sediment deposition within the impoundment, leading to incision when the dam is removed [20].

Upstream urbanization, with the increase in impervious surface, has been shown to alter the flow of streams, increasing the frequency of "flashiness" and bank incision [2]. Susceptibility to erosion and hydromodification from urbanization varies with stream bed composition, armoring, bank height, bed and bank materials, precipitation patterns, and other factors [21,22].

Sediment loading from bank erosion is a land management problem of global importance [23,24]. Sediment is one of the most common pollutants from non-point sources, with over 6,000 water bodies across the United States showing significant suspended sediments [25–27]. Different bank materials, aerial and subaerial weathering, variations in grain size, shear strength of the bank materials, bank angle, and water potential can influence river bank mass wasting, failure and fluvial entrainment [23–25]. The ability to predict bank failure and erosion, however, is often uncertain, especially for stream banks with a varied depositional history such as from anthropogenic disturbance [26].

Processes that determine channel geomorphology differ between first and second order streams and larger rivers with larger watersheds [27–29]. Few high order rivers have been studied [27,30]. In large rivers, conditions leading to river widening often are nonlinear, with energy adjustment resulting in different and sometimes opposite adjustment processes. This was seen in the North Fork Toutle River system in Washington State in the NW United States, where following the eruption of Mount St. Helens, one river was dominated by aggradation and widening, while another similar river was dominated by degradation [31]. Unstable channels continue to adjust following major disturbances, both anthropogenic and natural, until a stable floodplain is established with a progressive armoring of the channel bed [32,33].

Sediment and nutrient loading are a global water quality concern, and are significant issues in the Haw River, a high order river in the North Carolina Piedmont. Jordan Reservoir, a major drinking water supply, is formed by a dam on the Haw [34]. However, water in the reservoir is considered impaired due to algal blooms from excess nutrients. To improve water quality, nutrient reduction goals have been established. Understanding the patterns of geomorphic change of the river and contributions of the river bank to sediment and nutrient load may provide a model for water quality improvement in this and similar systems.

Similarly to other developing areas, the Haw River watershed has a history of profound disturbance through forest conversion to row crop agriculture, the construction of dams, and upstream urbanization. In 2014, it was identified as the 9th most threatened river in the United States by the organization American Rivers. The river today has little resemblance to the clear water and low banks from historical descriptions [35].

The primary objective of this study was to identify factors influencing bank geomorphic change and erosion of the Haw River using field measurements. Few reports on erosion of high order rivers have been published, with most based on model estimations. River traits included river slope, riparian soil type and slope, land cover, bank angle, surface protection, bank height, bankfull height and root density and depth. The Haw River has highly incised, unstable banks exhibiting extensive mass wasting, undercutting with bank collapse and fluvial entrainment. Understanding alluvial channel behavior and the channel response to disturbance will provide insight into understanding the factors controlling erosion patterns, shape and balance of the Haw and other high order rivers.

Materials and Methods

For study sites that were located on public lands, the field sites were located in local parks (town parks). Local government agencies were project partners and did not require permits for access. For sites that were located on private land, we obtained landowner permission for access. However, most landowners in our region are reluctant to allow access to their lands by strangers and requests for access may result in alienation of landowners. This research is being used as the foundation for development of the Haw River Trail - a project which uses recreation to achieve regional conservation goals. Because of the trust and contacts established with the research project described in the submitted paper, landowners have been willing to work with us and with local governments on this conservation/recreation project. Alienation of landowners would compromise progress on the conservation work we are developing on the Haw. When we made contact for this study, some of the landowners requested that their identities not be made public. Landowners highly value the personal connections that were established with this project, which is contributing to the success of the conservation work. However, landowners in this region also highly value their privacy and private property rights.

We will share an excel spreadsheet with the GPS coordinates through requests to the senior author. The authors will assist landowner contacts if requested.

The Haw River is located in the north central Piedmont region of North Carolina (Figure 1). This area is located between the coastal plain to the east, and the Appalachian Mountains to the west. North Carolina borders the Atlantic Ocean in the southeastern United States. The river is approximately 177 km long with a watershed of about 3952 km^2, including agricultural, forested, urban and suburban land cover. Nearly one million people live within the Haw River watershed, including the urbanized areas of Greensboro, Graham, Elon, and Burlington, North Carolina which are all upstream of the study reach.

The Haw River ends at the confluence with the Deep River near Moncure, North Carolina, forming the headwaters of the Cape Fear River. For most of its length, including the section studied, the river is relatively straight with few meanders. The watershed is characterized by low, gently rolling hills. Low naturally formed levees (typically $\leq$1.5 m in height) are frequently found on the river bank. Behind the river bank is a backswamp of varied width. Behind the backswamp is the toe slope, extending to the upland areas. The river channel bottom ranges from bedrock to mixed sand/silt/cobbles with large woody debris. The river is a riffle – pool system [36]. Mean discharge at the United States Geological Survey monitoring station near Bynum, North Carolina (1973–2012) immediately downstream of the study area was 34 m^3/s, with discharge ranging from 1642 m^3/s (1996) to 0.005 m^3/s (1983) [37]. Elevation ranged from 97m at the lower end of the study area to about 304 m at the top of the watershed upriver of the study section.

The study area extended 43.5 km, from the intersection of the Haw River and Interstate 40/85 in Burlington, North Carolina to the intersection of the river and US 15-501 in Bynum, North Carolina, flowing through the counties of Alamance, Orange, and

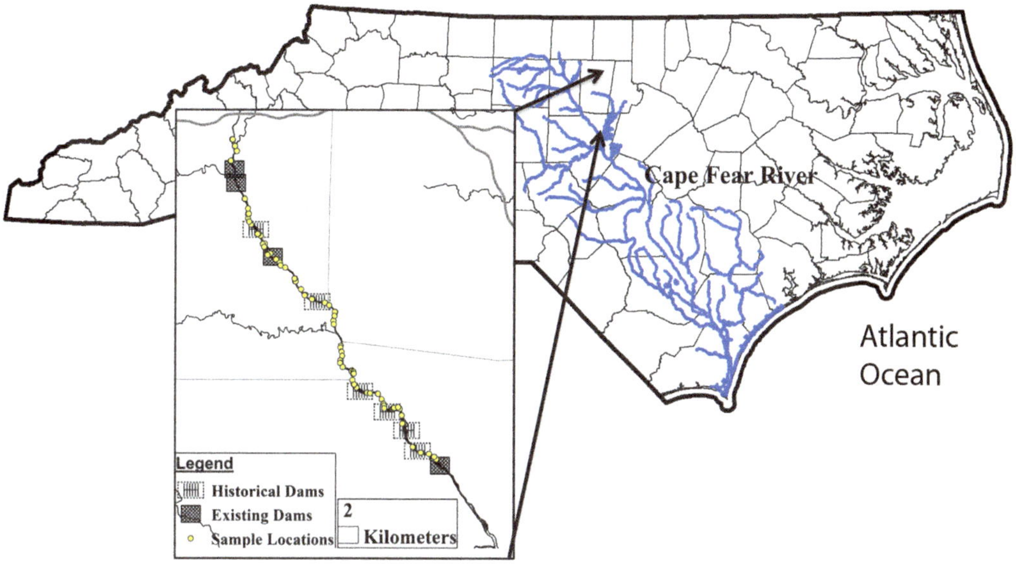

Figure 1. Location of the study area on the Haw River.

Chatham. The upriver location beginning the study area was approximately 112 km downstream from the headwaters and 65 km upstream from the confluence of the Haw River and the Deep River. The studied portion of the river is a fourth order river and higher [38]. Data were collected from 87 bank sites within the study area. Sites were accessed by kayak and land. (Figure 1).

An opportunistic sampling strategy for study site selection which was contingent upon landowner permission was used. Most study sites were on privately owned land. Study sites that were on public land were located in local parks where the local governments were project partners and no permits were required. Eighty-seven river sections that had more than 6 m in length of relatively homogenous slope, bank angle, bank height, and vegetation cover were selected for study. There were no endangered or protected species present at any of the study sites. Measurements were collected during 2006–2007, primarily during spring, summer and fall.

The geographic locations of the bank sites were documented using a Magellan handheld GPS. The lengths of the banks with homogenous characteristics were measured using a Leica Laser Distometer and heights were determined using a telescoping measuring rod and the distometer. The sites were also photographed and sketched to record the appearance of toppled trees and bare soil.

Bank Erosion Hazard Index Methods

A Bank Erosion Hazard Index (BEHI) was used to assess the potential stability of the Haw River's banks [39,40]. This method estimates and predicts erosion potential of rivers and streams, providing documentation of geomorphic field parameters and channel adjustment predictions. While the BEHI methodology is often a small component of more extensive river geomorphic assessments, it is a convenient and useful tool to rapidly create an inventory of river bank characteristics and to assess river bank and river adjustments. Rosgen's BEHI method is normally used to assess lower order rivers and is not often used on large rivers, such as the Haw River [39].

The BEHI is based on five measurements related to bank erosion potential (Table 1). Measurements were conducted according to methods outlined in Rosgen [39,40]. The bank height was measured at the top of the vertical or sloped area rising from the channel bed. Rooting depth was measured across the bank surface. Root density was determined from a visual estimation of the surface coverage across the bank surface. Surface protection was determined from the percentage of the bank with root, vegetation or hardened structure coverage. From these field measures, bank height ratio (bank height/bankfull height) and root depth ratio (root depth/bank height) were calculated [39].

Table 1 lists criteria used for calculation of the BEHI measurements (Table 1). In the Very Low erosion risk category, a value of between 1–2 is assigned. For example in a low risk stream, the Bank Height Ratio is low, there is extensive rooting in the bank (80–100% of the bank), there is high root density, low bank angle, and substantial surface protection, Summing the BEHI index values for the physical and biological traits gives a total of 5–10, indicating a low potential for erosion. In contrast, attributes that would suggest extreme risk of erosion (the bottom row) include high banks, few roots extending down the bank with low root density, an undercut riverbank with high angle, and little surface protection.

A BEHI value was calculated at each studied site on the river based on Rosgen's BEHI method [40]. A value of 1 (very low) to 10 (extreme) was assigned to each of the bank metrics, as indicated in Table 1. Numbers for all traits were summed for each site. A total value of <10 was considered a very low bank erosion hazard index. A value$>$ 45 was considered an extreme bank erosion hazard index.

Identification of the bankfull stage can be difficult to determine for rivers with an unstable channel. Rather than presenting a stable state, rivers are open systems which respond to variations in energy and materials, making bankfull assessment sometimes uncertain [31]. In cases where the bank height exceeded the bankfull height, the bankfull height was measured at the place on the river bank with visible vegetation changes indicating plants experienced flooding, such as a change in root morphology or root washout [26,27,41]. Where roots were not present, the bankfull height was measured from the point on the bank where bank

Table 1. River bank erosion metric ranking scores for calculation of the BEHI.

Category		Bank Ht. Ratio (m/m)	Root Depth Ratio (%)	Root Density (%)	Bank Angle (Degrees)	Surface Protection (%)	Total Index
Very Low	Value	1.0–1.1	100–80	100–80	0–20	100–90	
	Index	1–2	1–2	1–2	1–2	1–2	≤10
Low	Value	1.1–1.2	80–55	80–55	20–60	90–50	
	Index	2–4	2–4	2–4	2–4	2–4	10–20
Moderate	Value	1.2–1.5	55–30	55–30	60–80	50–30	
	Index	4–6	4–6	4–6	4–6	4–6	20–30
High	Value	1.5–2.0	30–15	30–15	80–90	30–15	
	Index	6–8	6–8	6–8	6–8	6–8	30–40
Very High	Value	2.0–2.8	15–5	15–5	90–120	15–5	
	Index	8–9	8–9	8–9	8–9	8–9	40–45
Extreme	Value	>2.8	<5	<5	>120	<5	
	Index	10	10	10	10	10	>45

materials showed evidence of saturation or shear force stress from flowing water.

Soils were sampled for nutrient analysis by bulk density sampler from the top 20 cm of the river bank at 31 locations throughout the study area. Duff was removed and soils were air dried and analyzed for bulk density and nutrient content by the NC Department of Agriculture and Consumer Services.

River Width and Slope Calculation Methods

The river width was measured using a geographic information system (ArcGIS, ESRI, Inc.) to analyze aerial photographs of the study area. Aerial photographs for Alamance, Orange (2005) and Chatham counties (2007) were obtained from the county GIS offices. Aerial photographs for Alamance and Orange (2011) were from DigitalGlobe (supplied through ESRI, Inc., Redlands, CA) and for Chatham Co. (2013) through NCOneMap. Widths were determined from the aerial photographs by measuring a line placed perpendicular to the center line of the river. The river width measurements were taken of the visible water surface, representing the river width at base flow. The river was at base flow at all dates of image acquisition. Digital resolution was 0.27 m. Locations of dams which are still present in the river were recorded from these aerial photographs. To confirm consistency of measurements, lengths of 10 hardened structures (dams, bridges, buildings) were measured in 2005/2007 and 2011/2013 images. There were no differences in measurements between the two image sets for hardened structures.

The slope of the river was estimated from Digital Elevation Models obtained from the NC Floodplain Mapping Program, 2013. Digital vertical resolution was 0.25 m and horizontal resolution was 6 m. A center line was established on the river image and points were placed every 500 meters. The river slope was measured for the entire length of the study section of the river.

Historical Dam Locations

The locations of historical dams, which are no longer present on the river, were determined by comparing the GIS data layers with photographs and topographic maps. Locations of historical dams were estimated from U.S. Geological Survey topographic quadrangles (scale 1:24,000) and historical records [42–44](Figure 1).

Soil and Land Cover Methods

Aerial photographs from 2005/2007 were used to digitize different types of land cover within 153 meters (500 ft.) of both of the river banks for evaluation of the riparian corridor over the entire river length studied. Within the GIS, polygons were drawn around each type of designated land cover. The areas for these polygons were then calculated and summed for each land cover category. The land cover classes were:

Forest – areas with evident canopy coverage of mature trees

Open - areas lacking trees or shrubs

Shrubland – areas dominated with small-canopied vegetation

Impervious – areas such as roads and buildings

Water – streams, ponds and other water features

Soil type and traits at each study site were determined from the Alamance, Orange and Chatham Co. Soil Survey. Soil maps (SSURGO) were obtained from the Natural Resource Conservation Service Web Soil Survey [45]. Locations for each study site were identified in the Soil Surveys using GIS. Soil types were determined for soils immediately adjacent to each study location, using the GIS shapefiles for soils in each county. Slopes which are characteristic for each soil type adjacent to the study sites were recorded based on the SSURGO soil type descriptions.

Statistical Analysis

Statistical analyses were conducted with SAS Enterprise Guide 4.3 (SAS Institute, Cary, NC). A Pearson product moment correlation analysis was calculated to determine independence of the river attributes and erosion potential. Analysis of variance or a Whitney Mann Rank Sum Test was calculated to determine effect of riparian slope and historical dams. Data used in these analyses is freely available at http://www.elon.edu/e-web/academics/elon_college/environmental_studies/macfall-plos-one.xhtml.

Results

Bank Erosion Hazard Index

A majority, 84% percent, of the studied banks had a BEHI value of moderate to high erosion potential (Table 2). The mean bank height was 11.8±0.5 m and the mean bankfull height was 5±0.1 m. The mean bank angle was 53°, with a maximum of 90°. The river had widened (bank retreat) by 2.3 m over the six year period from 2005/2007 to 2011/2013, suggesting rapid change to the river bank and ongoing erosion (Table 2).

River Width and Slope

The slope of the river varied throughout the study area, with some sections showing little drop and others a much steeper gradient over a short run (Figure 2). Upstream of the Saxapahaw Dam (9.1 m in height), the highest dam in the river segment studied, the river was 300 m wide, its widest point in the study segment (Table 2).

A number of significant correlations between the river bank geomorphology and erosion potential with attributes of the studied sites were noted (Table 3). There was a significant negative correlation between river width with bank height and BEHI, suggesting that as the river widens, the bank height and BEHI decrease.

One of the most significant results was that the slope of the riparian soils adjacent to the river bank was negatively correlated with bank height, BEHI and bank retreat. The BEHI, bank retreat and the bank heights were significantly higher at study sites with low riparian slope (<2%) compared to riparian areas with greater slopes (Table 4), suggesting greater erosion and erosion potential in areas with low riparian slope. The flatter the land adjacent to the river banks, the more the river had widened, the greater the height of the river banks, the greater the erosion potential (BEHI) and the more erosion had occurred through river widening.

Rooting patterns were negatively correlated with bank heights. There are likely two reasons for this observation. First is the protective effect from erosion which has been described from vegetation. A second reason, however, is if bank collapse or slumping had occurred, trees on the river bank would have also moved downward with the soil, increasing both the root depth ratio and root density lower in the river bank face.

In contrast to the correlation with bank height, root density and root depth ratio were not correlated with bank retreat. Areas with measureable erosion and river widening were also independent of river physical attributes including bank height, bank angle and the BEHI (Table 3).

Based on river bank height and bank retreat measurements, the amount of soil lost through erosion can be estimated for the six year period between 2005/2007 and 2011/2013. The mean value for annual soil loss from the 43.5 km section of the river which was studied was 205,320±23,000 m^3 per year. The bulk density measurement of the riparian soil adjacent to the river was 1.01. Based on this bulk density measure, $2.3 * 10^8$ kg of soil are lost annually from the studied river segment. The P concentration from soil nutrient analyses was measured to be 17.6±1.0 mg/L. Since P generally does not leach through the soil profile and is primarily retained in the surface, if P loading to the river with erosion is only from the top 20 cm of soil; approximately 3759 kg (8286 lb.) of P enters the river annually with sediment [46].

Historical Dams

Historically, there have been ten dams along the river reach studied, three of which still exist. The only two dams still providing hydroelectric power are the Saxapahaw Dam and the Bynum Dam (Table 5). There were low BEHI values and bank heights upriver of and closest to the Bynum and Saxapahaw Dams, as the sites were adjacent to upper reaches of the pools formed behind the dams. BEHI values and bank heights indicating severe erosion were measured downstream of the Saxapahaw dam. No measurements were made downstream of the Bynum dam, the downstream limit of this study.

Dams which are no longer present on the Haw River had little effect on river bank height, BEHI estimates of erosion potential or bank retreat. While it is generally accepted that sediment is deposited upstream of dams in the impoundments, the presence of historical dams on the Haw River does not appear to be a strong factor influencing current conditions (Table 6). In addition, the heights of the dams (with the exception of Saxapahaw Dam) were all lower than the mean values for bank heights measured in the current study. Impacts will likely be much greater with the remaining dams in Saxapahaw and Bynum, however, which are larger.

Riparian Land Cover and Soil Types

Most of the riparian areas adjacent to the study sites were forested with mature trees, primarily hardwoods, at the time of this study (Table 7). There was no relationship between BEHI values or bank heights with land cover in 2005/2007, as most of the riparian lands were forested. However, much of the land had been in agriculture for the past two centuries, so current geomorphic patterns may be a legacy from past land use.

Agriculture was a major land use for the first half of the past century in the 3 study area counties within the watershed (Table 8). Both farm number and percentage of agricultural land decreased, with the greatest loss in the last half of the twentieth century. Counties in the Haw River watershed had 80–85% of the land area in agriculture at the beginning of the twentieth century. The 2007 agricultural census indicates this area had shrunk to 24–32% of each county [47–49].

There were twenty one different soil types at the 87 BEHI study sites (Table 9). The most common soil type was Riverview Silt Loam, followed by Buncomb loamy fine sand.

Discussion

The banks of the Haw River, located in the central Piedmont of North Carolina, are deeply incised (11.8 m mean bank height), with steep banks in the straight reaches, inside bends, and outside bends of the river channel. There is overhanging vegetation at the top of banks, with mass wasting, bank failures, and collapse of trees into the river. The high banks, BEHI indices and bank retreat indicate active erosion is occurring and the potential for future erosion is high. Over half of the study sites had a BEHI of moderate to high erosion potential, supporting observations that the banks along the Haw River in the study area are unstable. The average difference between bank height and bankfull height was 6.6 meters.

Table 2. Summary statistics for bank characteristics of the Haw River.

Attribute	Mean	SE	Min	Max	Median
River Width 2005/2007 (m)	84.3	5.7	27.4	300.5	64.8
River Width 2011/2013 (m)	86.7	5.6	27.4	300.2	68.6
Bank retreat (m)	2.3	0.3	−1.4	9.5	1.5
Riparian slope (%)	6.6	0.9	1	30	1
Bank angle	53.2	2.5	3	90	55
Elevation (m)	122.7	1.4	97	145.7	120.4
River bed slope (%)	0.08	0.01	0	0.37	0
Bank height (m)	11.8	0.5	1.8	29.8	12.1
Bankfull height (m)	5.0	0.1	1.8	9.2	5
Bank height ratio	2.4	0.09	1	4.2	2.4
Root depth ratio	0.59	0.04	0	1.0	0.6
Surface protection (%)	35	2.4	4.3	83.4	30
Root density (%)	43	2	0	100	45
BEHI	24.3	0.7	9.1	39.6	24.1
BEHI Category	**Very low**	**Low**	**Moderate**	**High**	**Very High**
Number of Sites	1	13	58	15	0
Percent of Sites	1.2	14.9	66.7	17.2	0
Total sites	87				

The slopes of the riparian areas were significantly correlated with the BEHI, bank height and bank retreat, primarily through an increase in bank height, BEHI and bank retreat as the slope of the riparian areas decreased (Tables 3 and 4). This suggests that river banks adjacent to riparian areas with low slope are more eroded and erodible than river reaches with more steeply sloped riparian areas.

The riparian areas with low slope are mostly alluvial (Table 9), likely including migrating soils from the surrounding agricultural uplands and deposited sediment from past floods. Gross floodplain sediment trapping potential has been shown to be a function of floodplain area, with larger floodplains having greater trapping potential [50]. If sediment migrating from the uplands was deposited in riparian areas with low slope, the deposited sediment would likely be more erodible than the original base. Riparian areas with high slope adjacent to the river would be less likely to have extensive sediment deposition as the waters carrying the sediment have comparatively more energy than areas which were more flattened.

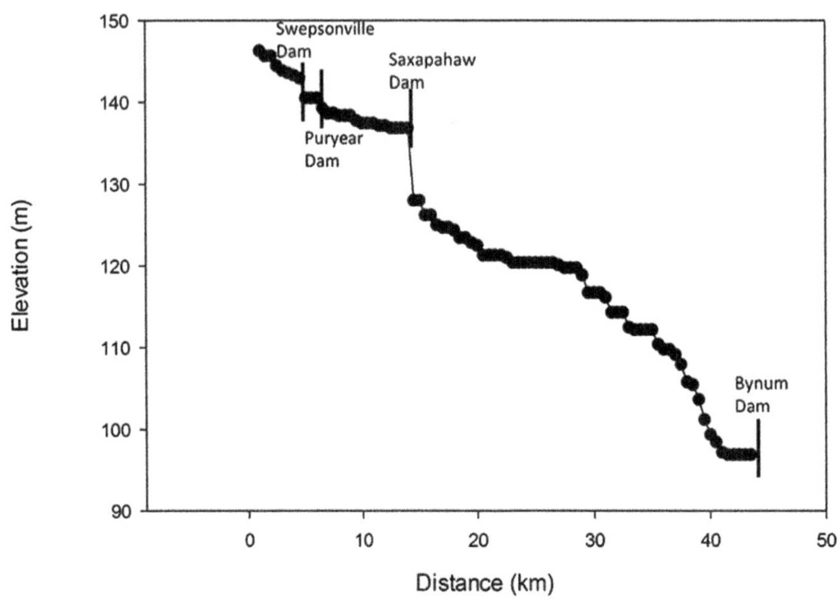

Figure 2. Longitudinal profile of the Haw River study site.

Table 3. Pearson Product-Moment Correlation Analysis of Erosion Factors on the Haw River.

Attribute		Bank height (m)	Bankfull height (m)	BEHI	Bank Retreat (m)
Bank angle	r	−0.04	−0.04		0.11
	p	0.7	0.7		0.29
Bank height (m)	r		**0.55**		−0.04
	p		**<0.001**		0.7
Bank height ratio	r	**0.76**	−0.03		−0.07
	p	**<0.001**	0.76		0.49
Bank retreat (m)	r	−0.04	0.10	0.11	
	p	0.7	0.38	0.34	
River width 2005/2007 (m)	r	**−0.68**	**−0.48**	**−0.40**	−0.09
	p	**<0.001**	**<0.001**	**<0.001**	0.37
Riparian slope (%)	r	**−0.341**	−0.15	**−0.50**	**−0.25**
	p	**<0.001**	0.16	**<0.001**	**0.02**
River bed slope (%)	r	0.02	**0.21**	−0.04	0.01
	p	0.9	**0.05**	0.73	0.9
Surface protection (%)	r	**−0.19**	−0.10		−0.002
	p	**0.08**	0.35		0.98
Root density (%)	r	**−0.36**	−0.17		0.004
	p	**<0.001**	0.12		0.96
Root depth ratio	r	**−0.29**	**−0.26**		−0.16
	p	**0.006**	**0.02**		0.14

Riparian sediment deposition from agriculture on the Haw is consistent with descriptions of sediment deposition layers between 1 and 6 m in depth which have been found adjacent to other streams of the eastern United States. A history of row crop agriculture has caused a loss of 14 cm of topsoil from the North Carolina Piedmont, with sediments migrating from the upland fields to riparian areas and associated rivers and streams [3,4]. These sediment deposit layers are likely an outcome of land clearing for development, agriculture within the watershed, deposition behind dams, and deposits from past flood events, showing the potential for substantial sediment migration to storage areas along riparian river borders [17]. Recent and continued sediment deposition from current agriculture into riparian areas has been documented in the upper Midwest of the United States [51], similarly to patterns on the Haw River.

The riparian areas were mostly forested at the time of the study, with root density and root depth ratios negatively correlated with bank height, but not with bank retreat. Although mature trees were present in the riparian zone throughout the length of the study site, roots rarely extended to the base of the bank. Root

Table 4. Analysis of Variance comparing bank characteristics between areas with low (<2%) and high (≥ 2%) riparian slopes.

Traits/Location	n	Median values (m)	Mean values (m)	SE	P*
Bank retreat					
Low slope	61	2.1	2.7	0.31	0.03
High slope	26	0.8	1.5	0.45	
Bank height					
Low slope	61	12.7	12.8	0.42	0.002
High slope	26	10.4	9.6	1.23	
Bank angle					
Low slope	61	60.0	56.1	2.60	0.06
High slope	26	40.0	46.1	5.35	
BEHI					
Low slope	61	26.5	26.3	0.70	<0.001
High slope	26	20.6	19.5	1.08	

*Mann Whitney Rank Sum Test was used for Bank retreat and Bank height analyses. ANOVA was used for BEHI and Bank Angle analyses.

Table 5. Historical and current dams on the Haw River listed in order on the river.

Dam Name	Dam Height (m)	Years
Virginia Falls Dam	3.0	1874- present, removed 2013
Puryear Dam	2.4	1763 - present
Cedar Cliffs Dam	1.5	1860–1910
Saxapahaw Dam	9.1	1938 - present
Dark's Dam	-	1790–1875
Elliot's Falls	1.2	1778–1810
Love's Dam	-	1790–1920
Pace's Dam	2.4	1789–1924
Burnett-Powell Dam	-	1776–1880
Bynum Dam	3.0	1874 - Present

density in river banks has been shown to reduce scour through both mechanical root reinforcement and matric suction [52]. However, on the Haw, the bank height frequently exceeded rooting depth (Table 2). The absence of roots at the base of the river bank would mean that roots are not present to provide reinforcement and protection from shear. In contrast to the protective effect of roots on bank height, the lack of correlation with bank retreat suggests vegetation provided no protection from erosion of the bank toe slope at the river's edge, a significant concern in river management [53].

The influence of dams on river geomorphology has been well documented [10,12,13]. In the present study, dams were lower than the heights of banks measured in the study area and the eroded bank areas extended well upriver beyond dam impoundments. The highest dam currently on the Haw is the Saxapahaw Dam, with a height of 9.1 m. The mean height for all dams in the studied segment was 2.9 m, including both historical and present dams. Yet the mean bankfull measurement was 5 m, and the mean bank height was 11.8 m, higher than all the dams. If bank erosion was primarily through legacy sediments from dam impoundments, it would be expected that the bankfull and bank heights would not exceed the heights of the dams. The Haw River bed is rocky, lined with rocks from 0.2 m diameter to large boulders and bedrock, so the contribution to bank heights from erosion of the river bed is likely to be minimal. Comparison of the bank heights, BEHI values and bank retreat upstream of the legacy dam sites with other study sites showed no difference between locations (Table 6). Although low dams can have significant ecological effects in many cases, the impact to higher order rivers such as the Haw appears to be small.

The lack of a correlation between bank height and the BEHI with bank retreat suggests that different but related processes are occurring. Bank retreat often occurs following bank failure and collapse, with the river bank collapsing downward towards the water's edge then eroding away as the river flows past. River bank stability can be highly variable, with many factors contributing to the structural integrity. The negative correlation between bank height and BEHI with river width suggests that many of the tall banks are stable, reflective of the wide range of conditions present in the river corridor. However, both high banks and bank retreat were correlated with low riparian slope, suggesting this condition increases risk from both patterns of erosion [54].

Geomorphic patterns of the Haw River are consistent with conceptual models describing changes in river geomorphology following disturbance such as the removal of a small dam. Following dam removal, the sequence would be: a) lowered water surface, b) degradation, c) degradation and widening d) aggradation and widening, ending with e) quasi-equilibrium [32,55,56]. This process usually happens quickly after removal of small dams, reaching quasi-equilibrium within a few years. A similar sequence of geomorphological changes following disturbance seems to be

Table 6. Effect of historical dams on river bank geomorphology.

Traits/location	n	Median values (m)	Mean values (m)	SE	p
Bank retreat					
Behind dam	8	1.8	2.3	0.87	0.94
Not behind dam	79	1.5	2.2	0.28	
Bank height					
Behind dam	8	11.7	11.4	1.5	0.83
Not behind dam	79	12.1	11.9	0.52	
BEHI					
Behind dam	8	23.2	24.1	2.5	0.65
Not behind dam	79	24.4	27.2	0.69	

*Mann Whitney Rank Sum Test was used for Bank retreat and Bank height analyses. ANOVA was used for BEHI analysis.

Table 7. Land cover within the 153 m zone adjacent to BEHI study sites.

Land Use	Percentage
Forest	78.4
Open	15.3
Shrubland	5.5
Impervious	0.8

occurring on the Haw River, with the river currently at stages b, c and d.

Disturbances on the Haw River include historical dams which are no longer in place, three dams which are currently in place, a history of row crop agriculture throughout the watershed and upstream urbanization. This is similar to the history of dams and land use on other rivers of the Piedmont and around the world [18,57]. Erosion of legacy sediments from agriculture followed by urbanization is has also been documented on a smaller stream in the Piedmont region, leaving similarly incised banks [36].

Reaching dynamic equilibrium does not always occur quickly. Following an eruption of Mount St. Helens in 1857, stable floodplains and re-vegetation of riparian zones had not yet been re-established at the time of the 1980 eruption [33]. In this case, dynamic equilibrium had not been reached over a century after the earlier disturbance. Similar patterns have been observed in other regions [5,58,59]. For the Haw River, re-connection with flood plains and equilibrium will likely take centuries or longer, if it all, compared with the decadal time scale described with dam removal.

In addition to the effects from dams and past agriculture, changes in flow can affect soil loss from erosion and sediment load. Data from a U.S.G.S. monitoring station just downstream of the study area were examined from 2005–2010, with flow and sediment measures collected on the same day [37,60]. As flow increased, so did sediment loading to the river, as reflected in the concentration of suspended solids (Figure 3). River discharge ranged from 2 m^3/sec to 659 m^3/sec, averaging 31 m^3/sec. The suspended solids in the river ranged from 1 mg/L to 187 mg/L, averaging 15 mg/L. As the energy in the river increased with flow, the concentration of suspended solids also increased exponentially, a relationship well documented in the literature [2]. As the region upstream of the study area continues to urbanize with increasing impervious surface, high flow events are likely to also increase with increasing erosion, a significant management concern.

Sediment loading from bank erosion to the Haw River will also contribute P. The analyses of riparian soil indicated approximately 3759 kg (8286 lb.) of P enters the river annually from sediment. This is about 1.5% of the total target P load from non-point sources in the Haw River flowing to Jordan Reservoir (of a targeted 106,884 kg), and potentially 6% of the total P load to the Haw River from river bank erosion (the study area was about 25% of the river length). Currently this is an unaccounted non-point source of P, with no Best Management Practices or nutrient loading targets assigned to this P source [34].

Like many other high order rivers globally, the geomorphology of the Haw River in North Carolina is changing following agricultural and other disturbances, with significant sediment and P loading to the river. The river is undergoing a process of reshaping with river widening and the formation of very high river banks. Occasional high flow events further contribute to erosion with the concentration of suspended solids in the water column increasing with discharge (Figure 3).

Future study of the Haw River and similar systems could be directed at understanding the patterns of erosion seen on this river. One major area of investigation would be to determine the origin and age of sediments deposited along the river, especially those in low slope areas which are experiencing the most erosion. Another study would be to evaluate the impacts of past land use and land cover on contemporary geomorphology, flow and erosion. Comparison with commonly used models for erosion estimates would also be valuable.

Conclusion

Few studies on the patterns of bank erosion and hydrogeomorphic change following disturbance have been published for high order rivers such as the Haw River. However, differences in geomorphology have been described between high and low order segments of rivers, with factors impacting the river attributes changing throughout the river's length. These studies show processes and attributes of low order rivers may not be applicable to larger river systems [52].

For most studies estimating erosion and erosion potential of larger rivers, a modeling approach has been used. Common methods are the Universal Soil Loss Equation and the Soil and Water Assessment Tool [61,62]. In contrast, field measurements of large systems are seldom the major approach used to identify factors influencing erosion and geomorphic change.

For the Haw River, regions of high erosion potential (as indicated by the BEHI and bank height) are negatively correlated with river width, suggesting regions high BEHI values and bank heights are have narrow river width and are relatively stable. On this river, erosion as measured by bank retreat is independent of most physical features such as bank height, BEHI, bank angle and

Table 8. History of farming within the studied counties of the Haw River watershed.

Year		Alamance	Chatham	Orange
1910	Number of farms	2508	3640	1957
	% of county in agriculture	80	85	84
1950	Number of farms	2940	2977	2038
	% of county in agriculture	79	66	70
2007	Number of farms	753	1089	604
	% of county in agriculture	32	24	24

Table 9. Soil types at BEHI study sites.

Soil ID	Soil Type	% slope	# sites
AdE	Appling, sandy loam, steep phase	20	2
Ba	Buncomb loamy fine sand, frequently flooded	1	14
BaE	Badin Nanford Complex	23	6
CbE	Cecil fine sandy loam, moderately steep phase	17	2
Cg	Congaree fine sandy loam, frequently flooded	1	2
ChA	Chewacla and Wehadkee soils, frequently flooded	1	1
Cp	Congaree fine sandy loam, frequently flooded	1	4
GaE	Georgeville silt loam, moderately steep phase	17	1
GbE3	Georgeville silt loam, severely eroded, moderately steep phase	20	1
GcE	Goldston slaty silt loam, moderately steep phase	17	5
GkE	Georgeville Badin complex	23	2
LbE	Lloyd loam, moderately steep phase	17	1
Mc	Mixed alluvial land, poorly drained	1	6
Md	Mixed alluvial land, well drained	1	6
NaD	Nanford Badin complex	5	1
RvA	Riverview silt loam, frequently flooded	1	28
TaD	Tirza silt loam, strongly sloping phase	30	1
WcE	Wilkes stony soils, moderately steep phase	17	4

bankfull height, suggesting different erosional processes are occurring. Although these physical features are often used to predict erosion potential, such as through use of the BEHI, that does not appear to be the case in this study. Independence of bank physical features with bank retreat has been observed in a few other systems [65].

Areas with low riparian slope (<2%) appear to have the highest erosion risk, experiencing high bank heights, BEHI values and bank retreat. This relationship is in contrast to some common models used for erosion estimation, which rank high slope as an erosion risk factor [61,62]. In the Haw River system, a history of extensive soil loss from upland agriculture suggests erosion of agricultural legacy sediments deposited on areas with low slope beside the river has occurred, with high erodibility. Increasing sediment concentration with flow suggests the river is still changing, reshaping as an outcome of past and present distur-

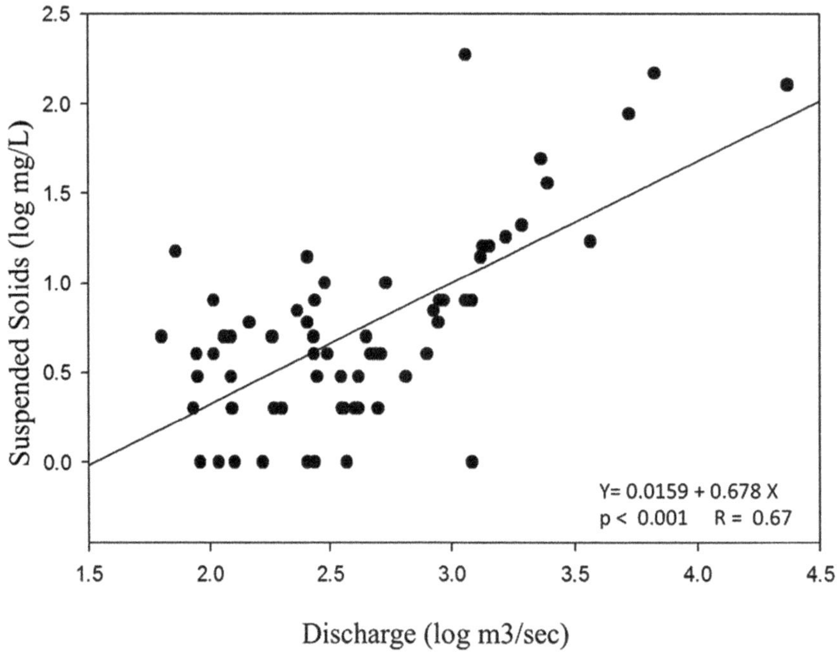

Figure 3. Relationship between log of river discharge and log of total suspended solids in the Haw River.

bance. The potential for further impairment as upstream urban development increases should be a management concern for the river health and water quality.

Rate of bank retreat and river widening is similar to the rate reported for other high order rivers. In rivers of southern Minnesota, in the Midwest of the United States, LIDAR studies have indicated widening rate of 0.57–5 m/yr since European settlement [63]. In our study, the Haw has widened 0.38 m/yr, slightly less than in Minnesota. But both studies indicate the rivers are not at equilibrium.

The Haw River is an unstable system, with river bank erosion and geomodification potential influenced by disturbance, riparian slope and episodes of high flow. The greatest erosion, measured by bank height and bank retreat occurred in regions with low riparian slope, usually with alluvial soils, suggesting erosion of deposited sediments. Historical dams were not a significant factor in influencing current conditions on the Haw River. This study provides a model for high order rivers, identifying factors driving erosion and changes to channel morphology which will help in management of these and other high order river systems.

Acknowledgments

We would like to thank Dr. Greg Jennings for his assistance and guidance throughout development of this project. We would also like to sincerely thank the reviewers of this manuscript for their thoughtful and insightful comments.

Author Contributions

Conceived and designed the experiments: JM PR DW. Performed the experiments: JM PR DW. Analyzed the data: JM PR DW. Contributed reagents/materials/analysis tools: JM PR DW. Wrote the paper: JM PR DW. Obtained landowner permission for access: JM PR DW.

References

1. Zaimes G, Emanuel R (2006) Stream processes Part I - Basic. Arizona Watershed Stewardship Guide.
2. Leopold LB (1994) A View of the River. Cambridge, Ma. Harvard University Press.
3. Trimble SW (1974) Man-induced soil erosion on the southern Piedmont, 1790–1970. Ankeny, Iowa: Soil and Water Conservation Society.
4. Trimble SW (1975) A volumetric estimate of soil eroson on the southern Piedmont. USDA Agricultural Research Service Publication S-40: 142–154.
5. Allen JD (2004) Landscapes and Riverscapes: The Influence of Land Use on Stream Ecosystems. Annual review of ecological and evolutionary systematics 35: 257–284.
6. Bennett HH (1928) The geographic relation of soil erosion to land productivity. Geographical Review 18: 579–605.
7. Jacobson RB, Coleman DJ (1986) Stratigraphy and recent evolution of Maryland Piedmont flood plains. American Journal of Science 286: 617–637.
8. Richter BD, Baumgartner JV, Powell J, Braun DP (1996) A method for assessing hydrologic alteration within ecosystems. Conservation Biology 10: 1163–1174.
9. Florsheim JL, Pellerin BA, Oh NH, Ohara N, Bachand PAM (2011) From deposition to erosion: Spatial and temporal variability of sediment sources, storage, and transport in a small agricultural watershed. Geomorphology 132: 272–286.
10. Smith SV, Renwick WH, Bartley JD, Buddenmeier RW (2002) Distribution and significance of small articial water bodies across the United States landscape. Science of the Total Environment 299: 21–36.
11. U.S. Army Corps of Engineers (2013) National Inventory of dams. http://geo.usace.army.mil/pgis/f?p=397:1:479164877021486. Accessed November, 2013
12. Beyer PJ (2005) Introduction to the special issue: Dams and geomorphology. Geomorphology 71: 1–2.
13. Walter RC, Merritts DJ (2008) Natural streams and the legacy of water-powered mills. Science 319: 299–304.
14. Lawson JM (1925) Effect of Rio Grande storage on river erosion and deposition. Engineering news Record: 327–334.
15. Williams GP, Wolman MG (1984) Downstream effects of dams on alluvial rivers. Geological Survey Professional Paper 1286: 1–83.
16. Petts G, Gurnell A (2005) Dams and geomorphology: Research progress and future directions. Geomorphology 71: 27–47.
17. Hupp CR, Schenk ER, Richter JM, Peet RK, Townsend PA (2009) Bank erosion along the dam-regulated lower Roanoke River, North Carolina. The Geological Society of America Special Paper 451: 97–108.
18. Schenk ER, Hupp CR (2009) Legacy effects of colonial millponds on floodplain sedimentation, bank erosion and channel morphology, Mid-Atlantic, USA. Journal of the American Water Resources Association 45: 596–606.
19. Ligon FK, Dietrick WE, Thrush WJ (1995) Downstream ecological effects of dams: a geomorphic perspective. BioScience 45: 183–192.
20. Doyle MW, Stanley EH, Orr CH, Selle AR, Sethi SA, et al. (2005) Stream ecosystem response to small dam removal: Lessons from the heartland. Geomorphology 71: 227–244.
21. Bledsoe BP, Stein E, Hawkley RD, Booth D (2012) Framework and tool for rapid assessment of stream susceptibility to hydromodofication. Journal of the American Water Resources Association 48: 788–808.
22. Paul MJ, Meyer JL (2001) Streams in the urban landscape. Annual review of ecological and evolutionary systematics 32: 333–365.
23. Thorne CR, American Society of Civil Engineers Task Committee on Hydraulics (1998) River width adjustment I: processes and mechanisms. Journal of Hydraulic Engineering 124: 881–902.
24. Thorne CR, American Society of Civil Engineers Task Committee on Hydraulics (1998) River width adjustment II: modeling. Journal of Hydraulic Engineering 124: 903–917.
25. Sass CK, Deane TD (2012) Application of rosgen's bancs model for NE Kansas and the development of predictive streambank erosion curves. Journal of the American Water Resources Association 48: 774–778.
26. Jennings GD, Harman WH. Measurement and stabilization of stream bank erosion in North Carolina. ASAE. In: Ascough JC, Flanagam DC, editors; 2001; Honolulu, Hawaii.
27. Rosgen D (1996) Applied River Morphology. Pagosa Springs, Colorado: Wildland Hydrology Books.
28. Abernethy B, Rutherford ID (1998) Where along a rivers length will vegetation most effectively stabilize stream banks? Geomorphology 23: 55–75.
29. Anderson RJ, Bledsoe BP, Hession WC (2004) Width of streams and rivers in response to vegetation, bank material and other factors. Journal of the American Water Resources Association 40: 1159–1172.
30. Rosgen D (1996) The use of infrared photography for the determination of sediment production. Fluvial Process and Sedimentation. Edmonton, Alberta: Canadian National Research Council. pp. 381–402.
31. Simon A, Doyle MW, Kondolf M, Shields FD, Rhoads B, et al. (2007) Critical evaluation of how the Rosgen classification and associated "natural channel design" methods fail to integrate and quantify fluvial processes and channel response. Journal of the American Water Resources Association 43: 1117–1131.
32. Doyle MW, Stanley EH, Orr CH, Selle AR, Sethi SA, et al. (2005) Stream ecosystem response to small dam removal: Lessons from the Heartland. Geomorphology 71: 227–244.
33. Simon A, Thorne CR (1996) Channel adjustment of an unstable coarse-grained stream: opposing trends of boundary and critical shear stress, and the applicability of extremal hypotheses. Earth Surface Processes and Landforms 21: 155–180.
34. General Assembly of North Carolina (2009) Jordan Water Supply Nutrient Strategy Rules. Session Law 2009-216, House Bill 239, 15A NCAC 02B . 0263 -. 0273, .0311(p).
35. Cassebaum AM (2011) Down along the Haw: A history of a North Carolina river. Raleigh, NC: McFarland and Company.
36. Hupp CR, Noe GB, Schenk ER, Bentham AJ (2013) Recent and historic sediment dynamics along Difficult Run, a suburban Virginia Piedmont Stream. Geomorphology 180–181: 156–169.
37. U.S. Geological Survey (2013) U.S.G.S. Water Information System. Available: http://waterdata.usgs.gov/nc/nwis/uv/?site_no=02096960&PARAmeter_cd=00065,00060. Accessed 2013 May 12.
38. Strahler AN (1952) Hypsometric (area altitude) analysis of erosional topology. Geological Society of America Bulletin 63: 1117–1142.
39. Rosgen DL (2001) A practical method of computing streambank erosion rate. pp. 9–15.
40. U.S. Dept. of Agriculture (2007) Rosgen Geomorphic Channel Design. Part 654 Stream Restoration Design National Engineering Handbook: Natural Resources Conservation Service. pp. 1–76.
41. Harman WH, Jennings GD, Patterson JM, Clinton DR, Slate LO (1999) Bankfull hydraulic geometry relationships for North Carolina Streams. In: Olsen DS, Potyondy JP, editors; Bozeman, Mt.
42. Spoon WL (1893) Map of Alamance County, NC. Cleveland, OH: H.B. Granahan and Co. Engravers.
43. Spoon WL (1928) Map of Alamance Co., NC. Cleveland, Ohio: H.B. Stranahan and Co. Engravers.
44. Chilton M (2008) A historical atlas of the Haw River. Carrboro, NC: Private printing.
45. U.S. Dept. of Agriculture (2014) Web Soil Survey. Natural Resource Conservation Service. Available: http://websoilsurvey.sc.egov.usda.gov/App/HomePage.htm. Accessed 2014 Apr.

46. Jobbagy EG, Jackson RB (2001) The distribution of soil nutrients with depth: Global patterns and the imprint of plants. Biogeochemistry 53: 51–77.
47. U.S. Dept. of Agriculture (1913) Thirteenth census of the United States taken in the year 1910. Agriculture 1909 and 1910, Reports by States with statistics for counties.
48. U.S. Dept. of Agriculture (1952) United States Census of Agriculture: 1950. Counts and State Economic Areas. North Carolina and South Carolina.
49. U.S. Dept. of Agriculture (2009) Census of Agriculture 2007, North Carolina State and County Data. Geographic Area Series, Part 33.
50. Schenk ER, Hupp CR, Gellis A, Noe G (2013) Developing a new stream metric for comparing stream function using a bank-floodplain sediment budget: a case study of three Piedmont streams. Earth Surface Processes and Landforms 38: 771–784.
51. Knox JC (2001) Agricultural influence on landscape sensitivity in the Upper Mississippi River Valley. Catena 42: 193–224.
52. Pollen-Bankhead N, Simon A (2010) Hydrologic and hydraulic effects of riparian root networks on streambank stability: Is mechanical root-reinforcement the whole story? Geomorphology 116: 353–362.
53. Docker BB, Hubble TCT (2008) Quantifying root-reinforcement of river bank soils by four Australian tree species. Geomorphology 100: 401–418.
54. Atkinson PM, German SE, Sear DA, Clark MJ (2003) Exploring the Relations between River bank Erosion and Geomorphological Controls using Geographically Weighted Logistic Regression. Geographical Analysis 35: 58–82.
55. Schumm A, Watson C, Harvey M (1986) Incised channels, morphology, dynamics and controls. Littleton, Co: Water Resources Publications.
56. Simon A, Hupp CR (1986) Channel evolution in modified Tennessee channels; Las Vegas, Nv. Federal Interagency Sedimentation Conference. pp. 71–82.
57. Cruise JF, Laymon CA, Al-Hamdan OZ (2010) Impact of 20 Years of Land-Cover Change on the Hydrology of Streams in the Southeastern United States1. Journal of the American Water Resources Association 46: 1159–1170.
58. Gregory S, Swanson F, McKee A, Cummins K (1991) An ecosystem perspective of riparian zones. BioScience 42: 540–551.
59. Parker C, Simon C, Thorne CR (2008) The effects of variability in bank material properties on river bank stability: Goodwin Creek, Mississippi. Geomorphology 101: 533–543.
60. (2011) Cape Fear River Basins Water Monitoring Coalition. Available: http://www.cormp.org/CFP/. Accessed 2011 Jul 11.
61. University T&M (2014) Soil and Water Assessment Tool. Available: http://swat.tamu.edu/. Accessed 2014 Jun.
62. Soupios M, Vallianaatos F (2009) Soil erosion prediction using the Revised Universal Soil Loss Equation (RUSLE) in a GIS Framework. Chania, Northwestern Crete, Greece. Environmental Geology 57: 483–497.
63. Kessler AC, Gupta SC, Brown MK (2013) Assessment of river bank erosion in Southern Minnesota rivers post European settlement. Geomorphology 201: 312–322.

9

Behavioral Response of *Corophium volutator* to Shorebird Predation in the Upper Bay of Fundy, Canada

Elizabeth C. MacDonald[1]*, Elisabeth H. Frost[1¤], Stephanie M. MacNeil[2], Diana J. Hamilton[1], Myriam A. Barbeau[2]

1 Department of Biology, Mount Allison University, Sackville, New Brunswick, Canada, **2** Department of Biology, University of New Brunswick, Fredericton, New Brunswick, Canada

Abstract

Predator avoidance is an important component of predator-prey relationships and can affect prey availability for foraging animals. Each summer, the burrow-dwelling amphipod *Corophium volutator* is heavily preyed upon by Semipalmated Sandpipers (*Calidris pusilla*) on mudflats in the upper Bay of Fundy, Canada. We conducted three complementary studies to determine if adult *C. volutator* exhibit predator avoidance behavior in the presence of sandpipers. In a field experiment, we monitored vertical distribution of *C. volutator* adults in bird exclosures and adjacent control plots before sandpipers arrived and during their stopover. We also made polymer resin casts of *C. volutator* burrows in the field throughout the summer. Finally, we simulated shorebird pecking in a lab experiment and observed *C. volutator* behavior in their burrows. *C. volutator* adults were generally distributed deeper in the sediment later in the summer (after sandpipers arrived). In August, this response was detectably stronger in areas exposed to bird predation than in bird exclosures. During peak predator abundance, many *C. volutator* adults were beyond the reach of feeding sandpipers (>1.5 cm deep). However, burrow depth did not change significantly throughout the summer. Detailed behavioral observations indicated that *C. volutator* spent more time at the bottom of their burrow when exposed to a simulated predator compared to controls. This observed redistribution suggests that *C. volutator* adults move deeper into their burrows as an anti-predator response to the presence of sandpipers. This work has implications for predators that feed on burrow-dwelling invertebrates in soft-sediment ecosystems, as density may not accurately estimate prey availability.

Editor: Jan M. Hemmi, University of Western Australia, Australia

Funding: Funding was provided by Natural Sciences and Engineering Research Council of Canada (Discovery Grants to DJH and MAB and Undergraduate Summer Research Award to EHF), the Canada Summer Jobs Program, Mprime (a Network of Centres of Excellence for the mathematical sciences), New Brunswick Wildlife Trust Fund, the Science Horizons Youth Internship Program (Environment Canada), Mount Allison Biology Department Ruggles-Gates Undergraduate Research Fellowship (to field assistant C.E. Gilroy) and Mount Allison University. The funders had no role in study design, data collection and analysis, decision to publish, or preparation of the manuscript.

Competing Interests: The authors have declared that no competing interests exist.

* Email: ecmacdonald@mta.ca

¤ Current address: Acadia Entrepreneurship Centre, Acadia University, Wolfville, Nova Scotia, Canada

Introduction

Predation is a key process contributing to the organization of ecological communities [1,2,3,4,5]. Predators have direct consumptive effects on prey populations while also eliciting strong indirect effects [6,7,8]. Prey can adopt avoidance strategies in response to predation, including biochemical, morphological and behavioral defenses [9,10,11,12,13]. Habitat structure can facilitate predator avoidance by providing refuges for prey [14,15]. Even in structurally simple environments, such as soft-sediment ecosystems, prey can take advantage of three-dimensional structure [16]. Soft-sediment invertebrates may burrow into the substratum [17,18,19], or create tubes to evade predators [20,21].

The amphipod *Corophium volutator* is abundant on intertidal mudflats in the upper Bay of Fundy, Canada [22,23]. *C. volutator* construct "U"–shaped burrows, which reduce predation risk [24] and dehydration [25,26]. Based on cursory observation, burrows typically occur in the upper 5 cm of sediment [27,28,29] but may extend 1–10 cm into the mud [30]. Most individuals remain burrowed as the tide recedes [29,31]; however, some will crawl on the surface for a short time after exposure [32,33,34].

C. volutator are vulnerable to predation by benthic fish, polychaetes, and other intertidal invertebrates [35,36,37], as well as disturbances and predation by mud snails (*Nassarius obsoletus* [*Ilyanassa obsoleta*]) [38,39]. Their greatest predation risk is concentrated during the late summer when hundreds of thousands of migrating Semipalmated Sandpipers (*Calidris pusilla*) stopover on mudflats in the upper Bay of Fundy [40]. Individual birds stay for 2–4 weeks [40,41], feeding intensively on *C. volutator*, as well as other invertebrates and biofilm [42]. Birds can consume thousands of amphipods per day [43], and typically double their mass during their short stay in this region [43,44,45].

Crawling *C. volutator* are highly vulnerable to predation and sharp declines in crawling behavior coincide with the arrival of sandpipers [24,46]. Boates et al. [46] found that surface-crawling activity by adult males fell by 97% after sandpipers arrived, but was concurrently unchanged at another mudflat that birds did not visit. They also noted, over several years, that cessation of crawling coincided directly with arrival of birds, and could not be fully explained by loss of individuals through predation. In Europe, foraging Common Redshanks (*Tringa totanus*) caused the number of crawling *C. volutator* individuals in close proximity to decline significantly, suggesting *C. volutator* may detect, and respond to, changes in substratum pressure [47]. Similarly, Boates et al. [46] observed that fewer individuals crawled on sediment visited by sandpipers than on adjacent areas where birds had not walked. This suggests that *C. volutator* responded directly to sandpipers rather than by using general seasonal cues.

Sandpipers can probe into burrows to obtain prey; however, their success is limited by bill length [48,49], which, in New Brunswick, averages 21.5 mm for females and 19.6 mm for males [50]. Thus, *C. volutator* may be able to escape predation by burrowing beyond the reach of foraging birds [51]. This avoidance mechanism, sometimes termed prey depression [52], temporarily reduces available prey [47,52,53], and has been inferred from declines in prey crawling activity [47,53] and reduced feeding rates for redshanks in areas recently occupied by conspecifics [54,55]. It has also been speculated to influence shorebird behavior on Bay of Fundy mudflats [56,57]. Coulthard and Hamilton [38] found proportionally fewer *C. volutator* adults near the sediment surface in the presence of high densities of mud snails at one mudflat, further suggesting that *C. volutator* have some capacity to adjust their position in the sediment in response to disturbances and/or predation.

We conducted three complementary studies examining behavioral responses of *C. volutator* to Semipalmated Sandpiper predation on mudflats in the Bay of Fundy. First, we designed a field experiment to test if the presence of shorebird predators affects the vertical distribution of *C. volutator* adults. We quantified their distribution in the sediment before and during the period when shorebirds were present, in plots where birds were either present or excluded. Next, we monitored *C. volutator* burrow depth by making polymer casts in the field before, during and after the period of peak shorebird abundance. Finally, we conducted a lab experiment where we simulated sandpiper pecking to test if *C. volutator* females moved deeper in their burrows in response to perceived predation risk. We expected that *C. volutator* would exhibit signs of predator avoidance in the presence of shorebirds. Specifically, we predicted they would be lower in the sediment when exposed to shorebird predation or perceived predation. This could occur through construction of deeper burrows and/or increasing time spent lower in their burrows. In summary, this study allowed us to examine the effect of shorebird predators on *C. volutator* vertical distribution in the sediment, and to determine if adult *C. volutator* exhibit predator avoidance behavior in the presence of foraging sandpipers.

Materials and Methods

Experiment 1: Predator exclusion in the field

We conducted a predator exclusion experiment, using a block design, to compare vertical distribution of *C. volutator* adults in areas with and without shorebird predators. We set up the experiment on 8 July 2007 at Pecks Cove in the Bay of Fundy, New Brunswick, Canada (45°45'N, W 64°28'W). Pecks Cove has an intertidal mudflat extending ~850 m from shore and supports migrating sandpipers each summer. Pecks Cove is not privately owned or protected, thus we did not require specific permission to sample at this site. We established 8 spatial blocks, each consisting of two treatment levels: a netted exclosure preventing shorebird predation and a control plot allowing free access to shorebirds (Figure 1). These treatment plots (caged and uncaged) were 1.6 m×1.2 m and were spaced 3 m apart within a block. These dimensions were chosen as they allowed us to sample an entire plot without disrupting the sediment within it, and previous work indicated that cages of this size were appropriate for detecting predator effects [58]. Exclosures consisted of bamboo stakes with a top made of clear plastic mesh (12.5-mm openings). The stakes were pounded 0.3 m into the mud, leaving the top of the exclosure 15 cm above the sediment. The sides of the exclosures were open, which avoided disruption of water flow [37]. Flagging tape streamers were tied to the top frame of the exclosure to help deter birds. Additional cage controls, typically consisting of mesh coverage of the top and some sides [59], were not possible or necessary because the exclosures already had open sides. We did not expect shading from the mesh top to be problematic as similar cages constructed with a finer mesh generated minimal reductions in photon flux density [37]. We marked control plots with bamboo stakes.

We regularly estimated shorebird density in a 400-m radius of the experiment within 2-h of a daytime low tide starting on 1 July, and increased the frequency of these bird counts to at least 3 times/week starting on 25 July 2007 (when shorebirds started to arrive at Pecks Cove). Additionally, beginning on 9 August, we estimated percent shorebird footprint cover in exclosures and control plots as an index of habitat use [60] ~4 times/week at low tide.

From 10 July to 19 August, we collected samples from exclosures and control plots at ~10-d intervals, totaling 5 sampling dates. In each plot, we counted crawling *C. volutator* individuals in a 0.25 m×0.25 m quadrat and recorded sediment temperatures at the surface and 5 cm depth (corresponding to the top and bottom of our sampling device) with a digital thermometer. We then collected sediment cores using a stratified sampling device divided into four horizontal layers (described in Coulthard and Hamilton [38]): 0–0.5, 0.5–1.5, 1.5–3.0, and 3.0–5.0 cm from the sediment surface. On each sampling date, we collected two replicate cores from exclosures and control plots in each spatial block immediately after mud exposure (time 0) and again 45 min later (Figure 1). Samples were not collected from a 10-cm buffer around the edge of plots, and different portions of the plot were sampled on each subsequent round. It is possible that individuals moved vertically within their burrows in response to sampling. However, the process was quick (<30 s to collect a sample and section the layers) and the disruptive nature of sampling was consistent among treatment levels and sampling dates, so any observed differences should be independent of effects due to sampling method or researcher presence.

We rinsed each layer of a core separately through a 250-µm sieve [61], and preserved the contents in 95% ethanol. We measured all *C. volutator* individuals (body length from rostrum to telson) in each layer, and grouped the adults (>4 mm body length) into two categories: 4–6 and >6 mm. Individuals <4 mm are considered juvenile and not preferred shorebird prey (data presented in Wallace [62]), while 4–6 mm amphipods are completing sexual development but are potential prey, and the fully developed >6 mm adults are considered preferred shorebird prey [22,37]. This protocol was developed in accordance with the policies of the Canadian Council on Animal Care and was approved by the Mount Allison Animal Care Committee (Permit Number: 07–12). We did not sample any protected species.

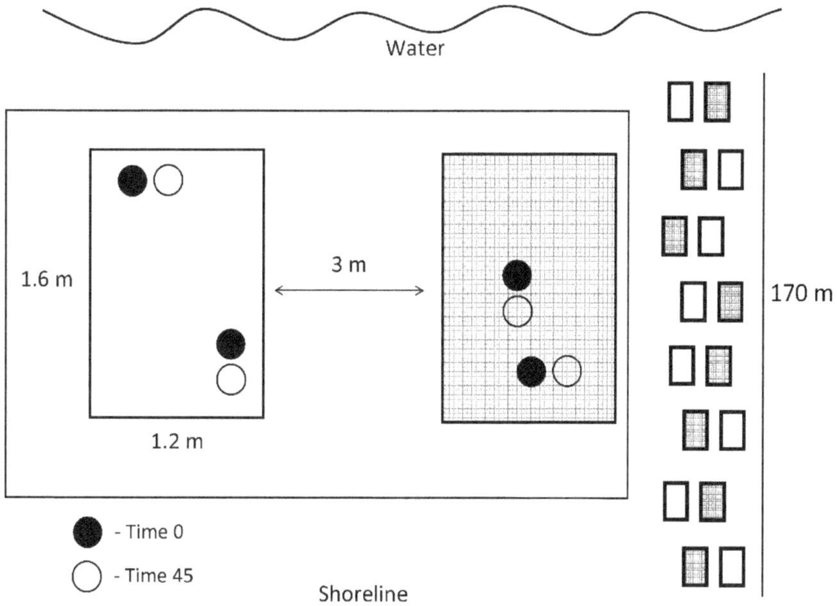

Figure 1. Design of the predator exclusion experiment with spatial blocks at Pecks Cove in summer 2007 (not to scale). Each pair of small rectangles represents a spatial block. Blocks were set up between 130 and 300 m from shore. Each block consisted of two treatment levels: a netted bird exclosure (with hatching) and a control plot (without hatching). Two replicate samples were taken within each treatment plot immediately following the ebb tide (time 0), and a second pair of samples were taken ~5 cm from the first 45 min later. Locations from which samples were taken within each plot were randomly selected in advance.

Field study: Burrow casts

We measured *C. volutator* burrow depths by making polymer resin casts of areas with *C. volutator* burrows on the Pecks Cove mudflat in summer 2011, following a protocol adapted from Gingras et al. [63]. We randomly selected 3 locations (between 100 and 300 m from shore, separated by at least 100 m) three times: 25 July, 19 August and 30 August, corresponding with shorebird arrival, and near peak abundance and departure, respectively. We attempted two casts per location per time; however, not all casts were successful. We have data from three casts (one from each location) on 25 July, five casts on 19 August and all six casts on 30 August. We made casts using polyester boat-repair resin catalyzed by 8% methylethyl ketone peroxide (MEKP). Before adding the catalyst on location at low tide, we diluted the resin by 50% with acetone to decrease viscosity, enabling the resin mix to penetrate small burrow openings. Immediately after adding the catalyst, we poured the resin mix into a circular plastic form (18 cm diameter) partly pushed into the mud, very slowly so that it seeped into burrows without trapping air and water. We collected the form and hardened cast 12–18 h later at low tide.

On each cast, we sampled burrows using a stratified random design, by first categorizing all *C. volutator* burrows in three burrow width classes (i.e., strata): small ($\leq$1 mm diameter), medium ($>$1 to $\leq$2 mm), and large ($>$2 mm). We attempted to measure 10 randomly selected burrows per width class per cast; however, on 40% of the width class-cast combinations, there were fewer than 10 burrows available (overall average $\pm$SD: 8.1$\pm$2.9 burrows per width class per cast). For each selected burrow, we measured burrow depth and width (in mm).

Experiment 2: Predator simulation in the laboratory

We conducted a laboratory experiment in summer 2011 to examine behavior of *C. volutator* females in the presence and absence of simulated shorebird pecking. We chose to use only females ($\geq$6 mm body size) because females (i) dominate adult densities in the field [23,64], (ii) are the main burrow-making adults [33,65], and (iii) previous work found no difference in burrow behavior of females when alone versus when with males, but male behavior varied in the presence of females (S.M. MacNeil and M.A. Barbeau, unpubl. data). For our experiment, we constructed narrow cages using a cedar wood frame (30 cm long$\times$30 cm high) with grooves to hold two 63-mm thick sheets of glass 0.3 cm apart (Figure S1). We filled the cages with 20 cm of mud collected from Pecks Cove and sieved through a 125-µm mesh to remove existing animals. We added nine *C. volutator* females to each cage (collected from Pecks Cove $<$24 h prior to set up), creating a density of 10,000 individuals/m^2 (similar to natural conditions [23]). We conducted the experiment in three temporal blocks (15–18 August, 23–26 August and 2–5 September), each consisting of four cages, with two replicate cages per treatment level (pecking and non-pecking). All cages experienced a simulated tidal cycle (emulating the 24-h natural cycle) divided into four observation stages: immediately high (first 20 minutes following submersion), high, immediately low (first 20 minutes following emersion), and low tide. A cage had a draining hole covered with 250-µm mesh at the level of the mud surface, and tides were created by lowering or elevating the rack of cages in a large salt water tank (recirculated salt water, kept at 13.8$\pm$0.9°C and 32.5$\pm$1.2 PSU, average $\pm$SD). The light:dark regime throughout the experiment was 15:9 h, with light levels at 1595 $\pm$342 lux during the day.

We designed the pecking treatment to mimic natural shorebird predation behavior. Pecking took place only during the first five minutes of immediately low tide at a rate of one peck per second [57] with a probe (2-mm diameter) inserted into the mud to a maximum depth of 5 mm. We did not simulate pecking at other times during low tide, because the area in a cage is small (9 cm^2), and the likelihood is almost zero that such a small area of mudflat would be visited more than once during a tidal cycle. Our pecking

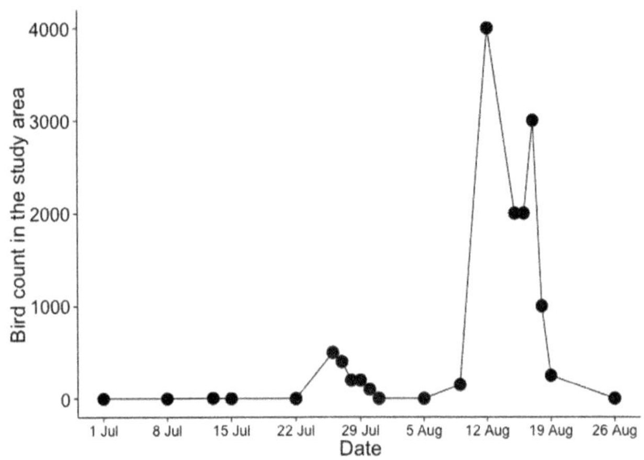

Figure 2. Estimates of Semipalmated Sandpiper density at Pecks Cove in 2007. Estimates of number of Semipalmated Sandpipers present within a 400 m radius of the blocked predator exclusion experiment. Estimates were taken approximately once per week until birds arrived (25 July) at which point density was estimated approximately 3 times/week for the remainder of the study.

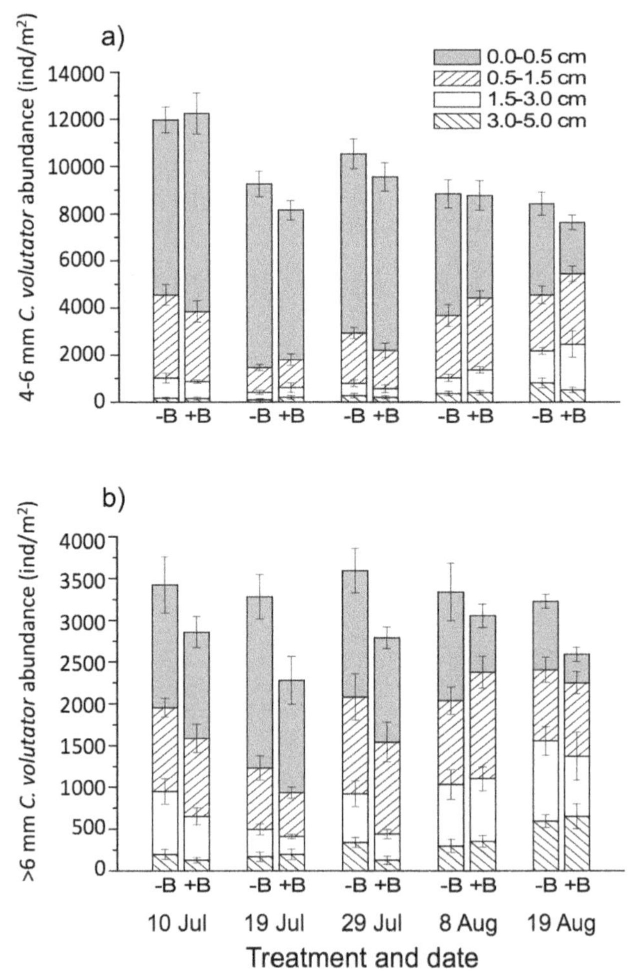

Figure 3. Abundance and vertical distribution of *Corophium volutator* adults sampled from the predator exclusion experiment at Pecks Cove in 2007. Abundance (individuals/m^2) of *C. volutator* sized A) 4–6 mm and B) >6 mm in bird exclosures (−B) and control plots (+B) immediately following sediment exposure (0 min after tidal recession). Each stack represents a layer of sediment; values are mean ±1 SE (n = 8) for each combination of layer, treatment and date. Note that scale of the y-axes changes between size classes.

treatment had no apparent confounding behavioral effects on *C. volutator* individuals (Figures S2.1, S2.2).

We observed each cage for four 20-min observation periods (1 observation period per tidal stage) per day over a 4-d trial (for a total of 80 minutes per tidal stage per cage per trial). Each day, we randomly selected the order in which the cages were observed. For an observation period for a cage, a burrow that was fully visible and that contained a *C. volutator* individual was randomly selected as the focal burrow with the focal amphipod. We recorded the proportion of time that the focal amphipod spent at the bottom of its burrow, defined as a part of its body touching the bottom of the "U". Other burrow behaviors were also recorded (Figures S2.1, S2.2). Animal care approval was not required at the University of New Brunswick as the experiment only involved invertebrates.

Statistical analyses

We conducted statistical analyses in SPSS version 15.0 [66] and R version 2.10.1 [67]. We applied appropriate transformations when parametric test assumptions were violated. We evaluated main effects at $p = 0.05$ and investigated interactions at $p \leq 0.10$ [58]. Means are reported as ±1SE, unless otherwise indicated.

We evaluated vertical distribution of *C. volutator* (density of individuals per sediment layer) in the predator exclusion experiment in the field for each size class separately using linear mixed effects models (repeated measures design). Sediment layer (4 levels), exposure time (2 levels: 0 and 45 min after tidal recession), treatment (2 levels: presence or exclusion of shorebirds) and date (5 levels) were fixed factors, and spatial block was a random factor. We used an autoregressive (AR1) covariance structure with homogenous variances [68] and a restricted maximum likelihood method of parameter estimation. Denominator degrees of freedom are calculated using the Satterthwaite method [69], producing an approximate but unbiased F-test. To avoid pseudoreplication, we averaged the two core samples within a treatment and exposure time. When there were significant interactions with date, we conducted separate analyses for each date [70]. For all other significant interactions, we either split by one of the interacting factors or compared all combinations of treatments [68,71] using posthoc testing with Bonferroni-adjusted p-values. For each size class, we also assessed the effect of date and shorebird treatment on overall *C. volutator* abundance (i.e., pooling over sediment layers) using linear mixed effects models.

Density of crawling *C. volutator* was also analyzed using linear mixed models with an AR1 covariance structure (repeated measures design, as above). To control for possible effects of sediment temperature on crawling behavior, we ran two additional models, one with surface temperature as a covariate and the other with temperature at 5 cm depth as a covariate. We used AIC model selection [72] to determine which of these models best predicted crawling behavior.

We compared burrow depths of *C. volutator* in the field, measured from casts, over three sampling rounds (before, during and near the end of the shorebirds' staging period). First, burrow depth was standardized for *C. volutator* body size by performing a linear regression of burrow depth versus burrow width (a proxy for *C. volutator* size). We then used the residuals from this regression in an ANOVA where sampling round (3 levels) was a fixed factor,

Table 1. Significant ($\alpha = 0.1$) and approaching significant interactions from global models assessing effects of sampling date, layer (vertical position in the sediment), length of exposure (time after tidal recession) and treatment (shorebird present or excluded) on abundance (individuals/m^2; square root transformed) of *Corophium volutator* adults sized 4-6 mm and >6 mm in an experiment at Pecks Cove in 2007.

C. volutator size	Interaction	df	F	p
4–6 mm	Treatment x date x exposure x layer	12, 428.0	1.67	0.07
>6 mm	Treatment x date	4, 217.6	1.80	0.13
	Treatment x layer	3, 461.2	2.57	0.05
	Date x exposure	4, 288.9	2.28	0.06
	Date x layer	12, 429.9	17.8	<0.0001

Results are from AR1 linear mixed models. The numerator and denominator degrees of freedom (df) are presented. For 4–6 mm *C. volutator*, main effects and simple interactions could not be interpreted due to a significant four-way interaction. For >6 mm *C. volutator*, numerous two-way interactions among all four factors made interpretation difficult. Higher-order interactions not presented were all non-significant (p>0.24).

location (3 levels) a random factor nested in round, and individual burrows were the error term. We pooled casts within a location, when both were successful, so that the unit of replication to test for the effect of round was location.

For the lab experiment, we analyzed the proportion of time spent by *C. volutator* females on the bottom of their burrow in the presence and absence of simulated pecking using a linear mixed model (repeated measures design). Treatment (2 levels: pecking, no pecking) and tidal stage (4 levels) were fixed factors, and tank was a random factor (split by tidal stage). We pooled data over the three temporal blocks, since the effect of block or interactions with block were highly non-significant (p>0.20 [73]). We applied an arcsine square-root transformation prior to analysis [71].

Results

Patterns in sandpiper abundance and foraging activity

Semipalmated Sandpipers began to arrive on 25 July in 2007, but foraging flocks were not present until early August. Abundance peaked at ~4,000 individuals on 12 August (Figure 2) and shorebirds remained in the study area until the end of the experiment.

From 9 to 26 August, shorebird footprint cover within exclosures was very low (mean cover $= 4.2 \pm 1.1\%$, n = 72) and was limited to the periphery, indicating that shorebird predators were effectively excluded. As well during this time, shorebird footprint cover in control plots, which were accessible to birds, was $43.6 \pm 4.1\%$ (n = 72).

Seasonal trends in *C. volutator* abundance and vertical distribution

Overall abundance of small *C. volutator* adults (4–6 mm) varied with date ($F_{4,133} = 25.7$, $p<0.0001$; interactions with date were non-significant, $p>0.59$), Specifically, abundance declined between 10 and 19 July, moderately increased by 29 July, and subsequently declined slightly over the last three sampling dates (Figure 3A). The effect of date on overall abundance of large *C. volutator* adults (>6 mm) approached significance ($F_{4,133} = 2.28$, $p = 0.06$; interactions with date were non-significant, $p>0.18$), but no clear temporal trend could be detected (Figure 3B). In general, the population of both size classes remained high throughout the experiment (Figure 3).

We could not interpret main effects or some simple interactions from global models examining vertical distribution of *C. volutator*, because differences among layers varied with other factors (Table 1). Consequently, we conducted separate analyses for each sampling date (Table 2). This prevented statistical comparison among dates, but allowed us to examine treatment, layer and exposure effects (presented below) without elevating the risk of a type I error [70]. Visual examination of Figures 3 and S3 indicate that *C. volutator* adults tended to be deeper in the sediment layer in the summer, following arrival of shorebird predators, both in control plots where shorebirds were able to forage and in predator exclosures.

Treatment and exposure effects on *C. volutator* abundance and vertical distribution

Overall abundance (all layers combined) varied with treatment level for both small ($F_{1,133} = 3.87$, $p = 0.05$) and large ($F_{1,133} = 21.8$, $p<0.0001$) *C. volutator* adults. In both size classes, abundance was lower in control plots relative to bird exclosures (Figure 3). There were slightly more small *C. volutator* adults in samples taken after 45 min of tidal exposure than samples taken immediately following exposure ($F_{1,133} = 4.74$, $p = 0.03$), while abundance of large *C. volutator* adults did not vary significantly with tidal exposure ($F_{1,133} = 4.74$, $p = 0.12$). All interactions were non-significant ($p>0.18$).

We observed a shorebird treatment effect for 4–6 mm *C. volutator* on both August sampling dates (Figures 3A, S3A), when shorebirds were present (Figure 2). Specifically, abundances among layers varied with treatment (treatment x layer interaction, Table 2). On 8 August, there was a pronounced vertical pattern in bird exclosures, with significantly more 4–6 mm amphipods in layer 1 than 2 (post-hoc testing, $p<0.0001$), more in layer 2 than 3 ($p<0.0001$), but no detectable difference between layers 3 and 4 ($p = 0.41$); conversely, this vertical separation was less pronounced in control plots (indicating movement downwards), with layers 1 and 2 being similar, although layer 2 still contained more amphipods than layer 3 ($p<0.0001$) and layer 3 more than layer 4 ($p = 0.041$). We observed similar vertical profiles on 19 August: most amphipods were in the top layer (layer 1> layer 2, $p = 0.015$) and numbers decreased with depth (layer 2>layer 3, $p = 0.009$; layer 3> layer 4, $p = 0.004$) in bird exclosures, whereas abundance did not vary significantly between the top two layers, but layer 2 had more amphipods than 3 ($p = 0.035$) and layer 3 had more than 4 ($p<0.0001$) in control plots.

For >6 mm *C. volutator*, the treatment effect was not as consistent as for small adults; during the period when shorebirds

Table 2. Effects of layer (vertical position in the sediment), length of exposure (time after tidal recession; Expos) and treatment (shorebird present or excluded; Treat) on abundance (individuals/m^2; square root transformed) of *Corophium volutator* adults sized 4–6 mm and >6 mm in an experiment at Pecks Cove in 2007.

		10 Jul			19 Jul			29 Jul			8 Aug			19 Aug		
C. volutator size	**Source of variation**	**df**	**F**	**p**	**df**	**F**	**p**	**df**	**F**	**p**	**df**	**F**	**p**	**df**	**F**	**p**
4–6 mm	Layer	3, 81.1	397.7	**<0.0001**	3, 81.6	380.5	<0.0001	3, 81.3	334.7	**<0.0001**	3, 80.4	167.2	<0.0001	3, 79.9	72.6	<0.0001
	Exposure	1, 59.8	1.5	0.22	1, 69.9	3.3	0.072	1, 64.4	0.5	0.49	1, 69.5	1.5	0.22	1, 48.9	1.8	0.19
	Treat	1, 42.4	0.1	0.75	1, 49.5	0.2	0.66	1, 46.5	2.3	0.14	1, 51.1	1.0	0.31	1, 29.7	2.69	0.12
	Expos x Layer	3, 82.1	0.4	0.72	3, 84.3	5.2	**0.003**	1, 82.3	0.3	0.81	3, 81.9	0.4	0.74	3, 77.4	0.4	0.78
	Treat x Layer	3, 82.5	0.4	0.76	3, 83.3	0.7	0.54	3, 82.4	0.8	0.51	3, 81.8	4.5	**0.006**	3, 82.4	3.9	**0.011**
	Treat x Expos	1, 50.4	0.1	0.73	1, 59.0	0.01	0.93	1, 55.0	0.6	0.45	1, 60.1	0.01	0.93	1, 37.4	0.2	0.66
>6 mm	Layer	3, 79.7	63.8	**<0.0001**	3, 81.8	93.8	**<0.0001**	3, 79.3	55.7	**<0.0001**	3, 78.4	24.5	<0.0001	3, 84.4	13.3	<0.0001
	Exposure	1, 59.6	0.2	0.68	1, 66.3	8.8	**0.004**	1, 52.6	0.3	0.61	1, 55.5	0.7	0.40	1, 56.9	1.9	0.17
	Treat	1, 42.0	0.3	0.59	1, 45.9	4.9	**0.032**	1, 35.4	2.3	0.14	1, 38.1	0.01	0.94	1, 39.6	13.2	**0.001**
	Expos x Layer	3, 81.0	0.6	0.63	3, 84.4	0.7	0.57	3, 79.4	1.1	0.34	3, 79.6	0.1	0.96	3, 84.8	3.6	**0.017**
	Treat x Layer	3, 81.1	0.6	0.60	3, 84.2	1.4	0.26	3, 81.2	0.7	0.55	3, 80.3	2.4	**0.078**	3, 86.2	2.0	0.13
	Treat x Expos	1,50.1	1.2	0.28	1, 55.1	0.1	0.73	1, 43.0	2.0	0.16	1, 45.9	0.01	0.96	1, 47.3	0.2	0.69

Results are from an AR1 linear mixed model. The numerator and denominator degrees of freedom (df) are presented. When interaction terms were significant, main effects were not interpreted. Significant main effects (i.e. not part of an interaction) are in bold.

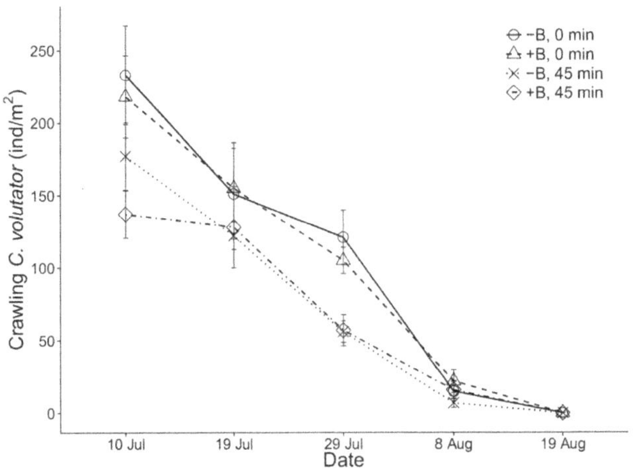

Figure 4. Abundance of *Corophium volutator* individuals crawling on the sediment surface throughout a predator exclusion experiment at Pecks Cove in 2007. Each symbol represents a different combination of treatment level (bird exclosures [−B] and control plots [+B]) and tidal exposure (0 min and 45 min). Values are presented as mean ±1 SE (n = 8).

were present and actively foraging (Figure 2) the treatment x layer interaction was only detected on 8 August, and not on 19 August (Table 2). Within bird exclosures on 8 August, abundances were similar among the top three layers, with the fourth layer containing significantly fewer individuals than the first ($p < 0.0001$, Figure 3B). In controls, amphipods were deeper in the sediment (Figure 3B, S3B); layer 2 had the most amphipods and significantly more than layer 4 ($p < 0.0001$), but other layers did not differ significantly from one another.

A main effect of treatment (with non-significant treatment x layer interaction) was significant for large *C. volutator* adults on 19 July (just before shorebirds arrived on the mudflat) and on 19 August (when shorebirds were present), reflecting that there were generally fewer amphipods in all layers in control plots than in exclosures (Table 2). The 19 July result could not have been due to shorebirds, and did not persist to the next sampling date. On 19 August, although a treatment x layer interaction was not detected statistically, there were relatively fewer amphipods in the top-most layer compared to other layers in the presence of shorebirds than in their absence (Figure 3B, S3B).

We also observed an exposure x layer interaction for small *C. volutator* adults on 19 July and for large adults on 19 August, indicating that differences among layers varied with tidal exposure (Table 2). On 19 July, however, post-hoc testing revealed that vertical profiles at 0 and 45 min exposure were similar with more small adults in layer 1 than 2 ($p < 0.0001$) and layer 2 than 3 ($p < 0.0001$), while layers 3 and 4 did not differ significantly ($p > 0.13$). On 19 August, abundance of large adults did not differ significantly among layers immediately following exposure (time 0, all comparisons between layers, $p > 0.05$). However, 45 min later, amphipods tended to concentrate in middle layers with significantly more residing in the second and third layer than the first layer or the fourth layer ($p \leq 0.01$ for these comparisons).

Crawling activity of *C. volutator* and temperature

Crawling declined over the summer (Figure 4), despite persistence of an adult *C. volutator* population (Figure 3). This effect varied with time since exposure (date x exposure interaction, $F_{4,137} = 3.39$, $p = 0.01$). Crawling activity declined significantly over the 45-min period ($F_{1,21} = 10.0$, $p = 0.005$) on 10 July, but there was no difference ($F_{1,21} = 0.81$, $p = 0.38$) on July 19. On 29 July (when the birds have recently arrived), crawling again declined in the 45 min after tidal recession ($F_{1,21} = 29.8$, $p < 0.0001$). Due to the almost absence of crawling amphipods later in the summer (Figure 4), we did not further examine the final two sampling dates. Crawling activity did not differ significantly between treatment levels ($F_{1,137} = 0.04$, $p = 0.85$, Figure 4). Changes in crawling activity were independent of sediment temperature, since the model that best predicted *C. volutator* crawling did not include a temperature variable (Table 3). Sediment temperatures were highest during late July and early August (Table 4), when crawling activity was intermediate (Figure 4).

Burrow casts

Depth of *C. volutator* burrows increased with tube width (linear regression, adjusted $R^2 = 0.75$, df = 340, $p < 0.0001$, Figure 5). However, once variation in width was accounted for, burrow depth did not vary with date (ANOVA, $F_{2,6} = 0.26$, $p = 0.78$). The maximum burrow depth measured was 58 mm. The *C. volutator* population structure at Pecks Cove was generally similar among sampling dates (Figure S4).

Predator simulation experiment

C. volutator females spent significantly more time on the bottom of their burrows when exposed to simulated shorebird pecking compared to controls without pecking ($F_{1,10} = 7.62$, $p = 0.02$, Figure 6). This effect did not vary with tidal stage (treatment x tide interaction, $F_{3,30} = 0.87$, $p = 0.47$). Even though amphipods were exposed to the predator simulation for only the first 5 minutes of low tide, the effect persisted throughout low and high tide. However, time spent on the bottom of a burrow did vary with tide

Table 3. AIC model selection results for models predicting *Corophium volutator* crawling activity during an experiment in Pecks Cove in summer 2007.

Model	AIC	ΔAIC	w_i
date + treatment + exposure + date*treatment + date*exposure + exposure*treatment	189.10	0.00	0.87
date + treatment + exposure + date*treatment + date*exposure + exposure*treatment + surface temp	195.59	6.49	0.03
date + treatment + exposure + date*treatment + date*exposure + exposure*treatment +5 cm temp	193.55	4.45	0.09

Number of crawling individuals (per 0.25×0.25 m quadrat) was log transformed to improve normality and an AR1 covariance structure was used to accommodate the repeated measures sampling design. Models included the fixed effects given below and block as a random effect. "Surface temp" is the sediment temperature (°C) on the surface of the mud and "5 cm temp" is the sediment temperature at 5 cm depth. ΔAIC is the difference in Akaike's information criterion from the top model and w_i is the AIC weight, which represents the likelihood that a particular model is the best of the options available.

Table 4. Mean (±1 SE; n = 16, per combination of date and depth) temperature (°C) at surface and 5.0 cm sediment depths during a predator exclusion experiment at Pecks Cove in 2007.

Location	Tidal Exposure	8 July	19 July	29 July	8 Aug	19 Aug
Surface	0 min	17.7±0.2	18.3±0.05	22.5±0.2	23.4±0.2	18.8±0.1
	45 min	19.3±0.4	18.8±0.05	25.9±0.1	25.4±0.1	20.4±0.2
5 cm depth	0 min	16.3±0.05	17.9±0.03	19.7±0.05	20.0±0.08	18.8±0.1
	45 min	17.5±0.08	18.1±0.03	21.4±0.08	22.0±0.08	19.2±0.1

Measures were taken immediately following receding tide (0 min) and 45 min later.

($F_{3,30}$ = 11.6, $p<0.0001$); *C. volutator* females spent the greatest amount of time at the bottom of their burrows during immediately low tide. Mean (±SD) burrow depth was 18.4±6.7 mm and 16.4±7.5 mm for the non-pecking and pecking treatment levels, respectively.

Discussion

Predator avoidance strategies can affect the outcome of predator-prey interactions [19,74,75,76]. For example, if prey relocate to an area with reduced risk of detection and capture [74], the result is a reversible decline in prey availability. Predator avoidance in shorebird-*Corophium volutator* interactions has been primarily identified through cessation of crawling where shorebirds are present and behavioral responses of the predator [54,57,77,78,79]. By quantifying vertical position in the sediment in the presence of predators (or simulated predators), our study provides direct evidence that *C. volutator* in the Bay of Fundy employ a predator avoidance mechanism beyond simple cessation of crawling.

Seasonal trends versus predator effects

In our predator exclusion experiment, we observed a distinct, temporal vertical redistribution of adult *C. volutator*. Although amphipods moved deeper into the sediment later in the summer, which happens to correspond with arrival of shorebirds, this redistribution was detectably stronger in areas where birds were able to forage compared to areas where birds were excluded. We also observed a concurrent cessation of crawling activity by amphipods, consistent with previous findings in the Bay of Fundy [46]. The observed vertical shift in the sediment was not simply due to selective predation of amphipods on or near the sediment surface, because absolute adult abundances in deeper layers increased after birds arrived and in areas where birds were able to forage, while the overall number of adults remained relatively consistent throughout the experiment (Figure 3). Sediment temperature was also not associated with the observed redistribution. High sediment temperatures can cause *C. volutator* to retreat deeper into burrows [80]; however, the highest temperature we recorded was on 29 July, after birds arrived but before they began foraging in flocks and prior to the most obvious redistribution by

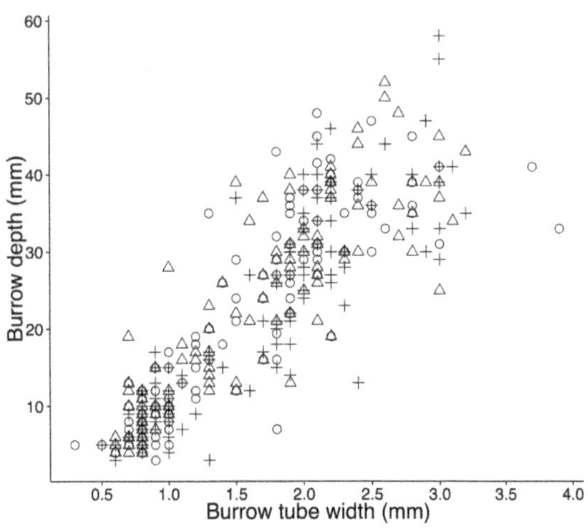

Figure 5. Depth (mm) versus tube width (mm) of *Corophium volutator* burrows in Pecks Cove in summer 2011. Burrow measurements were taken from polymer resin casts of naturally occurring burrows on three sampling dates, with n = 3, 5 and 6 casts, respectively. These dates were selected to roughly coincide with before arrival, peak abundance and departure of shorebirds. The linear regression coefficients for depth (mm) vs. width (mm) of burrows are 15.4±0.5 (SE) for the slope and −2.7±0.9 for the intercept.

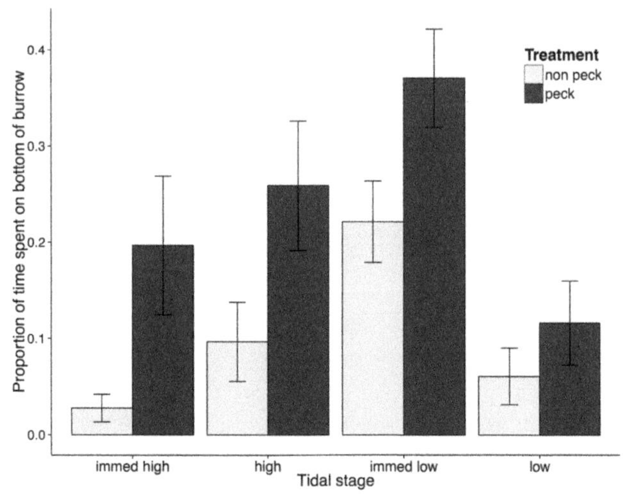

Figure 6. Proportion of time spent by focal *Corophium volutator* females on the bottom of their burrows in the presence and absence of simulated shorebird predation in laboratory. Clustered bars represent paired treatment levels (pecking and non-pecking) for each of four simulated tidal stages (immediately high, high, immediately low, and low). Proportion of time was calculated from 80 min of observation time per experimental unit and tidal stage. Experimental units in the pecking treatment level experienced 5 min of simulated pecking (1 peck per s) at the start of immediately low tide. Values are mean ±1 SE (n = 6 experimental units).

amphipods. Moreover, sediment temperatures in our study were lower than those tested by Gilroy [80].

In the field it is difficult to separate seasonal and bird effects to establish a causal relationship on vertical redistribution of infaunal prey. However, using appropriate controls in the field we did detect a significant difference in vertical distribution between areas with and without shorebird foraging activity on both August sampling dates for small *C. volutator* adults and on 8 August for large *C. volutator* adults. Further, the results of our predator simulation experiment corroborate our field observations; in the presence of perceived predator activity, *C. volutator* individuals spent more time deeper in their burrows. Minderman et al. [55] found that amphipod prey retreated into burrows in response to increased predation risk. Similarly, various species of bivalves have been shown to increase survival by burrowing deeper into the sediment in response to predators [19,76]. Our data suggest that *C. volutator* adults respond to the risk of predation, both by ceasing crawling and burrowing into the sediment beyond the reach of probing sandpipers.

Predator avoidance mechanisms

To avoid predation, *C. volutator* individuals could occupy lower portions of the sediment either by moving lower into existing burrows or by constructing deeper burrows. We found that *C. volutator* burrow depth did not increase later in the season when shorebirds were present, nor was it deeper for *C. volutator* exposed to simulated predator activity than those in the non-pecking treatment of the lab experiment. Rather, amphipods increased time spent at the bottom of their burrows in the presence of simulated predation risk. However, individuals would not necessarily have to go all the way to the bottom of their burrow to evade a predator. We observed that adult *C. volutator* burrows extend to ~6 cm in the sediment at Pecks Cove, while depths of up to 10 cm have been reported in the literature [30]. Sandpipers foraging in the Bay of Fundy have bills ~20 mm in length [50]; thus, it would be difficult for sandpipers to obtain prey burrowed deeper than 2 cm. Based on this, time at the bottom of a burrow for amphipods occupying deep burrows is a conservative metric for quantifying predator avoidance. An amphipod would not necessarily need to go to the bottom of its burrow to avoid a predator; even slight adjustments in its depth could reduce the risk of being eaten [81].

Trade-off between foraging and predation risk

Although we observed a vertical redistribution of prey, we still found amphipods in the top sediment layer in our predator exclusion experiment. This incomplete predator avoidance response is probably related to the fact that prey must deal with the competing pressures of avoiding predation and foraging for growth [82,83,84]. To manage this growth/predation risk trade-off, prey must make decisions regarding allocation of time to foraging and antipredator behavior [82,85,86]. Sandpipers forage with consistent success rates throughout low tide [87,88] and *C. volutator* individuals would presumably not be able to avoid foraging for that entire period. For *C. volutator*, food access is better near the sediment surface [29,31,89], which may explain why, in spite of the increased predation risk, we observed individuals near the top of their burrows.

Spatial scale and magnitude of the response by *C. volutator*

C. volutator may be able to broadly detect and respond to predators. This is suggested by our observation that, in addition to the response in areas exposed to predation, *C. volutator* individuals in bird exclosures also reduced crawling and appeared deeper in the sediment when shorebirds were present compared to before their arrival. However, based on our predator exclusion experiment, *C. volutator*'s response is stronger when directly exposed to foraging shorebirds. Stillman et al. [79], in modeling predator avoidance in a *C. volutator*-Redshank system based on observations by Yates et al. [54], estimated that *C. volutator* individuals respond to shorebirds within a 0.6-m radius (range 0.1 to 1 m). Thus, shorebirds foraging within our study area could potentially be detected by *C. volutator* in an exclosure, and so it is not surprising that we also observed a response in our exclosures. In addition, *C. volutator* exhibit considerable swimming movement during high tide [31,90,91], and could conceivably redistribute among bird exclosures and control plots. We observed that the effect of a perceived predator persisted throughout the tidal cycle in our lab experiment. Thus, amphipods collected from bird exclosures could have been responding to previously experienced predation risk. Given all of these factors, which could minimize differences between exclosures and control plots, our field experiment was a conservative test of predator avoidance. Indeed, we suggest that observing treatment effects on vertical distribution at the scale we tested is indicative of a strong response. Additional experiments with larger exclosures and/or enclosures to restrict emigration of swimming *C. volutator* are required to identify the scale of the response and to isolate the cues that trigger vertical redistribution and cessation of crawling.

Implications and future work

Prey availability for Semipalmated Sandpipers is typically assessed based on absolute density of *C. volutator* [46,92,93,94]. However, recent work on foraging behavior [95] and stable isotopes [42] indicates that sandpipers staging in the Bay of Fundy actually have a much broader diet. Our study suggests that traditional methods of sampling prey may misrepresent actual availability, which could contribute to the recent observation that *C. volutator* comprises a smaller portion of the sandpiper's diet than previously assumed [42]. As Bay of Fundy mudflats represent critical stopover habitat for migrating shorebirds [40,96], an accurate estimate of prey availability is needed to understand habitat use. Future studies in this system and others should consider vertical distribution of prey.

The behavioral response to predation we have observed in *C. volutator* could have community-wide implications. Predator avoidance limits prey availability [13,52,53] and, among shorebird-prey interactions, has been linked to changes in feeding [57] and activity rates [55], as well as interference competition among foraging conspecifics [54,55,79]. By burrowing out of reach of shorebird predators, *C. volutator* could displace predation pressure onto other, more accessible, prey items as has been observed in burrowing bivalves [19]. It has also been suggested that when foraging shorebirds increase activity rates in response to behavioral changes in their prey, they themselves experience a heightened risk of predation [55]. The non-consumptive effects of predation on community dynamics are considerable [7]. Studies of predator avoidance strategies are a key element to understanding the cascading impacts of predator-prey interactions.

Supporting Information

Figure S1 Experimental setup of the laboratory experiment examining effects of simulated shorebird predation on *Corophium volutator* behavior. Pictures showing the A) laboratory set up, B) thin glass cages with cedar frame used to observed behaviors of *Corophium volutator* adults in their burrow and C) burrows with a *C. volutator* individual inside, in one of the glass cages.

Figure S2 S2.1: Proportion of time *Corophium volutator* individuals spent in each of seven behaviors while in their burrow in the presence and absence of simulated shorebird pecking in a laboratory experiment. We monitored behavior over the course of a tidal cycle (20-min observation periods in each tidal stage per day conducted over 4 days; tidal stages: immediately high, high, immediately low, and low). To mimic natural conditions of shorebird predation, pecking occurred for the first 5 min of immediately low tide at a rate of one peck per s. Each section of a stack represents the mean proportion of time spent engaged in the corresponding behavior. Stacks do not sum to 1, because individuals were not always visible for behavior to be recorded. Error bars represent ±1 SE, n = 6 experimental units. **S2.2: Counts per 20 min of observation that *Corophium volutator* individuals spent in each of five types of movement in the presence and absence of simulated shorebird pecking in a laboratory experiment.** We monitored behavior over the course of a tidal cycle (20-min observation periods in each tidal stage per day conducted over 4 days; tidal stages: immediately high, high, immediately low, and low). To mimic natural conditions of shorebird predation, pecking occurred for the first 5 min of immediately low tide at a rate of one peck per s. Bars represent the mean number of times *C. volutator* individuals were observed engaging in a particular movement; error bars represent ±1 SE, n = 6 experimental units.

Figure S3 Proportional vertical distribution of *Corophium volutator* adults sampled from the predator exclusion experiment at Pecks Cove in 2007. Proportion of *C. volutator* sized A) 4–6 mm and B) >6 mm in bird exclosures (−B) and control plots (+B) found in each vertical layer of sediment. Layers 1–4 are 0–5, 0.5–1.5, 1.5–3.0, and 3.0–5.0 cm from the sediment surface, respectively.

Figure S4 Population structure of *Corophium volutator* at Pecks Cove in summer 2011. Sample sizes represent the number of individuals measured (pooling over cores) to generate the size frequency distribution in each sampling round.

Data S1 Raw data from three studies examining predator avoidance behavior of adult *Corophium volutator* in the presence of shorebirds.

Acknowledgments

We thank B. Cameron, M.E. Coulthard, W. Bell, W. Anderson and M. Marcoux (in 2007), and M.A. Hebert, C.E. Gilroy, L.K. Bursey, R. O'Keefe, I. Thomas and M. Keays (in 2011) for lab and field assistance, D.M. Green for helping to design and build the thin lab cages for observing *C. volutator* behavior, and M.K. Gingras for help with the resin cast protocol. We also thank M.R.S. Coffin, D. Drolet, A.V. Cheverie, M.G. Ginn, D. Shutler and his lab at Acadia University, M. Thiel, J.M. Hemmi and two anonymous reviewers for useful discussions and comments.

Author Contributions

Conceived and designed the experiments: EHF SMM DJH MAB. Performed the experiments: EHF SMM. Analyzed the data: ECM EHF DJH MAB. Contributed reagents/materials/analysis tools: MAB DJH. Wrote the paper: ECM EHF SMM DJH MAB.

References

1. Paine RT (1966) Food web complexity and species diversity. Am Nat 100: 65–75.
2. Menge BA (1976) Organization of the New England rocky intertidal community: role of predation, competition and environmental heterogeneity. Ecol Monogr 46: 355–393.
3. Menge BA, Sutherland JP (1976) Species diversity gradients: synthesis of the roles of predation, competition, and temporal heterogeneity. Am Nat 110: 351–369.
4. Huntly N (1991) Herbivores and the dynamics of communities and ecosystems. Ann Rev Ecol Syst 22: 477–503.
5. Vieira EA, Duarte LFL, Dias GM (2012) How the timing of predation affects composition and diversity of species in a marine sessile community? J Exp Mar Biol Ecol 412: 126–133.
6. Werner EE, Peacor SD (2003) A review of trait-mediated indirect interactions in ecological communities. Ecology 84: 1083–1100.
7. Preisser EL, Bolnick DI, Benard MF (2005) Scared to death? The effects of intimidation and consumption in predator-prey interactions. Ecology 86: 501–509.
8. Paterson RA, Pritchard DW, Dick JTA, Alexander ME, Hatcher MJ, et al. (2013) Predator cue studies reveal strong trait-mediated effects in communities despite variation in experimental designs. Anim Behav 86: 1301–1313.
9. Lindquist N, Hay ME (1996) Palatability and chemical defense of marine invertebrate larvae. Ecol Monogr 66: 431–450.
10. Honma A, Oku S, Nishida T (2006) Adaptive significance of death feigning posture as a specialized inducible defence against gape-limited predators. Proc Biol Sci 273: 1631–1636.
11. Domenici P, Turesson H, Brodersen J, Brönmark C (2008) Predator-induced morphology enhances escape locomotion in crucian carp. Proc Biol Sci 275: 195–201.
12. Manzur T, Vidal F, Pantoja JF, Fernandez M, Navarrete SA (2014) Behavioural and physiological responses of limpet prey to a seastar predator and their transmission to basal trophic levels. J Anim Ecol doi: 10.1111/1365-2656.12199.
13. Backwell PRY, O'Hara PD, Christy JH (1998) Prey availability and selective foraging in shorebirds. Anim Behav 55: 1659–1667.
14. Menge BA, Lubchenco J (1981) Community organization in temperature and tropical rocky intertidal habitats: prey refuges in relation to consumer pressure gradients. Ecol Monogr 51: 429–450.
15. Meerhoff M, Iglesias C, De Mello FT, Clemente JM, Jenson E, et al. (2007) Effects of habitat complexity on community structure and predator avoidance behavior of littoral zooplankton in temperate versus subtropical shallow lakes. Freshw Biol 52: 1009–1021.
16. Wilson WH (1990) Competition and predation in marine soft-sediment communities. Ann Rev Ecol Syst 21: 221–241.
17. Holohan BA, Klos EG, Oviatt CA (1998) Population density, prey selection, and predator avoidance of the burrowing anemone (*Ceriantheopsis americanus*) in Narragansett Bay, Rhode Island. Estuaries 21: 466–469.
18. Tallqvist M (2001) Burrowing behaviour of the Baltic clam *Macoma balthica*: effects of sediment type, hypoxia and predator presence. Mar Ecol Prog Ser 212: 183–191.
19. Griffiths CL, Richardson CA (2006) Chemically induced predator avoidance behaviour in the burrowing bivalve *Macoma balthica*. J Exp Mar Biol Ecol 331: 91–98.
20. Dill LM, Fraser AHG (1997) The worm re-turns: hiding behaviour of a tube-dwelling marine polychaete, *Serpula vermicularis*. Behav Ecol 8: 186–193.
21. Kicklighter CE, Hay ME (2007) To avoid or deter: interactions among defensive and escape strategies in sabellid worms. Oecologia 151: 161–173.
22. Peer DL, Linkletter LE, Hicklin PW (1986) Life history and reproductive biology of *Corophium volutator* (Crustacea: Amphipoda) and the influence of shorebird predation on population structure in Chignecto Bay, Bay of Fundy, Canada. Netherlands J Sea Res 20: 359–373.
23. Barbeau MA, Grecian LA, Arnold EE, Sheahan DC, Hamilton DJ (2009) Spatial and temporal variation in the population dynamics of the intertidal amphipod *Corophium volutator* in the upper Bay of Fundy, Canada. J Crustacean Biol 29: 491–506.
24. Boates JS, Smith PC (1989) Crawling behaviour of the amphipod *Corophium volutator* and foraging by Semipalmated Sandpipers, *Calidris pusilla*. Can J Zool 67: 457–462.
25. Vader W (1964) A preliminary investigation into the reactions of the infauna of the tidal flats to tidal fluctuations in water level. Netherlands J Sea Res 2: 189–222.
26. Mills A, Fish JD (1980) Effects of salinity and temperature on *Corophium volutator* and *C. arenarium* (Crustacea: Amphipoda), with particular reference to distribution. Mar Biol 58: 153–161.
27. Mathews SL, Boates JS, Walde SJ (1992) Shorebird predation may cause discrete generations in an amphipod prey. Ecography 15: 393–400.

28. Limia J, Raffaelli D (1997) The effects of burrowing by the amphipod *Corophium volutator* on the ecology of intertidal sediments. J Mar Biol Assoc UK 77: 409–423.
29. De Backer A, Van Ael E, Vincx M, Degraer S (2010) Behaviour and time allocation of the mud shrimp, *Corophium volutator*, during the tidal cycle: a laboratory study. Helgol Mar Res 64: 63–67.
30. McCurdy DG, Boates JS, Forbes MR (2000) Reproductive synchrony in the intertidal amphipod *Corophium volutator*. Oikos 88: 301–308.
31. Meadows PS, Reid A (1966) The behaviour of *Corophium volutator* (Crustacea: Amphipoda). J Zool Lond 150: 387–399.
32. Fish JD, Mills A (1979) The reproductive biology of *Corophium volutator* and *C. arenarium* (Crustacea: Amphipoda). J Mar Biol Assoc UK 59: 355–368.
33. Forbes MR, Boates SJ, McNeil NL, Brison AE (1996) Mate searching by males of the intertidal amphipod *Corophium volutator* (Pallas). Can J Zool 74: 1479–1484.
34. Lawrie SM, Raffaelli DG (1998) In situ swimming behaviour of the amphipod *Corophium volutator* (Pallas). J Exp Mar Biol Ecol 224: 237–251.
35. Raffaelli D, Milne H (1987) An experimental investigation of the effects of shorebird and flatfish predation on estuarine invertebrates. Estuar Coast Shelf Sci 24: 1–13.
36. McCurdy DG, Forbes MR, Logan SP, Lancaster D, Mautner SI (2005) Foraging and impacts by benthic fish on intertidal amphipod *Corophium volutator*. J Crustacean Biol 25: 558–564.
37. Cheverie AV, Hamilton DJ, Coffin MRS, Barbeau MA (2014) Effects of shorebird predation and snail abundance on an intertidal mudflat community. J Sea Res, in press.
38. Coulthard ME, Hamilton DJ (2011) Effects of *Ilyanassa obsoleta* (Say) on the abundance and vertical distribution of *Corophium volutator* (Pallas) on mudflats of the upper Bay of Fundy. J Exp Mar Biol 397: 161–172.
39. Coffin MRS, Barbeau MA, Hamilton DJ, Drolet D (2012) Effect of the mud snail *Ilyanassa obsoleta* on vital rates of the intertidal amphipod *Corophium volutator*. J Exp Mar Biol Ecol 418: 12–23.
40. Hicklin PW (1987) The migration of shorebirds in the Bay of Fundy. Wilson Bull 99: 540–570.
41. Neima SG (2014) Movement patterns, duration of stay, and diet of Semipalmated Sandpipers (*Calidris pusilla*) during migratory stopover in the upper Bay of Fundy. BSc Honours Thesis. Mount Allison University, Sackville, New Brunswick, Canada.
42. Quinn JT, Hamilton DJ (2012) Variation in diet of Semipalmated Sandpipers (*Calidris pusilla*) during stopover in the upper Bay of Fundy, Canada. Can J Zool 90: 1181–1190.
43. Boates JS (1980) Foraging Semipalmated Sandpipers *Calidris pusilla* L., and their major prey, *Corophium volutator* (Pallas) on the Starrs Point mudflat, Minas Basin. MSc Thesis, Acadia University, Wolfville, Nova Scotia, Canada.
44. Stoddard PK, Marsden JE, Williams TC (1983) Computer simulation of autumnal bird migration over the western North Atlantic. Anim Behav 31: 173–180.
45. Hicklin PW, Smith PC (1984) Selection of foraging sites and invertebrate prey by migrant Semipalmated Sandpipers, *Calidris pusilla* (Pallas), in Minas Basin, Bay of Fundy. Can J Zool 62: 2201–2210.
46. Boates JS, Forbes M, Zinck M, McNeil N (1995) Male amphipods (*Corophium volutator* [Pallas]) show flexible behaviour in relation to risk of predation by sandpipers. Ecoscience 2: 123–128.
47. Goss-Custard JD (1970) The responses of Redshank (*Tringa totanus* (L.)) to spatial variations in the density of their prey. J Anim Ecol 39: 91–113.
48. Gratto GW, Thomas MLH, Gratto CL (1984) Some aspects of the foraging ecology of migrant juvenile sandpipers in the outer Bay of Fundy. Can J Zool 62: 1889–1892.
49. Durell SEA (2000) Individual feeding specialisation in shorebirds: population consequences and conservation implications. Biol Rev 75: 503–518.
50. Gratto-Trevor CL, Morrison RIG, Mizrahi D, Lank DB, Hicklin P, et al. (2012) Migratory connectivity of Semipalmated Sandpipers: winter distribution and migration routes of breeding populations. Waterbirds 35: 83–95.
51. Hill C, Elmgren R (1987) Vertical distribution in the sediment in the co-occurring benthic amphipods *Pontoporeia affinis* and *P. femorata*. Oikos 49: 221–229.
52. Charnov EL, Orians GH, Hyatt K (1976) Ecological implications of resource depression. Am Nat 110: 247–259.
53. Goss-Custard JD (1980) Competition for food and interference among waders. Ardea 68: 31–52.
54. Yates MG, Stillman RA, Goss-Custard JD (2000) Contrasting interference functions and foraging dispersion in two species of shorebird (Charadrii). J Anim Ecol 69: 314–322.
55. Minderman J, Lind J, Cresswell W (2006) Behaviourally mediated indirect effects: interference competition increases predation mortality in foraging redshanks. J Anim Ecol 75: 713–723.
56. Robar NDP (2006) Fine-scale effects of prey density and interference on Semipalmated Sandpipers (*Calidris pusilla*) in the upper Bay of Fundy. BSc Honours Thesis, Mount Allison University, Sackville, New Brunswick, Canada.
57. Beauchamp G (2007) Competition in foraging flocks of migrating semipalmated sandpipers. Oecologia 154: 403–409.
58. Hamilton DJ, Diamond AW, Wells PG (2006) Shorebirds, snails, and the amphipod (*Corophium volutator*) in the upper Bay of Fundy: top-down vs. bottom-up factors, and the influence of compensatory interactions on mudflat ecology. Hydrobiologia 567: 285–306.
59. Boudreau MR, Hamilton DJ (2012) Seasonal variation in effects of multiple predators on an intertidal mussel bed: implications for interpretation of manipulative experiments. Mar Ecol Prog Ser 465: 137–153.
60. Robar NDP, Hamilton DJ (2007) A method for estimating habitat use by shorebirds using footprints. Waterbirds 30: 116–120.
61. Crewe TL, Hamilton DJ, Diamond AW (2001) Effects of mesh size on sieved samples of *Corophium volutator*. Estuar Coast Shelf Sci 53: 151–154.
62. Wallace EH (2008) Effects of predation by Semipalmated Sandpipers (*Calidris pusilla*) on vertical distribution of the amphipod *Corophium volutator*. BSc Honours Thesis. Mount Allison University, Sackville, New Brunswick, Canada.
63. Gingras MK, Pickerill R, Pemberton SG (2002) Resin cast of modern burrows provides analogs for composite trace fossils. Palaios 17: 206–211.
64. Drolet D, Barbeau MA (2012) Population structure of resident, immigrant, and swimming *Corophium volutator* (Amphipoda) on an intertidal mudflat in the Bay of Fundy, Canada. J Sea Res 70: 1–13.
65. Campbell JI, Meadows PS (1974) Gregarious behavior in a benthic marine amphipod (*Corophium volutator*). Experientia 30: 1396–1397.
66. SPSS Inc. (2006) SPSS version 16.0. SPSS, Chicago, Illinois.
67. R Development Core Team (2009) R: A language and environment for statistical computing. R Foundation for Statistical Computing, Vienna, Austria. Available: https://www.R-project.org. Accessed 8 October 2014.
68. Leech NL, Barrett KC, Morgan GA (2008) SPSS for Intermediate Statistics: use and interpretation, 3rd ed. New York: Taylor and Francis Group. 270 p.
69. Norušis MJ (2008) SPSS 16.0 advanced statistical procedures companion. New Jersey: Prentice Hall. 418 p.
70. Keppel G (1991) Design and analysis: a researcher's handbook, 3rd ed. New Jersey: Prentice-Hall. 594 p.
71. Zar JH (1999) Biostatistical analysis, 4th ed. New Jersey: Prentice Hall. 663 p.
72. Burnham JS, Anderson DR (2002) Model selection and multi-model inference: a practical information-theoretic approach, 2nd ed. New York: Springer. 488 p.
73. Winer BJ, Brown DR, Michels KM (1991) Statistical Principles in Experimental Design, 3rd ed. Boston: McGraw Hill. 1057 p.
74. Zwarts L, Esselink P (1989) Versatility of male curlews *Numenius arquata* preying upon *Nereis diversicolor*: deploying contrasting capture modes dependent on prey availability. Mar Ecol Prog Ser 56: 255–269.
75. Kruse I, Buhs F (2000) Preying at the edge of the sea: the nemertine *Tetrastemma melanocephalum* and its amphipod prey on high intertidal sandflats. Hydrobiologia 426: 43–55.
76. Flynn AM, Smee DL (2010) Behavioral plasticity of the soft-shell clam, *Mya arenaria* (L.), in the presence of predators increases survival in the field. J Exp Mar Biol Ecol 383: 32–38.
77. Goss-Custard JD (1977) Predator responses and prey mortality in Redshank, *Tringa totanus* (L.), and a preferred prey, *Corophium volutator* (Pallas). J Anim Ecol 46: 21–35.
78. Selman J, Goss-Custard JD (1988) Interference between foraging redshanks *Tringa totanus*. Anim Behav 36: 1542–1544.
79. Stillman RA, Goss-Custard JD, Alexander MJ (2000) Predator search pattern and the strength of interference through prey depression. Behav Ecol 11: 597–605.
80. Gilroy CE (2012) The effect of temperature and tidal exposure on behaviour and survival of *Corophium volutator*. BSc Honours Thesis, Mount Allison University, Sackville, New Brunswick, Canada.
81. Myers JP, Williams SL, Pitelka FA (1980) An experimental analysis of prey availability for sanderlings (Aves: Scolopacidae) feeding on sandy beach crustaceans. Can J Zool 58: 1564–1574.
82. Lima SL, Dill LM (1990) Behavioral decisions made under the risk of predation: a review and prospectus. Can J Zool 68: 619–640.
83. Werner EE, Anholt BR (1993) Ecological consequences of the trade-off between growth and mortality rates mediated by foraging activity. Am Nat 142: 242–272.
84. McPeek MA (2004) The growth/predation risk trade-off: so what is the mechanism? Am Nat 163: E88–E111.
85. Brodin T, Johansson F (2004) Conflicting selection pressures on the growth/predation-risk trade-off in a damselfly. Ecology 85: 2927–2932.
86. Strobbe F, McPeek MA, De Block M, Stoks R (2011) Fish predation selects for reduced foraging activity. Behav Ecol Sociobiol 65: 241–247.
87. Beauchamp G (2005) Low foraging success of Semipalmated Sandpipers at the edges of groups. Ethology 111: 785–798.
88. Beauchamp G (2006) Spatial, temporal and weather factors influencing the foraging behavior of migrating Semipalmated Sandpipers. Waterbirds 29: 221–225.
89. Riisgård HU, Schotge P (2007) Surface deposit-feeding versus filter-feeding in the amphipod *Corophium volutator*. Mar Biol Res 3: 421–427.
90. Drolet D, Barbeau MA (2009) Diel and semi-lunar cycles in the swimming activity of the intertidal, benthic amphipod *Corophium volutator* in the upper Bay of Fundy, Canada. J Crustacean Biol 29: 51–56.
91. Bringloe TT, Drolet D, Barbeau MA, Forbes MR, Gerwing TG (2013) Spatial variation in population structure and its relation to movement and the potential for dispersal in a model intertidal invertebrate. PLoS ONE 8(7): e69091. doi: 10.1371/journal.pone.0069091.
92. Wilson WH (1989) Predation and the mediation of intraspecific competition in an infaunal community in the Bay of Fundy. J Exp Mar Biol Ecol 132: 221–245.

93. Hamilton DJ, Barbeau MA, Diamond AW (2003) Shorebirds, mud snails, and *Corophium volutator* in the upper Bay of Fundy, Canada: predicting bird activity on intertidal mudflats. Can J Zool 81: 1358–1366.
94. Beauchamp G (2009) Functional response of staging semipalmated sandpipers feeding on burrowing amphipods. Oecologia 161: 651–655.
95. MacDonald EC, Ginn MG, Hamilton DJ (2012) Variability in foraging behavior and implications for diet breadth among Semipalmated Sandpipers staging in the upper Bay of Fundy. Condor 114: 135–144.
96. Hicklin PW, Chardine JW (2012) The morphometrics of migrant Semipalmated Sandpipers in the Bay of Fundy: evidence for declines in the eastern breeding population. Waterbirds 35: 74–82.

10

Insights into Deep-Sea Sediment Fungal Communities from the East Indian Ocean Using Targeted Environmental Sequencing Combined with Traditional Cultivation

Xiao-yong Zhang, Gui-ling Tang, Xin-ya Xu, Xu-hua Nong, Shu-Hua Qi*

Key Laboratory of Tropical Marine Bio-resources and Ecology/RNAM Center for Marine Microbiology/Guangdong Key Laboratory of Marine Material Medical, South China sea Institute of Oceanology, Chinese academy of sciences, Guangzhou, China

Abstract

The fungal diversity in deep-sea environments has recently gained an increasing amount attention. Our knowledge and understanding of the true fungal diversity and the role it plays in deep-sea environments, however, is still limited. We investigated the fungal community structure in five sediments from a depth of ~4000 m in the East India Ocean using a combination of targeted environmental sequencing and traditional cultivation. This approach resulted in the recovery of a total of 45 fungal operational taxonomic units (OTUs) and 20 culturable fungal phylotypes. This finding indicates that there is a great amount of fungal diversity in the deep-sea sediments collected in the East Indian Ocean. Three fungal OTUs and one culturable phylotype demonstrated high divergence (89%–97%) from the existing sequences in the GenBank. Moreover, 44.4% fungal OTUs and 30% culturable fungal phylotypes are new reports for deep-sea sediments. These results suggest that the deep-sea sediments from the East India Ocean can serve as habitats for new fungal communities compared with other deep-sea environments. In addition, different fungal community could be detected when using targeted environmental sequencing compared with traditional cultivation in this study, which suggests that a combination of targeted environmental sequencing and traditional cultivation will generate a more diverse fungal community in deep-sea environments than using either targeted environmental sequencing or traditional cultivation alone. This study is the first to report new insights into the fungal communities in deep-sea sediments from the East Indian Ocean, which increases our knowledge and understanding of the fungal diversity in deep-sea environments.

Editor: Vishnu Chaturvedi, California Department of Public Health, United States of America

Funding: The authors are grateful for the financial support provided by the National Basic Research Program of China (2010CB833803), National High Technology Research and Development Program of China (863 Program, 2012AA092104), Natural Science Foundation of China (41206139, 41376160), regional innovation demonstration project of Guangdong Province marine economic development (GD2012-D01-002), National Marine Public Welfare Research Project of China (201305017) and National Key Technologies R&D Program (2011BAE06B04-03). The funders had no role in study design, data collection and analysis, decision to publish, or preparation of the manuscript.

Competing Interests: The authors have declared that no competing interests exist.

* Email: shuhuaqi@scsio.ac.cn

Introduction

Although once thought to be an uninhabitable milieu owing to its extreme conditions, the deep-sea is now recognized as a home to rich and largely microbial communities [1]. Whitman et al. [2] reported that deep-sea-derived microbial communities, mainly composed of bacteria and archaea, accounted for a total cellular carbon content of approximately 3×10^{17} g. Besides bacteria and archaea [3–5], fungi in deep-sea environments have been extensively studied [6–8]. The isolation of deep-sea fungi was first reported approximately 50 years ago from the Atlantic Ocean at a depth of 4450 m [9]. Recently, an increasing number of fungal species were found in several deep-sea environments, e.g. sediments from Gulf of Mexico [1] and Mariana Trench at 11500 m depth [10], calcareous sediments [11], the Chagos Trench at a depth of 5500 m [12] and the Central Indian Basin at about 5000 m depth [8]. Orsi et al. [13] revealed that the deep-sea sediments are vast habitats for fungal life where cell may live on geologic timescale and these active fungi have an overlooked role in organic carbon turnover, which provide a direct evidence for active fungal metabolism in the deep-sea environments. Despite recent advances, the distribution and diversity of fungal communities in deep-sea environments are still largely unknown. With the recent development of more advanced instruments designed for sampling and researching life at greater depths, there has been more interest in evaluating the diversity and ecological role of fungi from deep-sea environments [14–20].

Traditionally, fungal diversity studies on deep-sea environmental samples have been based on cultivation techniques [21–24]. Fungi from deep-sea environments do not necessarily require extreme culture conditions, and many studies on deep-sea fungi describe using cultivation methods under standard laboratory conditions [21,22]. Currently, more than 120 fungal species have been isolated from deep-sea environments [6,18,21–24]. Targeted environmental sequencing analyses, however, have indicated that cultivable fungi are only a small fraction of the total number of fungi inhabiting deep-sea environments [7]. The molecular phylogenetic analysis of clone libraries constructed from environmental samples has become the gold-standard in fungal diversity research, and this technique is thought to be able to detect a wider range of fungi that more accurately represent the investigated environment [25]. To date, several unknown novel phylotypes including DSF-group with the phylum Ascomycota [17,26], KML11 clade and *Rozella* in the new described Cryptomycota [26,27] and BCGI clade [17], have been discovered by molecular phylogenetic analysis.

This technique of molecular phylogenetic analysis, however, can be easily biased at many steps of the process, such as PCR primer selection and the DNA extraction method used [20]. Previous studies have shown that the 18S rRNA technique is a valuable tool for assessing the global diversity of eukaryotes [28,29]. This approach, however, is limited because identification is often restricted to the genus or family level [30]. Buchan et al. [31] reported that the internal transcribed spacer (ITS) regions in fungal rDNA exhibit a high degree of polymorphism between species and are thought to be highly conserved within species. ITS regions can provide better taxonomic resolution than 18S rDNA sequences [16].

To identify the maximum amount of fungal species in the five deep-sea sediments collected from the East Indian Ocean and to find new sequences for phylogenetic studies, environmental gene libraries were constructed after amplifying the sediment DNA using an ITS rRNA gene primer set. Furthermore, the culturable fungi in these sediments were also isolated and identified by amplifying and sequencing the ITS rRNA gene.

Materials and Methods

Ethics statement

All the five sampling locations in this study were included in high seas. Permits for sediment sampling were provided by Ministry of Foreign Affairs of the People's Republic of China. No specific permissions were required for these locations and the field studies did not involve endangered or protected species.

Study site and sample collection

Five deep-sea sediment samples (A–E) were collected using a Remote Operated Vehicle (ROV) during the East Indian Ocean Open Cruise in March 2013 (Fig. S1 in File S1). The coordinates of Sample A–E were shown in Table 1. The collected sediment samples were mostly undisturbed and compact. The average length of the sediment cores collected from these locations was approximately 30 cm. Sub-cores of these samples were collected from a box corer using an alcohol-sterilized PVC cylinder that had a 5 cm inner diameter. Subsections of these samples were cut from the sediment sub-cores and immediately stored in sterile plastic bags to avoid any aerial contamination [18]. The bags were closed with rubber bands and transported to the laminar flow hood in the laboratory on board. A portion of sediment from the middle of each sub-sample that had not been in contact with the PVC cylinder wall was removed using an alcohol-flame-sterilized spatula and placed in a sterile vial for fungal isolation [21]. Fractions of these sediments were immediately stored at −20°C for direct DNA extraction after fungal isolation.

Fungal isolation and identification

The cultivation and isolation methods for fungi from deep-sea environments do not differ fundamentally from the methods used for fungi from shallow marine environments. Physiological analyses have demonstrated that deep-sea-derived fungi are able to grow in deep-sea salinity and at low temperatures [32]. Three different methods were used for fungal isolation in this study, including the particle plating method [33], dilution plating method [21] and low temperature (10°C) incubation method. These fungal isolation methods are described in more detail in a recently published paper by Zhang et al. [18].

Fungal isolates were identified using a combination of morphology characteristics and the internal transcribed spacer (ITS) sequences. Total genomic DNA was extracted from all of the selected fungal strains using a method described by Lai et al. [16]. From the genomic DNA, nearly full-length ITS sequences were amplified by polymerase chain reaction with the primers ITS1 (5′-TCCGTAGGTGAACCTGCGG-3′) and ITS4 (5′-TCCTCCGCTTATTGATATGC-3′) [34]. These fungal ITS gene sequencing and identification methods are described in more detail in a previously published paper by Toledo-Hernandez et al. [35].

DNA extraction, PCR amplification and clone library construction

Environmental DNA was isolated from two grams of sediment sample from each frozen subsection of the sediment cores using a soil DNA extraction kit (Omega Bio-Tek, Inc., Norcross, GA, USA) according to the manufacturer's instructions and sterile techniques to avoid cross contamination. The DNA samples from the five sediments were amplified using the primers ITS1 and ITS4. The polymerase chain reaction mixture (20 µl) consisted of 2 µl 10× PCR buffer (500 mM KCl, 100 mM Tris-HCl, 15 mM 1% (*w*/*v*) $MgCl_2$, and Triton X-100), 1.6 µl of 2.5 mM dNTP, 0.8 µl of each primer, 0.2 µl of 5 U Taq DNA polymerase (Takara Biotechnology Co., Ltd., Dalian, China), 13.6 µl of water, and 1.0 µl of template DNA (10–100 ng). PCR was conducted using

Table 1. The coordinates of Sample A–E.

Samples	Latitude	Longitude	Depth (m)
A	−2°57′N	95°19′E	4810
B	00°00′N	90°57′E	4532
C	00°30′N	82°03′E	4530
D	7°57′N	89°27′E	4614
E	10°00′N	84°33′E	4571

Table 2. Phylogenetic affiliations of uncultured fungi obtained from deep-sea sediment samples A–E.

OTU	**Closest identified relative**			**The number of clones**				
no.	**Taxon (Fungal phylum)**	**GenBank accession no.**	**Similarity %**	**A**	**B**	**C**	**D**	**E**
OTU-01	*Phoma glomerata* (Ascomycota)	EU273521	100			2		5
OTU-02	*Cryptococcus curvatus* (Basidiomycota)	KF472136	99			57		
OTU-03	*Phoma herbarum* (Ascomycota)	KC311476	99					1
OTU-04a	*Rhizoscyphus ericae* (Ascomycota)	JQ711893	98	1				
OTU-05a	*Sporobolomyces lactosus* (Basidiomycota)	HQ914907	99					1
OTU-06a	*Trichoderma asperellum* (Ascomycota)	KC479819	100	1				
OTU-07a	*Trichosporon moniliiforme* (Basidiomycota)	AF444415	100					27
OTU-08a	Uncultured fungus clone (Ascomycota)	GU211938	100				27	
OTU-09a	*Xeromyces bisporus* (Ascomycota)	GU733338	99	9				
OTU-10a	Uncultured ascomycete (Ascomycota)	EU046087	99	1				
OTU-11	Uncultured *Geomyces* (Ascomycota)	JQ346989	99	1				
OTU-12	*Basidiomycete* sp. (Basidiomycota)	EU871524	99	2				
OTU-13	Uncultured soil fungus (Basidiomycota)	DQ420877	100				14	
OTU-14	*Cryptococcus fragicola* (Basidiomycota)	AB035588	99		5			
OTU-15	*Cryptococcus podzolicus* (Basidiomycota)	FN428930	99	1				
OTU-16a	*Guehomyces pullulans* (Basidiomycota)	AF444418	100			2		
OTU-17	*Basidiomycete* sp. (Basidiomycota)	EU871524	99			2		
OTU-18	*Rhodotorula slooffiae* (Basidiomycota)	AB566328	99		14			
OTU-19	*Sterigmatomyces halophilus* (Basidiomycota)	NR073302	100	37				
OTU-20a	Uncultured compost fungus (Basidiomycota)	DQ365334	99		5			
OTU-21	Uncultured *Mortierella* (Zygomycota)	JF831505	100	1				
OTU-22	Uncultured *Mortierella* (Zygomycota)	JF831503	99	25				
OTU-23a	Uncultured soil fungus (Ascomycota)	EU826926	99	2				
OTU-24a	*Alternaria alternata* (Ascomycota)	GQ916545	99					39
OTU-25a	*Cladosporium tenuissimum* (Ascomycota)	AJ300331	100					3
OTU-26	*Alternaria* sp. (Ascomycota)	KF888649	99					2
OTU-27	*Aspergillus penicillioides* (Ascomycota)	AY373862	97	2				
OTU-28	Uncultured fungus clone (Ascomycota)	HQ143117	99				32	
OTU-29	*Candida etchellsii* (Ascomycota)	JQ653271	99		24			
OTU-30	*Candida inconspicua* (Ascomycota)	AB179767	99			6		
OTU-31	*Candida sake* (Ascomycota)	AJ549822	99				2	
OTU-32	*Candida xylopsoci* (Ascomycota)	FM178339	100				2	
OTU-33a	*Cladophialophora chaetospira* (Ascomycota)	EU035404	93	2				
OTU-34	*Cladosporium cladosporioides* (Ascomycota)	GU932679	99					2

Table 2. Cont.

OTU	Closest identified relative			The number of clones				
no.	Taxon (Fungal phylum)	GenBank accession no.	Similarity %	A	B	C	D	E
OTU-35a	*Cladosporium sphaerospermum* (Ascomycota)	GU017501	100	2				
OTU-36a	Uncultured soil fungus (Basidiomycota)	DQ420860	89	2		1		
OTU-37a	*Eurotium rubrum* (Ascomycota)	AY373891	99	1				
OTU-38a	*Fusarium solani* (Ascomycota)	JQ910159	100		5			
OTU-39a	*Galactomyces candidum* (Ascomycota)	JN974290	100		22	19		11
OTU-40a	*Dipodascus australiensis* (Ascomycota)	HQ115737	99		4			
OTU-41	*Geomyces pannorum* (Ascomycota)	DQ189228	99	1				
OTU-42	*Hortaea werneckii* (Ascomycota)	GQ334385	99					2
OTU-43a	*Hypocrea virens* (Ascomycota)	GU130297	100	1			12	
OTU-44	*Phoma sp.* (Ascomycota)	HQ630999	98			1		1
OTU-45	*Leptosphaeria* sp. (Ascomycota)	AB752252	99					1
Total				92	79	90	89	95

OTUs marked by a letter (a) are new reports for deep-sea environments. Bolded and italicized OTUs are affiliated with yeasts and filamentous fungi, respectively, and the remaining OTUs are affiliated with unidentified yeasts or filamentous fungi.

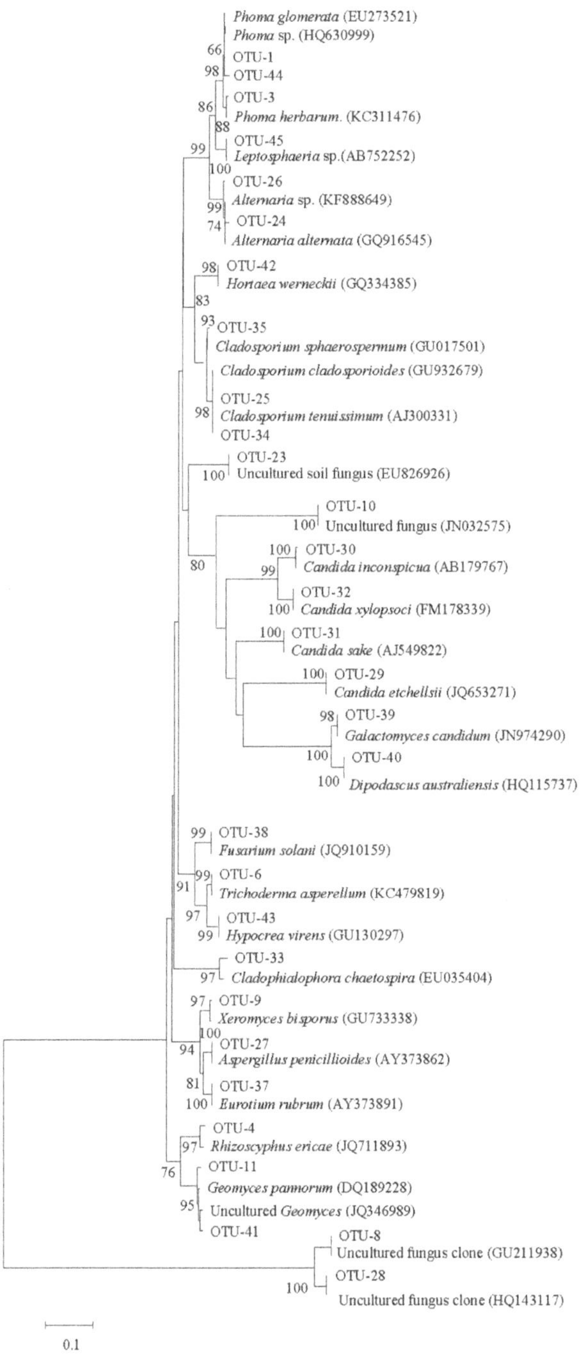

Figure 1. Neighbor-joining phylogenetic tree from analysis of ITS rDNA from 29 Ascomycota representative sequences in five libraries. The numbers at the nodes are the percentages indicating the level of bootstrap support based on a neighbor-joining analysis of 1000 resampled data sets. Only values >50% are shown. The scale bar represents 0.1 substitutions per nucleotide position.

an Eppendorf Mastercycler (Eppendorf German Co., Ltd., Hamburger, German) and the following program: denaturation at 95°C for 5 min, 25 cycles of 30 s at 95°C, 30 s at 55°C, and 90 s at 72°C, and a final extension at 72°C for 10 min. Reaction mixtures lacking template DNA were used as negative controls. Amplified products were gel-purified, ligated with pMD18-T easy vector (Takara Biotechnology Co., Ltd., Dalian, China) and transformed into *Escherichia coli* cells following the manufacturer's instructions. Transformants were grown overnight at 37°C on Luria-Bertani agar containing 100 μg/ml ampicillin. The presence of insert was confirmed by PCR with M13 forward and reverse primers. One microliter of broth containing the clone was added to 25 μl of PCR reaction mixture. The PCR protocol included an initial hot start incubation (5 min at °C) followed by 34 cycles of denaturation at 94°C for 30 s, annealing at 55°C for 30 s and extension at 72°C for 1 min, and then a final extension at 72°C for 5 min [8]. Clones containing the positive insert were further processed for plasmid isolation and purification using the Millipore plasmid preparation Kit (Millipore, USA). Sequencing of the PCR products from the plasmids was conducted by Invitrogen (China). A total of five environmental gene libraries were constructed from the DNA samples from the five deep-sea sediments. Approximately 100 clones were screened from each library.

Phylogenetic analyses

All of the vector sequences from the sequenced fungal clones were analyzed using the rRNA Database Project CHECK_CHIMERA program to detect and eliminate the potential chimeric sequences. Pairwise alignment of the sequences was conducted using Clustal W in the MEGA software version 5.0. Conserved motifs were identified, and the sequences were trimmed manually. Clones were grouped into operational taxonomic units (OTUs) using a sequence similarity cut-off value of 98% and the Mothur software version 1.32.1 [36]. Rarefaction curves for the number of observed OTUs were calculated with fungal assemblage at each dataset. A representative sequence from each OTU was queried against an NCBI-GenBank BLASTN search.

Nucleotide sequence accession number

The ITS sequences for 20 culturable fungal isolate representatives and 45 uncultured fungal clone representatives obtained in this study were deposited in GenBank under accession numbers KJ173524–KJ17352 and KJ173554–KJ173590.

Results

Phylogeny of environmental fungal ITS-rDNA sequences

A total of 515 clones from five deep-sea sediment samples A–E (Fig. S1 in File S1) from the East Indian Ocean were sequenced. Of the resulting sequences, 445 sequences were found to be fungal, and a total of 45 operational taxonomic units (OTUs) (Table 2) were identified after clustering based on a 98% sequence identity criterion. The other 70 clones (~13.6%) were eukaryotic or chimeric in nature and were excluded from this study. Rarefaction curves (Fig. S2 in File S1) were constructed for the ITS clone libraries from samples A–E. Rarefaction curves for three samples (B, C and D) demonstrated a plateau, which indicates that the number of sequences analyzed may sufficiently represent the fungal diversity in these samples. While the rarefaction curves of Sample A and E did not reach a plateau. It is likely that the fungal diversity of the two samples is higher than what was detected in this study.

Most of the ITS sequences from the 45 OTUs demonstrated ≥ 98% similarity with sequences from their closest relative taxa in GenBank. Three new sequence types, however, only demonstrated 89%–97% similarity with the existing database. Therefore, these results from the phylogenetic analysis suggest that OTU-27, 33 and 36 are novel fungal taxa that are not closely related to previous identified fungal ITS sequences in public databases (Table 2). The composition indicates that a majority (419/445) of

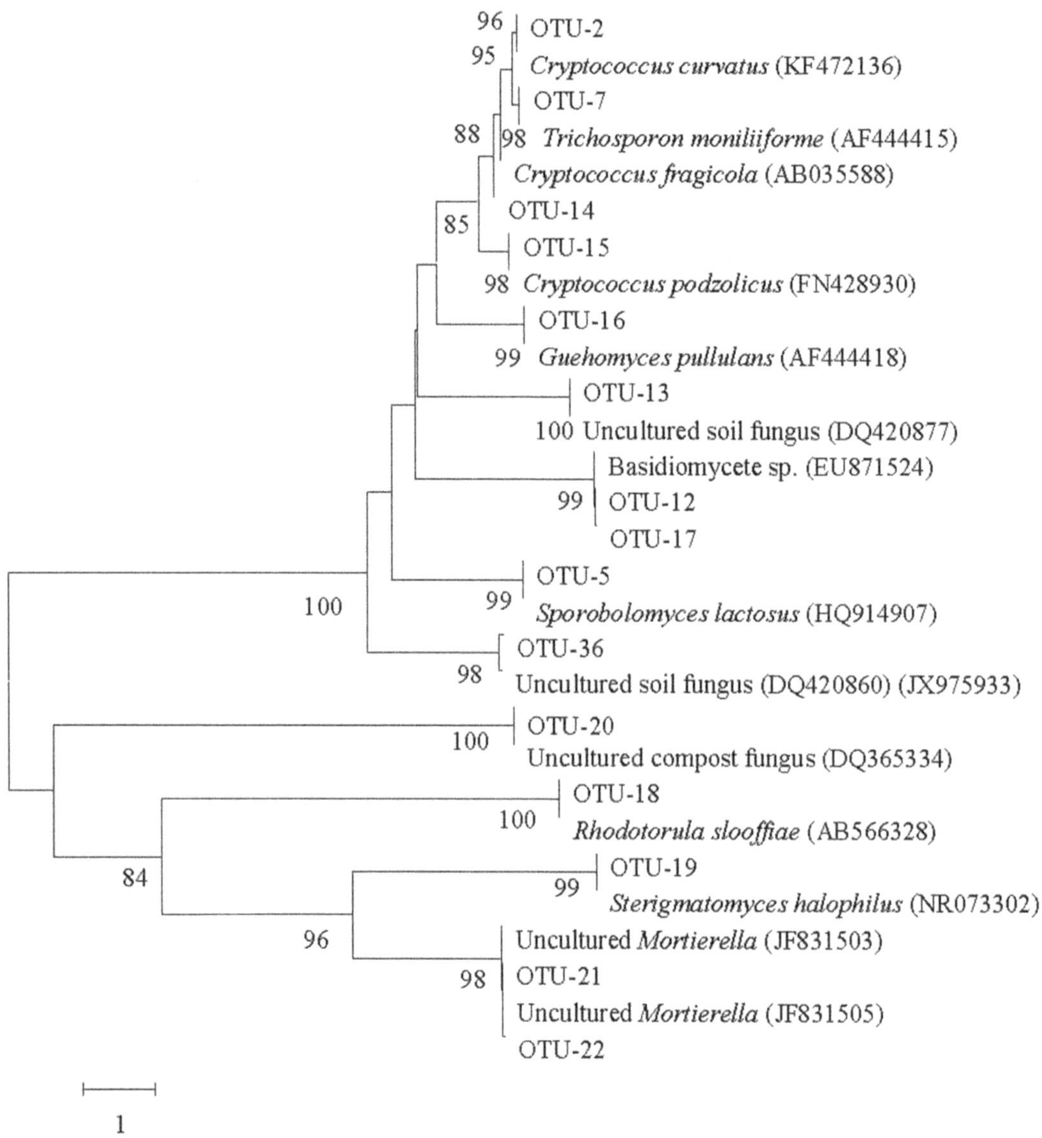

Figure 2. Neighbor-joining phylogenetic tree from analysis of ITS rDNA from 26 Basidomycota and Zygomycota representative sequences (OTU-21 and 22) in five libraries. The numbers at the nodes are percentages indicating the level of bootstrap support based on a neighbor-joining analysis of 1000 resampled data sets. Only values >50% are shown. The scale bar represents 1 substitution per nucleotide position.

these amplified ITS sequences belong to the phyla Ascomycota (276 clones from 29 OTUs) and Basidomycota (143 clones from 24 OTUs) (Fig. 1 and 2). The other remaining sequences belong to the phylum Zygomycota (26 clones from OUT-21 and 22) (Fig. 2).

Furthermore, the phylogenetic analyses revealed that 240 clones from 17 OTUs were most closely related to cultivable yeast forms, including three species of genus *Cryptococcus* (63 clones), *Galactomyces candidum* (52 clones), *Sterigmatomyces halophilus* (37 clones), four species of genus *Candida* (34 clones), *Trichosporon moniliiforme* (27 clones), *Rhodotorula slooffiae* (14 clones), *Basidiomycete* sp. (4 clones), *Dipodascus australiensis* (2 clones), *Guehomyces pullulans* (2 clones), *Hortaea werneckii* (2 clones) and *Sporobolomyces lactosus* (one clone) (Fig. 1 and Table 2). Another 120 clones from 20 OTUs were closely related to filamentous fungi, including *Aspergillus*, *Alternaria*, *Cladophialophora*, *Cladosporium*, *Eurotium*, *Fusarium*, *Geomyces*, *Hypocrea*, *Leptosphaeria*, *Mortierella*, *Phoma*, *Rhizoscyphus*, *Trichoderma* and *Xeromyces* (Fig. 1 and Table 2). The remaining 85 clones from 8 OTUs were closely related to uncultured fungi from soil or plant ecological systems that have been deposited in the NCBI database [37–41].

Culturable fungal isolates and species richness

A total of 78 fungal isolates belonging to 20 phylotypes (Fig. 3) were recovered using traditional cultivation. The ITS sequencing results showed that most of these fungal sequences demonstrated >97% similarity with sequences from their closet relative species. The one isolate EIODSF013 (accession number KJ173536), however, only demonstrates 95% similarity with the existing sequence (accession number JX981490) in the NCBI database (Table 3). A multigene analysis combined with detailed morphological and ultra structural studies, however, are needed to determine the novelty of this isolate. Most of the 78 fungal isolates belonged to Ascomycota, including two yeast and 16 filamentous

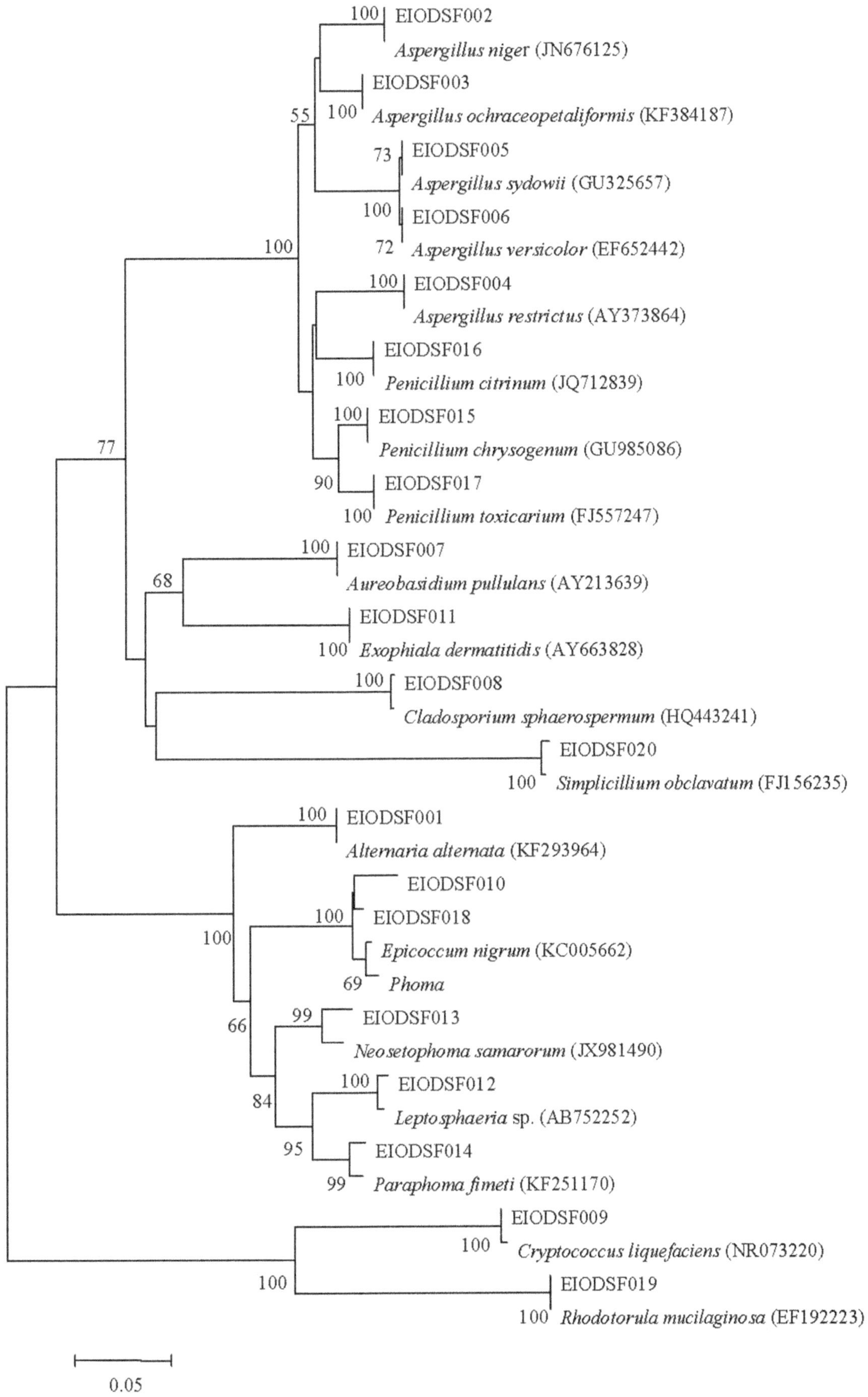
100
EIODSF002
Aspergillus niger (JN676125)
EIODSF003
55
100
Aspergillus ochraceopetaliformis (KF384187)
73
EIODSF005
Aspergillus sydowii (GU325657)
100
EIODSF006
100
72
Aspergillus versicolor (EF652442)
100
EIODSF004
Aspergillus restrictus (AY373864)
EIODSF016
100
Penicillium citrinum (JQ712839)
100
EIODSF015
77
Penicillium chrysogenum (GU985086)
90
EIODSF017
100
Penicillium toxicarium (FJ557247)
100
EIODSF007
68
Aureobasidium pullulans (AY213639)
EIODSF011
100
Exophiala dermatitidis (AY663828)
100
EIODSF008
Cladosporium sphaerospermum (HQ443241)
EIODSF020
100
Simplicillium obclavatum (FJ156235)
100
EIODSF001
Alternaria alternata (KF293964)
EIODSF010
100
EIODSF018
100
Epicoccum nigrum (KC005662)
69
Phoma
99
EIODSF013
66
Neosetophoma samarorum (JX981490)
100
EIODSF012
84
Leptosphaeria sp. (AB752252)
95
EIODSF014
99
Paraphoma fimeti (KF251170)
EIODSF009
100
Cryptococcus liquefaciens (NR073220)
100
EIODSF019
100
Rhodotorula mucilaginosa (EF192223)
0.05

Figure 3. Neighbor-joining phylogenetic tree from analysis of ITS sequences from fungi isolated from five deep-sea sediments from the East Indian Ocean. The numbers at the nodes are the percentages indicating the level of bootstrap support based on a neighbor-joining analysis of 1000 resampled data sets. Only values >50% are shown. The scale bar represents 0.05 substitutions per nucleotide position.

fungal species. Among these species, *Aspergillus* sp., *Penicillium* sp. and *Simplicillium obclavatum* were the most diverse and common, while *Alternaria alternata*, *Aureobasidium pullulans*, *Cryptococcus liquefaciens*, *Exophiala dermatitidis*, *Epicoccum nigrum* and *Neosetophoma samarorum* were the rarest fungal species with only single or double isolates. The remaining species occurred as several isolates.

Discussion

New insights into the fungal communities in deep-sea environments

The fungal diversity in deep-sea environments has recently gained an increasing amount of attention. Our knowledge and understanding of the true fungal diversity and the roles this diversity plays in deep-sea environments, however, is still limited. The aim of the present study was to obtain the maximum amount of fungal diversity from deep-sea sediments collected in the East Indian Ocean. A total of 45 fungal OTUs and 20 culturable fungal phylotypes were recovered in this study (Table 2 and 3), which revealed that there is a great amount of fungal diversity in the deep-sea sediments collected in the East Indian Ocean. Recently, many studies have shown that there is an increasing number of culturable fungal species and uncultured fungal clones in deep-sea sediments from the Central Indian Basin [8,21,42], South China Sea [16,18] and Eastern Equatorial Pacific [43]. Few studies, however, have specifically focused on the fungal diversity in the deep-sea environments from the East Indian Ocean. This study is the first to report new insight into the fungal communities in deep-sea sediments from the East Indian Ocean using a combination of targeted environmental sequencing and cultivation. These findings increase our knowledge and understanding of the fungal diversity in deep-sea environments.

In this study, most of the clones (94.2%) detected by targeted environmental sequencing and all of the culturable fungal isolates recovered using traditional cultivation belonged to the phyla Ascomycota and Basidiomycota (Table 2 and 3). The remaining 5.8% clones (OTU-21 and 22) belonged to the phylum Zygomycota (Table 2). In addition to these three phyla, the phylum Chytridiomycota has also been detected in many deep-sea environments, e.g. Izu-Ogasawara Trench [6]. In this study, however, no putative Chytridiomycota sequences were detected in any of the five sediment samples collected from the East Indian Ocean. We do not believe that our methods were incapable of detecting these higher taxonomic groups because the same primer set has been shown to be able to detect a diverse range of these taxa from deep-sea environments [16,44,45]. Therefore, these results suggest that the fungal communities in the deep-sea sediments from the East Indian Ocean tend to be dominated by Ascomycota and Basidomycota, while other fungal taxonomic groups are rare or absent.

Notably, 44.4% (20 out of 45 OTUs, Table 2) fungal OTUs and 30% (6/20) culturable fungal phylotypes (Table 3) identified in this study are new reports for deep-sea sediments (Table 2 and 3). Some of these species demonstrated phylogenetic similarity to fungal species and genera known to be found in shallow-sea environments. For examples, *A. alternata* was found in tropic sea grass *Enhalus acoroides* [45]; *E. rubrum*, *H. virens* and *Leptosphaeria* sp. were found in marine mangroves [46]; *A. ochraceopetaliformis* and *C. tenuissimum* were found in marine corals and sponges [36,47]; *C. inconspicua* may be found in crabs [48]. Other species were affiliated with culturable or uncultured fungi from soil or plant ecological systems. For examples, *R. ericae* was isolated from lodgepole pine in central British Columbia [49]; and *D. australiensis* was recovered from agricultural soil [50]; uncultured soil fungus (OTU-13) was detected in a genetically modified rice ecosystem [41]. These results may appear to be inconsistent, but previous studies have shown that while a majority of the fungal isolates from deep-sea environments were psychrotolerant, they grew more rapidly at 30°C than 5°C [42,51]. Moreover, most of the fungi isolated from deep-sea environments are halotolerant and do not absolutely require seawater for growth [24,32]. It is possible that there are no true indigenous fungi in deep-sea environments and that species from terrestrial environments have gradually adapted to deep-sea extreme conditions [6].

Yeast fungi detected by targeted environmental sequencing

Our targeted environmental sequencing results showed 53.9% fungal clones were closely related to yeast fungi. Among the yeast forms in this study, most of the clones represented phylotypes relevant to the genera *Cryptococcus* and *Galactomyces*. Some of *Cryptococcus*-related phylotypes are psychrotolerant and have been found in the majority of deep subseafloor samples from North Pond, Hydrate Ridge, Peru Margin and Eastern Equatorial Pacific [43], but *Galactomyces*-related phylotypes are rarely recovered from deep-sea environments. Four OTUs were closely related to *Candida* sp., which are considered to be associated with ecosystems in anaerobic environments [26].

In addition to the previously mentioned yeast phylotypes, *Rhodotorula*, *Sterigmatomyces* and *Trichosporon* yeasts were also found to be abundant in this study. Previous studies have shown that *Rhodotorula* spp. demonstrate remarkable ubiquity based on their presence in several different habitats, such as deep-sea sediments [52]. Phylotypes related to the genera *Trichosporon* and *Sterigmatomyces* are known pathogens or parasites of marine animals, which suggests that they may also be opportunistic pathogens or parasites of deep-sea animals [47,53]. Furthermore, OUT-12 and 17 are related to unidentified Basidiomycetious yeasts (EU871524) and were detected in a water column from the Equatorial Indian Ocean (unpublished). The remaining yeast forms were singletons or doubletons, indicating the low abundance of these clones.

Filamentous fungi dominated the fungal community based on traditional cultivation

In this study, filamentous fungi dominated the fungal community using traditional cultivation. The five genera *Aspergillus*, *Penicillium*, *Simplicillium*, *Cladosporium* and *Phoma* were distributed in more than three sediments from the East Indian Ocean (Table 2). Members of the mycelia genera *Aspergillus* and *Penicillium* are known to be globally distributed fungal taxa. It seems doubtful that these fungal species are indigenous to deep-sea environments, and evidence of physiological adaption of these species to deep-sea environments had been reported by Raghukumar et al. [54]. Moreover, *Aspergillus* spp. were frequently detected in anaerobic marine sediments and were shown to play

Table 3. Phylogenetic affiliations of culturable fungi obtained from deep-sea sediment samples A–E.

Isolates	Closest identified relative			The number of isolates				
no.	**Fungal genera or species (Phylum)**	**GenBank accession no.**	**Similarity %**	**A**	**B**	**C**	**D**	**E**
EIODSF 001a	*Alternaria alternata* (*Ascomycota*)	KF293964	99%	2				
EIODSF 002	*Aspergillus niger* (*Ascomycota*)	JN676125	99%				4	
EIODSF 003a	*A. ochraceopetaliformis* (*Ascomycota*)	KE384187	99%	1	1		4	1
EIODSF 004	*A. restrictus* (*Ascomycota*)	JX156352	99%	2	1			
EIODSF 005	*A. sydowii* (*Ascomycota*)	GU325657	99%	2				
EIODSF 006	*A. versicolor* (*Ascomycota*)	EF652442	99%	2			1	2
EIODSF 007	*Aureobasidium pullulans* (*Ascomycota*)	AY213639	99%					2
EIODSF 008	*Cladosporium sphaerospermum* (*Ascomycota*)	HQ443241	99%	2	1			3
EIODSF 009	*Cryptococcus liquefaciens* (*Basidiomycota*)	AF145331	99%	1				
EIODSF 010	*Epicoccum nigrum* (*Ascomycota*)	KC005662	98%					1
EIODSF 011	*Exophiala dermatitidis* (*Ascomycota*)	AY663828	99%				1	
EIODSF 012	*Leptosphaeria sp.* (*Ascomycota*)	AB752252	99%					6
EIODSF 013a	*Neosetophoma samarorum* (*Ascomycota*)	JX981490	95%					1
EIODSF 014a	*Paraphoma fimeti* (*Ascomycota*)	KF251170	99%			2		1
EIODSF 015	*Penicillium chrysogenum* (*Ascomycota*)	GU985086	99%			2		
EIODSF 016	*P. citrinum* (*Ascomycota*)	JN624897	99%	1		4	2	
EIODSF 017a	*P. toxicarium* (*Ascomycota*)	FJ557247	99%	3				
EIODSF 018	*Phoma sp.* (*Ascomycota*)	EF120404	99%	2	1			2
EIODSF 019	*Rhodotorula mucilaginosa* (*Basidiomycota*)	EF192223	99%	4				
EIODSF 020a	*Simplicillium obclavatum* (*Ascomycota*)	FJ156235	99%	5	3	2	2	1
Total				27	7	10	14	20

Species marked by a letter (a) are new reports for deep-sea environments. Bolded and italicized isolates are included in yeasts and filamentous fungi, respectively.

an important role in the denitrification process [20]. These findings suggest that these *Aspergillus* species may play a potentially versatile role for fungi in major ecological processes in the deep-sea environments. The genus *Simplicillium* was segregated from *Verticillium* and contains 10 species [55,56]. These *Simplicillium* species occur in a broad range of ecological niches, such diseased plant tissue, soil, human nails, dog tissue and mushrooms [55,57]. In this study, we report for the first time the presence of *Simplicillium* sp. in deep-sea sediments. *Phoma* species are known to be associated with not only land plants but also with marine plants. Previous studies have demonstrated that many terrestrial microorganisms could accumulate in deep-sea sediments [58,59]. One terrestrial fungal genus, *Cladosporium* was isolated from three deep-sea sediments below 3000 m in this study, which indicates that sedimentation may be an important factor responsible for the accumulation of facultative marine fungi in deep-sea sediments.

In addition, only two isolates belonging to *A. alternate* were recovered using traditional cultivation, but this species was also detected by targeted environmental sequencing in this study. Therefore, *A. alternata* should be abundant in these deep-sea sediments from the East Indian Ocean. Previous studies have shown that the genera *Alternaria* was found in nearly every survey of free-living fungal communities associated with biological soil crusts [60,61]. Few *Alternaria* sp., however, were detected in deep-sea environments.

Comparison of fungal community by targeted environmental sequencing and traditional cultivation

A distinct difference in the fungal community based on targeted environmental sequencing compared with traditional cultivation was that the Zygomycota spp. was not recovered by traditional cultivation but was detected by targeted environmental sequencing. Previous studies have also shown that Zygomycota spp. could be found in different deep-sea environments using targeted environmental sequencing [17,24]. Currently, however, there are no reports of isolating Zygomycota species from deep-sea environment cultures. After a further comparison of the fungal phylotypes recovered using these two methods, it was found that the majority of the fungal phylotypes recovered using targeted environmental sequencing could not be recovered using a traditional cultivation method. This finding is consistent with the findings published by Le Calvez et al. [24], who reported that there are striking differences in the deep-sea fungal diversity results when using targeted environmental sequencing compared with traditional cultivation. These findings suggest that a combination of targeted environmental sequencing and traditional cultivation will generate a more accurate assessment of the fungal diversity in deep-sea environments compared with using targeted environmental sequencing or traditional cultivation alone. Furthermore, to obtain an even greater abundance of deep-sea fungi, it is necessary to combine the methods used in this study with other methods, such as the microscopic observation of samples appropriately staining, FISH, measuring ergosterol, metagenomic methods, and other new powerful tools expected to be developed in the future [62].

Supporting Information

File S1 Contains Fig. S1 Map of the East Indian Ocean, location and depth of the sampling site and Fig. S2 Rarefaction curves constructed for ITS clone libraries from each of the five sampling sites. ITS, internal transcribed spacer.

Author Contributions

Conceived and designed the experiments: XZ, SQ. Performed the experiments: XZ. Analyzed the data: XX. Contributed reagents/materials/analysis tools: GT. Contributed to the writing of the manuscript: XZ. Helped perform the analysis with constructive discussions: XN.

References

1. Thaler AD, Dover CLV, Vilgalys (2012) Ascomycete phylotypes recovered from a Gulf of Mexico methane seep are identical to an uncultured deep-sea fungal clade from the Pacific. Fungal Ecol 5: 270–273.
2. Whitman WB, Coleman DC, Wiebe WJ (1998) Prokaryotes: the unseen majority. Proc Acad Natl Sci USA 95: 6578–6583.
3. DeLong EF, Pace NR (2001) Environmental diversity of bacteria and archaea. Syst Biol 50: 470–478.
4. Sogin M, Morrison HG, Huber JA, Welch DM, Huse SM, et al. (2006) Microbial diversity in the deep sea and underexplored "rare biosphere". Proc Natl Acad Sci USA103: 12115–12120.
5. Luna GM, Stumm K, Pusceddu A, Danovaro R (2009) Archaeal diversity in deep-sea sediments estimated by means of different terminal-restriction fragment length polymorphisms (T-RFLP) Protocols. Curr Microbiol 59: 356–361.
6. Nagano Y, Nagahama T (2012) Fungal diversity in deep-sea extreme environments. Fungal Ecol 5: 463–471.
7. Nagano Y, Nagahama T, Hatada Y, Nunoura T, Takami H, et al. (2010) Fungal diversity in deep-sea sediments-the presence of novel fungal groups. Fungal Ecol 3: 316–325.
8. Singh P, Raghukumar C, Verma P, Shouche Y (2012) Assessment of fungal diversity in deep-sea sediments by multiple primer approach. World J Microbiol Biotechnol 28: 659–667.
9. Roth FJ, Orpurt PA, Ahearn DJ (1964) Occurrence and distribution of fungi in a subtropical marine environment. Can J Bot 42: 375–383.
10. Takami H, Inoue A, Fuji F, Horikoshi K (1997) Microbial flora in the deepest sea mud of the Mariana Trench. FEMS Microbiol Lett 152: 279–285.
11. Raghukumar C, Raghukumar S (1998) Barotolerance of fungi isolated from deep-sea sediments of the Indian Ocean. Aquat Microb Ecol 15: 153–163.
12. Raghukumar C, Raghukumar S, Sheelu G, Gupta S, Nagendernath B, et al. (2004) Buried in time: culturable fungi in a deep-sea sediment core from the Chagos Trench, Indian Ocean. Deep Sea Res I 51: 1759–1768.
13. Orsi WD, Edgcomb VP, Christman GD, Biddle JF (2013) Gene expression in the deep biosphere. Nature 499: 205–210.
14. Bhadury P, Bik H, Lambshead JD, Austen MC, Smerdon GR, et al. (2011) Molecular diversity of fungal phylotype co-amplified alongside nematodes from coastal and deep-sea marine environments. PloS One 6(10): e26445.
15. Takishita K, Tsuchiya M, Reimer JD, Maruyama T (2006) Molecular evidence demonstrating the basidiomycetous fungus *Cryptococcus curvatus* is the dominant microbial eukaryotic in sediment at the Kuroshima Knoll methane seep. Extremophiles 10: 165–169.
16. Lai X, Cao L, Tan H, Fang S, Huang Y, et al. (2007) Fungal communities from methane hydrate-bearing deep-sea marine sediments in South China Sea. ISME J 1: 756–762.
17. Nagahama T, Takahashi E, Nagano Y, Abdel-Wahab MA, Miyazaki M (2011) Molecular evidence that deep-branching fungi are major fungal components in deep-sea methane cold seep sediments. Environ Microbiol 13: 2359–2370.
18. Zhang XY, Zhang Y, Xu XY, Qi SH (2013) Diverse deep-sea fungi from the South China Sea and their antimicrobial activity. Curr Microbiol 67: 525–530.
19. Zhang XY, Xu XY, Peng JX, Ma CF, Nong XH, et al. (2014) Antifouling potentials of eight deep-sea-derived fungi from the South China Sea. J Ind Microbiol Biotechnol 41: 741–748.
20. Jebaraj CS, Raghukumar C, Behnke A, Stoeck T (2010) Fungal diversity in oxygen-depleted regions of the Arabian Sea revealed by targeted environmental sequencing combined with cultivation. FEMS Microbiol Ecol 71: 399–412.
21. Damare S, Raghukumar C, Raghukumar S (2006) Fungi in deep-sea sediments of the central Indian basin. Deep-sea Res Part I 53: 14–27.
22. Burgaud G, Arzur D, Durand L, Cambon-Bonavita MA, Barbier G (2010) Marine culturable yeasts in deep-sea hydrothermal vents: species richness and association with fauna. FEMS Microbiol Ecol 73: 121–133.
23. Gadanho M, Sampaio JP (2005) Occurrence and diversity of yeasts in the mid-Atlantic ridge hydrothermal fields near the Azores Archipelago. Microb Ecol 50: 408–417.
24. Le Calvez T, Burgaud G, Mahe S, Barbier G, Vandenkoornhuyse P (2009) Fungal diversity in deep sea hydrothermal ecosystems. Appl Environ Microbiol 75: 6415–6421.

25. Pang KL, Mitchell JI (2005) Molecular approaches for assessing fungal diversity in marine substrata. Bot Mar 48: 332–347.
26. Bass D, Howe A, Brown N, Barton H, Demidova M, et al. (2007) Yeast forms dominate fungal diversity in the deep oceans. Proc Biol Sci 274: 3069–3077.
27. Lara E, Moreira D, Lopez-Garcia P (2010) The environmental clade LKM11 and *Rozella* Form the deepest branching cladeof fungi. Protist, 161: 116–121.
28. Stocek T, Taylor GT, Epstein S (2003) Novel eukaryotes from the permanently anoxic Cariaco Basin (Caribbean Sea). Appl Environ Microbiol 69: 5656–5663.
29. Stocek T, Epstein S (2003) Novel eukaryotic lineages inferred from small-subunit rRNA analyses of oxygendepleted marine environments. Appl Environ Microbiol 69: 2657–2663.
30. Anderson IC, Cairney JWG (2004) Diversity and ecology of soil fungal communities: increased understanding through the application of molecular techniques. Environ Microbiol 6: 769–779.
31. Buchan A, Newell SY, Moretam JIL, Moranm MA (2002) Analysis of internal transcribed spacer (ITS) regions of rRNA genes in fungal communities in a southeastern US salt marsh. Microb Ecol 43: 329–340.
32. Burgaud G, Calvez TL, Arzur D, Vandenkoornhuyse P, Barbier G (2009) Diversity of culturable marine filamentous fungi from deep-sea hydrothermal vents. Environ Microbiol 11: 1588–1600.
33. Bills GF, Polishook JD (1994) Abundance and diversity of microfungi in leaf litter of a lowland rain forest in Costa Rica. Mycologia 86: 187–198.
34. White TJ, Bruns TD, Lee SB, Taylor JW (1990) Amplification and direct sequencing of fungal ribosomal RNA genes for phylogenetics. In: Innis MA, Gelfand DH, Sninsky JJ, White TJ, editors. PCR Protocols: A Guide to Methods and Applications. San Diego: Academic. 315–322.
35. Toledo-Hernandez C, Zuluaga-Montero A, Bones-Gonzalez A, Rodriguez JA, Sabat AM, et al. (2008) Fungi in healthy and diseased sea fan (Gorgonia ventalina): is Aspergillus sydowii always the pathogen? Coral Reefs 27: 707–714.
36. Singh P, Raghukumar C, Verma P, Shouche Y (2011) Fungal community analysis in the deepsea sediments of the Central Indian Basin by culture-independent approach. Microb Ecol 61: 507–517.
37. Allison SD, McGuire KL, Treseder KK (2010) Resistance of microbial and soil properties to warming treatment seven years after boreal fire. Soil Biol Biochem 42: 1872–1878.
38. Urban A, Puschenreiter M, Strauss J, Gorfer M (2008) Diversity and structure of ectomycorrhizal and co-associated fungal communities in a serpentine soil. Mycorrhiza 18: 339–354.
39. Waldrop MP, Zak DR, Blackwood CB, Curtis CC, Tilman D (2006) Resource availability controls fungal diversity across a plant diversity gradient. Ecol Lett 9: 1127–1135.
40. Rao S, Hyde KD, Pointing SB (2013) Comparison of DNA and RNA, and cultivation approaches for the recovery of terrestrial and aquatic fungi from environmental samples. Curr Microbiol 66: 185–191.
41. Lee SH, Kim CG, Kang H (2011) Temporal dynamics of bacterial and fungal communities in a genetically modified (GM) rice ecosystem. Microb Ecol 61: 646–659.
42. Singh P, Raghukumar C, Verma P, Shouche Y (2010) Phylogenetic diversity of culturable fungi from the deep-sea sediments of the Central Indian Basin and their growth characteristics. Fungal Divers 40: 89–102.
43. Orsi W, Biddle JF, Edgcomb V (2013) Deep sequencing of subseafloor eukaryotic rRNA reveals active fungi across marine subsurface provinces. PLoS One, 8(2): e56335.
44. Wang Z, Nilsson RH, Lopez-Giraldez F, Zhuang WY, Dai YC, et al. (2011) Tasting soil fungal diversity with earth tongues: phylogenetic test of SATe alignments for environmental ITS data. PLoS One 6(4): e19039.
45. Sakayaroj J, Preefanon S, Supaphon O, Jones EBG, Phongpaichit S (2010) Phylogenetic diversity of endophyte assemblages associated with tropical seagrass *Enhalus acoroides* from Thailand. Fungal Divers 41: 1–19.
46. Liu T, Li ZL, Wang L, Tian L, Pei YH, et al. (2011) A new alkaloid from the marine-derived fungus *Hypocrea virens*. Nat Prod Res 25: 1596–1599.
47. Liu WC, Li CQ, Zhu P, Yang JL, Cheng KD (2010) Phylogenetic diversity of culturable fungi associated with two marine sponges: *Haliclona simulans* and *Gelliodes carnosa*, collected from the Hainan Island coastal waters of the South China Sea. Fungal Divers 42: 1–15.
48. De Araujo FV, Soares CA, Hagler AN, Mendonca-Hagler LC (1995) Ascomycetous yeast communities of marine invertebrates in a southeast Brazilian mangrove ecosystem. Antonie van Leeuwenhoek 68: 91–99.
49. Jones MD, Phillips LA, Treu R, Ward V, Berch SM (2012) Functional responses of ectomycorrhizal fungal communities to long-term fertilization of lodgepole pine (Pinus contorta Dougl. ex Loud. var. latifolia Engelm.) stands in central British Columbia. Appl Soil Ecol 60: 29–40.
50. Gorfer M, Blumhoff M, Klaubauf S, Urban A, Inselsbacher E, et al. (2011) Community profiling and gene expression of fungal assimilatory nitrate reductases in agricultural soil. ISME J 5: 1771–1783.
51. Damare S, Raghukumar C (2008) Fungi and macroaggregation in deep-sea sediments. Microb Ecol 56: 168–177.
52. Nagahama T, Hamamoto M, Nakase T, Takami H, Horikoshi K (2001) Distribution and identification of red yeasts in deep-sea environments around the northwest Pacific Ocean. Antonie van Leeuwenhoek 80: 101–110.
53. Edgcomb VP, Beaudoin D, Gast R, Biddle JF, Teske A (2011) Marine subsurface eukaryotes: the fungal majority. Environ Microbiol 13: 172–183.
54. Raghukumar C, Raghukumar S, Sheelu G, Gupta SM, Nagender NB, et al. (2004) Buried in time: culturable fungi in a deep-sea sediment core from the Chagos Trench, Indian Ocean. Deep-sea Res Part I 51: 1759–1768.
55. Zare R, Gams W (2001) A revision of *Verticillium* section *Prostrata*. IV. The genera *Lecanicillium* and *Simplicillium* gen. nov. Nova Hedwigia 73: 1–50.
56. Zhao D, Liu B, Li LY, Zhu XF, Wang YY, et al. (2013) *Simplicillium chinense*: a biological control agent against plant parasitic nematodes, Biocontrol Sci Technol 23: 980–986.
57. Zare R, Gams W (2008). A revision of *Verticillium* fungicola species complex and its affinity with the genus *Lecanicillium*. Mycol Res 112: 811–824.
58. Baross JA, Hanus FJ, Morita RY (1975) Survival of human enteric and other sewage microorganisms under simulated deep-sea conditions. Appl Microbiol 30: 309–318.
59. Pivkin MV (2000) Filamentous fungi associated with holothurians from the sea of Japan, off the primorye coast of Russia. Biol Bull 198: 101–109.
60. Bates ST, Garcia-Pichel F (2009) A culture-independent study of free-living fungi in biological soil crusts of the Colorado Plateau: their diversity and relative contribution tomicrobial biomass. Environ Microbiol 11: 56–67.
61. Green LE, Porras-Alfaro A, Sinsabaugh RL (2008) Translocation of nitrogen and carbon integrates biotic crust and grass production in desert grassland. J Ecol 96: 1076–1085.
62. Nagahama T, Nagano Y (2012) Cultured and uncultured fungal diversity in deep-sea environments. In: Raghukumar C, editor. Biology of Marine Fungi. Germany: Springer. pp. 173–187.

11

New Luminescence Ages for the Galería Complex Archaeological Site: Resolving Chronological Uncertainties on the Acheulean Record of the Sierra de Atapuerca, Northern Spain

Martina Demuro[1,2]*, Lee J. Arnold[1,2,3], Josep M. Parés[2], Alfredo Pérez-González[2], Ana I. Ortega[2], Juan L. Arsuaga[4,5], José M. Bermúdez de Castro[2], Eudald Carbonell[6]

1 Institute for Photonics and Advanced Sensing, School of Chemistry and Physics, The University of Adelaide, Adelaide, Australia, **2** Centro Nacional de Investigación sobre la Evolución Humana, Burgos, Spain, **3** The Environment Institute and School of Earth and Environmental Sciences, The University of Adelaide, Adelaide, Australia, **4** Centro Mixto Universidad Complutense-Instituto de Salud Carlos III de Evolución y Comportamiento Humanos, Madrid, Spain, **5** Departamento de Paleontología, Facultad de Ciencias Geológicas, Universidad Complutense de Madrid, Madrid, Spain, **6** Institut Català de Paleoecologia Humana i Evolució Social, Àrea de Prehistòria, Universitat Rovira i Virgili, Tarragona, Spain

Abstract

The archaeological karstic infill site of Galería Complex, located within the Atapuerca system (Spain), has produced a large faunal and archaeological record (*Homo sp. aff. heidelbergensis* fossils and Mode II lithic artefacts) belonging to the Middle Pleistocene. Extended-range luminescence dating techniques, namely post-infrared infrared stimulated luminescence (pIR-IR) dating of K-feldspars and thermally transferred optically stimulated luminescence (TT-OSL) dating of individual quartz grains, were applied to fossil-bearing sediments at Galería. The luminescence dating results are in good agreement with published chronologies derived using alternative radiometric dating methods (i.e., ESR and U-series dating of bracketing speleothems and combined ESR/U-series dating of herbivore teeth), as well as biochronology and palaeoenvironmental reconstructions inferred from proxy records (e.g., pollen data). For the majority of samples dated, however, the new luminescence ages are significantly (~50%) younger than previously published polymineral thermoluminescence (TL) chronologies, suggesting that the latter may have overestimated the true burial age of the Galería deposits. The luminescence ages obtained indicate that the top of the basal sterile sands (GIb) at Galería have an age of up to ~370 thousand years (ka), while the lowermost sub-unit containing Mode II Acheulean lithics (base of unit GIIa) was deposited during MIS 9 (mean age = 313±14 ka; $n = 4$). The overlying units GIIb-GIV, which contain the richest archaeopalaeontological remains, were deposited during late MIS 8 or early MIS 7 (~240 ka). Galería Complex may be correlative with other Middle Pleistocene sites from Atapuerca, such as Gran Dolina level TD10 and unit TE19 from Sima del Elefante, but the lowermost archaeological horizons are ~100 ka younger than the hominin-bearing clay breccias at the Sima de los Huesos site. Our results suggest that both pIR-IR and single-grain TT-OSL dating are suitable for resolving Middle Pleistocene chronologies for the Sierra de Atapuerca karstic infill sequences.

Editor: Michael D. Petraglia, University of Oxford, United Kingdom

Funding: JMP, LJA and MD were financially supported by a grant (CGL2010-16821) from the Ministerio de Economía y Competitividad (Spain). MD was supported by a Marie Curie Reintegration grant (PIRG08-GA-2010-276810) within the 7th European Community Framework Programme. LA received additional support from an Australian Research Council (ARC) Future Fellowship grant (FT130100195). AIO has been supported by a postdoctoral grant from Juan de la Cierva Subprogram (JCI-2010-06792). The funders had no role in study design, data collection and analysis, decision to publish, or preparation of the manuscript.

Competing Interests: The authors have declared that no competing interests exist.

* Email: martina.demuro@adelaide.edu.au

Introduction

The Sierra de Atapuerca archaeological complex, located in northern Spain, has been systematically excavated for the last 30 years and has provided a wealth of information for the Early to Middle Pleistocene palaeoanthropological record of Europe. Some of the most important findings at Atapuerca have included the unearthing of mandible ATE9-1 at Sima del Elefante, one of the oldest human remains of Western Europe (1.22±0.16 Ma) [1], the discovery of the largest accumulation of Middle Pleistocene hominin remains worldwide at Sima de los Huesos [2], which has also provided the oldest human mitochrondial DNA recovered thus far [3], and the excavation of >150 *Homo antecessor* remains from level TD6 at Gran Dolina [4]. These sites have also produced rich associated lithic artefacts and macro/micro mammal assemblages, which provide the necessary environmental and cultural context to reconstruct the nature of hominin occupation at Atapuerca [5,6]. Key among the Atapuerca palaeoanthropological sites is the mid-late Middle Pleistocene site of Galería Complex, which represents one of the karstic infill exposures found along the abandoned railway trench and is located 50 m to the south of Gran Dolina and 100 m to the north of Sima del Elefante (**Figure 1**). This site contains various archaeo-palaeontological levels that have produced ~12,000 large mammal remains, thousands of small vertebrates, ~1800 lithic tools and two human skeletal fragments tentatively assigned to *Homo sp. aff. heidelbergensis* [6–9]. Examination of the large collection of Mode II lithics and abundant Acheulean large cutting tools (handaxes, cleavers), together with use-wear analysis and taphonomic investigations of preserved faunal remains, indicate that Galería was visited by humans for the exploitation of carcasses of large herbivores that had fallen down the vertical shaft at the southern end of the cavity [9,10].

Numerical age constraint has been previously established at Galería using uranium-thorium (U-series) and electron spin resonance (ESR) dating of calcite/speleothems bracketing the sedimentary sequence [11–13], thermoluminescence (TL) and infrared stimulated luminescence (IRSL) dating of sedimentary silicates [14], and combined ESR/U-series dating of fossil teeth [15]. The existing luminescence chronologies, which are mostly based on TL dating, are systematically older than those obtained using the other methods. In particular, the TL/IRSL chronologies suggest that units GIIa-GIIIa (archaeological horizons TG7-TG10A) were deposited ~500–440 ka, which corresponds to MIS 13/12 [14]. However, the combined ESR/U-series ages of mammal teeth from these units, as well ESR and U-series ages on intervening speleothems, indicate that the fossil accumulation occurred during MIS 9 through to MIS 8, and perhaps during early MIS 7 [15,16]. The causes of the disagreement between the TL/IRSL ages and those ages obtained using alternative dating methods have not been clearly identified.

The importance of the Early-to-Middle Pleistocene archaeological records at Galería means there is a pressing need to resolve the chronological inconsistencies observed using different radiometric dating methods. This is also necessary because the polymineral IRSL/TL chronologies established at Galería and neighbouring sites (e.g., TD10; Gran Dolina) are often used as the basis for correlating the various archaeological and palaeontological records of Atapuerca [5], and are more broadly used to postulate hypotheses about the timing and nature of human occupation across Europe [17–20]. The episodic nature of sediment infilling and erosion in cave environments results in highly complex stratigraphic associations that are difficult to correlate temporally in the absence of direct numerical dating [21], as has been shown in recent studies [22,23]. Improved site-specific chronological frameworks are therefore necessary for reliably reconstructing the karst system dynamics of the Atapuerca complex, as well as for constraining the spatio-temporal relationships of palaeoenvironmental proxies and biological and human remains trapped in caves, and for understanding the types of diagenetic processes that have affected these sedimentary sequences [24,25]. For all these reasons, it is crucial to assess the accuracy, and expand the spatial coverage, of existing numerical chronologies established on the Atapuerca cave infills.

Luminescence dating of silicate minerals using visible light stimulation (optically stimulated luminescence, OSL) or infrared wavelengths (IRSL) offers advantages over many other chronological techniques because the material being dated is ubiquitously found in sedimentary environments [26]. In comparison to TL methods, OSL and IRSL techniques are also better suited for dating sedimentary deposits that have received relatively limited daylight exposure prior to burial because the signals employed can be reset by natural sunlight (either completely or to very low levels) in a matter of seconds to minutes [27]. The development of single-grain dating equipment, new measurement protocols, and improvements in statistical analyses of burial dose datasets [28,29] have further enhanced the applicability of OSL and IRSL techniques by enabling incomplete signal resetting (partial bleaching) or post-depositional mixing to be more readily detected and appropriately treated in age calculations [30–33]. Quartz OSL dating has been successfully applied to sediments from a range of caves and rock shelters that contain associated archaeological records [34–36]. However, OSL is commonly only applicable to sediments that are <200 ka because the signal used for dating rapidly reaches saturation level (i.e., it does not continue to grow with additional radiation exposure) over moderate natural doses ranges.

The emergence of alternative luminescence dating techniques, such as thermally-transferred OSL (TT-OSL) dating of quartz and post-infrared IRSL (pIR-IR) dating of K-feldspars [37–40], offers new potential for circumventing the dose saturation restrictions of conventional OSL dating and for extending the upper age range of luminescence dating techniques over Middle Pleistocene time-scales. Reliability assessments performed on these 'extended-range' luminescence techniques using known-age comparisons have produced encouraging results across a range of depositional contexts [33,37,39,41–44]. There have also been several reliable TT-OSL and pIR-IR dating studies in potentially complex cave or rock shelter settings [35,45–47]. However, a number of complications can potentially affect the accuracy of TT-OSL and pIR-IR ages, necessitating careful sample-specific suitability assessments in any dating study. In particular, there has been some debate over the long-term stability of the TT-OSL signal, with laboratory studies of electron-retention lifetimes suggesting that age under-estimations may be a potential problem when dating very old (Early rather than Middle Pleistocene) samples [48]. There is also evidence to suggest that different types of pIR-IR measurement conditions may be required to obtain accurate ages for certain samples [42,47,49]. An additional complication with pIR-IR and TT-OSL dating is the slower rate at which these signals are optically bleached in comparison with conventional OSL signals [39,50–52] (i.e., on the order of minutes-to-hours and weeks-to-months, respectively). These slower bleaching rates may increase the potential for insufficient signal resetting prior to deposition, particularly in the case of the very slowly bleaching TT-OSL signal. However, it has been recently shown that it is feasible to apply TT-OSL dating at the single-grain scale of analysis at the Atapuerca sites [33,47], which permits detailed assessments of bleaching and post-depositional histories in the present context.

Here we report on results obtained using single-grain TT-OSL and multi-grain pIR-IR dating applied in tandem at the Middle Pleistocene site of Galería Complex. The aim of the paper is twofold. First, to generate new, systematic ages for the karstic infill sediment layers that have previously been dated in an attempt to resolve existing chronological inconsistencies. Second, to establish the first numerical age constraint on previously undated units at Galería and Covacha de los Zarpazos that underlie the archaeo-palaeontological levels. This extended-range luminescence study builds on similar comparative assessments undertaken at other Atapuerca sites [47]. We evaluate our findings with respect to these related studies and discuss the broader applicability of novel TT-OSL and pIR-IR techniques in similar cave settings.

1. Study Site - Galería

The Sierra de Atapuerca palaeoanthropological sites are located within a Cretaceous limestone karstic complex found 15 km east of

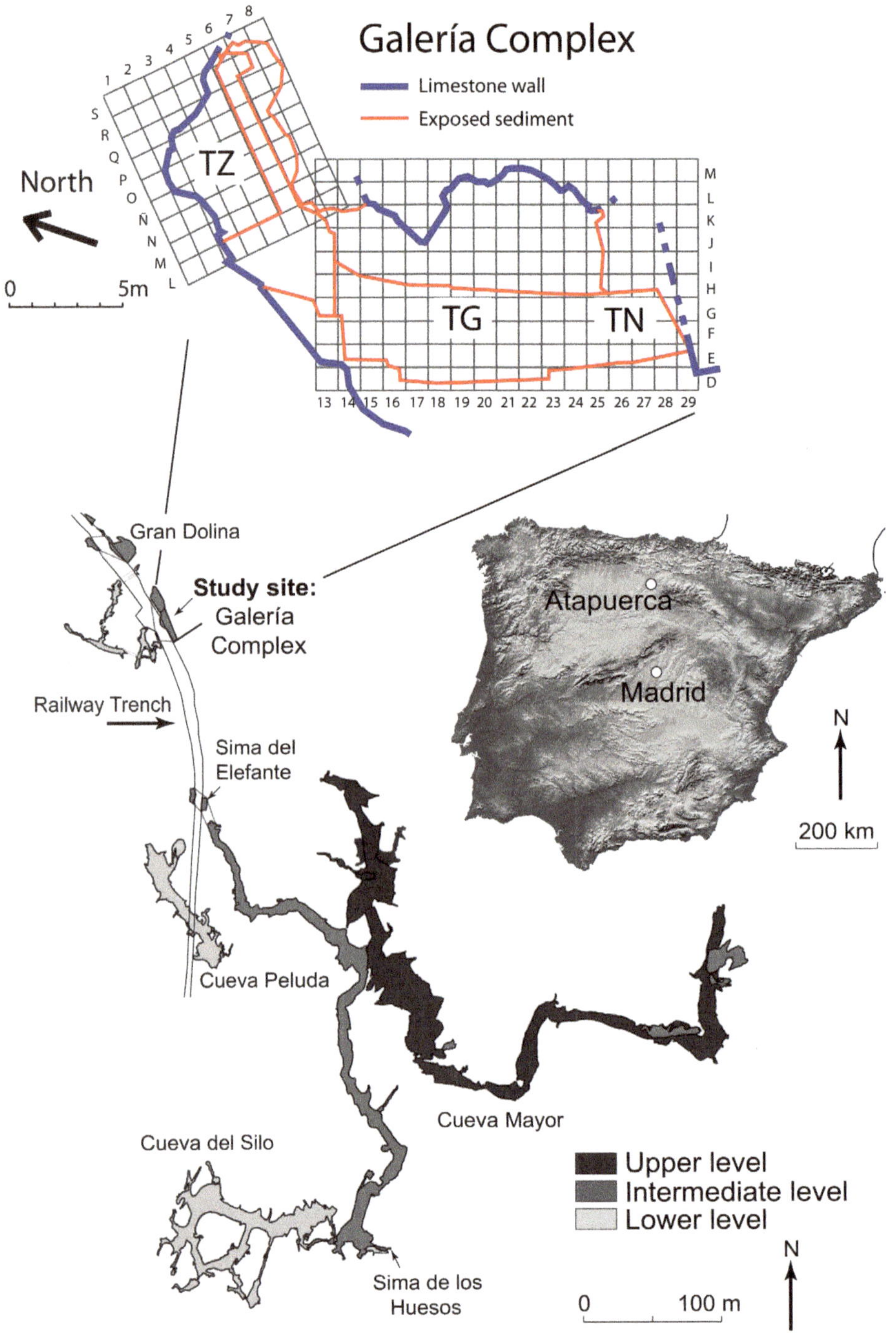

Figure 1. Geographical location of the Galería site within the abandoned railway trench at Atapuerca, Spain. The plan view of the Galería karstic complex shows the three conduits (Zarpazos: TZ, Galería: TG, and Tres Simas norte: TN) that have contributed to the infilling of the cavity.

Burgos, northern Spain (**Figure 1**). The sites are situated in the headwaters of the Duero basin and close to the Bureba corridor, which forms a connection with the Ebro basin to the east [53]. The Galería site (42° 21.08′N, 003° 31.19′W) is found within the intermediate zone (~995 masl) of a multi-level karst system that formed during the Early Pleistocene by phreatic groundwater circulation and subsequent river downcutting of local base levels [54,55] (**Figure 1**). Three separate conduits are thought to have contributed to the sedimentary infilling of the Galería Complex cavity, resulting in a complex sedimentary sequence characterised by strong lateral lithological variations. These conduits are associated with three sub-sections: Covacha de los Zarpazos (TZ) to the north; the main sub-horizontal gallery (Galería itself) in the centre (TG); and the northernmost infill segment (TN) of the neighbouring (and connected) Tres Simas cavity, which is located at the southern end of the Galería site [16,56] (**Figure 1**).

2. Stratigraphy

The 6 to 13 m-thick stratigraphic sequence has been divided into five allostratigraphic units according to well-defined unconformities [12,57] (**Figure 2**). From the base of the sequence upwards, the units are denoted as GI to GV (the uppermost unit contains an old edaphic horizon named TGVI). Unit GI is an internally-transported (endokarstic) sedimentary deposit, whereas GII to GV are allochthonous to the cave system. All of the palaeontological and archaeological finds have been collected from units GII and GIII; units GI and GIV-GV are sterile. A number of lithological levels have been identified in the north-central (TG) and southern (TN) parts of the cavity [58]. This stratigraphical framework was later used to delimit the main archaeo-palaeontological levels (TG7 to TG12 and TN2 to TN10) of Galería [10,16,59], though their exact relationships can be difficult to determine due to the spatial complexity of the stratigraphic sequence. The nomenclature adopted in this study for the allostratigraphic units and lithological/archaeo-palaeontological levels of Galería is based on published descriptions [15]. A summary of the sedimentological properties for each unit is given below and in **Table 1** following [12,16,57].

Unit GI. A 5-m thick facies dominated by light brown/yellow laminated fine sands and silts of fluvial origin that display occasional planar- and cross-stratification [57]. Deposition of this unit occurred prior to the roof collapse event(s) that subsequently enabled infilling of the cavity by material sourced from the immediate exterior of the cave. As such, this unit does not contain any fossils or archaeological remains and is thus considered to be autochthonous in the sense that it was transported through the endokarst system rather than being sourced directly from a nearby cave opening (note: this interpretation does not preclude an exterior origin for the sediments). A polarity reversal has been identified within this unit in the central sector [12] and has been attributed to the Matuyama-Brunhes (MB) boundary [60]. The upper section of GI, which has a normal polarity, is denoted GIb [16], while the lower, with reversed polarity, is termed GIa. The upper part of GI contains interstratified and sporadic speleothems and <4 cm-thick guano layers that form irregular laminar sequences [57]. The high proportion of crandallite (3–30%) within the clay fraction (<2 μm) of GIb and the lower section of GII (i.e., GIIa), indicates leaching of phosphates from the overlying/surrounding heavily altered bone and guano layers [57,61].

Unit GII. This unit is separated from GI by an angular discontinuity and marks the beginning of the allochthonous sediment input. The infill has been sourced directly from the nearby cave exterior, as suggested by (i) the presence of collapsed limestone blocks towards the base of GIIa, corresponding with the opening of the TN shaft, (ii) a change in the colour of sediments towards reddish hues (5 YR 5/8), in agreement with the terra rossa deposits currently blanketing the surrounding landscape, and (iii) an increment in the amount of micro-morphological material originating from outside the cave [8,16,62]. The thickness of this unit ranges laterally from 1 m (to the north) to 2.5 m (to the south). Unit GII contains the first appearance of faunal remains and lithic artefacts at Galería and represents two phases of sediment infill. The first phase (GIIa) encompasses archaeo-palaeontological levels TG7 to TG9 and TN2B to TN2A, while the second phase (GIIb) contains levels TG10B-D and TN5 to TN6 [6,16,59]. The fragment of an adult human mandible containing M2 and M3 and assigned to *Homo sp. aff. heidelbergensis* [8] was uncovered from GII, although its exact stratigraphical position is unknown.

Sub-unit GIIa is distinguished by high levels of post-depositional alteration that are readily observable at the micro-morphological scale and that are absent in GIIb [62]. Within GIIa, high levels of crandellite point to lixivation of bat guano, which have resulted in the formation of weathering rings on limestones and significant bone destruction. The base of GIIa (level TG7) is laterally continuous and is composed of brown/yellow fine silts and clays devoid of clasts. The aformentioned alteration processes have prevented the preservation of much of the biological remains in this horizon, though numerous lithic tools attest to the presence of hominins during this time period [6,10]. In the southern and central sectors of Galería, TG7 is overlain by two 5–30 cm-thick dark grey organomineral layers (containing bat guano) separated by red clays and silts (TN2A and TN2B) and a localised speleothem (TG8). In the northern sector, level TG7 is directly overlain by TG9, which is composed, from the base upwards, of brown-to-red fine silts and clays and an increasing abundance of small to medium size (<64 cm) limestone boulders. The second phase of GII (GIIb) was formed by an accumulation of clastic gravity deposits (i.e., debris fall) and contains abundant archaeo-palaeontological remains [16,59]. Unit GIIb is composed mainly of boulders and gravels encased in red sandy silts and clays, or small nodules encased in a matrix of red silts and clays.

Prior to the opening of the TN shaft, which occurred sometime during the formation of GIIa, the environment inside the cave would have been characterised by semi-darkness and waterlogging, and was thus unsuitable for permanent hominin occupation [10]. However, the recovery of a lithic assemblage consisting of ~200 pieces from GIIa (and ~400 from GIIb; [63]) point to the presence and activities of hominins. It is inferred that hominins entered the cave for the attainment of animal parts (via a passageway), and subsequently transported and consumed these remains outside the cavity. The low number of refits and disproportionate abundance of shaped tools, medium-size flakes and percussion material throughout GII suggest that most of the knapping activities occurred outside the cave [6], further supporting the idea that the cave was visited for the collection of trapped animals.

A recent study of the lithic assemblage of unit GII [63] shows that both large cutting tools (e.g., handaxes and cleavers) and small- and medium-sized tools (e.g., denticulates and scrapers) are present in the assemblage (comprising ~20% and ~80%, respectively). Quartzite and Neogene chert are predominantly used in GIIa, but an increase in the exploitation of limestone, sandstone and Cretaceous chert is recorded in GIIb. Sub-unit GIIa is characterised by an almost exclusive use of cobbles as blanks for the production of large cutting tools, most of which are in the sharpening stages of configuration. Knapping techniques are mainly longitudinal and multipolar [63]. In contrast, sub-unit GIIb shows an increment in centripetal knapping techniques and the use of flakes as blanks, as well as a higher frequency of handaxes at the finishing stage of configuration. These trends indicate the use of longer operative chains to produce more extensively elaborated tools when compared to GIIa [63].

Unit GIII. This unit is separated from GII by a gentle unconformity and was deposited via gravitational mass movement and slopewash processes. The preserved sedimentary features indicate that water transportation and alluvial deposition were more dominant with this unit. The base of GIII is composed of a massive fine sand and silt-clay layer that displays a higher frequency of gravels and blocks towards the southern and northern margins. This is overlain by a deposit characterised by marked lateral variations owing to influences of different sedimentary inputs from the northern (TZ), central (TG) and southern (TN) sector conduits (see **Table 1** for details). GIII has been divided into GIIIa (lower) and GIIIb (upper) sub-units based on the presence of calcitic sands in the upper half of the deposit [62].

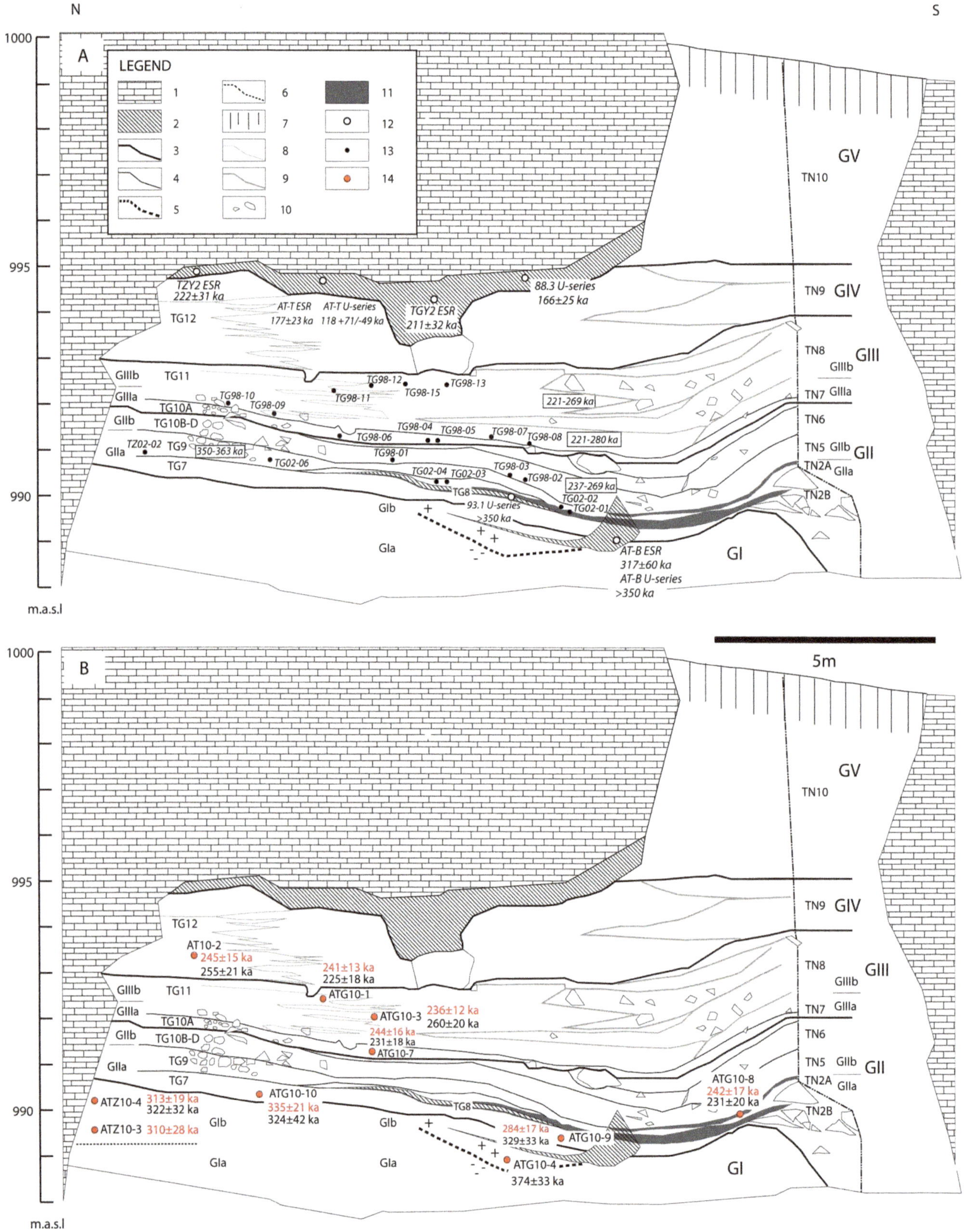

Figure 2. Stratigraphic sequence of the cave deposits at Galería Complex. (a) Ages/sample positions of teeth previously dated by combined ESR/U-series [15] and ages/sample positions of speleothem samples previously dated using U-series and/or ESR [11,15,104]. (b) Location of the luminescence dating samples and ages obtained in this study. Legend: (1) Cretaceous limestone; (2) Speleothem; (3) Main stratigraphic uncomformity – allostratigraphic unit (e.g., GII); (4) Lithologic/archaeo-palaeontological level (e.g., TG7); (5) Palaeomagnetic reversal; (6) Zarpazos – tentative GIa/GIb boundary; (7) Soil; (8) Clayey silt/small gravel boundary; (9) Gravels/breccia boundary; (10) Limestone clasts and boulders; (11) Organomineral layer

(includes bat guano); (12) U-series and/or ESR speleothem samples; (13) ESR/U-series teeth samples; (14) Luminescence samples (this study).

GIIIa corresponds to archaeo-palaeontological levels TG10A, the base of TG11 and TN7. The upper sub-unit of GIII (GIIIb) is composed of archaeo-palaeontological levels TN8 and upper TG11. Twelve occupation floors have been identified within unit GIII, indicating repeated exploitation of the cave for the acquisition of pitfall carcasses. These floors contain abundant paleontological and archaeological remains and collectively form archaeo-palaeontological level TG11 (GSU1-12). A human neurocranial fragment was uncovered during 1995 from the base of GIII [7]. In terms of the lithic assemblage [6], most of the Acheulean implements from GIII were derived from large flakes, and the raw materials were predominantly limestone, sandstone and Cretaceous chert. As with GII, small- and medium-sized shaped tools (denticulates, scrapers, etc) make up a substantial part of the identifiable lithics recovered from GIII (~80–85%; see Tables 10 and 12 of [6]).

Units GIV to GV. These remaining units do not contain any archaeological or palaeontological remains. Unit GIV sits unconformably on GIII and is of similar lithology. This unit corresponds to lithological levels TG12 and TN9. Unit GV represents the final infilling of the vertical conduit to the south and comprises 6 or 7 clastic mass movement episodes. The sediments of GV are primarily subangular gravels separated by fine sandy silt and silty clay layers [57]. The top of unit GV contains one of the oldest edaphic phases preserved in the Atapuerca hills.

3. Previous chronological work

3.1. Radiometric dating and palaeomagnetism. The ages obtained previously at Galería are summarised in **Figure 2a** and **Table 2**. The entire archaeo-palaeontological sequence is bracketed by the MB geomagnetic reversal in the lowermost level (GI) [60], and an U-series and ESR dated speleothem crust (stalagmitic floor) overlying GIV. This stalagmitic floor has produced the following radiometric ages: 118+71/−49 (U-series, [11]), 166±25 ka (U-series, cited in [15]), 177±23 ka (ESR, [11]) and 211±32 ka (ESR, [13]). Another stalagmite attached to the Zarpazos cave ceiling, which has been correlated with the capping crust of unit GIV, has also produced an age of 222±31 ka (ESR, [13]). Three additional speleothems have been dated from the horizons associated with units GI to GIII at Galería. A speleothem from a section now absent at the neighbouring Tres Simas site (labelled TS [15]), which has been stratigraphically correlated with the boundary between GIIIa and GIIIb, has been dated to 256±33 ka (ESR, [13]). Another speleothem located in unit GIIa (TG8) was found to be in isotopic equilibrium and yielded a minimum U-series age of >350 ka (Bischoff unpublished data; reported in [15]). Finally, in the upper section of GI, a third speleothem has been dated to >350 ka and 318±60 ka by U-series and ESR, respectively [11]. Relatively few details have been provided about the ESR and U-series speleothem ages obtained on units GI to GIV at Galería. Consequently, it remains difficult to assess any potential variations in methodological reliability from the published literature.

Multiple aliquot, polymineral fine-grained luminescence dating results for the main sedimentary sequence at Galería have been reported [14]. The TL and IRSL ages obtained in this study showed a wide range of values, some of which were stratigraphically reversed and/or in disagreement with associated palaeomagnetic, U-series and ESR chronologies (see Table 2 of [14]). A number of these results (5 of the 15 ages) were excluded from the final age summary as they were deemed to be affected by signal saturation (in the case of IRSL signals) or partial bleaching (in the case of slower-bleaching TL signals). The resultant luminescence chronological framework suggested that units GIV and GIIIb were deposited ~190–260 ka, which is in broad agreement with the aforementioned intervening speleothem ages for units GIV and GIIIb. The TL samples that were considered to be reliable for the lower units GIIIa and GII yielded ages of between ~420 ka and ~500 ka (**Table 2**).

Most recently, 20 ESR/U-series ages have been presented for large herbivore teeth from various levels of units GII and GIII [15]. Bayesian modelling was applied to the resulting ages to produce a chronological framework for the stratigraphical units at Galería that contain archaeo-palaeontological remains. The combined age range calculated for GIIIb, GIIIa and GIIb were 269–221 ka (*n* = 4), 280–221 ka (*n* = 7) and 269–237 ka (*n* = 3), respectively. For unit GIIa, the modelled ages increase significantly to 363–350 ka (*n* = 2), indicating a temporal hiatus between GIIa and GIIb. However, four additional teeth ages of between ~270 and ~211 ka were obtained for unit GIIa (TN2B and TG8) in close contact with the guano layer. These younger ages were excluded from the Bayesian age modelling as they were deemed to exhibit complex uranium leaching and uptake histories. The final ESR/U-series chronologies (minus the few young ages) indicate that units GIII and GIIb were deposited between 220–280 ka, while unit GIIa was deposited substantially earlier at 350–363 ka [15].

3.2. Biochronology. The micromammal assemblages from GII and GIII are ascribed to Atapuerca Faunal Unit 6 (FU6) [64]. This faunal unit is represented by *Terricola atapuerquensis*, *Iberomys brecciencis*, *Microtus arvalis*, *Microtus argestis jansoni* and *Pliomys lenki*. The Galería assemblage can therefore be confidently placed within the I. brecciensis Biozone for the Spanish Pleistocene [64], which is defined by the presence of *I. brecciencis* and *T. atapuerquensis*. This micromammal biozone covers a relatively broad age range from 780 to 100 ka. Biometric analysis of *I. brecciencis* molars indicates the presence of a discontinuity between the lower levels of GII with respect to upper GII and GIII [65]. The large mammal assemblage is composed of species typical of the European Middle Pleistocene, such as the carnivores *Panthera leo*, *Vulpes vulpes*, *Canis lupus* and *Cuon alpinus europaeus*, and the hervibores *Dama dama clactoniana*, *Cervus elaphus*, *Stephanorhinus hemitoechus* and *Equus ferus* [5]. The composition and relative abundance of large mammal species are similar throughout the various levels of the Galería sequence (GII-GIII), suggesting that infilling of the cavity and accumulation of the fossils occurred within a relatively short lapse of time (i.e., within the same glacial-interglacial cycle) [16,66].

Materials and Methods

1. Luminescence dating sample details

Access and permission to collect samples in Atapuerca was granted by the Junta de Castilla y León, Spain. Ten samples were collected from units GI to GIV, as shown in **Figure 2b** and **Figure 3** and **Table 1**. Units, or lenses within units, dominated by homogenous sandy silts and clays were targeted for sampling. We specifically avoided locations that were affected by cementation, carbonate/calcite growth and (where possible) those with non-homogenous compositions/grain sizes to minimise the potential effects of beta-dose heterogeneity (see Table S1 in File S1 for lithological descriptions of each sample). One sample was

Table 1. Allostratigraphic units and lithological levels (same nomenclature used to describe archaeo-palaeontological levels) for Galería, together with the position of the luminescence dating samples analysed in this study.

Allostra-tigraphic unit	Thickness (m)	Description	Sub-units description	Lithological/Archaeo-palaeontological level		Luminescence dating Samples
				NORTH	SOUTH	
GV	5	Breccia composed of subangular gravels and blocks within a fine sandy silt and silty clay matrix [57]. Capped by a 1-1.5-m thick edaphic phase.		Sterile	Sterile	
GIV	1.5 to 2	Southern sector: Breccia deposits made up of angular limestone clasts supported by fine gravels and silt (<10%). Central sector: Dominated by well-sorted very fine subangular gravels (<3 cm) with <10% silt matrix. These deposits alternate with silts and clays to the north. Northern sector: Laminated sandy silts and clays with layers of small gravels or cemented mud balls.		TG12	TN9	AT10-2
GIII	1 to 2	The base displays an apparently massive fine sand and silt-clay layer up to 35 cm-thick that contains, at the extremities, gravels and up to 9–12 cm clasts (to the north). This deposit is overlain by three distinct infill facies: Southern sector: Breccia composed of angular limestone clasts supported by fine gravels and silty clay (<10%). Central sector: Succession of layers (up to15 cm-thick) composed of small (<3 cm) limestone gravels supported by a silty clay matrix, which are overlain by massive but occasionally laminated silty clays (up to 10 cm thick). Northern sector: Sandy silts and clays.	GIIIb: Presence of highly calcitic sands [62]	TG11	TN8	ATG10-1 (GSU 2), ATG10-3 (GSU 7)
			GIIIa: Encompasses the lower half of GIII and is characterised by the absence of calcitic sands.	TG11, TG10A	TN8, TN7	ATG10-7
GII	1 to 2.5	Yellowish sandy silts and clays with a 20–30 cm-thick dark grey organomineral horizon (containing bat guano). The organimineral layers (TN2A and TN2B) are intercalated with localised speleothems and clays. This horizon is overlain by breccias of rounded clasts (eroded and with double patinas) and laminated yellowish sands and reddish clays and silts. The breccias are overlain by clasts supported by gravels and clays	GIIb: Allochthonous sediments of reddish colour displaying significantly less post-depositional alternation of bones and sedimentary structures compared to GIIa.	TG10B-D	TN6, TN5	

Table 1. Cont.

Allostra-tigraphic unit	Thickness (m)	Description	Sub-units description	Lithological/Archaeo-palaeontological level		Luminescence dating Samples
				NORTH	SOUTH	
		.	GIIa: Encompasses fine grained sediments and guano layers that display substantial post-depositional alteration and evidence of in situ lixivation (as in GIb), resulting in a lack of bone preservation in level TG7. Sub-unit contains microfacies composed of humate formation or peat, evidence of high biological activity and the presence of crandellite [62].	TG9, TG8, TG7	TN2A, TN2B	ATG10-8, ATG10-9, ATG10-10
GI	5	Internal facies (infill transported through the cave system rather than sourced from the immediate cave exterior). Laminated sands and silts with occasional planar and cross-stratification [57]. Deposition occurred prior to the roof collapse; as such this unit does not contain any fossils or archaeological remains.	GIb: Sands and silts with interstratified, sporadic speleothems and <4 cm-thick guano layers forming irregular laminar sequences [57]. High proportion of crandallite (3–30%) indicative of phosphate leaching from heavily altered bone and guano layers [57,61].	Sterile	Sterile	ATG10-4, ATZ10-4, ATZ10-3
			GIa: Similar to GIb but separated by an unconformity. Palaeomagnetism indicates reversed polarity orientation.	Sterile	Sterile	

Based on [12,15,57–59,62].

Table 2. Final luminescence ages obtained for the Galería samples in this study.

Allostratigraphic unit		Lithological/Archaeo-palaeontological level	Speleothems			Polymineral fine grain-Additive dose *				ESR/U-series on teeth		This study		
			Sample	Age (ka)		Sample	Age (ka)			Sample	Age (ka)	Sample	Age (ka)	
				ESR	U-series		IRSL	TL	IRSL and TL				SG TT-OSL	pIR-IR$_{225}$
Stalagmite			AT-T	177±23 **	118+71/−49 **									
			TGY2	211±32[#]										
			TZY2	222±31[#]										
			88.3		166±25 **									
GIV		TG12				98–19	1340±170							
		TG12				97–14	185±26	510±100				AT10-2	255±21	245±15
GIII	GIIIb	TG11 (GSU2)				98–17			269±27			ATG10-1	225±18	241±13
		TG11 (GSU3)								TG9815	221 +15/−12			
		TG11 (GSU3)								TG9813	234 +20/−19			
		TG11 (GSU3)								TG9812	238 ±19			
		TG11 (GSU4)								TG9811	269 +51/−44			
		TG11 (GSU7)				97–24		224±42				ATG10-3	260±20	236±12
Speleothem	TS-S[a]		TSY12	256±33[#]										
	GIIIa	TG11 (GSU11A)				98–15		480±48		TG9810	280 +30/−27			
		TG10 (GSU12)								TG9807	231 +25/−24			
		TG10 (GSU12)								TG9809	233 +30/−28			
		TG10 (GSU12)								TG9808	244 +49/−26[b]			
		TG10A				97–19		439±66		TG9806	256 +26/−25[b]	ATG10-7	231±18	244±16
		TG10A								TG9805	227 +34/−34			
		TG10A								TG9804	239 +26/−24			
GII	GIIb	TG10D								TG9803	237 +26/−24			
		TG10D								TG9802	262 +35/−34			
		TG10D								TG9801	269 +26/−24			
	GIIa	TN2A										ATG10-8	231±20	242±17
TZ		(TG9)								TZ0202	363 +44/−42			
	GIIa	TG9				98–12		422±55		TG0206	350 +47/−46			
		TN2B								TG0201	217 +31/−28			
		TG8								TG0203	244 +56/−47			
		TG8								TG0202	274 +28/−26[b]			
		TG8								TG0204	211 +18/−16			
		TG8	93.1		>350 **									
		TG7				97–6		503±95				ATG10-9	329±33	284±17

Table 2. Cont.

Allostratigraphic unit	Lithological/Archaeo-palaeontological level	Speleothems			Polymineral fine grain-Additive dose *				ESR/U-series on teeth		This study		
		Sample	Age (ka)		Sample	Age (ka)			Sample	Age (ka)	Sample	Age (ka)	
			ESR	U-series		IRSL	TL	IRSL and TL				SG TT-OSL	pIR-IR_{225}
	TG7										ATG10-10	324±42	335±21
TZ											ATZ10-4	322±32	313±19
											ATZ10-3		310±28
Speleothem		AT-B	318±60 **	>350 **									
GI	GIb				97-2		3000±1800				ATG10-4	374±33	
	GIa												

Also shown are the ages obtained for the infill sequence using TL and IRSL dating of (*) polymineral fine grains [14], (**) U-series and ESR dating of speleothems [11,15], (#) ESR dating of speleothems [104], and combined ESR/U-series dating of herbivore teeth [15].
[a]TS-S refers to the location of the speleothem at Tres Simas Sur (see [15] for details).
[b]Samples having potentially complex uranium leaching and uptake histories according to [15]. Errors on ages are reported at 1σ.

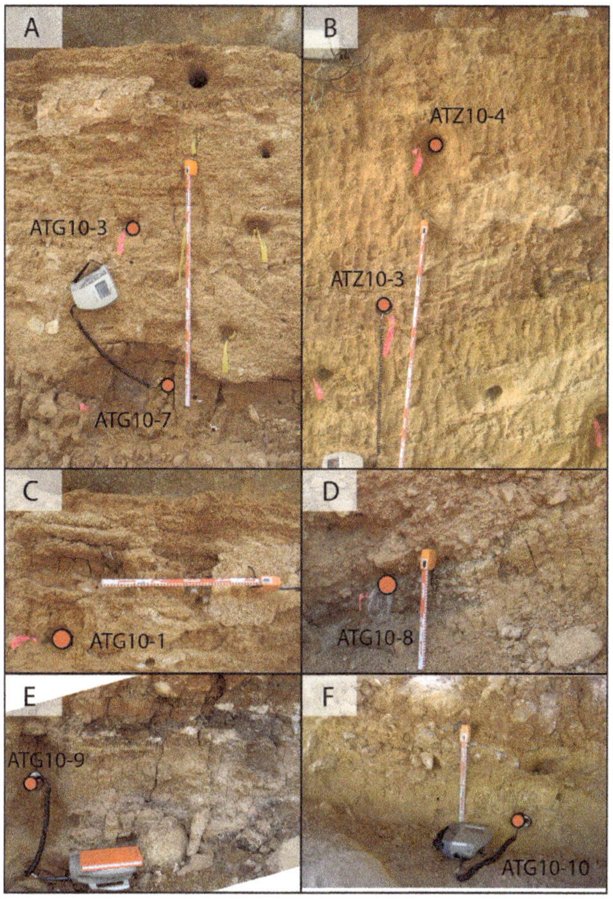

Figure 3. Photos showing the location of the luminescence dating samples and the types of deposits investigated at Galería. (a) and (c) show the sampling positions in unit GIII; (b) shows the stratigraphic sequence sampled at Covacha Zarpazos (northern wall); (d) sampling of the upper guano in unit GIIa (level TN2A); (e) and (f) shows sampling of the lower part of unit GIIa (level TG7).

collected from unit GIV (AT10-2) and two were collected from GIIIb (ATG10-1 and ATG10-3), within fine silts and clays towards the northern (Zarpazos) side of the profile (**Figure 3a, c**). During sampling of sub-unit GIIIb, care was taken to avoid layers that were predominantly composed of small limestone gravels and cemented calcitic sands. Specifically, samples ATG10-1 and ATG10-3 were collected from two thin (<10 cm) layers of sandy silts and clays corresponding to GSU 2 and GSU 7, respectively, which intercalate gravelly layers towards the central northern sectors. The base of GIII (GIIIa) was sampled within a massive horizon of homogenous fine silt and clay preserved in the centre of the profile (ATG10-7; **Figure 3a**). Three samples were collected from GIIa; one from the upper guano layer in sub-unit TN2A, which is underlain directly by red clays and fine silts (ATG10-8; **Figure 3d**), and another two samples from the base of GIIa within TG7, which is composed of brown/yellow sandy silt and clay (samples ATG10-9 and ATG10-10; **Figure 3e-f**). Two samples were collected from a level at Zarpazos that lies directly below the archaeo-palaeontological levels at Galería (i.e., below GII) and is stratigraphically correlated as lying at the top of unit GI (samples ATZ10-3 and ATZ10-4; **Figure 3b**). The upper part of unit GI (GIb) was sampled in the central sector of Galería (sample ATG10-4) within a horizon bracketed by the palaeomag-

Table 3. Environmental dose rate values for the quartz and K-feldspar fractions measured in this study.

				Environmental dose rate (Gy/ka)[b]							
Sample	Sample depth (m)	Grain fraction (μm)	Water content[a]	Gamma dose rate[c]	Beta dose rate[d]	Cosmic dose rate[e]	Internal dose rate for Qz (U+Th)[f]	Internal dose rate for K-feldspars (U+Th)[f]	Internal dose rate for K-feldspars (K+Rb)[g]	Total dose rate for Qz (Gy/ka)[h]	Total dose rate for K-feldspars (Gy/ka)[h]
AT10-2	7.01	90–125	16.0	0.69±0.03	1.53±0.07	0.07±0.01	0.03±0.01	0.06±0.03	0.43±0.03	2.32±0.12	2.78±0.13
ATG10-1	7.75	90–125	19.2	0.65±0.02	1.53±0.08	0.06±0.01	0.03±0.01	0.06±0.03	0.43±0.03	2.27±0.13	2.73±0.14
ATG10-3	8.15	90–125	16.7	0.59±0.02	1.52±0.07	0.06±0.01	0.03±0.01	0.06±0.03	0.43±0.03	2.20±0.12	2.66±0.12
ATG10-7	9.00	90–125	22.1	0.95±0.03	1.57±0.08	0.06±0.01	0.03±0.01	0.06±0.03	0.43±0.03	2.60±0.16	3.06±0.16
ATG10-8	10.10	90–125	30.8	0.95±0.03	1.34±0.02	0.05±0.01	0.03±0.01	0.06±0.03	0.43±0.03	2.37±0.17	2.83±0.18
ATG10-9	10.80	90–125	22.0	1.11±0.04	1.63±0.09	0.05±0.01	0.03±0.01	0.06±0.03	0.43±0.03	2.81±0.17	3.27±0.18
ATG10-10	10.60	90–125	22.0	0.96±0.04	1.47±0.08	0.05±0.01	0.03±0.01	0.06±0.03	0.43±0.03	2.51±0.15	2.97±0.16
ATZ10-4	10.00	90–125	24.9	1.00±0.03	1.83±0.10	0.05±0.01	0.03±0.01	0.06±0.03	0.43±0.03	2.91±0.19	3.37±0.19
ATZ10-3[i]	10.40	90–125	26.0	1.06±0.03	2.05±0.12	0.05±0.01	-	0.06±0.03	0.43±0.03	-	3.65±0.22
ATG10-4[i]	11.10	90–125	20.4	1.00±0.03	1.48±0.08	0.05±0.01	0.03±0.01	-	-	2.56±0.25	-

[a] Long-term water content, expressed as % of dry mass of sample and assigned a relative uncertainty of ±20%. Long-term water contents are calculated as 60% of saturated values following assessments made in deeper parts of the endokarst system [47].
[b] Radionuclide concentrations and specific activities have been converted to dose rates using the conversion factors given in [68,105], making allowance for beta-dose attenuation [106,107].
[c] Gamma dose rates were calculated from *in situ* measurements made at each sample position with a NaI:Tl or $LaBr_3$:Ce detector using the 'energy windows' method [67].
[d] Beta dose rates were calculated using a Risø GM-25-5 low-level beta counter [108], after making allowance for beta dose attenuation due to grain-size effects and HF etching [107].
[e] Cosmic-ray dose rates were calculated following published procedures [69] and assigned a relative uncertainty of ±10%.
[f] Assumed internal (alpha plus beta) dose rate for the quartz fractions are based on published ^{238}U and ^{232}Th measurements for etched quartz grains from a range of locations [109–112] and an alpha efficiency factor (a-value) of 0.04±0.01 [113,114]. For K-feldspar grains, the internal alpha and beta dose rate contributions from ^{238}U and ^{232}Th were calculated using assumed concentrations of 0.15±0.03 ppm and 0.35±0.07 ppm, respectively, based on modal values [112] and similar published values [115–117]. An a-value of 0.09±0.03 was used to estimate the internal alpha dose rate contributions from these ^{238}U and ^{232}Th concentrations based on published estimates obtained for a wide range of K-feldspar samples [14,113,118–121].
[g] Internal dose rate of feldspar grains arising from ^{40}K and ^{87}Rb concentrations were calculated using assumed values of 12.5±0.5% [122] and 400±100 ppm [123], respectively.
[h] Mean ± total uncertainty (68% confidence interval), calculated as the quadratic sum of the random and systematic uncertainties.
[i] TT-OSL measurements have not been made on sample ATZ10-3 and pIR-IR measurements have not been made on sample ATG10-4. Hence, only the relevant quartz or K-feldspar dose rates are shown for these two samples.

Table 4. SAR protocols used in this study to obtain ages from quartz (protocol A) and K-feldspars (protocol B).

A			B		
Step	**Single-grain TT-OSL SAR protocol**		**Step**	**SAR pIR-IR multi-grain aliquots**	
1[a]	Give dose		1[a]	Give dose	
2	Preheat to 260°C for 10 s		2	Preheat to 250°C for 60 s	
3	Stimulate with green laser at 125°C for 2 s (90% power)		3	Stimulate with infrared diodes 50°C for 200 s (90% power)	
4	Preheat to 260°C for 10 s		4[b]	Stimulate with infrared diodes 225°C for 200 s (90% power)	pIR-IR L_n or L_x
5	Stimulate with green laser at 125°C for 2 s (90% power)	TT-OSL L_n or L_x	5	Give test dose (100 Gy)	
6	Stimulate with blue LEDs at 280°C for 400 s		6	Preheat to 250°C for 60 s	
7	Give test dose (200–300 Gy)		7	Stimulate with infrared diodes 50°C for 200 s (90% power)	
8	Preheat to 260°C for 10 s		8[b]	Stimulate with infrared diodes 225°C for 200 s (90% power)	pIR-IR T_n or T_x
9	Stimulate with green laser at 125°C for 2 s (90% power)		9[c]	Return to 1	
10	Preheat to 260°C for 10 s				
11	Stimulate with green laser at 125°C for 2 s (90% power)	TT-OSL T_n or T_x			
12	Stimulate with blue LEDs at 290°C for 400 s				
13	Return to 1				

L_n and L_x refer to the natural and regenerative-dose signal measurements, respectively. T_n and T_x refer to the test dose signals measured after the L_n and L_x signals, respectively. Each SAR measurement cycle was repeated for the natural-dose, three to four different sized regenerative doses, a 0 Gy regenerative-dose (to measure OSL signal recuperation) and a replicate of the first regenerative-dose cycle (to assess the suitability of the test-dose sensitivity correction). In the case of the single-grain TT-OSL SAR procedure, the OSL IR depletion ratio [124] was measured separately and used to check for the presence of feldspar contaminants.

[a] Step omitted when measuring the natural signal (L_n).

[b] Aliquots were held at pIR-IR stimulation temperatures for 10 s prior to switching on the IR diodes to avoid any potential isothermal TL contamination of the pIR-IR signal.

[c] A high temperature IRSL stimulation was not included at the end of each pIR-IR SAR measurement cycle because these samples did not display significant signal recuperation (the 0 Gy regenerative dose pIR-IR signal was consistently <3% of the sensitivity-corrected natural pIR-IR signal for all measured aliquots).

netic reversal and an overlaying speleothem that has been dated to >350 ka and 318±60 ka by U-series and ESR, respectively.

2. Luminescence dating methods

Luminescence dating was carried out at the CENIEH Luminescence Dating Laboratory, Burgos, Spain. Pure coarse-grained quartz (90–125 μm) and K-feldspar fractions (90–125 μm) were extracted using standard preparation procedures [26]. The outer (~10 μm) alpha-irradiated rinds of the quartz extracts were etched with a treatment of 48% hydrofluoric (HF) acid for 45 minutes. For K-feldspars, etching was done using 10% HF for 10 minutes. The etched fractions were treated with 30% HCl for 45 minutes to eliminate any acid-soluble fluoride precipitates, and sieved again (using a 63 μm sieve) to remove any disaggregated or partially etched grains.

The external dose rates were calculated using a combination of *in situ* gamma-ray spectrometry and low-level beta counting (**Table 3**). The beta dose rate measurements were made on dried and homogenised material collected directly from the luminescence sampling positions. Beta dose rate measurements were made using a Risø GM-25-5 beta counter. Gamma dose rates were determined from *in situ* gamma-ray spectrometry measurements to account for any spatial heterogeneity in the surrounding gamma radiation field. Radionuclide concentrations of U, Th and K were derived from the gamma-ray spectra using the approach described in [67] and published conversion factors [68]. Cosmic-ray dose rates were calculated according to theoretical calculations [69] after taking into consideration site altitude, geomagnetic latitude, and density, thickness and geometry of sediment and bedrock overburden. The state of secular equilibrium in the ^{238}U and ^{232}Th decay series was assessed via high resolution gamma-ray spectrometry (HRGS) measurements (Table S2 in File S1). The daughter-parent isotopic ratios for ^{238}U, ^{226}Ra, ^{210}Pb, ^{228}Ra and ^{228}Th indicate that the ^{238}U and ^{232}Th chains of the measured samples are in present-day secular equilibrium (ratios are consistent with unity at either 1σ or 2σ).

Luminescence signals were measured using a Risø TL-DA-20 reader [70]. Quartz and K-feldspar stimulations were performed with blue LED units (470±20 nm, maximum power of 84 mW cm^{-2}) and an array of infrared (IR) LEDs (875 nm, maximum power of 151 mW cm^{-2}), respectively. Single-grain stimulations were performed using a 10 mW Nd:YVO4 single-grain laser attachment emitting at 532 nm (maximum power of ~50 W cm^{-2}). Luminescence signals were detected using an EMI 9235QA photomultiplier tube. The ultraviolet quartz emissions were measured through a 7.5 mm-thick Hoya U-340 filter. Blue emissions from K-feldspar were measured through a filter combination pack composed of a 3 mm-thick Schott BG39 and a 4 mm-thick Corning 5–79. Samples were irradiated using a calibrated $^{90}Sr/^{90}Y$ beta source mounted on the reader. For single-grain measurements, the beta source had been calibrated to administer known doses to each grain-hole position across the disc plane.

TT-OSL and pIR-IR equivalent dose (D_e) values were measured using the single-aliquot regenerative-dose (SAR) proto-

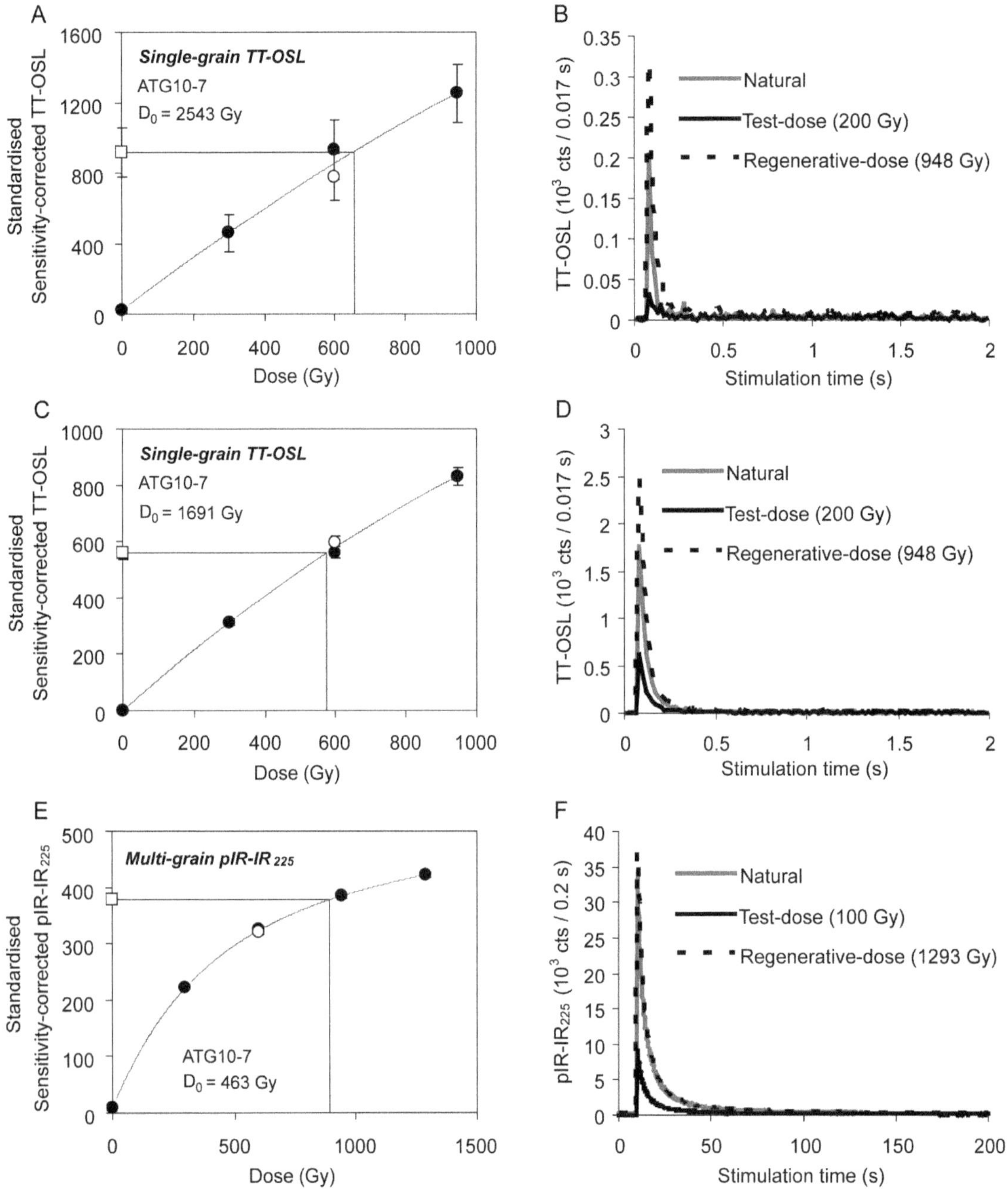

Figure 4. Selected examples of dose-response curves and signal decay curves. (a-d) TT-OSL single-grain measurements and (e-f) pIR-IR_{225} measurements of the Galería samples.

cols shown in **Table 4**. TT-OSL measurements were made on individual quartz grains rather than on multi-grain aliquots, following the successful application of this approach elsewhere at the Atapuerca sites [33,47]. Quartz D_e estimates were determined using a published TT-OSL SAR protocol [71], which has been modified to enable measurements at the single-grain scale of analysis (**Table 4a**). K-feldspar D_e values were measured using a multi-grain aliquot pIR-IR SAR protocol [72]. This D_e measurement protocol (herein denoted pIR-IR_{225}) involves pIR-IR stimulation temperatures of 225°C and regenerative-/test-dose preheat temperatures of 250°C for 60 s (**Table 4b**). Further details of the D_e measurement procedures, including dose-recovery suitability tests and the rejection criteria used to eliminate unreliable grains and aliquots, are provided in the Supporting Information (see results in Table S4 and Table S5).

Results

1. TT-OSL and pIR-IR signal characteristics

Representative TT-OSL decay curves and sensitivity-corrected dose-response curves are shown in **Figure 4a-d** for two accepted grains from unit GIIIa (sample ATG10-7). These two grains were measured from the same single-grain disc but display very different luminescence sensitivities; one has a relatively bright TT-OSL signal (T_n intensity >1500 cts/0.02 s) (**Figure 4d**), while the other displays a dimmer T_n luminescence intensity of ~200 cts/

Table 5. Summary of single-grain TT-OSL and multi-grain pIR-IR$_{225}$ D_e values for the Galería samples.

Luminescence signal	TT-OSL						pIR-IR$_{225}$						
Mineral	Quartz						K-feldspar						
Resolution	'Pseudo' single-grain						160-grain aliquot						
Grain size (μm)	90–125						90–125						
Sample	**Accepted/measured**	**OD (%)[a]**	**D_e (Gy)[b, c]**	**Weighted skewness value[d, e]**	**Critical skewness 68% C.I.[d]**	**Critical skewness 95% C.I.[d]**	**Accepted/measured**	**OD (%)[a]**	**D_e (Gy)[f]**	***g*-values (%/decade)**	**Weighted skewness value[d, e]**	**Critical skewness 68% C.I.[d]**	**Critical skewness 95% C.I.[d]**
AT10-2	33/1000	32±6	591±37 (FMM: k_2)	−0.604	0.426	0.853	6/6	7±3	679±23	1.39±0.04 (n=4)	0.40	1	2
ATG10-1	46/800	22±5	511±25 (CAM)	−0.007	0.362	0.722	6/6	0±0	658±7	1.43±0.10 (n=4)	0.06	1	2
ATG10-3	43/800	23±5	572±29 (CAM)	−0.401	0.374	0.747	6/6	2±2	629±8	1.47±0.10 (n=4)	0.45	1	2
ATG10-7	93/1100	31±4	601±27 (FMM: k_2)	−0.076	0.254	0.508	5/6	6±3	745±23	1.26±0.07 (n=4)	−0.06	1.1	2.19
ATG10-8	67/1200	20±4	546±21 (CAM)	−0.106	0.299	0.599	6/6	5±2	688±17	1.42±0.16 (n=4)	0.57	1	2
ATG10-9	17/600	12±11	925±71 (CAM)	−0.730	0.594	1.188	6/6	4±2	929±19	1.51±0.12 (n=3)	0.94	1	2
ATG10-10	12/700	24±11	813±90 (CAM)	−0.201	0.707	1.414	6/6	6±2	996±27	1.86±0.22 (n=3)	0.01	1	2
ATZ10-4	21/600	19±8	937±66 (CAM)	−0.133	0.534	1.069	6/6	0±0	1054±15		0.00	1	2
ATZ10-3							6/6	14±5	1131±72		−0.07	1	2
ATG10-4	18/1700	12±9	957±62 (CAM)	−0.744	0.561	1.123							

Also shown are the g-values (fading rates) measured on a subset of aliquots and normalised to a delay time of 2 days.
[a] OD = overdispersion.
[b] CAM = central age model; FMM = finite mixture model.
[c] The FMM was fitted by varying the common overdispersion parameter (σ_k) between 5 and 30% and incrementally increasing the specified number of k_n components. The FMM D_e values shown here were obtained from the optimum FMM fit (i.e., the fit with the lowest BIC score; [32]), which corresponded to a σ_k value of 15% for both samples (consistent with the lowest empirical overdispersion values obtained for other D_e datasets at Galería). Using this approach, the D_e datasets of sample AT10-2 and ATG10-7 are shown to contain two and three discrete dose populations, respectively. The main dose components (those containing the majority of individual D_e estimates) have been used to derive the final burial doses for these two samples.
[d] Weighted skewness scores have been calculated on log-transformed D_e values using Eq. 7–8 [32]. Critical skewness scores have been calculated using Eq. 16 [125].
[e] The D_e distribution are considered to be significantly positively skewed at the 68% C.I. or 95% C.I. if the weighted skewness value is greater than the corresponding critical skewness value.
[f] pIR-IR D_e values have been calculated using the CAM.

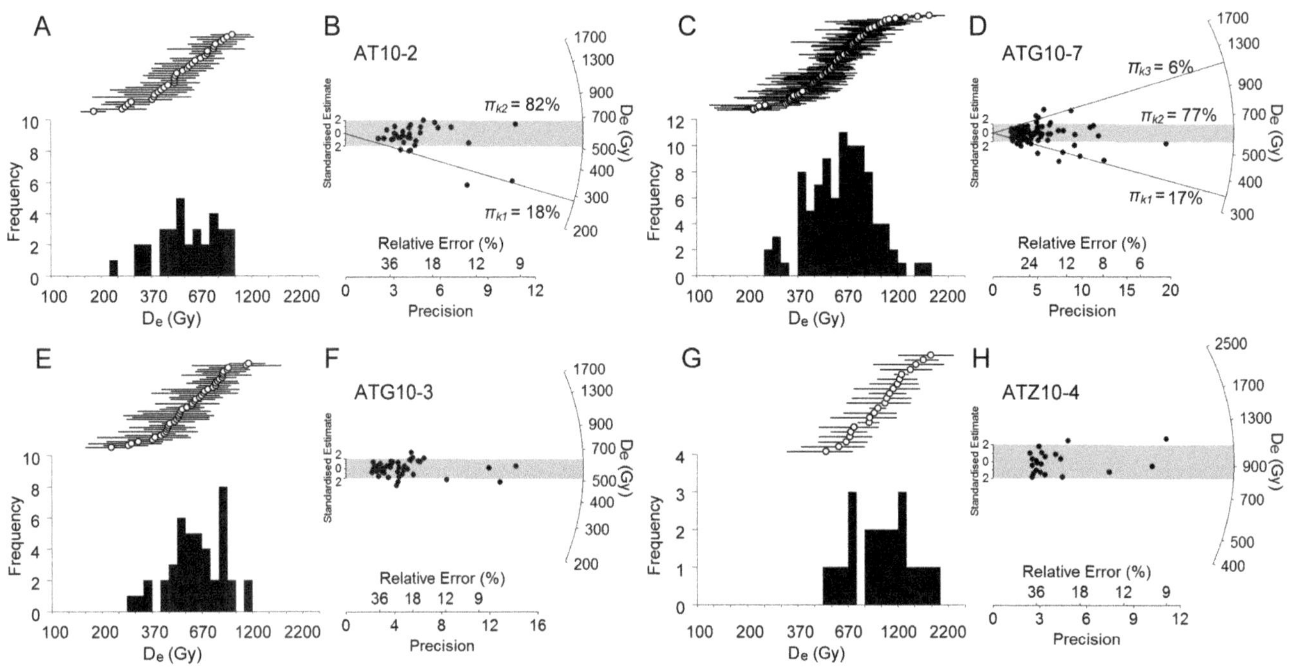

Figure 5. Examples of single-grain TT-OSL D_e distributions plotted as histograms (logarithmic x-axis) and radial plots. Data is shown for selected Galería samples. See Figure S4 in File S1 for the D_e datasets of all other samples.

0.02 s (**Figure 4b**). The latter is representative of the majority of grains measured for these samples. In both cases, the TT-OSL signals decay rapidly to background and the dose-responses are well represented by a single-saturating exponential function (**Figure 4a, c**). Samples collected from the lowermost units at Galería, i.e., unit GI (ATG10-4), layer TG7 at the base of GIIa (ATG10-9 and ATG10-10) and the base of TZ (ATZ10-4), display distinctly different TT-OSL brightness characteristics. Further details of these differences are provided in the Supporting Information. The dose-saturation properties of the accepted grains permit finite D_e determination over very high dose ranges. A pooled distribution of characteristic saturation dose (D_0) values for the accepted grains of all nine samples is shown in Figure S3 in File S1. Grains with dose-response curves that are best characterised by a single saturating exponential function ($n = 270$) yielded D_0 values ranging between 250 Gy and 10,000 Gy. Additionally, ~23% of the accepted grains produced linear dose-response curves over the administered dose range (highest dose-point = 900–1200 Gy). All of the accepted grains had D_e values that are lower than their corresponding 2^*D_0 value; i.e., they are well below the commonly cited threshold for precise burial dose determination.

A representative pIR-IR_{225} decay curve and sensitivity-corrected dose-response curve is shown in **Figure 4e-f**. The pIR-IR_{225} decay curves of these samples typically decrease by >90% within the first 50 s of stimulation and most are optimally fitted with a single saturating exponential plus linear function. All the D_e values were obtained from the region of the dose-response that was not in saturation when using this type of fitting function (for example **Figure 4e**).

To examine the potential for athermal loss of K-feldspar pIR-IR signals over burial timescales [73], we performed anomalous fading assessments on subsets of aliquots used to derive D_e values. Fading tests were conducted on three to four aliquots per sample for storage times of up to 24 hours [74]. Mean *g*-values normalised to 2 days [75] were calculated from these results, and used to quantify the expected percentage of signal loss per decade of storage time. The mean fading rates for individual samples range between 1.26±0.07 and 1.86±0.22%/decade (**Table 5**) and the combined weighted average *g*-value is 1.39±0.07%/decade. These empirical fading rates are similar to those published previously for pIR-IR_{225} and higher temperature ("pIR-IR_{290}") signals [36,39,42,72,76–80], and agree with pIR-IR *g*-values reported for other Atapuerca karstic infill deposits [47]. Elsewhere, low fading rates (on the order of 1-2%/decade) have been shown to be potential artefacts of laboratory procedures following comparisons made with independent age control, observations of natural signal saturation, and measurements of similarly sized *g*-values for quartz [39,42,49]. Consequently, the low *g*-values recorded here are not usually considered indicative of the need for pIR-IR age corrections [39].

2. D_e distributions

The vast majority of Galería samples display similar single-grain TT-OSL D_e characteristics (**Table 5**). With the exception of AT10-2 and ATG10-7, all samples have overdispersion values below 25%, which is consistent with values obtained for ideal (well-bleached and unmixed) samples at the neighbouring Atapuerca site of Sima de los Huesos [47]. The log-transformed D_e distributions of these seven samples are normally distributed according to published skewness assessments [32] (**Table 5**, **Figure 5** and Figure S4 in File S1) and they do not display pronounced 'leading-edge' shapes or elongated high dose tails. The D_e datasets are also well-represented by a single dose population, and most of the individual D_e values are consistent with the weighted mean burial dose estimate at 2σ (they fall within the grey shaded bands of the radial plots shown in Figures 5F, 5H and Figure S4 in File S1). These observations, together with the relatively high doses measured for these samples, indicate that any D_e scatter arising from partial bleaching is likely to be insignificant

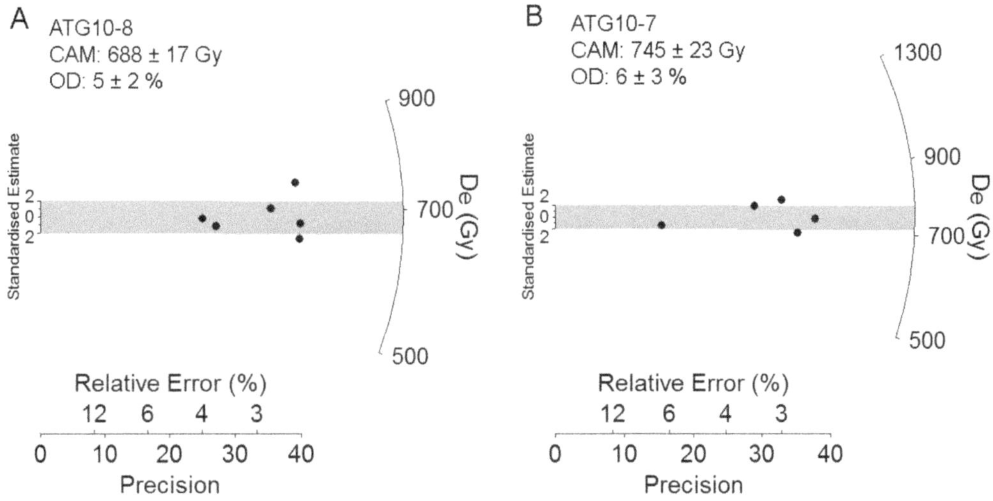

Figure 6. Radial plots showing pIR-IR$_{225}$ D$_e$ distributions of representative Galería samples. See Figure S5 in File S1 for the D$_e$ datasets of all other samples.

or effectively absent in these datasets. As such, we have used the weighted mean D$_e$ values, calculated using the central age model (CAM) [81], to calculate representative burial dose estimates. The final single-grain TT-OSL D$_e$ estimates for these seven Galería samples range between ~510 Gy and ~960 Gy (**Table 5**).

Samples AT10-2 and ATG10-7 display more noticeable D$_e$ scatter (**Figure 5a-d**) and their overdispersion values (32±6% and 31±4%, respectively) are systematically higher than the weighted mean estimate obtained for the other Galería samples (20±2%) and the ideal Sima de los Huesos samples (21±1%) [47]. Inspection of the radial plots shown in **Figure 5** reveals that sample AT10-2 contains two precise low-dose grains (**Figure 5a**) and that sample ATG10-7 displays a wider spread in D$_e$ values at both the lower and upper ends of the D$_e$ distribution (**Figure 5d**). For both of these samples, however, the distribution is not positively skewed (**Table 5**). Application of the FMM model [82] confirms that these two D$_e$ datasets contain multiple dose populations. The dose components identified by the optimum FMM fit are shown in **Figure 5b and 5d**. These optimum fits were determined according to the BIC criterion [33], using a common overdispersion value of 15% (consistent with the lowest empirical overdispersion values obtained for other D$_e$ datasets at Galería). The lowest dose components for the AT10-2 and ATG10-7 datasets yield D$_e$ values of 287±34 Gy and 348±35 Gy, respectively (**Figure 5**), and corresponding ages of 123±18 ka and 134±16 ka. These ages are significantly younger than the replicate pIR-IR$_{225}$ ages for AT10-2 and ATG10-7 (**Table 2**). In the case of the latter sample, the age of the low dose component also is not in stratigraphic agreement with the TT-OSL ages obtained on surrounding samples (**Table 2**). The highest dose component identified for sample ATG10-7 contains a very small population of grains ($n = 6$) and its corresponding age (429±71 ka) is not consistent with the associated TT-OSL and IR-IR ages of neighbouring samples (all of which display homogeneous D$_e$ distributions). This high dose component is therefore not considered reliable for burial dose determination.

The very young and stratigraphically inconsistent ages obtained for the low dose components of samples AT10-2 and ATG10-7 suggests that partial bleaching is not the primary cause of the above-average overdispersion in these D$_e$ datasets. Post-depositional intrusion of younger grains can probably also be ruled out since the host deposits are either capped by well-preserved contacts/speleothems or retain primary bedding features. Beta dose heterogeneity may be a possible cause of the minor, low dose populations identified for AT10-2 and ATG10-7. However, these particular samples were collected from homogenous silt and clay sediments, and were devoid of limestone clasts, gravels or carbonates that could have acted as beta-emitting cold spots or beta-particle attenuators. The minor dose components, which comprise <20% of grains, could therefore be caused by intrinsic sources of D$_e$ dispersion, such as grain-to-grain variations in TT-OSL response to fixed measurement conditions or grain-to-grain variations in the thermal stability properties of the TT-OSL signals. For both samples, the main dose components (FMM component 2) contain the vast majority (~80%) of grains and the resulting TT-OSL ages are in stratigraphic agreement with those obtained for the surrounding samples. We have therefore used these main FMM dose components to derive the final ages for samples AT10-2 and ATG10-7 (**Table 5**).

The pIR-IR$_{225}$ D$_e$ distributions are shown in **Figure 6** and Figure S5 in File S1. The majority of these D$_e$ datasets cover a narrow range and all are normally distributed when assessed using a log weighted skewness test (**Table 5**). The pIR-IR$_{225}$ overdispersion values are generally low (<10% at 1σ; **Table 5**) and are consistent at 2σ with the mean value of 5±1% obtained for well-bleached, unmixed samples elsewhere at Atapuerca [47]. These D$_e$ characteristics suggest that any dose dispersion originating from extrinsic or intrinsic sources, including partial bleaching, are relatively insignificant in relation to the size of the measurement uncertainties. As such, we have derived the final pIR-IR$_{225}$ burial doses estimates using the weighted mean (CAM) D$_e$ values (**Table 5**).

3. Ages

Table 2 and **Figure 7** show the final TT-OSL and pIR-IR$_{225}$ ages for the Galería Complex sequence. Both sets of extended-range luminescence ages are stratigraphically consistent and the replicate TT-OSL and pIR-IR$_{225}$ ages of individual samples are in agreement at 1σ. The internal consistency of the two chronological datasets for the entire sequence provides support for the reliability of the TT-OSL and pIR-IR$_{225}$ approaches adopted at this site. The TT-OSL age obtained for sample ATG10-4 suggests

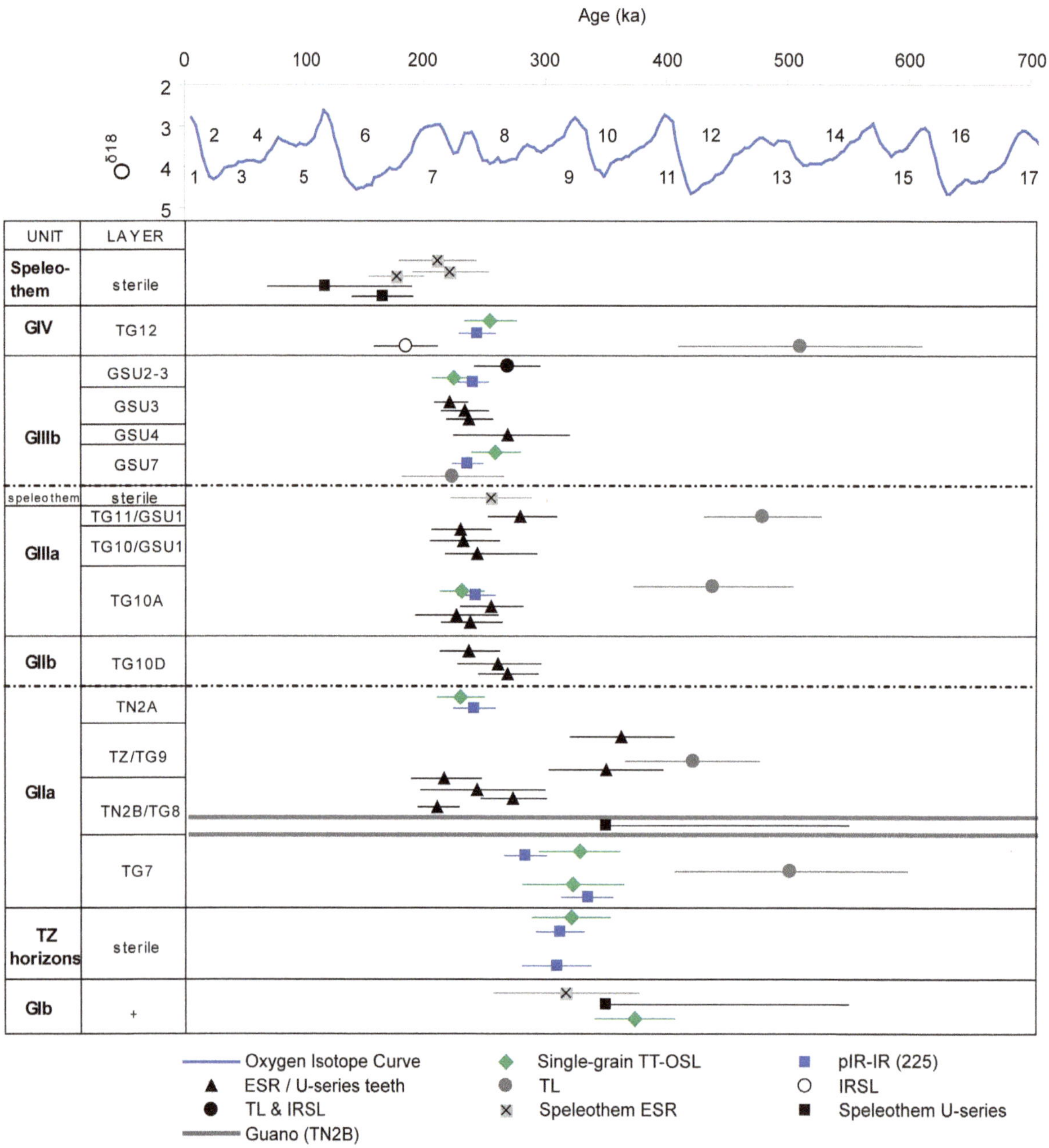

Figure 7. Published ages obtained using different dating methods (including this study) for the various allostratigraphic units and lithologic/archaeo-palaeontological levels at Galería. The data included here has been derived from combined ESR/U-series dating of teeth [15], TL and IRSL dating of sediment [14], ESR dating of calcite/speleothems [11,104], U-series dating of calcite/speleothems [11,15]. The numerical age estimates are shown against the Marine Oxygen Isotope curve record [94]. Two additional TL ages of 1340±170 ka (unit GIV) and 3000±1800 ka (unit GI) [14] are not shown on this plot as they lie beyond the x-axis limits (see Table 2 for a full summary of the dataset in [14]).

that the upper part of GI was deposited 374±32 ka. The two samples collected from the exposure in Zarpazos (ATZ10-3 and ATZ10-4), which are stratigraphically correlated as laying directly below GIIa, produced pIR-IR$_{225}$ and TT-OSL ages of 310–322 ka. The four TT-OSL and pIR-IR$_{225}$ ages for the lowest archaeological level (TG7) range between 284±16 ka and 335±20 ka (samples ATG10-9 and ATG10-10; **Table 2**), and produced a combined weighted mean age of 313±14 ka ($n = 4$) for the first appearance of lithic tools at Galería. Finally, the five samples from the overlying archaeological units (upper GIIa, GIIIa, GIIIb) and sterile GIV horizon (ATG10-8 to AT10-2) produced replicate pIR-IR$_{225}$ and TT-OSL ages ranging between ~220 and ~260 ka. These ages are all statistically indistinguishable at 2σ (**Figure 7**), suggesting relatively rapid infilling of the Galería cavity over a timeframe that is comparable to our empirical error ranges. With the possible exception of the ATG10-9 pIR-IR$_{225}$ age, all of the extended-range luminescence ages obtained in this study are in good agreement with the previously published ESR and U-series ages for speleothems [11,13,15] and combined ESR/U-series teeth ages [15] at either 1σ or 2σ (**Table 2**, **Figure 7**). The pIR-IR$_{225}$ and TT-OSL ages are also in agreement with the TL and combined TL/IRSL ages obtained on unit GIIIb [14]. However, the other published TL ages for units GIIa, GIIIa and GIV are systematically older than our extended-range luminescence chronologies (**Table 2**, **Figure 7**).

Discussion

1. Suitability of extended-age luminescence techniques in karstic-infill environments

Previous luminescence dating studies of cave deposits have shown that quartz OSL and K-feldspar pIR-IR signals can be completely reset prior to deposition [36,45,83,84]. However, it is less clear whether slower-bleaching TT-OSL signals [50] are able to be sufficiently reset to low levels for sediments that have been transported into caves via alluvial and colluvial processes. To some extent, the adequacy of TT-OSL bleaching in this context is likely to depend on the exposure and recycling histories of grains before they were ultimately washed into the cave. Quartz grains that had experienced prolonged and direct sunlight exposure immediately prior to entrainment and transportation into the cave and/or experienced relatively short residency times in well-bleached transient sediment bodies may still be amenable for TT-OSL dating, in spite of the unfavourable bleaching conditions they potentially experience during their incorporation into the cave deposit. Samples from the nearby Sima de los Huesos site, which is part of the same karst system, display seemingly well-bleached single-grain TT-OSL D_e distributions (i.e., low overdispersion values, single dose populations, non-skewed D_e distributions) [47]. The resultant weighted mean TT-OSL ages for these samples are also in good agreement with independent chronologies established at the site using U-series analysis [85]. In the same study it was shown that surface deposits located near the cave entrance have a very low mean TT-OSL residual of ~7 Gy (<1% of the mean D_e value for the Sima de los Huesos samples), which is similar to the lowest values measured elsewhere for modern-age aeolian samples [37,51,52,71]. This modern sample provides a useful analogue for the potentially well-bleached nature of the sediments found preserved in the cave environment. The majority of single-grain TT-OSL D_e distributions obtained in the present study are consistent with those reported for the Sima samples and the resultant TT-OSL ages are also in agreement with existing ESR, U-series and combined ESR/U-series chronologies. Together, the Galería Complex and Sima de los Huesos TT-OSL results indicate that the sediments blanketing the surrounding hills of the Atapuerca karst complex (i.e., the external sources of the karstic infill sediments) probably experienced prolonged bleaching periods (in the order of weeks to months) prior to entering the cave, as suggested previously [14]. At Galería, grains from units GIIa (upper) to GIV are also likely to have experienced additional sunlight exposure within the cave chamber prior to burial owing to their proximity to the nearby cave entrance (TN shaft).

Encouragingly, the sediments that were not directly sourced from the nearby cave opening (unit GI and the Zarpazos deposits, i.e., those transported through the endokarst system) yielded TT-OSL ages in agreement with replicate pIR-IR ages and the existing ESR and U-series (minimum) ages established on bracketing speleothems (**Figure 7**). Unit GIb was initially considered to be autochthonous to the cave system [12,57] and possibly sourced from dissolution of the host limestone bedrock. However, this interpretation is not consistent with the reasonable TT-OSL chronologies we obtain for samples ATG10-4, ATZ10-3 and ATZ10-4. If the dated grains had been directly derived from weathered bedrock, it is likely that they would have yielded either "infinite" D_e values or stratigraphically incoherent ages, as well as disagreement with replicate pIR-IR ages. Moreover, truly autochthonous sediments from this karst system are unlikely to have yielded the quantities of coarse-grained quartz grains found in our samples. Dissolution of limestone fragments collected from the Galería site as part of the present study showed that the Cretaceous bedrock contains very limited concentrations of quartz (<1%; [86]) and that the available quartz is exclusively <50 µm. In contrast, the luminescence dating samples collected from Zarpazos and unit GIb at Galería are composed primarily of coarse sand and silt grains that are >60 µm in size and the mineralogies are strongly dominated by quartz [14,86]. These observations, together with the consistency of the resultant luminescence dating results, suggest that the original sediment source was most likely external to the karst system.

The samples from unit GIb in Galería and Zarpazos also display distinctly different quartz luminescence brightness properties in comparison with the overlying samples (units upper GIIa to GIV; see File S1), indicating a potentially different sedimentary origin or geomorphic reworking history. This finding is consistent with the lack of exokarstic material (i.e., large mammal remains) in unit GIb and suggests that, at the time of deposition, the cave was closed-off from the immediate surrounding, and that these sediments were derived from an external source located some distance away. Consequently, it would be inappropriate to assume that sediments from GIb were fully bleached immediately prior to their final deposition, as was demonstrated [47] for sediments sourced from the immediate cave entrance (GII-GIV). In this regard, the TT-OSL and pIR-IR ages obtained for samples ATG10-4, ATZ10-3 and ATZ10-4 could conservatively equate to maximum age estimates for the emplacement of the host deposits at Galería. The similarity between the TT-OSL ages and the bracketing (minimum) ages of associated speleothems suggests that the sands of Zarpazos and unit GI were deposited relatively quickly after their initial incorporation into the karst system (i.e., over a time span similar to our analytical uncertainties). However, further luminescence studies are needed on similar "internal" facies to better constrain transportation times within the deeper cave system.

In assessing the suitability of the TT-OSL and pIR-IR chronologies obtained in this study, it is worth considering how our ages might change if we were to apply additional corrections for potential athermal and thermal signal losses over burial timescales. The need for application of a TT-OSL thermal instability correction is not clear-cut from existing known-age field assessments (see synthesis in [87]), though laboratory lifetime predictions suggest that an empirical correction may be prudent for some samples at least, particularly when dating over Early Pleistocene timescales. In the present context, it is also not known whether existing (multi-grain aliquot) laboratory lifetime predictions are of direct relevance to the specific grain populations isolated in our single-grain analysis. Nevertheless, if we assume a worst-case scenario and apply a first order TT-OSL thermal stability correction based on published laboratory lifetime calculations [48] and a sample burial temperature of 10°C (based on the mean current annual air temperature recorded inside the deeper chambers of the Atapuerca karst system [88] and outside the cave at the nearby Villafría (INM-2331) weather station), the TT-OSL ages of the Galería samples increase by only 6–16 ka. The difference between the original and corrected TT-OSL ages equate to <4% and is statistically insignificant at 1σ for all samples. Any potential systematic biases arising from signal instability are therefore largely inconsequential over the Middle Pleistocene timescales considered here.

The application of an anomalous fading correction to our pIR-IR_{225} ages (based on the empirically determined g-values) causes the resultant chronologies to increase by 12–20% but none of the ages change beyond their original 2σ error ranges. Importantly, the fading corrected pIR-IR_{225} ages are still in broad agreement with existing ESR, U-series and combined ESR/U-series chronologies at 2σ and they are still systematically younger than the TL ages [14] for units GIIa, GIIIa and GIV. Hence, our overall findings remain unchanged regardless of whether we opt for the original or fading corrected pIR-IR_{225} ages.

2. Comparison with previous luminescence dating results

As already noted, the majority of polymineral fine-grain TL ages [14] are significantly older than the chronologies established using other dating approaches, as well as those reported here (**Figure 7** and **Table 2**). Only the TL ages of 269±27 ka and 224±42 ka obtained for TG11 (upper GIIIb) are in agreement with our TT-OSL and pIR-IR ages at 1σ. The TL age of 503±95 ka for sublevel TG7 of lower GIIa is systematically older than the four ages obtained for this same unit using TT-OSL and pIR-IR_{225} (mean age 313±14 ka); though this TL age is in agreement with our ages at 2σ owing to its large relative uncertainty (19%). Given the systematic differences with the TL results, it seems plausible that this technique may not have provided representative burial ages for the Galería sequence. The most likely sources of inaccuracy with the TL ages include problems of partial bleaching, the use of 2π- instead of 4π-geometry gamma spectrometry in heterogeneous sedimentary matrices, and limitations with the chosen D_e measurement and data analysis procedures. TL dating was performed at Galería using a multiple-aliquot additive-dose (MAAD), total-bleach method, for which tens of thousands of grains were measured simultaneously and a single D_e value was computed per sample via extrapolation of the dose-response curve [89]. Though our single-grain TT-OSL results suggest that these samples were exposed to prolonged sunlight before transportation and burial, the adequacy of TL signal bleaching was not specifically tested in the earlier study [14]. Indeed, the TL plateau tests presented for some samples showed an increase in D_e with temperature (see Figure 6 and S2 of [14]), which may indicate insufficient bleaching of deeper TL traps. Though IRSL residual doses were shown to be very low for nearby surface sediments, the same assessments were not performed on the TL signals; hence the likelihood of MAAD TL resetting remains unknown in this environment. Additional D_e inaccuracies could have been introduced by the use of poorly defined dose-response curve fits coupled with significant extrapolation of the MAAD responses [89,90], particularly as many of the measured dose-response curves lay within 50% of saturation [91]. In light of these analytical uncertainties and the systematic TL age offsets seen throughout the sequence (**Figure 7**), we consider that the TL chronologies developed for Galería, and by extension those from other Atapuerca sites (e.g., TD10 Gran Dolina), should be viewed with a degree of caution (see also [92]).

It is also worthwhile noting that five additional IRSL ages were not reported [14] because they diverged from replicate TL ages at Galería and they ceased to increase with stratigraphic depth beyond ~250 ka. These trends were taken to be indicative of IRSL signal saturation but a similar non-incremental age-depth profile is apparent for both the ESR/U-series dataset [15] (GIIb to GIIIb) and for our extended-range luminescence chronologies (units upper GIIa to GIV). The similarities between these independent geochronological results suggests that the allochthonous Galería sequence genuinely accumulated over a relatively short timescale and thus the rejected IRSL ages [14] may have been more reliable than originally thought.

3. Chronostratigraphic implications of the new extended-range luminescence ages and comparison with recent ESR/U-series dating of teeth

Our new TT-OSL and pIR-IR results enable us to quantitatively evaluate the chronostratigraphic model proposed previously for the Galería sequence [16]. According to micromorphological data and pollen analysis undertaken on units GII to GIII, the allochthonous Galería sequence records a climatic deterioration from warm and wet condition (GIIa) to a cool and temperate phase (GIII-GIV) [9,62,93], and may correspond to a single interglacial–glacial cycle [8,10]. The upper section of the preceding sterile level (GIb) also records interglacial-like conditions characterised by high biological activity, degradation and bioturbation [9,12,16,57]. On the basis of these palaeoenvironmental indicators, it was suggested that (i) the beginning of the Galería sequence (GIb) accumulated during either MIS 9 or MIS 11, (ii) the lower archaeo-palaeontological levels (GIIa) formed within MIS 9, and (iii) the upper units (GIIb-GIIIb) were deposited during MIS 8 and possibly early MIS 7. The TT-OSL age for GIb (374±33 ka) aligns closely with the MIS 11/MIS 10 boundary but is also consistent with an MIS 9 age at 2σ. The mean luminescence age for the base of GIIa is 313±14 ka (i.e., TG7; $n = 4$) and indicates that this horizon accumulated during MIS 9. Together, our TT-OSL ages suggest that levels GIb and GIIa were either deposited continuously from early MIS 11/late MIS 10 through to MIS 9, or, if we consider the intervening finite stalagmite age (318±60 ka; **Table 2**) and the relatively warm conditions recorded in GIb, that both levels were deposited over a shorter time period within MIS 9. The shared sedimentological properties of the lower GIIa horizons and the upper part of unit GI [16] suggests that the two depositional episodes could have occurred during a closely spaced time period (i.e., within the same interglacial cycle), hence the latter interpretation may be the most feasible.

Our numerical dating results provide broad support for earlier chronostratigraphic interpretations [16]. We note, however, that the age of unit GIb (sample ATG10-4) is based on a relatively low number of TT-OSL D_e estimates. Further single-grain TT-OSL measurements would ideally be needed to improve the precision (and potentially accuracy) of this particular age estimate and to better constrain the chronological relationship between GIb and the base of GII. Regardless, we do not observe any evidence to suggest that the base of the archaeo-palaeontological levels at Galería Complex is ~500 ka (MIS 13), as indicated by the earlier TL dating results. Instead, our combined TT-OSL and pIR-IR datasets reveal that the earliest Acheulean lithics at Galería (i.e., those recorded in level TG7) were most likely formed during MIS 9 (i.e., 337–300 ka; [94]).

The 20 ESR/U-series teeth ages [15] reveal a potential temporal discontinuity between sub-units GIIa and GIIb at Galería (**Figure 7**). This chronological divergence occurs between archaeo-palaeontological levels TG10D and TG9, and is marked by an increase in age from 237–269 ka ($n = 4$) to 350–363 ka ($n = 2$) (see Figure 11 of [15]). It is worth highlighting that this apparent shift in the ESR/U-series ages is potentially negated by four additional teeth ages of 211–270 ka obtained on the underlying TG8 horizon of GIIa. These four ages were deemed unreliable on geochemical grounds [15], though it was acknowledged that there is no direct evidence for post-depositional uranium leaching in at least three of these specimens. A similar chronological shift is apparent in our luminescence dataset; however this potentially occurs at a lower stratigraphic position, between levels TG7 and TN2A of unit GIIa (**Figure 7**). This temporal discontinuity in unit GIIa appears to coincide with a major alteration in the local depositional environment at Galería, as evidenced by a change in the intrinsic TT-OSL characteristics of samples collected from TN2A upwards. The increased frequency of TT-OSL-producing grains and higher signal sensitivities observed for samples ATG10-8 through to AT10-2 (Figure S1 in File S1 and Table S3 in File S1) indicates an apparent shift in local sediment sourcing in comparison with units TG7, lower Zarpazos and GI. These changes in luminescence characteristics are consistent with lithostratigraphic evidence from unit GIIa, which suggest that the Galería chamber was opened-up to the cave exterior at this stage and started to be filled by large quantities of locally derived terra rossa sediment. The field-based evidence for a major change in sedimentary regime includes (i) the presence of large limestone blocks above layer TG7 at TN, which indicates extensive roof collapse at the southern end of the chamber; (ii) the appearance and increasing dominance of exokarstic material in layers TN2B, TG8 and TG9, as documented by the micro-morphological, microfacies and litho-logical analysis [9,16,62]; (iii) the deposition of a large guano layer and evidence of peat formation [62] immediately above layer TG7; and (iv) the occurrence of speleothem formations in layer TG8, which lies between levels TG7 and TN2A. Both the guano layer and the speleothem formation are indicative of a potential hiatus in sediment deposition prior to the cave opening event. The new TT-OSL and pIR-IR_{225} ages suggest that this depositional hiatus occurred between 313±14 ka (the mean age of TG7) and the opening of the cave chamber at (or slightly before) ~250 ka (the age of TN2A).

In comparing the luminescence ages from layer TN2A (sample ATG10-8) with the two combined ESR/U-series ages from levels TZ and TG9 (samples TZ02-02 and TG02-06), it is worth bearing in mind that the lateral associations between these horizons is somewhat ambiguous. Level TN2A is only preserved in the southern section of the profile, whereas level TG9 is found exclusively in the northern sector. It has been reported that TN2A is either located above or is correlative with level TG9 [59], and that both layers lie above TG8. It has also been acknowledged, however, that the true relationship between these levels is not easy to distinguish because they are laterally discontinuous and have undergone some post-depositional alterations [59]. Furthermore, one of the herbivore teeth dated by ESR/U-series was obtained from the entrance of the TZ chamber, which is disconnected from the TN section and is not likely to be directly representative of the GIIa levels preserved ~10 m away at the southern end of the Galería exposure. Added to this complexity is the fact that the luminescence and ESR/U-series samples were collected from sections that accumulated through different sediment conduits (TZ and TN, respectively). The difference in the ages obtained for TG9 and TN2A suggests that these deposits are not directly correlative, and they may have been deposited non-synchronously. Resolving the complex stratigraphic relationships observed in sub-unit GIIa will require a higher resolution dating study using complementary methods, as well as additional, quantitative sedimentological analyses.

It is interesting to compare the new Galería ages with those obtained for other Middle Pleistocene infill sequences at Atapuerca, namely Gran Dolina level TD10, the uppermost units of Sima del Elefante and the clay breccia infills at Sima de los Huesos. The TD10 level of Gran Dolina (located 30 m away) contains an abundant Mode II Acheulean assemblage and has been dated to between 337±29 ka and 418±63 ka using combined ESR/U-series analysis of teeth [95] (although see below for further discussion on contradictory TL/IRSL ages by [14]). The new weighted mean luminescence age of 313±14 ka for the lowermost archaeo-palaeontolgical level at Galería (TG7) may be contemporaneous with the age range of the TD10 sub-levels (TD10.1 and TD10.2), indicating that the two sites could have been occupied broadly simultaneously or over closely spaced time frames. Similarly, our ages for the earliest hominin presence at Galería appear to be correlative with the deposition of Upper Red Unit TE19 at the nearby site of Sima del Elefante, which contain sparse Mode II lithics and has been dated to 350–200 ka using biochronological associations [96]. Recently, new luminescence ages were presented [47] for the Sima de los Huesos "café con leche" bone breccia found directly above a rich *H. heidelbergensis* fossil assemblage and a single, well-shaped Acheulean hand-axe. The TT-OSL and pIR-IR_{225} mean age for this bone breccia is 427±12 ka ($n = 6$), which provides a firm minimum age for the hominin fossils. The combined luminescence chronology for the Sima de los Huesos assemblage reveals that there is no temporal overlap with the archaeological layers at Galería; the evidence for hominin presence at these two sites is separated by at least ~100 ka.

The new luminescence ages for Galería units GII and GIII confirm the presence of the Acheulean tradition in north-central Iberia within MIS 9 and MIS 8, at a time when Middle Palaeolithic technologies were appearing elsewhere in western Europe (Organc 3 [97,98]; see [99] for overview). Our chrono-logical results therefore support previous evidence pointing to the co-existence of these two lithic traditions during the Middle-Pleistocene [99–103]. Interestingly, the presence of Levallois-like methods at the Gran Dolina TD10.1 (upper) sub-level, as well as the progressive decrease in the number of large cutting tools within the entire 4 m-thick TD10 level, have been interpreted as representing a local Mode 2 to Mode 3 transitional phase [6] (similar observations have also been reported for the 'top of Galería' [19], albeit with limited information regarding tool description and their exact stratigraphic position). ESR/U-series dating of teeth from TD10.1 places this local transition [6] at ~350 ka (337±29 ka and 379±57 ka; [5,95]), but younger ages of ~240 ka have been obtained for bracketing layers TD11 and TD10.2 using polymineral fine-grain IRSL and TL dating, respectively [14]. The ESR/U-series ages for TD10.1 clearly pre-date our new ages and the published ESR/U-series ages [15] for units GIIb and GIII, which contain Mode II Acheulean artefacts. If the ESR/U-series ages for TD10.1 are correct, then there is an apparent temporal contradiction between the TD10.1 and Galería lithic records. New radiometric dating efforts at Gran Dolina, which are currently underway, will prove crucial for resolving this discrepancy, and will lead to an improved understanding of the pattern of technological change that occurred at Atapuerca during the Middle Pleistocene.

Conclusion

The extended-range luminescence chronologies obtained in this study enable us to refine and extend the existing numerical age constraint at the Middle Pleistocene archaeological site of Galería. Our combined TT-OSL and pIR-IR ages reveal that the basal sterile sands (unit GIb) underlying the archaeo-palaeontological sequence were deposited ~370 ka. The first hominin exploitation of Galería appears to have taken place during MIS 9 at ~313 ka, as recorded by the accumulation of Mode II lithics in layer TG7 of sub-unit GIIa. This period of hominin activity occurred prior to the opening of the main TN shaft. The new luminescence ages for horizon TN2A, which overlies the organomineral and speleothem formations (TG8), suggest that the main chamber became directly exposed to the exterior by ~240 ka. Hominin exploitation of animal carcasses (pitfall trap remains) continued during MIS 8 and early MIS 7 and was accompanying by a relatively rapid sediment infilling phase, as suggested by the indistinguishable ages obtained for the upper part of unit GIIa to unit GIV.

The new ages for the Galería sequence also provide insight into the origin of chronological disagreements published previously using different radiometric dating methods. The TT-OSL and pIR-IR ages obtained in this study are in overall agreement with the published ESR and U-series speleothems ages and combined ESR/U-series teeth ages at 2σ. However, the published TL ages for units GIIa, GIIIa and GIV are systematically older than both our luminescence chronologies and the ages obtained using other dating techniques (by ~200 ka in most instances). Given the significant offset with these TL ages, we suggest that chronologies established elsewhere at Atapuerca using this approach should be interpreted with caution.

The suitability of the extended-range chronologies presented in this study is supported by laboratory quality assurance tests (dose-recovery tests), assessments of single-grain D_e datasets, agreement between replicate TT-OSL and $pIR\text{-}IR_{225}$ ages, and the stratigraphic consistency of the resultant chronologies. Our data are consistent with the recent results presented for the neighbouring site of Sima de los Huesos [47] and suggest that TT-OSL and pIR-IR techniques offer good potential for constraining human occupation and depositional histories at the Middle Pleistocene Atapuerca sites. We emphasise, however, that the applicability of these extended-range luminescence dating techniques may vary at other localities and that sample-specific suitability assessments should always be performed in pIR-IR and TT-OSL dating studies. This study also demonstrates the potential for using luminescence sensitivity characteristics as a proxy for identifying changes in local depositional conditions and sediment sourcing. This novel approach may provide useful ancillary information for lithostratigraphic assessments of autochthonous versus allochthonous infill horizons at other Atapuerca sites.

Supporting Information

File S1 This file contains additional text about methodology and results, as well as associated figures and tables. Explanations are provided about measurement conditions and data analysis used for both the TT-OSL and $pIR\text{-}IR_{225}$ signals (e.g., D_e calculation and rejection criteria). Dose recovery test results for the pIR-IR and the TT-OSL signals are shown in Table S4. File S1 also contains the TT-OSL signal brightness characteristics and single-grain rejection statistics obtained for all samples. This file also contains Table S1-S5 and Figure S1-S5. Table S1, Details of the Galería luminescence dating samples analysed in this study. Table S2, Radionuclide activities (Bq kg-1) and daughter-to-parent ratios obtained from high resolution gamma spectrometry (HRGS) measurements of the 238U and 232Th decay chains. Table S3, Single-grain TT-OSL classification statistics for the Galería samples. Table S4, Results obtained for the dose recovery tests performed using protocol A of Table 4 for the TT-OSL signal. Table S5, Rejection statistics from the single-grain TT-OSL dose recovery test performed on samples ATG10-3 corresponding to Table S4. Figure S1, (A) Cumulative light-sum plots for the Galería samples constructed from the ranked net natural test dose signal (Tn) (using the first 0.24 s of laser stimulation minus a background-subtraction from the final 0.25 s). Plot (B) shows ranked signal counts normalised to the given Tn dose (200–300 Gy). Data shown is for single-grain TT-OSL measurements made using the 90–125 µm grain fraction (~18 grains per hole; Arnold et al., 2012). Figure S2, Single-grain TT-OSL dose recovery test results (~18 grains per hole). Figure S3, Histogram showing the distribution of single-grain TT-OSL D0 values for the Galería samples. Figure S4, Single-grain TT-OSL De distributions for the Galería samples, shown as histograms (left column) and radial plots (right column). Figure S5, Radial plots showing the pIR-IR225 De distributions of the remaining Galería samples not included in Figure 6 of the main text.

Acknowledgments

LA and MD thank Carlos Pérez Garrido for his assistance with preparing and measuring the luminescence dating samples and Virgil Drăguşin for his assistance with sample collection. We thank chief excavator Isabel Cáceres for her advice during the course of this study. We are also grateful to Sébastien Huot for providing Excel macros to calculate K-feldspar fading rates and age corrections. Access and permission to collect samples in Atapuerca was granted by Junta de Castilla y León.

Author Contributions

Conceived and designed the experiments: MD LJA JMP. Performed the experiments: MD LJA. Analyzed the data: MD LJA. Contributed reagents/materials/analysis tools: MD LJA JMP APG. Wrote the paper: MD LJA JMP AIO. Obtained permission for site excavation: JLA JMB EC. Provided sample locations: APG AIO.

References

1. Carbonell E, Bermúdez de Castro JM, Parés JM, Pérez-González A, Cuenca-Bescós G, et al. (2008) The first hominin of Europe. Nature 452: 465–470.
2. Arsuaga JL, Martínez I, Garcia A, Carretero JM, et al. (1997) Sima de los Huesos (Sierra de Atapuerca, Spain). The site. J Hum Evol 33: 109–127.
3. Meyer M, Fu Q, Aximu-Petri A, Glocke I, Nickel B, et al. (2014) A mitochondrial genome sequence of a hominin from Sima de los Huesos. Nature 505: 403–406.
4. Bermúdez de Castro JM, Arsuaga JL, Carbonell E, Rosas A, Martínez I, et al. (1997) A hominid from the Lower Pleistocene of Atapuerca, Spain: possible ancestor to Neandertals and modern humans. Science 276: 1392–1395.
5. Rodríguez J, Burjachs F, Cuenca-Bescós G, García N, Van der Made J, et al. (2011) One million years of cultural evolution in a stable environment at Atapuerca (Burgos, Spain). Quat Sci Rev 30: 1396–1412.
6. Ollé A, Mosquera M, Rodríguez XP, Lombera-Hermida A, García-Antón MD, et al. (2013) The Early Middle Pleistocene technological record from Sierra de Atapuerca (Burgos, Spain). Quat Int 295: 138–167.
7. Arsuaga JL, Gracia A, Lorenzo C, Martínez I, Pérez PJ (1999) Resto cranial humano de Galería/Cueva de los Zarpazos (Sierra de Atapuerca, Burgos). In: Carbonell Roura E Rosas González A, Díez Fernández-Lomana JC, editors, Atapuerca: Ocupaciones Humanas y Paleoecología del Yacimiento de Galería. Arqueología en Castilla y León 7. Valladolid. pp. 233–235.
8. Rosas A, Bermúdez de Castro JM (1999) Descripción y posición evolutiva de la mandíbula AT76-T1H del yacimiento de Galería (Sierra de Atapuerca). In: Carbonell Roura E, Rosas González A, Díez Fernández-Lomana JC, editors. Atapuerca: Ocupaciones Humanas y Paleoecología del Yacimiento de Galería. Arqueología en Castilla y León 7. Valladolid. pp. 237–244.
9. Rosas A, Carbonell E, Ollé A, Pérez-González A, Vallverdú J, et al. (1999) Contribución del yacimiento de Galería (Sierra de Atapuerca) al Cuaternario Ibérico. In: Carbonell Roura E, Rosas González A, Díez Fernández-Lomana JC, editors. Atapuerca: Ocupaciones Humanas y Paleoecología del Yacimiento de Galería. Arqueología en Castilla y León 7. Valladolid. pp. 377–390.
10. Ollé A, Cáceres I, Vergès JM (2005) Human Occupation at Galería site (Sierra de Atapuerca, Burgos, Spain) after the technological and taphonomical data. In: Molinea N, Moncel MH, Monnier J-L, editors. Les premiers peuplements en Europe. Oxford: John and Erika Hedges Ltd. pp. 269–280. BAR International Series, S1364.
11. Grün R, Aguirre E (1987) Datación por ESR y por la serie del U, en los depósitos cársticos de Atapuerca. In: Aguirre E. Carbonell E, Bermúdez de Castro JM, editors. El Hombre Fósil de Ibeas y el Pleistoceno de la Sierra de Atapuerca. Junta de Castilla y León. Valladolid. pp. 201–204.
12. Pérez-González A, Parés JM, Gallardo J, Aleixandre T, Ortega AI, et al. (1999) Geología y estratigrafía del relleno de Galería de la Sierra de Atapuerca (Burgos). In: Carbonell Raura E, Rosas González A, Díez Fernández-Lomana

JC, editors. Atapuerca: Ocupaciones Humanas y Paleoecología del Yacimiento de Galería. Arqueología en Castilla y León 7. Valladolid. pp. 31–42.

13. Falguères C, Bahain J-J, Yokoyama Y, Bischoff JL, Arsuaga JL, et al. (2001) Datation par RPE et U-Th des sites pléistocènes d'Atapuerca: Sima de los Huesos, Trinchera Dolina et Trinchera Galería. Bilan géochronologique. L'Anthropologie 105: 71–81.
14. Berger GW, Pérez-González A, Carbonell E, Arsuaga JL, Bermúdez de Castro JM, et al. (2008) Luminescence chronology of cave sediments at the Atapuerca palaeoanthropological site, Spain. J Hum Evol 55: 300–311.
15. Falguères C, Bahain J-J, Bischoff JL, Pérez-González A, Ortega AI, et al. (2013) Combined ESR/U-series chronology of Acheulian hominid-bearing layers at Trinchera Galería site, Atapuerca, Spain. J Hum Evol 65: 168–184.
16. Rosas A, Carbonell E, Cuenca G, García N, Fernández Jalvo Y, et al. (1998) Cronología, bioestratigrafía y paleoecología del Pleistoceno Medio de Galería (Sierra de Atapuerca, España). Revista Española de Paleontología 13: 71–80.
17. Muttoni G, Scardia G, Kent DV (2010) Human migration into Europe during the late Early Pleistocene climate transition. Palaeogeogr Palaeoclimatol Palaeoecol 296: 79–93.
18. Arsuaga JL, Peris Fernández J, Gracia-Téllez A, Quam R, Carretero JM, et al. (2012) Fossil human remains from Bolomor Cave (Valencia, Spain). J Hum Evol 62: 629–639.
19. Mosquera M, Ollé A, Rodríguez XP (2013) From Atapuerca to Europe: Tracing the earliest peopling of Europe. Quat Int 295: 130–137.
20. Bermúdez de Castro JM, Martinón-Torres M, Blasco R, Rosell J, Carbonell E (2013) Continuity and discontinuity in the European Early Pleistocene human settlement: the Atapuerca evidence. Quat Sci Rev 76: 53–65.
21. White WB (2007) Cave sediments and paleoclimate. J Cave Karst Stud 69: 76–93.
22. Constantin S, Robu M, Munteanu C-M, Petculescu A, Vlaicu M, et al. (2013) Reconstructing the evolution of cave systems as a key to understanding the taphonomy of fossil accumulations; The case of Urşilor Cave (Western Carpathians, Romania). Quat Int 339–340: 25–40.
23. Hoffmann DL, Pike AWG, Wainer K, Zilhão J (2013) New U-series results for the speleogenesis and Palaeolithic archaeology of the Almonda karstic system (Torres Novas, Portugal). Quat Int 294: 168–182.
24. Karkanas P, Bar-Yosef O, Goldberg P, Weiner S (2000) Diagenesis in prehistoric caves: the use of minerals that form in situ to assess the complétense of the archaeological record. J Archaeol Sci 27: 915–929.
25. Goldberg P, Sherwood S (2006) Deciphering human prehistory through the geoarchaeology study of cave sediments. Evol Anthropol 15: 20–36.
26. Aitken MJ (1998) An Introduction to Optical Dating: The Dating of Quaternary Sediments by the Use of Photon-Stimulated Luminescence. Oxford: Oxford University Press. 267 p.
27. Godfrey-Smith DI, Huntley DJ, Chen WH (1988) Optical dating studies of quartz and feldspar sediment extracts. Quat Sci Rev 7: 373–380.
28. Duller GAT (2008) Single-grain optical dating of Quaternary sediments: why aliquot size matters in luminescence dating. Boreas 37: 589–612.
29. Galbraith RF, Roberts RG (2012) Statistical aspects of equivalent dose and error calculation and display in OSL dating: An overview and some recommendations. Quat Geochronol 11: 1–27.
30. Jacobs Z, Roberts RG (2007) Advances in optically stimulated luminescence dating of individual grains of quartz from archeological deposits. Evol Anthropol 16: 210–223.
31. Arnold LJ, Bailey RM, Tucker GE (2007) Statistical treatment of fluvial dose distributions from southern Colorado arroyo deposits. Quat Geochronol 2: 162–167.
32. Arnold LJ, Roberts RG (2009) Stochastic modelling of multi-grain equivalent dose (D_e) distributions: Implications for OSL dating of sediment mixtures. Quat Geochronol 4: 204–230.
33. Arnold LJ, Demuro M, Navazo M, Benito-Calvo A, Pérez-González A (2013) OSL dating of the Middle Palaeolithic Hotel California site, Sierra de Atapuerca, north-central Spain. Boreas 42: 285–305.
34. Jacobs Z, Wintle AG, Roberts RG, Duller GAT (2008) Equivalent dose distributions from single grains of quartz at Sibudu, South Africa: context, causes and consequences for optical dating of archaeological deposits. J Archaeol Sci 35: 1808–1820.
35. Demeter F, Shackelford LL, Bacon A-M, Duringer P, Westaway K, et al. (2012) Anatomically modern human in southeast Asia (Laos) by 46 ka. Proc Natl Acad Sci U S A 109: 14375–14380.
36. Gliganic LA, Jacobs Z, Roberts R (2012) Luminescence characteristics and dose distributions for quartz and feldspar grains from Mumba rockshelter, Tanzania. Archaeo Anthropol Sci 4: 115–135.
37. Wang XL, Wintle AG, Lu YC (2006) Thermally transferred luminescence in fine-grained quartz from Chinese loess: Basic observations. Radiat Meas 41: 649–658.
38. Thomsen KJ, Murray AS, Jain M, Bøtter-Jensen L (2008) Laboratory fading rates of various luminescence signals from feldspar-rich sediment extracts. Radiat Meas 43: 1474–1486.
39. Buylaert J-P, Jain M, Murray AS, Thomsen KJ, Thiel C, et al. (2012) A robust feldspar luminescence dating method for Middle and Late Pleistocene sediments. Boreas 41: 435–451.
40. Duller GAT, Wintle AG (2012) A review of the thermally transferred optically stimulated luminescence signal from quartz for dating sediments. Quat Geochronol 7: 6–20.
41. Sun X, Mercier N, Falgueres C, Bahain J-J, Despriee J, et al. (2010) Recuperated optically stimulated luminescence dating of middle-size quartz grains from the Palaeolithic site of Bonneval (Eure-et-Loir, France). Quat Geochronol 5: 342–347.
42. Roberts HM (2012) Testing Post-IR IRSL protocols for minimising fading in feldspars, using Alaskan loess with independent chronological control. Radiat Meas 47: 716–724.
43. Roberts DL, Karkanas P, Jacobs Z, Marean C, Roberts RG (2012) Melting ice sheets 400,000 yr ago raised sea level by 13 m: Past analogue for future trends. Earth Planet Sci Lett 357–358: 226–237.
44. Pickering R, Jacobs Z, Herries AIR, Karkanas P, Bar-Matthews M, et al. (2013) Paleoanthropologically significant South African sea caves dated to 1.1-1.0 million years using a combination of U-Pb, TT-OSL and palaeomagnetism. Quat Sci Rev 65: 39–52.
45. Rink WJ, Mercier N, Mihailović D, Morley MW, Thomspon JW, et al. (2013) New radiometric ages for the BH-1 hominin from Balanica (Serbia): implications for understanding the role of the Balkans in Middle Pleistocene human evolution. PLoS One 8: e54608.
46. Sun X, Lu H, Wang S, Yi S, Shen C, et al. (2013) TT-OSL dating of Longyadong Middle Plaleolithic site and paleoenvironmental implications from hominin occupation in Luonan Basin (central China). Quat Res 79: 168–174.
47. Arnold LJ, Demuro M, Parés JM, Arsuaga JL, Aranburu A (2014) Luminescence dating and palaeomagnetic age constraint on hominins from Sima de los Huesos, Atapuerca, Spain. J Hum Evol 67: 85–107.
48. Adamiec G, Duller GAT, Roberts HM, Wintle AG (2010) Improving the TT-OSL SAR protocol through source trap characterisation. Radiat Meas 54: 768–777.
49. Vasiliniuc Ş, Vandenberghe DAG, Timar-Gabor A, Panaiotu C, Cosma C, et al. (2012) Testing the potential of elevated temperature post-IR IRSL signals for dating Romanian loess. Quat Geochronol 10: 75–80.
50. Jacobs Z, Roberts RG, Lachlan TJ, Karkanas P, Marean CW, et al. (2011) Development of the SAR TT-OSL procedure for dating Middle Pleistocene dune and shallow marine deposits along the Cape coast of South Africa. Quat Geochronol 6: 491–513.
51. Porat N, Duller GAT, Roberts HM, Wintle AG (2009) A simplified protocol for TT-OSL. Radiat Meas 44: 538–542.
52. Tsukamoto S, Duller GAT, Wintle AG (2008) Characteristics of thermally transferred optically stimulated luminescence (TT-OSL) in quartz and its potential for dating sediments. Radiat Meas 43: 1204–1218.
53. Benito-Calvo A, Pérez-González A, Parés JM (2008) Quantitative reconstruction of Late Cenozoic landscapes: a case study in the Sierra de Atapuerca (Burgos, Spain). Earth Surf Proc Land 33: 196–208.
54. Ortega AI, Benito-Calvo A, Pérez-González A, Martín-Merino MA, Pérez-Martínez R, et al. (2013) Evolution of multilevel caves in the Sierra de Atapuerca (Burgos, Spain) and its relation to human occupation. Geomorphology 196: 122–137.
55. Ortega AI (2009) La evolución geomorfológica del karst de la Sierra de Atapuerca (Burgos) y su relación con los yacimientos Pleistocenos que contiene. Ph.D. Thesis, University of Burgos, 624 p.
56. Aguirre E (1987) Los yacimientos mesopleistocenos en cavidades de la Sierra de Atapuerca. In: Aguirre E, Carbonell E, Bermúdez de Castro JM, editors. El Hombre Fósil de Ibeas y el Pleistoceno de la Sierra de Atapuerca. Junta de Castilla y León Valladolid pp. 201–204.
57. Pérez-González A, Aleixandre T, Pinilla A, Gallardo J, Benayas J, et al. (1995) An approach to the Galería stratigraphy in the Sierra de Atapuerca trench (Burgos). In: Bermúdez de Castro JM, Arsuaga JL, Carbonell E, editors. Evolución Humana en Europa y los Yacimientos de la Sierra de Atapuerca. Junta de Castilla y León. Valladolid. pp. 99–122.
58. Gil E, Aguirre E, Hoyos M (1987) Contexto estratigráfico. In: Aguirre E, Carbonell E, Bermúdez de Castro JM, editors. El Hombre Fósil de Ibeas y el Pleistoceno de la Sierra de Atapuerca I. Junta de Castilla y León. Soria. pp. 47–54.
59. Ollé A, Huguet R (1999) Sequencia arqueoestratigráfica del yacimiento de Galería, Atapuerca. In: Carbonell Roura E, Rosas González A, Díez Fernández-Lomana JC, editors. Atapuerca: Ocupaciones Humanas y Paleoecología del Yacimiento de Galeria. Arqueología en Castilla y León 7. Valladolid. pp. 55–62.
60. Pérez-González A, Parés JM, Carbonell E, Aleixandre T, Ortega AI, et al. (2001) Géologie de la Sierra de Atapuerca et stratigraphie des remplissages karstiques de Galería et Dolina (Burgos, Espagne). L'Anthropologie 105: 27–43.
61. Shahack-Gross R, Berna F, Karkanas P, Weiner S (2004) Bat guano and preservation of archaeological remains in cave sites. J Archaeol Sci 31: 1259–1272.
62. Vallverdú J (1999) Microfacies y micromorfologia de las unidades GII y GIII de Galería (Sierra de Atapuerca). In: Carbonell Roura E, Rosas González A, Díez Fernández-Lomana JC, editors. Atapuerca: Ocupaciones Humanas y Paleoecología del Yacimiento de Galería. Arqueología en Castilla y León 7. Valladolid, pp. 43–54.
63. García-Medrano P, Ollé A, Mosquera M, Cáceres I, Díez C, et al. (2014) The earliest Acheulean at Atapuerca (Burgos, Spain): Oldest levels of the Galería site (GII unit). Quat Int. (In press).

64. Cuenca-Bescós G, Rofes J, López-García JM, Blain HA, De Marfá RJ, et al. (2010) Biochronology of Spanish Quaternary small vertebrate faunas. Quat Int 212: 109–119.
65. Cuenca-Bescós G, Canudo JI, Laplana C (1999) Análisis bioestratigráfico de los roedores del Pleistoceno medio del yacimiento de Galería (Sierra de Atapuerca, Burgos). In: Carbonell Roura E, Rosas González A, Díez Fernández-Lomana JC, editors. Atapuerca: Ocupaciones Humanas y Paleoecología del Yacimiento de Galería. Arqueología en Castilla y León 7. Valladolid. pp. 189–210.
66. Van der Made J (1999) Artiodactyla del yacimiento mesopleistoceno de Galería (Sierra de Atapuerca). In: Carbonell Roura E, Rosas González A, Díez Fernández-Lomana JC, editors. Atapuerca: Ocupaciones Humanas y Paleoecología del Yacimiento de Galería. Arqueología en Castilla y León 7. Valladolid. pp. 143–168.
67. Arnold LJ, Duval M, Falguères C, Bahain J-J, Demuro M (2012) Portable gamma spectrometry with cerium-doped lanthanum bromide scintillators: suitability assessments for luminescence and electron spin dating applications. Radiat Meas 47: 6–18.
68. Guérin G, Mercier M, Adamiec G (2011) Dose-rate conversion factors: update. Ancient TL 29: 5–8.
69. Prescott JR, Hutton JT (1994) Cosmic ray contributions to dose rates for luminescence and ESR dating: large depths and long-term time variations. Radiat Meas 23: 497–500.
70. Thomsen KJ, Bøtter-Jensen L, Jain M, Denby PM, Murray AS (2008) Recent instrumental developments for trapped electron dosimetry. Radiat Meas 43: 414–421.
71. Stevens T, Buylaert J-P, Murray AS (2009) Towards development of a broadly-applicable SAR TT-OSL dating protocol for quartz. Radiati Meas 44: 639–645.
72. Buylaert J-P, Murray AS, Thomsen KJ, Jain M (2009) Testing the potential of an elevated temperature IRSL signal from K-feldspar. Radiat Meas 44: 560–565.
73. Aitken MJ (1985) Thermoluminescence Dating. London: Academic Press. 365 p.
74. Auclair M, Lamothe M, Huot S (2003) Measurement of anomalous fading for feldspar IRSL using SAR. Radiat Meas 37: 487–492.
75. Huntley DJ, Lamothe M (2001) Ubiquity of anomalous fading in K-feldspars and the measurement and correction for it in optical dating. Can J Earth Sci 38: 1093–1106.
76. Buylaert J-P, Thiel C, Murray AS, Vandenberghe DAG, Yi S, et al. (2011) IRSL and post-IR IRSL residual doses recorded in modern dust samples from the Chinese loess plateau. Geochronometria 38: 432–440.
77. Thiel C, Coltorti M, Tsukaoto S, Frechen M (2010) Geochronology of some key sites along the coast of Sardinia (Italy). Quat Int 222: 36–47.
78. Lowick SE, Trauerstein M, Preusser F (2012) Testing the application of post IR-IRSL dating to fine grain waterlain sediments. Quat Geochronol 8: 33–40.
79. Sohbati R, Murray AS, Buylaert J-P, Ortuño M, Cunha PP, et al. (2012) Luminescence dating of Pleistocene alluvial sediments affected by the Alhama de Murcia fault (eastern Betics, Spain) – a comparison between OSL, IRSL and post-IR IRSL ages. Boreas 41: 250–262.
80. Kars RH, Busschers FS, Wallinga J (2012) Validating post IR-IRSL dating on K-feldspars through comparison with quartz OSL ages. Quat Geochronol 12: 74–86.
81. Galbraith RF, Roberts RG, Laslett GM, Yoshida H, Olley JM (1999) Optical dating of single and multiple grains of quartz from Jinmium rock shelter, northern Australia: Part I, experimental design and statistical models. Archaeometry 41: 339–364.
82. Galbraith RF, Green PF (1990) Estimating the component ages in a finite mixture. Nucl Tracks Rad Meas 17: 197–206.
83. Gaar D, Preusser F (2012) Luminescence dating of mammoth remains from northern Switzerland. Quat Geochronol 10: 257–263.
84. Clark-Balzan LA, Candy I, Schwenninger J-L, Bouzouggar A, Blockley S, et al. (2012) Coupled U-series and OSL dating of a Late Pleistocene cave sediment sequence, Morocco, North Africa: significance for constructing Palaeolithic chronologies. Quat Geochronol 12: 53–64.
85. Arsuaga JL, Martínez I, Arnold LJ, Aranburu A, Gracia-Téllez A, et al. (2014) Neandertal roots: cranial and chronological evidence from Sima de los Huesos. Science 344: 1358–1363.
86. Aleixandre T, Pérez-González A (1999) Facies mineralógicas de las arenas de los rellenos karsticos de la Sierra de Atapuerca (Burgos). In: Andreo B, Carrasco F, Duran JJ, editors. Contribución del Estudio científico de las Cavidades Kársticas al Conocimiento Geológico. Patronato de la Cueva Nerja. Nerja. Málaga. pp. 231–242.
87. Arnold LJ, Demuro M, Parés JM, Pérez-González A, Arsuaga JL, et al. (2014) Evaluating the suitability of extended-range luminescence dating techniques over Early and Middle Pleistocene timescales: Published datasets and case studies from Atapuerca, Spain. Quat Int. (In press).
88. Martín-Chivelet J, Muñoz-García MB, Edwards RL, Turrero MJ, Ortega AI (2011) Land surface temperature changes in Northern Iberia since 4000 yr BP, based on δ13C of speleothems. Global Planet Change 77: 1–12.
89. Wintle AG (1997) Luminescence dating: laboratory procedures and protocols. Radiat Meas 27: 769–817.
90. Foreman SL (1989) Applications and limitations of thermoluminescence to date Quaternary sediments. Quat Int 1: 47–59.
91. Wintle AG, Questiaux DG, Roberts RG, Spooner NA (1993) Dating loess up to 800 ka by thermoluminescence: Comment and Reply. Geology 21: 568–569.
92. Parés JM, Arnold LJ, Duval M, Demuro M, Pérez-González A, et al. (2013) Reassessing the age of Atapuerca TD6 (Spain): New palaeomagnetic results. J Archaeol Sci 40: 4586–4595.
93. Garcia-Antón M, Sainz-Ollero H (1991) Pollen records from the middle Pleistocene Atapuerca site (Burgos, Spain). Palaeogeogr Palaeoclimatol Palaeoecol 85: 199–206.
94. Lisiecki LE, Raymo ME (2005) A Pliocene-Pleistocene stack of 57 globally distributed benthic δ18O records. Paleoceanography 20: PA 1003.
95. Falguères C, Bahain J-J, Yokoyama Y, Arsuaga JL, Bermudez de Castro JM, et al. (1999) Earliest humans in Europe: the age of TD6 Gran Dolina, Atapuerca, Spain. J Hum Evol 37: 343–352.
96. López-García JM, Blain H-A, De Marfà R, García A, Martinell J, et al. (2011) Small-mammals from the Middle Pleistocene layers of the Sima del Elefante (Sierra de Atapuerca, Burgos, northwestern Spain). Geologica Acta 9: 29–43.
97. Moncel M-H, Moigne A-M, Sam Y, Combier J (2011) The emergence of Neanderthal technical behavior: new evidence from Orgnac 3 (level 1, MIS 8), southeastern France. Curr Anthropol 52: 37–75.
98. Michel V, Shen G, Shen C-C, Wu C-C, Vérati C, et al. (2013) Application of U/Th and $^{40}Ar/^{39}Ar$ dating to Orgnac 3, a Late Acheulean and Early Middle Palaeolithic site in Ardèche, France. PLoS One 8(12): e82394.
99. Villa P (2009) The Lower to Middle Paleolithic Transition. In: Camps M, Chauhan P, editors. Sourcebook of Paleolithic Transitions, Methods, Theories, and Interpretations. New York: Springer, pp. 265–270.
100. Santonja M, Villa P (2006) The Acheulian of western Europe. In: Goren-Inbar N, Sharon G, editors. Axe Age: Acheulian Toolmaking from Quarry to Discard. London: Equinox. pp. 429–478.
101. Monnier GF (2006) The Lower/Middle paleolithic periodization in western Europe: an evaluation. Curr Anthropol 47: 709–744.
102. Santonja M, Pérez-González A (2010) Mid-Pleistocene Acheulean industrial complex in the Iberian Peninsula. Quat Int 223–224: 154–161.
103. Santonja M, Pérez-González A, Domínguez-Rodrigo M, Panera J, Rubio-Jara S, et al. (2014) The Middle Paleolithic site of Cuesta de la Bajada (Teruel, Spain): a perspective on the Acheulean and Middle Paleolithic technocomplexes in Europe. J Archaeol Sci 49: 556–571.
104. Falguères C (1986) Datations de sites acheuléens et moustériens du Midi Méditerranéen par la méthode de Résonance de Spin Electronique. Muséum National d'Historie Naturelle, Paris. Ph.D. Thesis.
105. Readhead ML (2002) Absorbed dose fraction for ^{87}Rb β particles. Ancient TL 20: 25–28.
106. Mejdahl V (1979) Thermoluminescence dating: beta-dose attenuation in quartz grains. Archaeometry 21: 61–72.
107. Brennan BJ (2003) Beta doses to spherical grains. Radiat Meas 37: 299–303.
108. Bøtter-Jensen L, Mejdahl M (1988) Assessment of beta dose-rate using a GM multicounter system. Nucl Tracks Radiat Meas 14: 187–191.
109. Bowler JM, Johnston H, Olley JM, Prescott JR, Roberts RG, et al. (2003) New ages for human occupation and climate change at Lake Mungo, Australia. Nature 421: 837–840.
110. Jacobs Z, Duller GAT, Wintle AG, Henshilwood CS (2006) Extending the chronology of deposits at Blombos Cave, South Africa, back to 140 ka using optical dating of single and multiple grains of quartz. J Hum Evol 51: 255–273.
111. Pawley SM, Bailey RM, Rose J, Moorlock BSP, Hamblin RJO, et al. (2008) Age limits on Middle Pleistocene glacial sediments from OSL dating, north Norfolk, UK. Quat Sci Rev 27: 1363–1377.
112. Mejdahl V (1987) Internal radioactivity in quartz and feldspar grains. Ancient TL 5: 10–17.
113. Rees-Jones J (1995) Optical dating of young sediments using fine-grain quartz. Ancient TL 13: 9–14.
114. Rees-Jones J, Tite MS (1997) Optical dating results for British archaeological sediments. Archaeometry 39: 177–187.
115. Huntley DJ, Clague JJ (1996) Optical dating of tsunami-laid sands. Quat Res 46: 127–140.
116. Huntley DJ, Lian OB (1999) Using optical dating to determine when a sediment was last exposed to sunlight. In D.S. Lemmen and R.E. Vance (Eds.), Holocene Climate and Environmental Change in the Palliser Triangle: A Geoscientific Context for Evaluating the Impacts of Climate Change on the Southern Canadian Prairies. Geological Survey of Canada, Ottawa, Bulletin 534: 211–222.
117. Alappat L, Tsukamoto S, Singh P, Srikanth D, Ramesh R, et al. (2010) Chronology of Cauvery Delta sediments from shallow subsurface cores using elevated temperature post-IR IRSL dating of feldspar. Geochronometria 37: 37–47.
118. Lang A, Wagner GA (1997) Infrared stimulated luminescence dating of Holocene colluvial sediments using the 410 nm emission. Quat Sci Rev 16: 393–396.
119. Lang A, Hatté C, Rousseau DD, Antoine P, Fontugne M, et al. (2003) High-resolution chronologies for loess: comparing AMS ^{14}C and optical dating results. Quat Sci Rev 22: 953–959.
120. Banerjee D, Murray AS, Bøtter-Jensen L, Lang A (2001) Equivalent dose estimation using a single aliquot of polymineral fine grains. Radiat Meas 33: 73–94.

121. Feathers JK, Casson MA, Schmidt AH, Chithamboet ML (2012) Application of pulsed OSL to polymineral fine-grained samples. Radiat Meas 47: 201–209.
122. Huntley DJ, Baril MR (1997) The K content of the K-feldspars being measured in optical dating or in thermoluminescence dating. Ancient TL 15: 11–13.
123. Huntley DJ, Hancock RGV (2001) The Rb contents of the K-feldspar grains being measured in optical dating. Ancient TL 19: 43–46.
124. Duller GAT (2003) Distinguishing quartz and feldspar in single grain luminescence measurements. Radiat Meas 37: 161–165.
125. Bailey RM, Arnold LJ (2006) Statistical modelling of single grain quartz De distributions and an assessment of procedures for estimating burial dose. Quat Sci Rev 25: 2475–2502.

12

Heavy Metal Pollution Characteristics of Surface Sediments in Different Aquatic Ecosystems in Eastern China

Wenzhong Tang[1]*, Baoqing Shan[1]*, Wenqiang Zhang[1], Hong Zhang[1], Lishuo Wang[1,2], Yuekui Ding[1]

1 State Key Laboratory on Environmental Aquatic Chemistry, Research Center for Eco-Environmental Sciences, Chinese Academy of Sciences, Beijing, China, **2** China University of Mining & Technology (Beijing), Beijing, China

Abstract

Aquatic ecosystems in eastern China are suffering threats from heavy metal pollution because of rapid economic development and urbanization. Heavy metals in surface sediments were determined in five different aquatic ecosystems (river, reservoir, estuary, lake, and wetland ecosystems). The average Cd, Cr, Cu, Ni, Pb, and Zn concentrations were 0.716, 118, 37.3, 32.7, 56.6, and 204 mg/kg, respectively, and the higher concentrations were mainly found in sediment samples from river ecosystems. Cd was the most anthropogenically enriched pollutant, followed by Zn and Pb, indicated by enrichment factors >1.5. According to consensus-based sediment quality guidelines, potential ecological risk indices, and risk assessment codes, all five types of aquatic ecosystems were found to be polluted with heavy metals, and the most polluted ecosystems were mainly rivers. Cd was the most serious pollutant in all five aquatic ecosystems, and it was mainly found in the exchangeable fraction (about 30% of the total Cd concentration, on average). The results indicate that heavy metal contamination, especially of Cd, in aquatic ecosystems in eastern China should be taken into account in the development of management strategies for protecting the aquatic environment.

Editor: Manuel Reigosa, University of Vigo, Spain

Funding: This work was supported by the National Natural Science Foundation of China (No. 21107126), the National Water Pollution Control Program (No. 2012ZX07203-002), and the One-Three-Fiver Program of Research Center for Eco-Environmental Sciences (No. YSW2013B02). The funders had no role in study design, data collection and analysis, decision to publish, or preparation of the manuscript.

Competing Interests: The authors have declared that no competing interests exist.

* Email: wztang@rcees.ac.cn (WZT); bqshan@rcees.ac.cn (BQS)

Introduction

Heavy metal contamination has been a worldwide environmental problem for many years [1–4]. Heavy metals are widespread, persistent, and potentially toxic in aquatic ecosystems [5–7]. After being transported into aquatic ecosystems, heavy metals can be absorbed by suspended solids and accumulated in sediments. Sediment acts as a sink for heavy metals but can re-release the heavy metals, acting as a source of pollutants and degrading the quality of the aquatic ecosystems [8,9]. Therefore, the accumulation of heavy metals in sediment is a cause of growing interest and concern. Environmental problems caused by heavy metal pollution in aquatic ecosystems have recently been extensively studied so that procedures for effectively managing these ecosystems can be developed [10–12].

Heavy metal contamination in aquatic ecosystems usually accompanies rapid economic development, in both developed and developing countries [13]. Little attention was paid to environmental problems caused by heavy metal contamination in sediments in China by researchers or the government until 2000 [4,14]. Economic development, especially industrial development, has recently been continuously and rapidly intensifying in China, particularly in eastern China. Heavy metals discharged into the air and effluents by industrial activities have caused severe pollution in aquatic ecosystems in eastern China [4,6,15]. The State Council of China approved the "12th National Five Year Plan for Comprehensive Prevention and Control of Heavy Metal Pollution" in early 2011, with the aim of controlling heavy metal contamination in the environment [16]. Eastern China is the most developed and most densely populated region in China. Rapid economic development and urbanization have caused significant amounts of pollution, including heavy metal pollution, in different aquatic ecosystems (rivers, lakes, and wetlands) in eastern China [6,13,15,16]. It is important to comprehensively understand the characteristics of heavy metal contamination in different types of aquatic ecosystems in this area to provide a reference point for the development of better procedures for controlling and managing heavy metals in the affected ecosystems.

In the study presented here, we evaluated the heavy metal pollution status of five different types of aquatic ecosystems (river, reservoir, estuary, lake, and wetland ecosystems) in eastern China. The specific goals were: 1) to determine heavy metal (Cd, Cr, Cu, Ni, Pb, and Zn) concentrations in surface sediments in twelve study areas; 2) to use enrichment factors (EFs) to determine the importance of lithogenic and anthropogenic sources of the heavy metals; and 3) to assess the risks associated with the presence of the heavy metals in the aquatic ecosystems using consensus-based sediment quality guidelines (SQGs), potential ecological risk (PER) indices, and risk assessment codes (RACs).

Materials and Methods

Ethics Statement

No specific permits were required for the field studies described here. The study areas are not privately owned or protected in any way, and the field studies did not involve endangered or protected species.

Description of the Study Areas

Twelve study areas (Fig. 1) in eastern China that have been affected by rapid industrial development and urbanization were selected to represent five typical aquatic ecosystems (reservoirs, rivers, estuaries, wetlands, and lakes). The study areas were in Heilongjiang Province, Heibei Province, Beijing Municipality, Tianjin Municipality, Zhejiang Province, Anhui Province, Jiangxi Province, and Guangdong Province. The study areas were in the region between 113° 15′ 31.08″ E and 126° 30′ 52.43″ E and between 23° 50′ 26.85″ N and 45° 46′ 18.01″ N.

The river ecosystems were the Songhuajiang River (SR), two typical rivers in the Haihe River basin (HR1 and HR2), and the Zhujiang River (ZR). The reservoir ecosystem was the Miyun Reservoir (MRe). Estuarine ecosystems were represented by the estuaries of rivers HR1 (estuary HE1) and HR2 (estuary HE2). The lake ecosystems were Chaohu Lake (CL) and Poyanghu Lake (PL), and the wetland ecosystems were the Baiyangdian Wetland (BW) and two areas in the Shijiuyang Wetland (SW1 and SW2).

Sample Collection and Analysis

Surface sediment samples were collected from the twelve study areas (Fig. 1) using a Peterson grab sampler between August and October 2013. Five sampling stations were established in each study area, and three parallel samples were taken from each station (Fig. S1 in File S1). The samples (n = 180) were air-dried in the laboratory then dried in an oven at 40°C. The dried samples were ground and passed through a 100 mesh sieve before they were analyzed.

For the total heavy metal analysis, each sediment sample (0.100 g) was digested in a 5:1 mixture of hydrofluoric acid and perchloric acid [17] in a microwave digestion system (MARS XPress; CEM, Matthews, NC, USA) using the digestion conditions shown in Table S1 in File S1. The geochemical fractionation of the heavy metals in a sample was determined using the BCR three-step sequential extraction procedure [18], to determine the exchangeable (sediment solution, carbonates, and exchangeable metals), reducible (Fe/Mn oxides), oxidizable (organic matter and sulfides), and residual (remaining non-silicate bound) fractions of the heavy metals. All of the above solutions were stored at 4°C until instrumental analysis was performed. The Cd, Cr, Cu, Fe, Ni, Pb, and Zn concentrations in the extracts were determined using an inductively coupled plasma-mass spectrometry (ICP-MS) instrument (7500a; Agilent Technologies, Santa Clara, CA, USA), which gave detection limits of 0.015–0.120 μg/L, and an inductively coupled plasma optical emission spectrometry (ICP-OES) instrument (Optima 2000DV; Perkin Elmer, Waltham, MA, USA), which gave detection limits of 0.001–0.030 mg/L. The quality control procedures used in the laboratory were analyzing a sediment reference material (GBW07302a; National Institute of Metrology, Beijing, China) and analyzing samples in triplicate. The recoveries of the heavy metals varied but were all within the range 90–105%. The analytical method gave a good level of precision, the relative standard deviation being less than 5%. Each result that is presented is the average of the measurements made at the five sampling stations in each study area.

Figure 1. Locations of the five aquatic ecosystem types and the 12 areas that were studied.

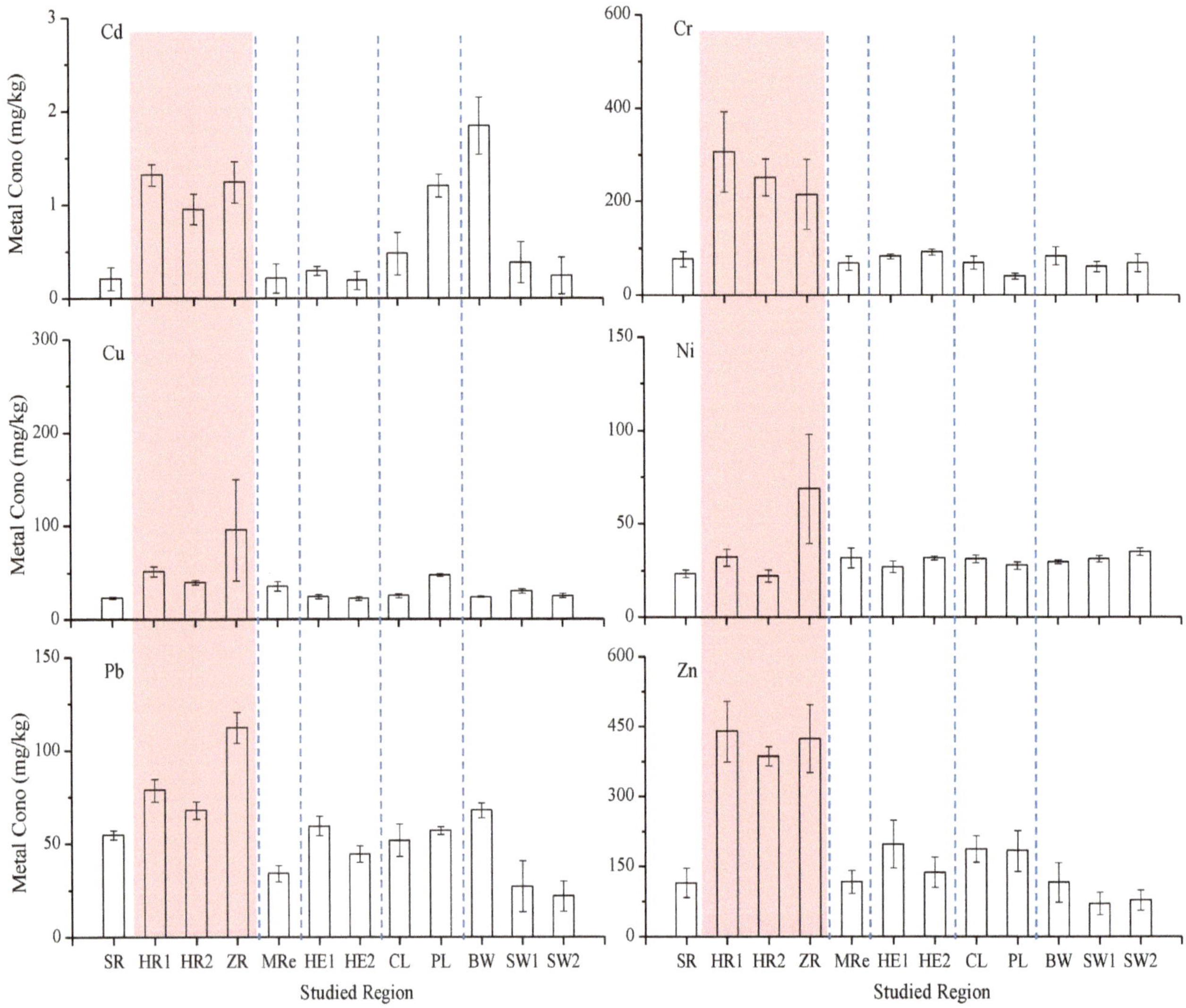

Figure 2. Heavy metal concentrations in the surface sediment samples from each of the study sites.

Enrichment Factors

The EFs for the heavy metals in the surface sediment samples were calculated using eq. 1, which was published by Zhang et al. [19], to provide information on the sources of the metal contaminants and temporal variations in the contributions of the different sources.

$$EF = [C_{n(sample)}/C_{Fe(sample)}]/[B_{n(baseline)}/B_{Fe(baseline)}] \quad (1)$$

where C_n is the metal concentration in a sediment, B_n is the background concentration of the metal, C_{Fe} is the Fe concentration in the sediment, and B_{Fe} is the background Fe concentration. The background metal concentrations in soils from each of the twelve study areas were used as the baseline concentrations [20].

Risk Assessment

The ecological risks posed by heavy metals in surface sediments were assessed using three methods, consensus-based SQGs [7], PER indices [15,21], and RACs [22]. Details of the assessment methods are provided in Table S2 in File S1.

Results and Discussion

Heavy Metal Concentrations in Surface Sediments

The heavy metal concentrations found in the surface sediments from the five types of aquatic ecosystems are shown in Fig. 2. The average Cd, Cr, Cu, Ni, Pb, and Zn concentrations in the twelve study areas were 0.716, 118, 37.3, 32.7, 56.6, and 204 mg/kg, respectively. The highest average Cd concentrations (>1.00 mg/kg) were found in surface sediments from the HR1, ZR, PL, and BW sites. The highest average Cr concentrations were found in surface sediments from the HR1, HR2, and ZR sites (307, 252, and 215 mg/kg, respectively). The highest average Cu, Ni, and Pb concentrations (96.1, 68.8, and 112 mg/kg, respectively) were found in surface sediments from the ZR site. The highest average Zn concentrations were found in surface sediments from the HR1, HR2, and ZR sites (440, 387, and 424 mg/kg, respectively). The highest total heavy metal concentrations (i.e., the sum of the Cd, Cr, Cu, Ni, Pb, and Zn concentrations) in surface sediments were mainly found in samples from river ecosystems (Fig. 1 and Fig. 2), especially samples from the HR1 and ZR sites, which contained total heavy metal concentrations of 911 and 918 mg/kg, respectively. The high heavy metal concentrations in the surface

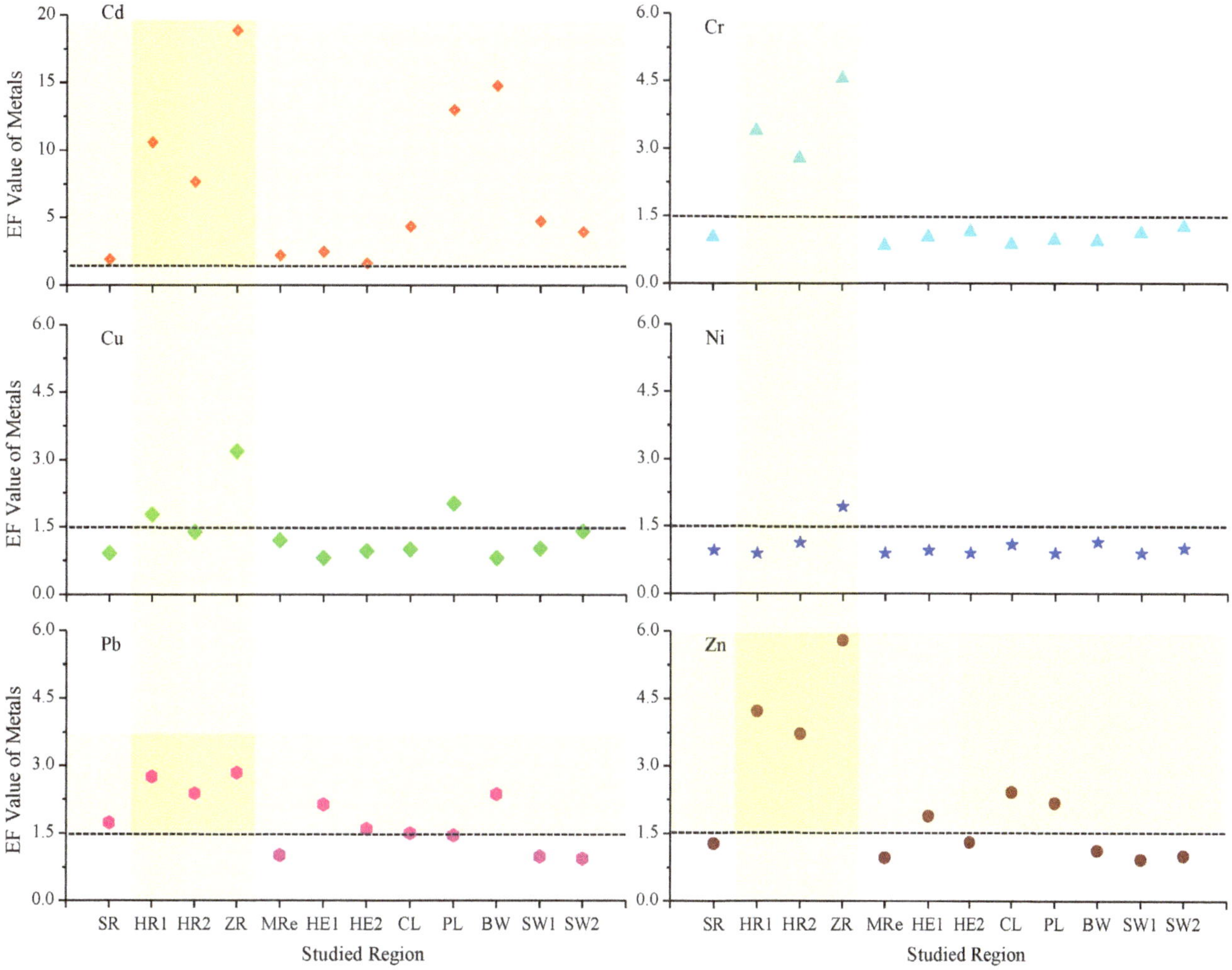

Figure 3. Heavy metal enrichment factors (EFs) in the sediment samples from the study sites.

sediments from the rivers implied that the rivers in the study area have suffered substantial inputs of heavy metals from anthropogenic sources [23].

The relationships between the concentrations of different heavy metals elements can provide information on the sources of the heavy metal [24]. Therefore, the Spearman correlation coefficients for the relationships between the heavy metal concentrations were determined (Table S3 in File S1). Positive correlations ($p \leq 0.01$) were found between the Cd and Pb concentrations and between the Pb and Zn concentrations. Significant positive correlations ($p \leq 0.05$) were also found between the Cd and Cu concentrations, the Cd and Zn concentrations, the Cr and Pb concentrations, and the Cr and Zn concentrations.

Heavy Metal Sources to the Surface Sediments

The EF is a normalization technique that is widely used to identify the proportion of a metal derived from natural environmental sources and the proportion of the metal derived from anthropogenic activities [16,25]. The EF for each metal was calculated, and is shown in Fig. 3, to determine the anthropogenic influences on the heavy metal concentrations in surface sediments in the five types of aquatic ecosystems that were studied in the study areas. The mean EF was highest for Cd (7.16), indicating that anthropogenic sources had had a stronger impact on Cd concentrations than the concentrations of the other heavy metals in the sediments in the study areas. The remaining EFs decreased in the order Zn (2.23), Pb (1.81), Cr (1.65), Cu (1.37), and Ni (1.05). The EFs were highest for the surface sediments from the river sites, and the average EFs were 3.93, 3.17, and 6.18 for the samples from the HR1, HR2, and ZR sites, respectively. The spatial distribution of the EFs for each metal was similar to the spatial distribution of the concentrations of that metal.

An EF of approximately 1 suggests that the metal may originate entirely from natural sources, such as crustal material and natural weathering processes. An EF slightly higher than 1 may not necessarily be caused by the addition of the metal from anthropogenic activities, but may be caused by natural differences in the metal concentrations in the sediment and the reference soil used for the EF calculation. Therefore, an EF of between 0.5 and 1.5 suggests that the metal may be entirely derived from crustal materials and natural weathering processes. However, an EF higher than 1.5 suggests that a significant proportion of the metal originates from anthropogenic processes [16,25,26].

The Cd, Pb, and Zn concentrations in the sediments from the aquatic ecosystems in the study areas positively correlated with each other (Table S3 in File S1) and these were found to be the most anthropogenically enriched of the heavy metals that were analyzed at most of the study sites (EF>1.5) (Fig. 3). The potential anthropogenic sources of these metals include industrial processes,

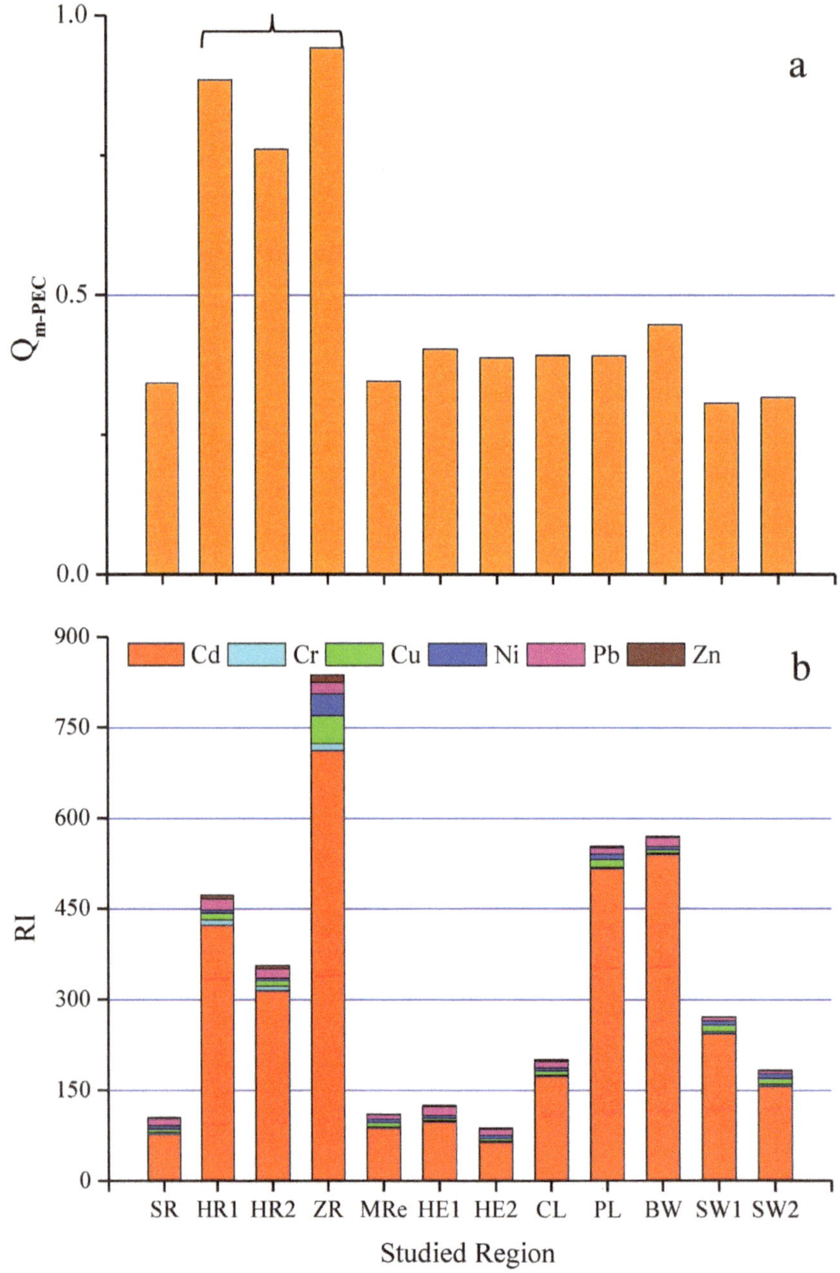

Figure 4. Heavy metal risk assessment indices for the sediment samples from each study site. (a) $Q_{m\text{-}PEC}$ and (b) comprehensive PER index (*RI*).

agriculture, and mining [16,27,28]. The mean EF for Cr in the sediments that were analyzed was slightly higher than 1.5, and this was mainly caused by high EFs at the HR1, HR2, and ZR sites (3.39, 278, and 4.55, respectively). This indicates that Cr was enriched in the sediments at the HR1, HR2, and ZR sites because of anthropogenic emissions. The mean EFs for Cu and Ni were less than 1.5, indicating that these two metals were mainly derived from natural sources, such as the underlying geological material. The EFs that were obtained may be useful indicators of the role of anthropogenic processes in the distributions of the metals in the study area. The mean EFs for Cd in the sediment samples from the five types of aquatic ecosystems were all greater than 1.5, and the Cd EFs were particularly high at the ZR (18.9), BW (14.8), and PL (13.0) sites (Fig. 3). These results indicate that Cd pollution has become a major environmental problem in aquatic ecosystems in eastern China, and this conclusion is consistent with the conclusions of other studies [4,6,16,27].

Risks Posed by Heavy Metals in Surface Sediments

The ecological risks posed to the five types of aquatic ecosystems that were studied by the heavy metals that were measured in surface sediments were assessed using consensus-based SQGs, PER indices, and RACs. The mean probable effect concentration quotients ($Q_{m\text{-}PEC}s$) for the sediments from sites HR1, HR2, and ZR were higher than 0.5 (0.885, 0.761, and 0.942, respectively), which indicates that these sediments were toxic and were seriously

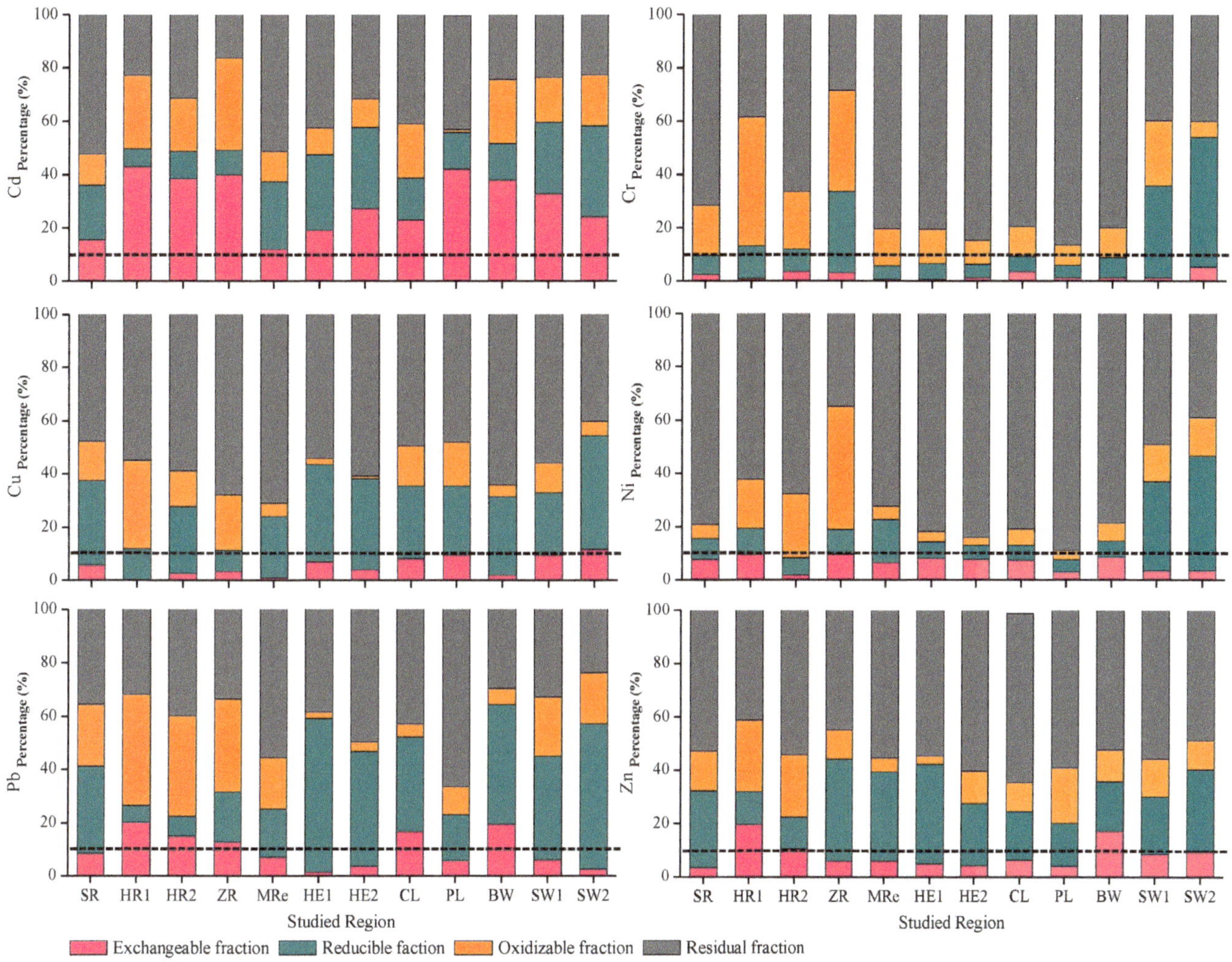

Figure 5. Proportions of the heavy metal fractions in sediments from the five aquatic ecosystem types.

contaminated with heavy metals (Fig. 4a). The PER index for an individual element (E^i_r) describes the sensitivity of a biological community to a given substance and can be used to illustrate the risk posed to that community by contaminants [4,28]. For the aquatic ecosystems in eastern China that we studied, the E^i_r values for the heavy metals decreased in the order Cd>Cu>Pb>Ni>Cr>Zn. All of the metals that we studied apart from Cd had low PER indices. The E^i_r values for Cd ranged from 64.2 (at HE2) to 711 (at ZR), and the average was 283, indicating that sediments in aquatic ecosystems in eastern China are highly contaminated with Cd, which is consistent with the conclusions we drew from the Cd EFs.

The comprehensive PER index (*RI*) distribution can be useful for identifying sites that need more attention than others because of risks associated with pollutants [16]. In this study, the mean RI values for the surface sediments were highest at the ZR, BW, and PL sites (836, 570, and 554, respectively), followed by the HR1 and HR2 sites (472 and 356, respectively), and the overall average RI was 322. According to the PER index criteria (Table S2 in File S1), the comprehensive PER for the surface sediments at the ZR site showed that this site suffers from very high-grade risks. The BW, PL, HR1, and HR2 sites all had high-grade risks, and the CL, SW1, and SW2 sites had moderate-grade risks. The sediments in the river ecosystems had higher PER indices than did the sediments from the other ecosystems, mainly resulting from Cd pollution, and this conclusion was similar to the conclusions we drew from the distributions of the heavy metal concentrations and EFs. Heavy metal pollution was found to be most severe in the river ecosystems, and this is probably because rivers, being open systems, are more vulnerable to contamination caused by anthropogenic processes. Pollutants are often discharged directly into rivers, and they can be quickly deposited in river sediments. The pollutants that are not deposited in the sediments will be transported to other ecosystems in the flowing water [6].

The Cd, Cr, Cu, Ni, Pb, and Zn speciation in the sediment samples was investigated (Fig. 5) to further evaluate the heavy metal contamination in the sediment samples. Cr, Cu, Ni, and Zn were predominantly found in the residual fraction in the studied sediments, and this fraction contained an average of more than 50% of the total metal content. In contrast, about 30% of the total Cd content and about 10% of the total Pb content were found in the exchangeable fraction of each sample. A high proportion of a metal in the exchangeable fraction indicates anthropogenic pollution [29,30]. According to the RAC classification (Table S2 in File S1), the risks posed by Cd in the surface sediments were high at the HR1, HR2, ZR, PL, BW, and SW1 sites and medium at the other sites. The risks posed by Pb in the surface sediments were found to be medium at the HR1, HR2, ZR, CL, and BW sites, and the risks posed by Zn were also found to be medium at the HR1, HR2, and BW sites. These results corroborate the EF and PER index results. Contaminants in the sediment may move into the foodchain, particularly if the contaminants are in bioavailable forms. Cd has a high transfer rate in the environment, so it can accumulate in relatively large amounts in plants without any apparent effects on the plants. Cd can, therefore, be found in most human foodstuffs, and this can cause health problems in humans [16,31,32]. Therefore, heavy metal contaminants, especially Cd, in aquatic ecosystems (particularly river ecosystems) in eastern China should be carefully monitored and controlled in the future [6,33].

Conclusions

We studied heavy metal pollutants in surface sediments from different aquatic ecosystems in eastern China. The heavy metal concentrations were usually higher in surface sediments in river ecosystems than in other ecosystems. The EFs showed that Cd, Zn, and Pb were enriched in the surface sediments more because of anthropogenic sources than were Cr, Cu, and Ni, which were predominantly derived from natural sources in most of the study areas. Using consensus-based SQGs, PER indices, and RACs, we found that all five types of aquatic ecosystems, especially river ecosystems, in the study areas have been polluted with heavy metals. The most serious pollutant in the study areas is Cd, which was mainly found in the exchangeable fraction of the sediment samples. This information could be useful for the development of effective management strategies for controlling heavy metal pollution in aquatic ecosystems in eastern China.

Supporting Information

File S1 Table S1. Microwave digestion conditions used for the sediment samples. **Table S2.** The evaluation criteria for different risk assessment methods. **Table S3.** Correlation coefficients for the heavy metals in the surface sediments from different types of aquatic ecosystems. **Figure S1.** Map showing the layouts used at the sampling stations in each type of aquatic ecosystem.

Author Contributions

Conceived and designed the experiments: WZT BQS WQZ. Performed the experiments: WZT LSW YKD. Analyzed the data: WZT BQS HZ. Contributed reagents/materials/analysis tools: WZT WQZ LSW. Wrote the paper: WZT BQS.

References

1. Iksandar IK, Keeney DR (1974) Concentration of heavy metals in sediment cores from selected Wisconsin lakes. Environ Sci Technol 8: 165–170.
2. Allengil SM, Gubala CP, Landers DH, Lasorsa BK, Crecelius EA, et al. (1997) Heavy metal accumulation in sediment and freshwater fish in US Arctic lakes. Environ Toxicol Chem 16: 733–741.
3. Kim IS, Kang KH, Johnson-Green P, Lee EJ (2003) Investigation of heavy metal accumulation in Polygonum thunbergii for phytoextraction. Environ Pollut 126: 235–243.
4. Ma Z, Chen K, Yuan Z, Bi J, Huang L (2013) Ecological risk assessment of heavy metals in surface sediments of six major Chinese freshwater lakes. J Environ Qual 42: 341–350.
5. Taweel A, Shuhaimi-Othman M, Ahmad AK (2013) Assessment of heavy metals in tilapia fish (*Oreochromis niloticus*) from the Langat River and Engineering Lake in Bangi, Malaysia, and evaluation of the health risk from tilapia consumption. Ecotox Environ Safe 93: 45–51.
6. Zhang LW, Shao HB (2013) Heavy metal pollution in sediments from aquatic ecosystems in China. Clean-Soil, Air, Water 41: 878–882.
7. MacDonald DD, Ingersoll CG, Berger T (2000) Development and evaluation of consensus-based sediment quality guidelines for freshwater ecosystems. Arch Environ Con Tox 39: 20–31.
8. Hakanson L (1980) An ecological risk index for aquatic pollution control: A sedimentological approach. Water Res 14: 975–1001.
9. Suresh G, Sutharsan P, Ramasamy V, Venkatachalapathy R (2012) Assessment of spatial distribution and potential ecological risk of the heavy metals in relation to granulometric contents of Veeranam lake sediments, India. Ecotox Environ Safe 84: 117–124.
10. Karak T, Bhattacharyya P, Paul RK, Das DK (2013) Metal accumulation, biochemical response and yield of Indian mustard grown in soil amended with rural roadside pond sediment. Ecotox Environ Safe 92: 161–173.
11. Buggy CJ, Tobin JM (2008) Seasonal and spatial distribution of metals in surface sediment of an urban estuary. Environ Pollut 155: 308–319.
12. Jiang M, Zeng G, Zhang C, Ma X, Chen M, et al. (2013) Assessment of Heavy Metal Contamination in the Surrounding Soils and Surface Sediments in Xiawangang River, Qingshuitang District. PLoS One 8: e71176.
13. Gao XL, Chen CTA (2012) Heavy metal pollution status in surface sediments of the coastal Bohai Bay. Water Res 46: 1901–1911.
14. He MC, Wang ZJ, Tang HX (1998) The chemical, toxicological and ecological studies in assessing the heavy metal pollution in Le An River, China. Water Res 32: 510–518.
15. Guo WH, Liu XB, Liu ZG, Li GF (2010) Pollution and potential ecological risk evaluation of heavy metals in the sediments around Dongjiang Harbor, Tianjin. Procedia Environ Sci 2: 729–736.
16. Tang WZ, Zhao Y, Wang C, Shan BQ, Cui JG (2013) Heavy metal contamination of overlying waters and bed sediments of Haihe Basin in China. Ecotox Environ Safe 98: 317–323.
17. Tessier A, Campbell PGC, Bisson M (1979) Sequential extraction procedure for the speciation of particulate trace metals. Anal Chem 51: 844–851.
18. Nemati K, Bakar NKA, Abas MR, Sobhanzadeh E (2011) Speciation of heavy metals by modified BCR sequential extraction procedure in different depths of sediments from Sungai Buloh, Selangor, Malaysia. J Hazard Mater 192: 402–410.
19. Zhang H, Shan BQ (2008) Historical records of heavy metal accumulation in sediments and the relationship with agricultural intensification in the Yangtze–Huaihe region, China. Sci Total Environ 399: 113–120.
20. China National Environmental Monitoring Center (CNEMC) (1990). The background concentrations of soil elements in China. Beijing: China Environmental Science Press.
21. Liu YJ, Shi GX, Mao L, Cheng G, Jiang SJ, et al. (2012) Direct and indirect influences of 8 yr of nitrogen and phosphorus fertilization on Glomeromycota in an alpine meadow ecosystem. New Phytol 194: 523–535.
22. Singh KP, Mohan D, Singh VK, Malik A (2005) Studies on distribution and fractionation of heavy metals in Gomti river sediments-a tributary of the Ganges, India. J Hydrol 312: 14–27.
23. Sondi I, Lojen S, Juračić M, Prohić E (2008) Mechanisms of land–sea interactions–the distribution of metals and sedimentary organic matter in sediments of a river-dominated Mediterranean karstic estuary. Estuar Coast Shelf S 80: 12–20.
24. Dragovic S, Mihailivic N, Gajic B (2008) Heavy metals in soils: distribution, relationship with soil characteristics and radionuclides and multivariate assessment of contamination sources. Chemosphere 72: 491–495.
25. Menge DNL, Hedin LO, Pacala SW (2012) Nitrogen and phosphorus limitation over long-term ecosystem development in terrestrial ecosystems. PLoS One 7: e42045.
26. Zhang J, Liu CL (2002) Riverine composition and estuarine geochemistry of particulate metals in China - weathering features, anthropogenic impact and chemical fluxes. Estuar Coast Shelf S 54: 1051–1070.
27. Tang WZ, Shan BQ, Zhang H, Mao ZP (2010) Heavy metal sources and associated risk in response to agricultural intensification in the estuarine sediments of Chaohu Lake Valley, East China. J Hazard Mater 176: 945–951.
28. Fu J, Zhao CP, Luo YP, Liu CS, Kyzas GZ, et al. (2014) Heavy metals in surface sediments of the Jialu River, China: Their relations to environmental factors. J Hazard Mater 270: 102–109.
29. Yang Z, Wang Y, Shen Z, Niu J, Tang Z (2009) Distribution and speciation of heavy metals in sediments from the mainstream, tributaries, and lakes of the Yangtze River catchment of Wuhan, China. J Hazard Mater 166: 1186–1194.
30. Weng H, Ma X, Fu F, Zhang J, Liu Z, et al. (2014) Transformation of heavy metal speciation during sludge drying: Mechanistic insights. J Hazard Mater 265: 96–103.
31. Lambert R, Grant C, Sauve S (2007) Cadmium and zinc in soil solution extracts following the application of phosphate fertilizers. Sci Total Environ 378: 293–305.
32. Satarug S, Garrett SH, Sens MA, Sens DA (2010) Cadmium, environmental exposure, and health outcomes. Environ Heal Persp 118: 182–190.
33. Cheng SP (2003) Heavy metal pollution in China: Origin, pattern and control. Environ Sci Pollut R 10:192–198.

13

Charlemagne's Summit Canal: An Early Medieval Hydro-Engineering Project for Passing the Central European Watershed

Christoph Zielhofer[1]*, Eva Leitholdt[1], Lukas Werther[2], Andreas Stele[3], Jens Bussmann[3], Sven Linzen[4], Michael Schneider[4], Cornelius Meyer[5], Stefanie Berg-Hobohm[6], Peter Ettel[2]

1 Institute of Geography, Leipzig University, Leipzig, Germany, **2** Chair of Prehistory and Early History, Friedrich-Schiller University, Jena, Germany, **3** Institute of Geography, Osnabruck University, Osnabruck, Germany, **4** Leibniz Institute of Photonic Technology (IPHT), Jena, Germany, **5** Eastern Atlas, Berlin, Germany, **6** Bavarian State Department of Cultural Heritage, Munich, Germany

Abstract

The Central European Watershed divides the Rhine-Main catchment and the Danube catchment. In the Early Medieval period, when ships were important means of transportation, Charlemagne decided to link both catchments by the construction of a canal connecting the Schwabian Rezat and the Altmühl rivers. The artificial waterway would provide a continuous inland navigation route from the North Sea to the Black Sea. The shortcut is known as Fossa Carolina and represents one of the most important Early Medieval engineering achievements in Europe. Despite the important geostrategic relevance of the construction it is not clarified whether the canal was actually used as a navigation waterway. We present new geophysical data and *in situ* findings from the trench fills that prove for the first time a total length of the constructed Carolingian canal of at least 2300 metres. We have evidence for a conceptual width of the artificial water course between 5 and 6 metres and a water depth of at least 60 to 80 cm. This allows a crossing way passage of Carolingian cargo scows with a payload of several tons. There is strong evidence for clayey to silty layers in the trench fills which reveal suspension load limited stillwater deposition and, therefore, the evidence of former Carolingian and post-Carolingian ponds. These findings are strongly supported by numerous sapropel layers within the trench fills. Our results presented in this study indicate an extraordinarily advanced construction level of the known course of the canal. Here, the excavated levels of Carolingian trench bottoms were generally sufficient for the efficient construction of stepped ponds and prove a final concept for a summit canal. We have evidence for the artificial Carolingian dislocation of the watershed and assume a sophisticated Early Medieval hydrological engineering concept for supplying the summit of the canal with adequate water.

Editor: John P. Hart, New York State Museum, United States of America

Funding: The authors are deeply grateful to the German Research Foundation (DFG) for funding the project (ZI 721/10-1 and ET 20/7-1). URL: http://www.dfg.de. The funders had no role in study design, data collection and analysis, decision to publish, or preparation of the manuscript.

* Email: zielhofer@uni-leipzig.de

Introduction

European Dimension of a Carolingian canal

Central Europe is covered by a dense network of navigable rivers. Along with overland routes they built the backbone of communication and commerce during the Middle Ages. Rivers formed the infrastructural link between traffic and economic systems of the Black Sea and Mediterranean basin on the one hand and of the North Sea and Baltic region on the other hand in this period [1]–[3]. Although generally disregarded in historical, archaeological and geoarchaeological research, ports and watersheds are the most important nodal points within this network [4].

During the Early to High Medieval period (cf. Table S1) the entire region between Denmark and Italy was controlled by powerful elites which were extraordinarily mobile, building up itinerant kingships and huge economic networks controlled by religious institutions [5]–[8]. Freund [9] highlights the important role of Central European river valleys for the communication networks of these groups. The basic work of Eckholdt [10] features methodological problems. Here, the role of the small rivers seems to be underrepresented [11]. However, until now there is poor knowledge about the location of inland ports, the explicit medieval navigability of the rivers and the bridging of watersheds between these rivers and their catchment areas [12]. So far there is mainly evidence for small and simple constructed medieval inland ports and hythes [13], [14]. But there is one extraordinary piece of engineering which highlights the enormous economic and human resources which have been mobilised for the building of navigable waterways: the Fossa Carolina which is the major focus of our geoarchaeological study presented here. The canal mirrors the first attempt to bridge the Central European watershed during the Early Medieval period [15]–[18]. The successful use of the canal would connect the Rhine-Main-catchment with the Danube catchment which means a continuous shipping lane between the

Atlantic and the Black Sea (Fig. 1a and b, Fig. 2). Hence, a successful bridging of the Central European watershed would be a performance of engineering of European dimension in the scope of Early Medieval infrastructure and economies [1], [19]. The demand on economic resources for the canal construction was enormous, but even bigger was the potential advantage of controlling this pivotal point in the inland navigation network of Europe, connecting such tremendous powers as the Carolingian and the Byzantine Empires [1]. However, political control was a prerequisite for the canal construction and the outstanding man who gathered this control in a long series of conflicts and wars during the late 8th century AD was Charlemagne [8], [20], [21].

Historical tradition

Decisive information about Charlemagne's canal derives from Early Medieval written sources. The annals of the Frankish empire record for the year 793 AD a major construction site at the eastern edge of the realm: *in autumn Charlemagne was coming per ship from Ratisbon to the big canal between the Altmühl and the Rezat rivers [...]* (Annales Regni Francorum, [22]). Numerous Carolingian sources describe the construction of the canal in contradictory details [23], [24]. In the Annals of Einhard the following description is found: *as the king [...] was convinced that it might be possible to build a navigable canal between the Altmühl River and the Rezat River arriving suitable from the Danube River into the Rhine River he betakes himself to the place, assembles a large amount of workers and was staying the entire autumn at the construction site [...] but without avail: due to ongoing rainstorms [...] it was not possible to achieve the progress. The material which was excavated by the workers during the day collapsed during the night* (Annales Regni Francorum, [22]). Legends about the construction of the Carolingian canal are mentioned in numerous more recent sources. This is the origin of the Ludwig-Danube-Main Canal which was dedicated in 1843 and the modern Rhine-Maine-Danube Canal which was finished in 1992 after 32 years of construction [25].

Different reasons hinder a clear interpretation of the written sources. Especially the real finishing of the construction and the effective use of the building as a waterway were contradictorily discussed [26]–[29]. But accomplished or not, Charlemagne's vision of the Fossa Carolina highlights the major importance of waterways for the transportation of people and goods in the supra-regional network of Early Medieval traffic and customs facilities [1], [5], [30], [31].

Building remains and previous knowledge about use and engineering concept of the canal

At the northern edge of Graben village impressive remnants of the Fossa Carolina are visible (Fig. 3). A first topographic documentation of the canal by C. L. Thomas reveals a length of 1230 m, an average breadth of the canal of 30 m and banks with maximum heights of 6.5 m and maximum breadths of 40 m [32]. G. Hock reconstructed the Carolingian excavation level by a first archaeological survey in 1910 [32]. Based on these results and the available written sources Schwarz [32] concluded that the building was never completed. First aerial photographs from the 1970s and 1980s indicate that remnants of the former banks can be tracked at least 1000 m further north in the zone of the Rezat fen [27]. Regarding the southern continuation of the canal, Koch et al. [27] assumed from oral tradition and poorly documented former casting pits that the present-day Graben main street follows the former course of the Carolingian canal in the direction to the Altmühl River.

Koch et al. [27] and Koch and Leininger [33] conducted a first drilling campaign in the fills of the central part of the canal. Fifteen cores feature a peat layer and unfortunately poorly documented

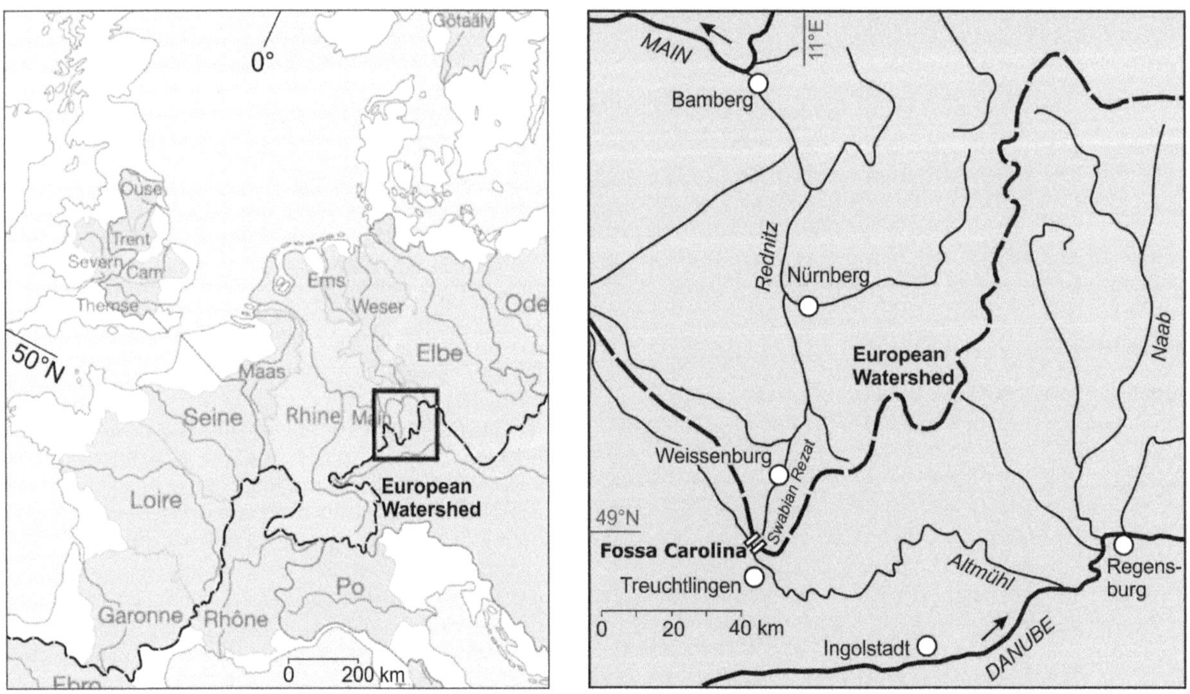

Figure 1. Fossa Carolina bridging the Central European Watershed. a) The outstanding role of the Fossa Carolina for Early Medieval shipping by bridging the Central European Watershed between the Rhine-Main and the Danube catchments. b) The Fossa Carolina was constructed by Charlemagne for linking the Swabian Rezat and the Altmühl Rivers. The figure is not similar to formerly published figures.

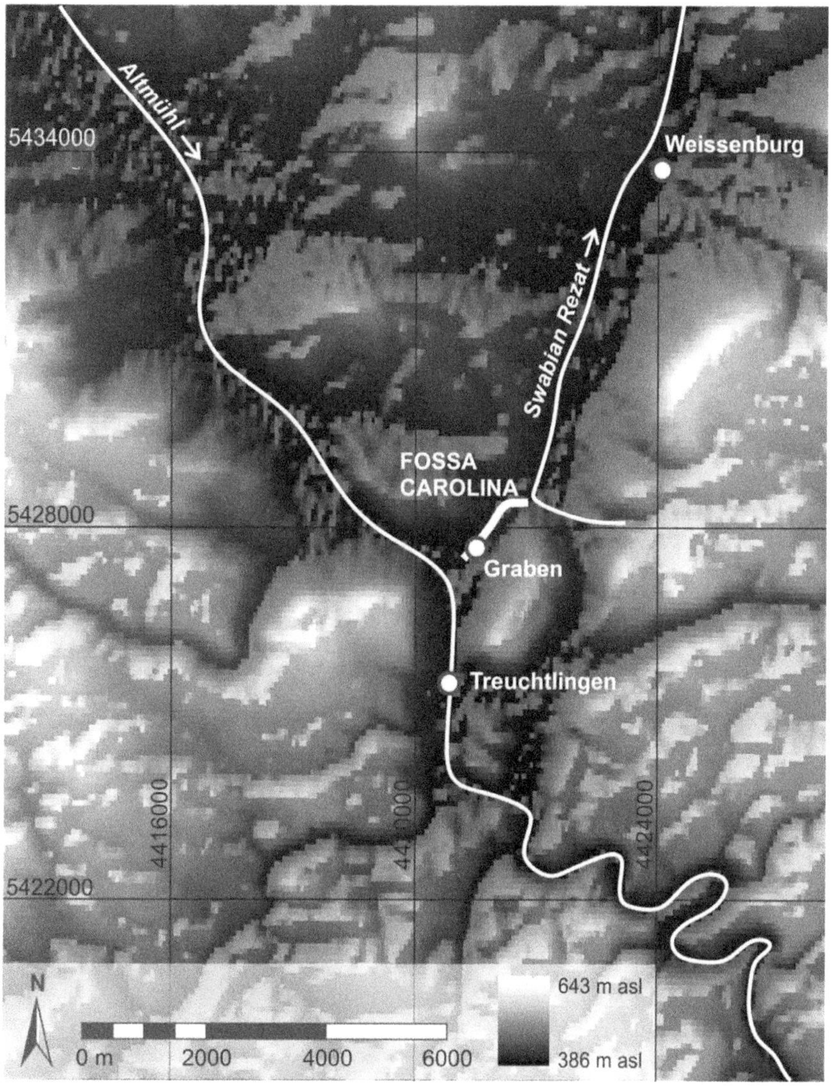

Figure 2. The local topographical setting of the Fossa Carolina. The canal is located across a valley watershed between the Altmühl River (Danube catchment) and the Swabian Rezat River (Rhine-Main catchment) within the Treuchtlingen valley junction at the piedmont of the Franconian Jura escarpment. Data source: Open access SRTM data (3 arc-second, 90 m raster, [72]) and OpenStreetMap [73].

first palynological evidence for former ponds [34]. Regarding the levels of the drilled peat layers, Küster [34] assumed that the level of excavation of the trench does not reach the necessary depth for a continuous canal. Hence, the building was never finished or was already initially planned as a summit canal. This corresponds with the descriptions of the Carolingian Annales Guelferbytani. Here, it is mentioned for 793 AD that the boats were towed through the water and overland (Annales et chronica aevi Carolini, [24]). However, there is no clear geoarchaeological evidence for a planned summit canal yet. Survey works by Koch et al. [27] and Koch and Leininger [33] indicate that the West-East Section (Fig. 3) of the canal does not follow the natural Rezat drainage system to the north. Rather, the canal might be connected in the zone of a noticeable rectangular bend with the uppermost course of the slightly diverted Rezat stream to subjoin the canal with additional water. Here, Koch [15] assumed that the embankment of the road between Grönhart and Dettenheim (Section V, Fig. 3) belongs to remnants of a Carolingian reservoir which might be constructed for supplying the Fossa Carolina with water from the Rezat spring. However, new ^{14}C data from *in situ* dating of the embankments (Fig. 3) reveal High Medieval ages (Tables 1 and 2, [35]) and give no evidence for a Carolingian dam. The lowermost strata of the embankment reveal an age of 1194–1246 cal AD, followed by a younger heightening during the 15th century AD (Table 2, [35]).

During a first survey our research group drilled seven cores in the Central Section of the canal. Radiocarbon dating feature for the first time a chronological model for the trench fills [36]. Our findings show a clear stratigraphic order of the fills. The oldest fills include wood remnants of Carolingian age. Therefore, the former hypotheses of an initial Roman age of the canal [37] must be discarded.

Aims of our study

In former times archaeological and historical research about the Fossa Carolina was only focused on a potential Carolingian use of the construction [15]. First ^{14}C data from the canal fills [17], [36] but also from the potential dam of the reservoir [35] do not

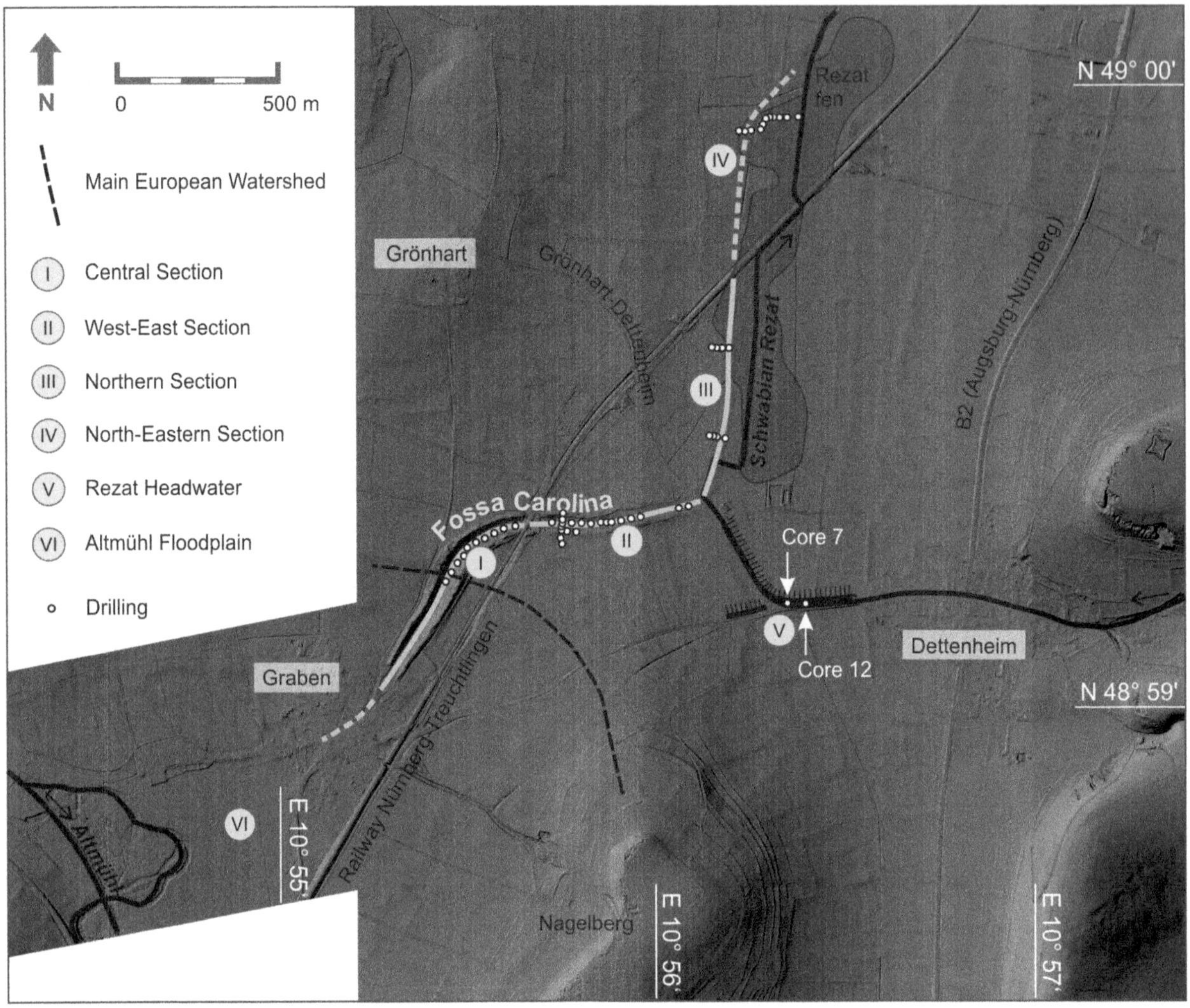

Figure 3. The course of the Fossa Carolina and the subdivision in six sections. I) Central Section, II) West-East Section, III) Northern Section, IV) North-Eastern Section, V) Rezat Headwater with cores 7 and 12 [35] within the former assumed dam for a Rezat reservoir [15], and VI) Altmühl Floodplain. For detailed drilling positions of the sections I, II, III, and IV see Figs 4, 8 and 12. Raw data source: LIDAR data are cordially provided by Bavarian Land Surveying Office for illustrative purposes.

support this concept. Hence, we have to start thinking about an alternative water engineering concept. In this study we focus on the probable course of the canal from Graben village to the Rezat fen. The course includes the Central Section, the West-East Section, the Northern Section and the North-Eastern Section (Fig. 3).

a) For the first time we want to obtain detailed chronostratigraphical data from the trench fills (Fig. 3), scrutinising the possible evidence for open water bodies and, therefore, the potential use of the canal for shipping.

b) We want to improve the stratigraphical model for a better reconstruction of the trench refilling process. Up to now, it is not clear, if the Carolingian canal collapsed due to the abrupt re-deposition of the bank material into the excavated trenches.

c) We generally want to improve the knowledge about the Carolingian progress of construction works, and we aim to clarify the advised minimum water depth of the completed canal getting an idea about minimum requirements for Early Medieval shipping routes. Here, multiple reconstructions of cross sections of the canal course are required.

d) The Fossa Carolina is an outstanding example of Early Medieval hydraulic engineering know-how. This study aims to reconstruct the longitudinal profile of the Fossa Carolina including the levels of excavation and the potential evidence for the artificial dislocation of the watershed. Here, our focus is to clarify the probable concept of a summit canal.

e) Regarding the European dimension of the project, the detection of the confluence of the Fossa Carolina with the Altmühl River (Section VI in Fig. 3) will be a challenging part of future geoarchaeological research. In this study, we want to obtain for the first time *in situ* evidence about the potential extension of the canal course in the northern direction. Assuming that there was a navigable connection between the canal and the Rezat River, we must find remnants of a canal in the North-Eastern Section (Fig. 3).

Table 1. Radiocarbon ages of the Fossa Carolina trench fill (Part 1).

Mean depth [cm below surface/m a.s.l.]	Core	Location	Lab No.	Material	^{14}C [BP]	Calibrated 1σ range [cal. AD]	$\delta^{13}C$ (‰)	Reference (if already published)
375/411.8	A	Trench fill (peat layer)	KIA36403	Wood	973±28	1024–1116	−26.1	Leitholdt et al. 2012
460/410.9	A	Trench fill (sandy layer)	KIA36404	Charcoal *Quersus sp.*	1267±27	692–762	−23.7	Leitholdt et al. 2012
610/409.4	A	Trench fill (sandy layer)	KIA36406	Charcoal *Quercus sp.*	1269±27	691–759	−24.5	Leitholdt et al. 2012
245/414.2	G	Trench fill (peat layer)	SU-GL-21	Wood	805±30	1207–1251	−27.5	Leitholdt et al. 2014
368/413.0	G	Trench fill (peat layer)	SU-GL-22	Wood	1180±30	793–879	−27.6	Leitholdt et al. 2014
238/414.2	H	Trench fill (peat layer)	SU-GL-23	Wood	765±30	1228–1266	−30.6	Leitholdt et al. 2014
386/412.8	H	Trench fill (peat layer)	SU-GL-24	Wood	1150±30	827–939	−27.1	Leitholdt et al. 2014
265/414.1	I	Trench fill (peat layer)	SU-GL-25	Wood	870±30	1078–1200	−28.2	Leitholdt et al. 2014
185/414.8	J	Trench fill	SUERC-42068	Wood *Quercus sp.*	834±25	1178–1232	−26.6	
275/413.9	J	Trench fill (peat layer)	SUERC-42069	Wood *Quercus sp.*	934±23	1048–1136	−25.1	
165/415.2	K	Trench fill (peat layer)	SUERC-42070	Caulis	874±25	1079–1195	−27.1	
225/414.6	K	Trench fill (peat layer)	SUERC-42074	Caulis	942±25	1043–1133	−27.4	
335/413.5	K	Trench fill	SUERC-42075	Wood	1253±26	701–773	−29.4	
355/413.3	K	Trench fill	SUERC-42076	Wood	1163±26	813–923	−27.9	
410/412.9	K	Trench fill	SUERC-42140	Wood *Alnus sp.*	1170±37	797–921	−28.7	
282/414.5	L	Trench fill (peat layer)	SUERC-42141	Wood *Salix sp.*	806±37	1193–1251	−27.4	
375/413.6	L	Trench fill (peat layer)	SUERC-42142	Wood	860±37	1087–1211	−28.2	
465/412.6	L	Trench fill	SUERC-42143	Wood	1110±37	893–969	−27.0	
230/415.3	M	Trench fill (peat layer)	SUERC-42147	Wood *Salix sp.*	782±37	1216–1260	−26.1	
474/412.9	M	Trench fill (trench bottom)	SUERC-42148	Wood *Salix sp.*	1223±37	723–851	−27.0	

Calibration (one sigma) of the conventional ages was performed using the Cologne Radiocarbon Calibration Program (quickcal2007 ver.1.5, [48]). KIA samples were conducted at Kiel AMS facility. SU-GL and SUERC samples were conducted at Glasgow AMS facility. MAMS samples were conducted at Mannheim AMS facility.

Table 2. Radiocarbon ages of the Fossa Carolina trench fill (Part 2).

160/415.4	B	Trench fill (peat layer)	KIA36407	Caulis	307±22	1524–1624	−27.3	Leitholdt et al. 2012
300/414.0	B	Trench fill (peat layer)	KIA36408	Wood	882±29	1067–1189	−29.5	Leitholdt et al. 2012
390/413.1	B	Trench fill	KIA36409	Wood	859±29	1149–1207	−30.9	Leitholdt et al. 2012
256/414.9	O	Trench fill (sapropel layer)	SUERC-44081	Wood *Salix sp.*	713±29	1266–1286	−28.4	
550/412.0	Q	Trench fill	SUERC-44082	Wood *Quercus sp.*	1271±29	689–761	−26.4	
87/416.7	S	Trench fill (sandy layer)	MAMS 18371	Wood *Juniperus sp.*	316±18	1523–1619	−20.3	
541/412.1	S	Trench fill (trench bottom)	MAMS 17461	Wood *Quercus sp.*	1312±17	666–708	−22.5	
482/413.0	T	Trench fill (peat layer)	SUERC-44083	Wood *Salix sp.*	1156±29	819–933	−29.4	
285/414.9	W	Trench fill (peat layer)	SUERC-44085	Seed *Sparganium sp.*	1021±29	990–1020	−24.1	
341/414.3	W	Trench fill (peat layer)	SUERC-44084	Wood *Salix sp.*	1186±26	792–870	−28.0	
369/414.5	Z2	Trench fill (sapropel layer)	SUERC-44089	Wood *Salix sp.*	1101±26	903–971	−28.9	
485/413.7	Z3	Trench fill	SUERC-44090	Wood *Salix sp.*	1234±29	716–836	−28.1	
75/421.2	7	Embankment	KIA38898	Charcoal	330±35	1498–1614	−21.0	Berg-Hobohm and Kopecky-Hermanns 2012
85/421.1	7	Embankment	KIA38899	Charcoal	445±30	1430–1454	−27.5	Berg-Hobohm and Kopecky-Hermanns 2012
138/420.6	7	Embankment	KIA38900	Charcoal	815±30	1194–1246	−23.2	Berg-Hobohm and Kopecky-Hermanns 2012
189/420.0	7	Buried A horizon	KIA38901A	Charcoal	1485±40	543–611	−24.7	Berg-Hobohm and Kopecky-Hermanns 2012
189/420.0	7	Buried A horizon	KIA38901B	Humic acid	1990±40	−38–46	−23.4	Berg-Hobohm and Kopecky-Hermanns 2012
74/422.3	12	Embankment	KIA38902A	Charcoal	340±30	1493–1609	−22.6	Berg-Hobohm and Kopecky-Hermanns 2012
74/422.3	12	Embankment	KIA38902B	Humic acid	320±35	1506–1616	−21.7	Berg-Hobohm and Kopecky-Hermanns 2012
367/414.3	QP-1	Trench fill (trench bottom)	MAMS 18372	Charcoal	1338±17	656–670	−30.7	
137/415.6	407-1	Trench fill (peat layer)	MAMS 18374	Charcoal	410±16	1444–1464	−28.5	
155/413.7	256-3	Trench fill (trench bottom)	MAMS 18375	Wood	1158±18	829–927	−23.6	

Geographical setting

The Fossa Carolina is located in the Southern Franconian Jura. It is surrounded by Upper Jurassic (Malm) carbonate rocks and Middle Jurassic Aalenian (Dogger Beta) sandstones [38] (Fig. S1, Fig. 2). The Swabian Rezat River (Rhine-Main catchment) and the Upper Altmühl River (Danube catchment) flow through undulating Middle Jurassic Aalenian (Dogger Alpha) and Lower Jurassic (Lias) foothills. The Fossa Carolina is not directly situated on the clayey sediments of the Middle Jurassic Aalenian (Dogger Alpha), but rather on fine clayey remains of Miocene lake deposits and Quaternary fills. The Miocene lake resulted from a meteorite impact that sealed the Miocene drainage system [38]–[40]. In the Central and West-East Sections limnic Miocene clays are covered by Quaternary fills consisting of fluvial sands down to a depth of 5 m [38], [39]. The Northern and North-Eastern Sections show different geological preconditions to the Central and West-East Sections. These are characterized by a natural, relatively wide valley floor with only slightly inclined valley edges. The groundwater table is naturally very high resulting in the development of the Rezat fen (Fig. 3). The Rezat fen comprises a part of the upper course of the Rezat stream, which originates in a V-shaped valley in the East of Dettenheim. The V-shaped valley is incised in the water-bearing Middle Jurassic Aalenian sandstone formation of the Franconian escarpment [38].

The location of the Fossa Carolina is closely associated with the position of the Altmühl and Swabian Rezat Rivers (Figs. 2 and 3), draining both parts of the European Watershed: the Altmühl River is a tributary of the Danube River that finally flows into the Black Sea. The modern mean water level of the Altmühl at its supposed confluence with the Fossa Carolina is 408.3 m a.s.l. [33]. Medieval navigability of the Altmühl close to the canal is documented in different Carolingian documents [33]. The Swabian Rezat River flows northward from Dettenheim to the east of the Fossa Carolina (Fig. 3) into the Main, a tributary of the Rhine that flows into the North Sea. During the Medieval period, the Swabian Rezat was at least navigable from Weissenburg downstream [33]. The Swabian Rezat was probably rerouted during the Carolingian period, turning it into an artificial channel. The present pond located in the Fossa Carolina is fed by a stream that originates from the western slopes of the Fossa Carolina, and is additionally filled up with seepage water. The outlet of the pond flows through Graben into the Altmühl.

Methods

LIDAR digital elevation model

High-resolution airborne laser scanning data (LIDAR) were supplied by the Bavarian land surveying office in Munich. These data were used to generate a digital elevation model (Fig. 3). We used the program ArcGIS 10 for data management and mapping.

Flux gate magnetic survey

In the West-East Section of the Fossa Carolina a Bartington Grad601 fluxgate magnetometer was applied for a magnetic survey [41], [42]. Due to the difficult topography in the narrow zone between the two banks we used the handheld fluxgate magnetometer for the precise detection of the canal course. Grid measurement was carried out using a Topcon HiPer II DGPS device. Wooden canes mark the corners of each grid. PE-cords with metre spacing indicate northern and southern grid baselines. Additional PE-cords equipped with metre distance marks spanned between these two baselines were used as indicator for walking speed and position. For a higher resolution line spacing of 0.5 m and 4 measuring intervals per metre were selected resulting in a pixel size of 0.25×0.5 m. Since archaeological features with usually low magnetic anomalies are subject of this survey, the logged data range was limited to 100 nT, resulting in a sensitivity of 0.03 nT [42]. Georeferencing and patching of magnetic survey grids were handled in ArcGIS 10. For detailed analysis, processing and visualization of the data Geoplot 3.00 v software [43] was applied. The findings of the magnetic survey were used to determine locations for drilling.

SQUID magnetic survey

To allow precise magnetic prospecting of large areas in the order of tens of hectares within reasonable time a motorized measurement system was developed [44]. The system based on SQUID (Superconducting Quantum Interference Device) sensors, which provide very high magnetic field resolution also during rapid traverses over the ground. The recorded magnetic maps have a centimetric resolution thanks to the SQUID sampling rates of 1000 measurement points per second, the used differential GPS technique (Trimble R8 GNSS) and data acquisition unit including very low noise and fast 24-bit A/D converters developed in-house. The detailed data post processing results in magnetic maps as well as a morphology map of high resolution. No data filtering has to be applied. The maps are *a priori* geo-referenced, which allows a precise and easy integration with aerial images and other GIS data.

The current setup of the system allows the synchronous recording of up to 12 SQUID gradiometer signals. Thus, several components of the gradient of the Earth's magnetic field up to the full tensor can be detected. This maximum of magnetic information is the basis of our approaches to calculate the possible subsoil distributions of the magnetic sources [45]. In this way, depth informations and three-dimensional shapes corresponding to the detected anomalies can be estimated in addition to the standard two-dimensional magnetic maps.

Drilling campaigns

The flux gate and SQUID magnetic data were used to locate the canal course and to determine drilling positions. Our research group carried out drilling campaigns within the Fossa Carolina trench fills in 2012 and 2013 (Fig. 3), using an Atlas Copco Cobra Pro hammer and a 60 mm diameter open corer. Using this technique, we obtained 60 cores with lengths between 2 and 9 m. Drillings were levelled using a Topcon HiPer II DGPS device. The District Office of Weißenburg issued the permit for the drilling campaign. The study was partially carried out on private land. All owners were contacted personally by phone call, email or directly in the field to attain entry rights. All owners permitted access to their private lands and to take samples by the authors. The permission were recorded in an internal excel file after oral or written confirmation. Please contact the cities of Treuchtlingen and Weissenburg for future permissions and for the list of registered proprietors. The field studies (see Fig. 3 for GPS coordinates) did not involve endangered or protected species.

Sedimentological analyses

We determined grain size distributions of each sediment layer to obtain information about the deposition process. This method is applicable to samples with a content of organic matter less than 5%. Bulk samples (10 g) were left in 50 ml 35-% hydrogen peroxide (H_2O_2) overnight, and heated to remove organic matter during the next day. Afterwards, the samples were dispersed using 10 ml 0.4 N sodium pyrophosphate solution ($Na_4P_2O_7$) and ultrasonic treatment for 45 minutes. Grain size analysis of the sand fraction was carried out by means of the dry-sieving technique (2000–630 µm: coarse sand, 630–200 µm: medium sand, 200–125 µm: fine sand, 125–63 µm: finest sand). Coarse silt (63–20 µm), medium silt (20–6.3 µm), fine silt (6.3–2.0 µm), coarse clay (2.0–0.6 µm), medium clay (0.6–0.2 µm) and fine clay (< 0.2 µm) were measured by X-ray granulometry (XRG) using a SediGraph III 5120 (Micromeritics) [46].

For calculating the content of organic matter, we measured the content of total carbon by using a CNS analyser vario EL cube (Elementar), and determined the content of inorganic carbon by calcimeter measurements (Scheibler method, Eijkelkamp). Resulting values of organic carbon were multiplied by 1.72 in order to obtain contents of organic matter [47].

Numerical dating

Chronological information was obtained using ^{14}C dating: Plant macrofossils (charcoal and wood) from clastic and peat layers were processed by accelerator mass spectrometry (AMS) (Tables 1 and 2). Radiocarbon ages were calibrated using quickcal2007 ver.1.5 [48].

Classification of the organic-rich layers

Following Eckelmann [47], organic-rich layers were classified into peat and sapropel layers. The classification of organic-rich layers offers significant information about local hydrological conditions during phases of growth/sedimentation (e.g. fen facies, limnic facies). The degree of peat decomposition was determined following Von Post [47]. Degrees of decomposition depend on the preservation of plant residues, peat colour, colour of the water that is squeezed out and the condition of the remains after squeezing.

Electrical resistivity tomography

Measurements of 2D and 3D electrical resistivity tomography were carried out in the West-East Section to detect boundaries of artificial ponds and potential Carolingian shipping chutes. Geophysical resistivity methods harness contrasts of the electrical conductivity σ or the electrical resistivity ρ ($\rho = 1/\sigma$) in sediments in order to detect different underground structures. Hence, buried archaeological features might be mapped under the precondition of existing resistivity contrasts in comparison to the surrounding sediment matrix. Generally, gravels and stones show increased resistivity whereas a clayey to silty texture is characterized by lower resistivity. To define the electrical characteristics of an underground volume metal probes (electrodes) are placed in the ground. Two probes are used in order to feed into the ground an electrical current. The resulting potential difference between two other probes is measured. By applying Ohm's law under consideration of the geometry of the electrode configuration the apparent electrical resistivity can be calculated. These values have to be transformed in a model of the spatial distribution of the resistivity by use of numeric inversion algorithms.

The electrical resistivity measurements were carried out using the multi-electrode system GeoTom MK1E100. Using this system up to 100 electrodes can be controlled. For data acquisition a pole-dipole configuration ("Half Wenner") was used [49]. The spacing of the elctrodes was 1 m. The longitudinal profile was processed by means of the numeric inversion software DC2DInvRes [50].

Results

Types of sediment texture

Regarding the grain size data we were able to separate generally three types of sediment texture with characteristic grain size distributions. a) The ubiquitous 'fluvial type' sediment texture (Fig. 4a) reveals a bimodal distribution with peaks in medium to fine sand fraction as well as in the fine clay fraction. The majority of sediment samples from the Central Section and the West-East Section show this quite similar type of grain size distribution. We suggest that the trenches of the Central Section and of the West-East Section were excavated within this one fluvial grain size type only. The maximum in the medium to fine sand fraction can be explained by fluvial sediments which were transported by saltation and surface creeping [51]. Hence, the Carolingian and post-Carolingian 'fluvial type' trench fills represent re-deposited material from the banks which are build up with the originally Quaternary fluvial fills of the valley divide. b) The 'ponding type' sediment texture shows a bimodal distribution with peaks in medium silt and fine clay (Fig. 4b). For this second group we assume a deposition process typical for open water bodies, since the grain size distribution is characteristic for suspended load [51]. The ubiquitous sandy peak from the 'fluvial type' group is missing here. c) A third type of sediment texture, the 'Miocene clay type'

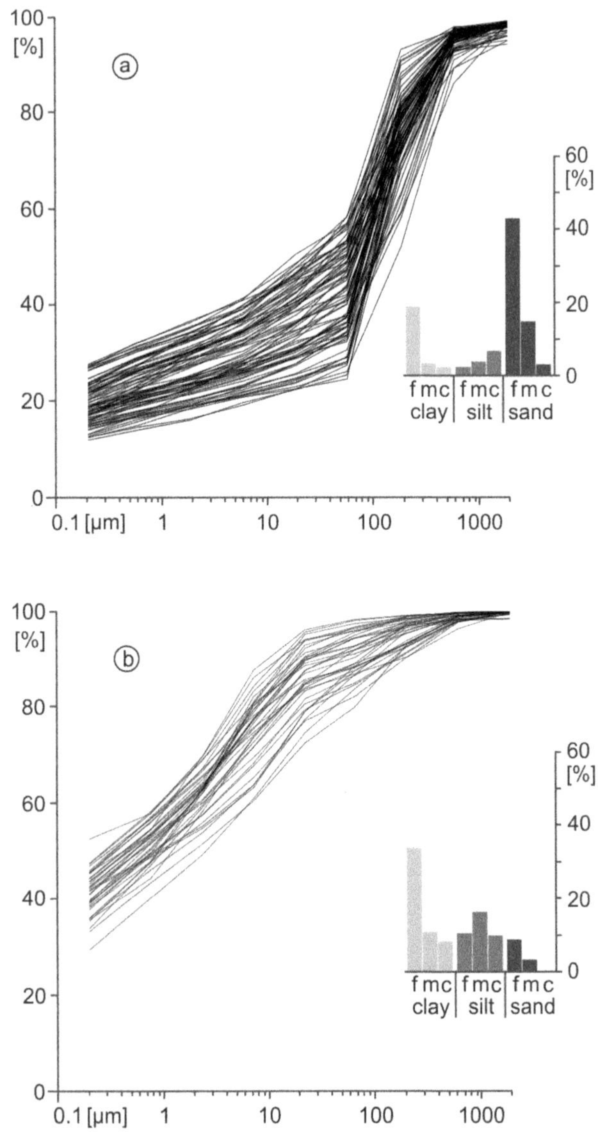

Figure 4. Types of sediment texture in the Central Section and the West-East Section. a) fluvial type and b) ponding type.

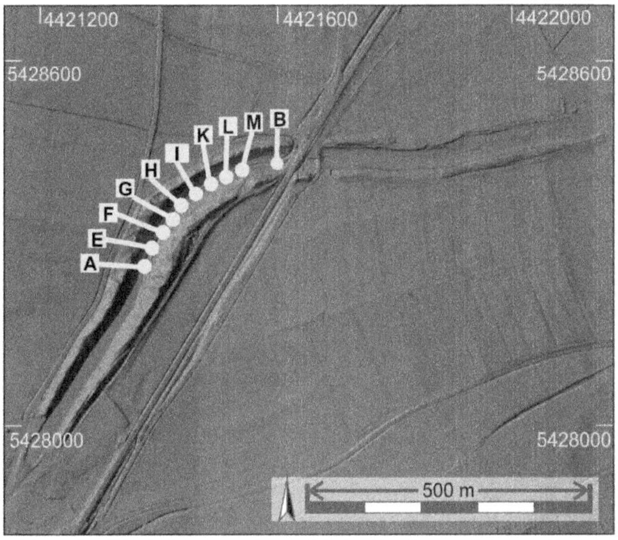

Figure 5. Digital elevation model of the Central Section including detailed drilling positions. Raw data source: LIDAR data are cordially provided by Bavarian Land Surveying Office for illustrative purposes.

indicates a strong maximum in the fine clay fraction. These thick and homogenous sediments were deposited during a long-term limnic phase. We detected this type of sediment texture only in drilling E (Fig. 5).

Central Section

On the basis of all core findings from the Central Section (Fig. 5), we are able to present the refill process of the Central Section in a synthetic chronostratigraphical form (Fig. 6). We can assume five principal phases of geomorphological or depositional relevance. ***I)*** The pre-Carolingian parent material in the Central Section consists of Miocene clays (drilling E only) and Quaternary fluvial fills of the valley divide. During the Quaternary the upper part of the valley floor was filled-up with well-stratified fluvial sediments poor in organic matter (File S1 and Fig. S2). These deposits feature generally a 'fluvial type' sediment texture (Fig. 4a). ***II)*** The 2nd phase represents the excavation of the Carolingian trench. Our radiocarbon ages clearly indicate that the excavation started during the Early Medieval period. A *Quercus sp.* charcoal sample taken from a brownish layer at the base of core A shows an age of 691–759 cal AD (Table 1). The oxidized brownish layer in core A might feature the walking horizon during the time of Carolingian excavation. In core G and H we assume that the Carolingian excavation depth is below the level of two radiocarbon samples which yield ages of 793–879 and 827–939 cal AD (Fig. 6 and Table 1). Comparing all excavation depths in the Central Section, a stepwise altitudinal increase from core A up to core M is detectable (Fig. 6). In core A we have evidence for an abrupt refill of the excavated trench bottom. A *Quercus sp.* charcoal sample taken 1.5 m above the brownish layer indicates an age of 692–762 cal AD which is comparable with the result of a ^{14}C dating from the brownish layer itself (691–759 cal AD, Table 1). This is quite similar to the observations in core K. Here, a probably re-deposited wood fragment features an age of 701–773 cal AD and covers ^{14}C dated wood remains which reveal younger ages of 813–923 and 797–921 cal AD (Fig. S2 and Table 1). The Early Medieval age inversion at the lowermost sequence of core K (Fig. S2) suggests an abrupt refill process shortly after the excavation of the trench. ***III)*** The 3rd phase indicates evidence for Carolingian ponds in the Central Section. In core H the sapropel layer between 412.28 and 412.35 m is located beneath a peat layer which reveals an age of 827–939 cal AD (Table 1). However, there is no evidence for Carolingian ponds in the majority of the cores of the Central Section. Indeed, at the trench bottoms of cores G and M we found Carolingian peat layers (793–879 cal AD and post 723–851 cal AD, Table 1) which show at least semi-terrestrial conditions but not clear evidence for Carolingian ponds. Numerous *Salix sp.* and *Alnus sp.* macro-remains indicate a local semi-terrestrial environment as well (Fig. S2). ***IV)*** The 4th phase represents the main ponding and peat-growing phase within the Central Section. The deposits do not feature Carolingian but younger High Medieval ages between the 11th and 13th centuries AD (Table 1). Compared to the lower Carolingian deposits, the High Medieval sediments provide stronger evidence for ponds as indicated by sapropels and ponding-type sediment textures in almost all cores (E, F, G, H, I, K, L, M) (Fig. 6). ***V)*** The overlying terrestrial deposits above the High Medieval organic-rich layers characterize the final 5thphase in the Central Section which is probably still continuing today. In the uppermost sequence of the cores A, E, F, G and H (Fig. 6) the grain size distributions are similar to those found below the organic-rich layers which are dominated by the 'fluvial type' grain size group. Here, the sediments of the 5thphase originate from the Quaternary valley fills, feature re-deposited material from the adjacent Carolingian banks which were built up by Quaternary valley deposits. The uppermost clayey to silty sequence of cores I, K, L and M mirrors a suspension-load and partially organic-rich deposition process starting in the 13th century AD (Fig. 6). We assume a High Medieval to modern ponding phases at this point.

West-East Section

The magnetogram of the West-East Section reveals a slight positive, elongate anomaly which is parallel to the direction of the banks but 2.5 m north of a current drainage trench. The anomaly is particularly noticeable in the western part of the West-East Section (Fig. 7) and became gradually more diffuse in the eastern part. In the western part the total width of the anomaly is between 6 to 10 m. Due to the elongate and narrow form of the anomaly we assumed here the course of the Carolingian canal. As the course of the canal should be constrained by *in situ* findings and ^{14}C dating, cores were drilled exactly in the centre of the anomaly with consistent intervals of 30 m. Of specific interest were an interruption and a slightly displacement of the anomaly between the cores S and U. At this point we assumed a subdivision of the canal, probably due to a favoured cross bank or shipping chute. Putting all core findings from the West-East Section together, we document the deposition processes of the West-East Section in a synthetic chronostratigraphical context (Fig. 8). Comparable to the Central Section we postulate five principal phases of geomorphological or depositional relevance for the West-East Section: ***I)*** the drilled pre-Carolingian parent material in the West-East Section consists of Quaternary fluvial fills of the valley divide. The stratified fluvial sediments are poor in organic matter and reveal a bimodal grain size distribution with a typical maximum in medium to fine sand and a secondary peak in fine clay. ***II)*** The 2nd phase reflects the excavation of the Carolingian trench. The trench bottom in the West-East Section is generally characterized by the distinct onset of plant macro-remains (Fig. S3) indicating earliest trench refill processes. Three wood remnants in cores T, W and Z3 were found slightly above the detected trench bottoms, and all ^{14}C ages are in the range of the Carolingian period (819–933, 792–870 and 716–836 cal AD, Table 2). Two wood remnants

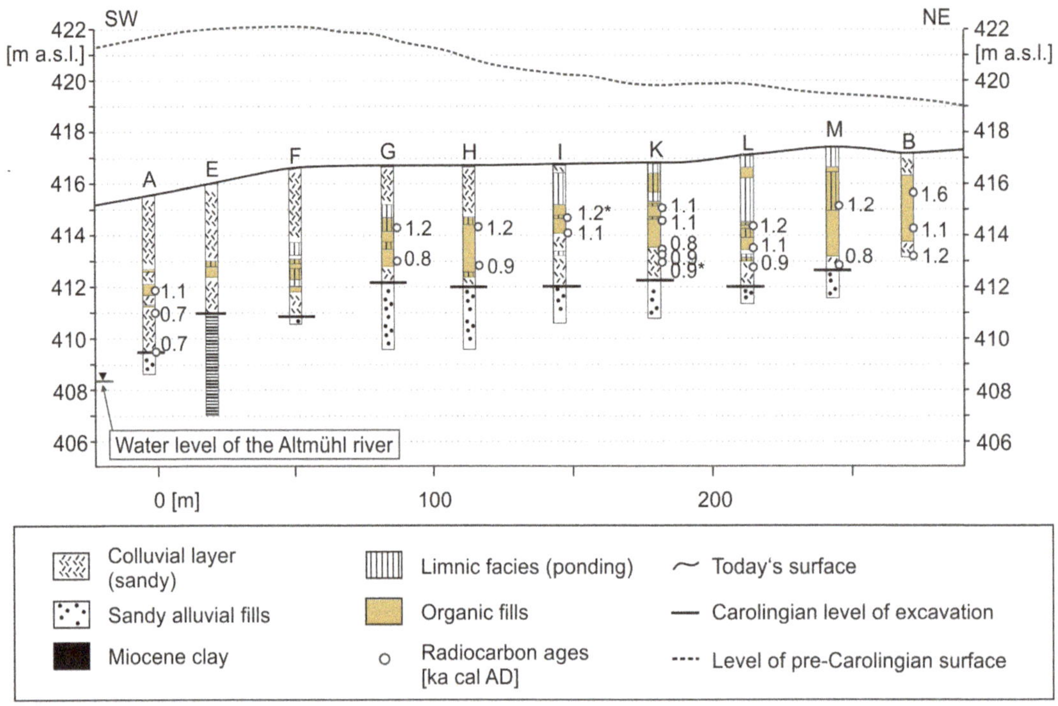

Figure 6. Chronostratigraphical documentation of the Central Section of the canal. Raw data from the cores A and B are published in Leitholdt et al. [17], those from the cores E, F, G, H and I are published in Leitholdt et al. [36]. The figure is not similar to formerly published figures.

from the trench bottom at cores Q and S reveal calibrated ^{14}C ages of 689–761 and 666–708 cal AD. Both ages are 70 to 100 years older than the historical documented onset of canal construction in 793 AD [24]. Two pieces of *Quercus sp.* suggest the use of timber for stabilizing the excavated trench. Comparing all excavation depths in the West-East Section, cores N to T show levels of 412.24 m with very small deviations of only ±0.31 m (Fig. 8). We interpret the small deviations as evidence for a planned construction of a pond at this site. The cores U to Z3 mirrors more shallow excavation depths between 413.00 and 413.30 m (Fig. 8). Hence, in the West-East Section there is evidence for one step-like increase of the trench bottom between cores T and U. Regarding the findings from the electrical survey, there is no evidence for a cross bank or shipping chute between the detected trench bottoms in core S and U (Fig. 8). ***III)*** The 3rd phase indicates evidence for Carolingian ponds in the West-East Section. On the one hand the presence of 'fluvial type' clastic deposits, with visible plant macro-remains at the base of the trench fills in cores N, O, P, Q, and W, points to an initial colluvial re-deposition process shortly after the time of excavation. On the other hand we have strong evidence for ponds in most of the lowermost trench fills of the West-East Section. Here, cores Q, T, W, X, Z2 and Z3 reveal 'pond type' sediment texture or sapropel layers at the canal bottom. The presence of alternating peat and sapropel layers and the abundance of *Salix sp.* and *Alnus sp.* macro-remains points to open water bodies with adjacent semi-terrestrial zones of fen and swamp forest during the Carolingian period. According to the ^{14}C data from the lower trench fills, we have evidence for this type of canal ecosystem from the Carolingian period to at least 990–1020 cal AD (core W, Table 2). The electrical survey features a continuous layer of low electrical resistivity between the cores N and V (green and blue colours, Fig. 8). In contrast to the different trench bottom levels in the western and eastern part of the West-East Section, there is no evidence for isolated ponds from the magnetic survey. However, a wooden Carolingian weir would not be necessarily visible in the electrical resistivity profile. ***IV)*** The 4th phase represents an enduring ponding and peat-growing phase within the West-East Section. Compared to the Carolingian deposits, the High Medieval sediments reveal even more evidence for ponds as indicated by sapropels and ponding-type sediment textures in all cores (N, O, P, Q, R, S, T, U, V, W, X, Z2 and Z3) (Fig. 8). Chronologically, we have evidence for this ponding-rich environment up to the 13th century AD (core O, Fig. 8, Table 2). ***V)*** The clastic sediments above the medieval peat and sapropel layers represent the final 5th phase in the West-East Section which is probably running until today. In the western part of the West-East Section the grain size distributions are equivalent to those found below the organic-rich layers which are dominated by the 'fluvial type' sediment texture. Here, the medium to fine sandy sediments of the 5th phase originate from the Quaternary valley fills, respectively feature re-deposited material from the adjacent Carolingian banks which were built up by Quaternary valley deposits. The sandy texture is also indicated by a high electrical resistivity (Fig. 8). In the eastern part of the West-East Section a final clayey to silty sequence of cores U, V, W, X, and Z2 indicates a suspension-load deposition process. Although the matrix of the final sequence is reduced in organic matter, we assume evidence for a former open water body due to the analysed 'ponding type' sediment texture which was identified in all sediment samples. This compact facies is also visible in the low electrical resistivity values (Fig. 8).

Fig. 9 shows a cross section of the West-East Section at core O site. The drilling positions are plotted in Fig. 7. The chronostratigraphical findings reveal evidence for a planned canal width of around 5 to 6 m. Only in this narrow zone sapropel layers and organic fills are detectable. Surprisingly, we have evidence for a narrow zone of ponding and semi-terrestrial conditions not only

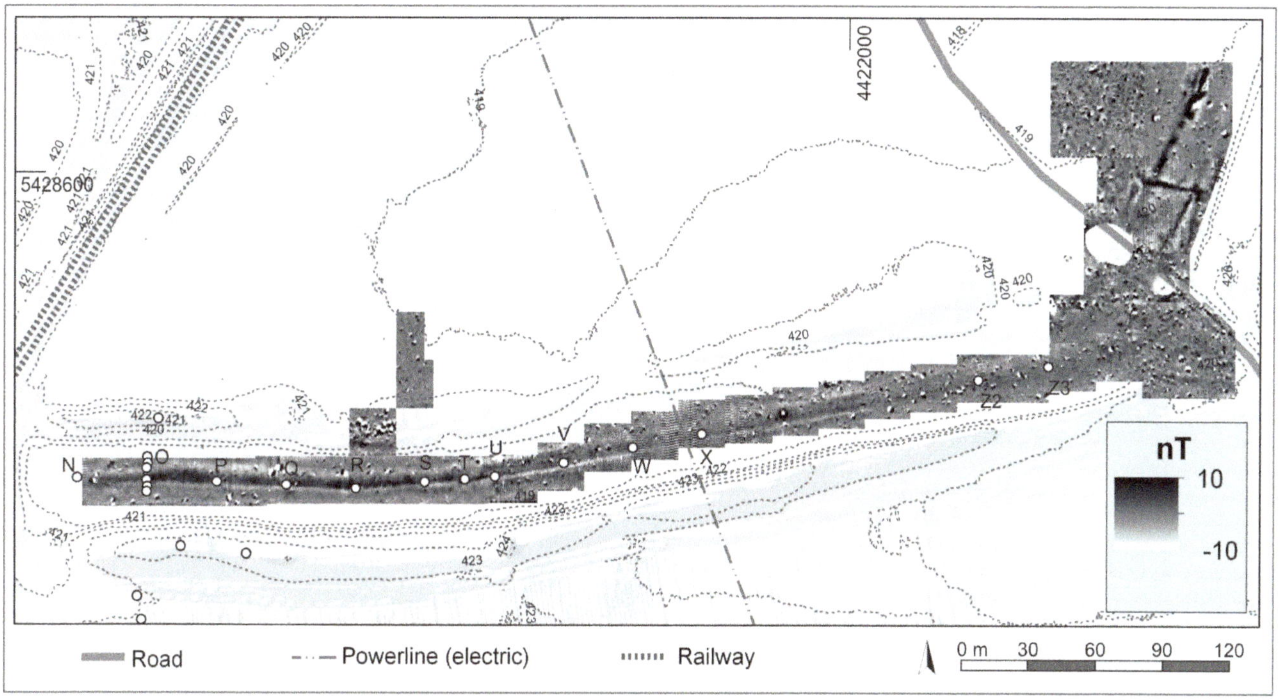

Figure 7. West-East Section of the Fossa Carolina. The figure shows a digital elevation model with detailed drilling positions, electrical transects, and the flux gate magnetic survey. White dots mark the drilling positions. The black lines show the geo electrical transects 1 and 2. Raw data source: LIDAR data are cordially provided by Bavarian Land Surveying Office for illustrative purposes.

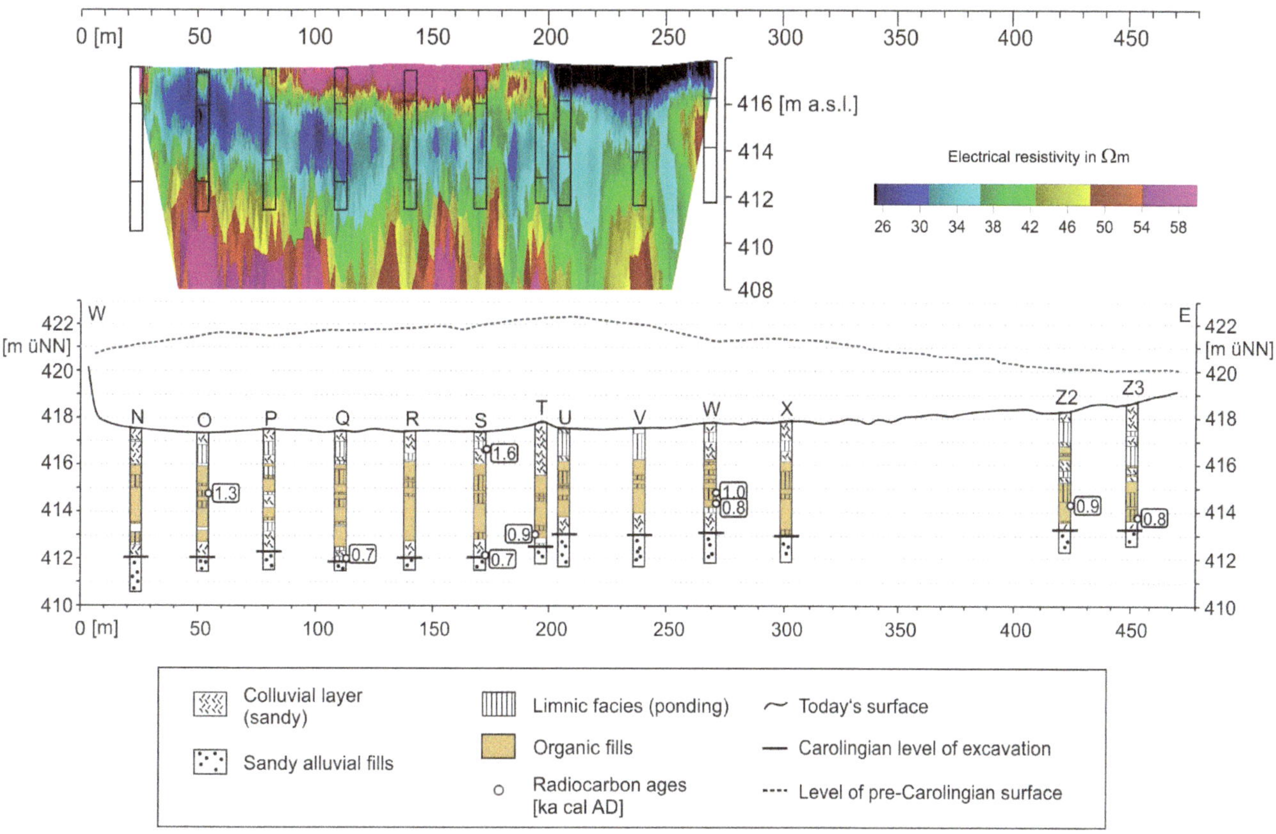

Figure 8. Longitudinal profile of the West-East Section including an electrical resistivity profile.

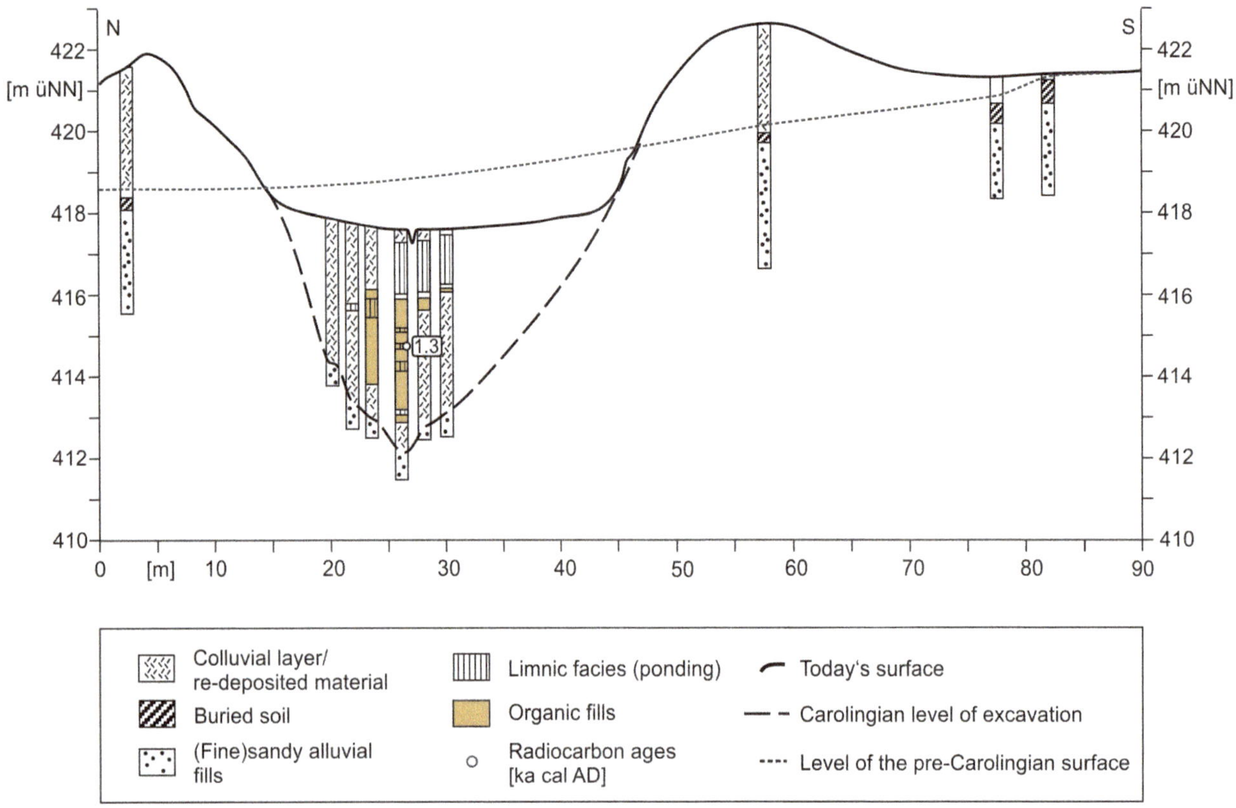

Figure 9. Cross section of the West-East Section at core O position.

for the Early Medieval strata but also for all younger sequences in the cross section (Fig. 9).

Northern Section

The Northern Section covers the probable course of the canal from the present-day connection road between Grönhart and Dettenheim to the second crossing of the Treuchtlingen-Nuremberg railway (Fig. 3). Hence, there must be a quite angular bend of the canal course in the transition from the West-East Section to the Northern Section (Fig. 3). The LIDAR digital elevation model from the Northern Section features two banks orientated in North-South direction (Fig. 10). The bank remnants are hardly visible in the field but the digital elevation model clearly documents evidence for an anthropogenic structure. The distance between the two bank crests varies only little between 25 and 30 metres (Fig. 10). Hence, this structure might reflect the extension of the Fossa Carolina in northern direction as already assumed by Röder [52] and later on by Koch and Leininger [33]. The SQUID geomagnetic data strongly support this assumption. Pronounced linear magnetic anomalies with nearly North-South direction are detectable within the complete North-South extension of the Northern Section (arrows in Fig. 11). The linear anomalies with the strongest signal of more than 80 nT/m (sharp black-white contrast in Fig. 11) and a total length of about 250 m occur between a sharp 90 degree turn of the current Rezat course in the south and a modern drainage trench in the north. However, these anomalies do not mark the principal canal course but they seem to flank the west side at a distance of 10 to 20 m. First calculations of the corresponding subsoil structure from the magnetic data show a narrow shape with a depth of about 1.5 m, which is in good agreement with the below presented drilling findings. Based on the detected magnetic anomalies and the LIDAR data we have chosen drilling sites for proving the course of the canal in the Northern Section. Two cross sections (Fig. 10 and 11) were selected for the stratigraphical documentation of the alluvial valley deposits, the trench bottoms, the banks and the trench fills. For the first time we sampled wood remnants at the trench bottom to prove the Carolingian age of the canal in the Northern Section.

QP cross section (Fig. 12) consists of four cores (QP1 to QP4). QP1 is exactly positioned in the assumed canal course. A charcoal sample (*Quercus sp.*) from the trench bottom features a calibrated ^{14}C age of 656–670 AD (Table 2). This age is around 130 years older than the historically documented canal construction at 793 AD [24]. However, keeping possibly matured oak wood in mind, the ^{14}C sample supports the hypothesis of the Carolingian age of the trench bottom and indicates evidence for the use of timber in an advanced Carolingian construction phase. Above the Carolingian trench bottom the QP1 core features a 220 cm long sequence of predominantly organic trench fills. The fills consist of peat and sapropel layers indicating local semi-terrestrial and limnic deposit conditions. The upper part of QP1 core is characterized by a covering organic-poor colluvial layer of 120 cm thickness. The lower part of QP3 and QP4 cores reveals sandy alluvial fills of the Rezat valley predominantly poor in organic matter. The sandy fills are covered by an *in situ* half-bog A horizon featuring the Carolingian surface and the edge of the close by Rezat fen. The A horizon is buried by 60 to 100 cm of excavated trench material. QP2 core is exactly positioned in the strong magnetic anomaly (Fig. 11). The lower part of the QP2 sequence represents organic-poor sandy Rezat alluvial fills. The alluvial fills are covered by a black and a red (2.5 R) clayey to silty layer (130 to 170 cm below surface). These layers reveal extremely high values of volume

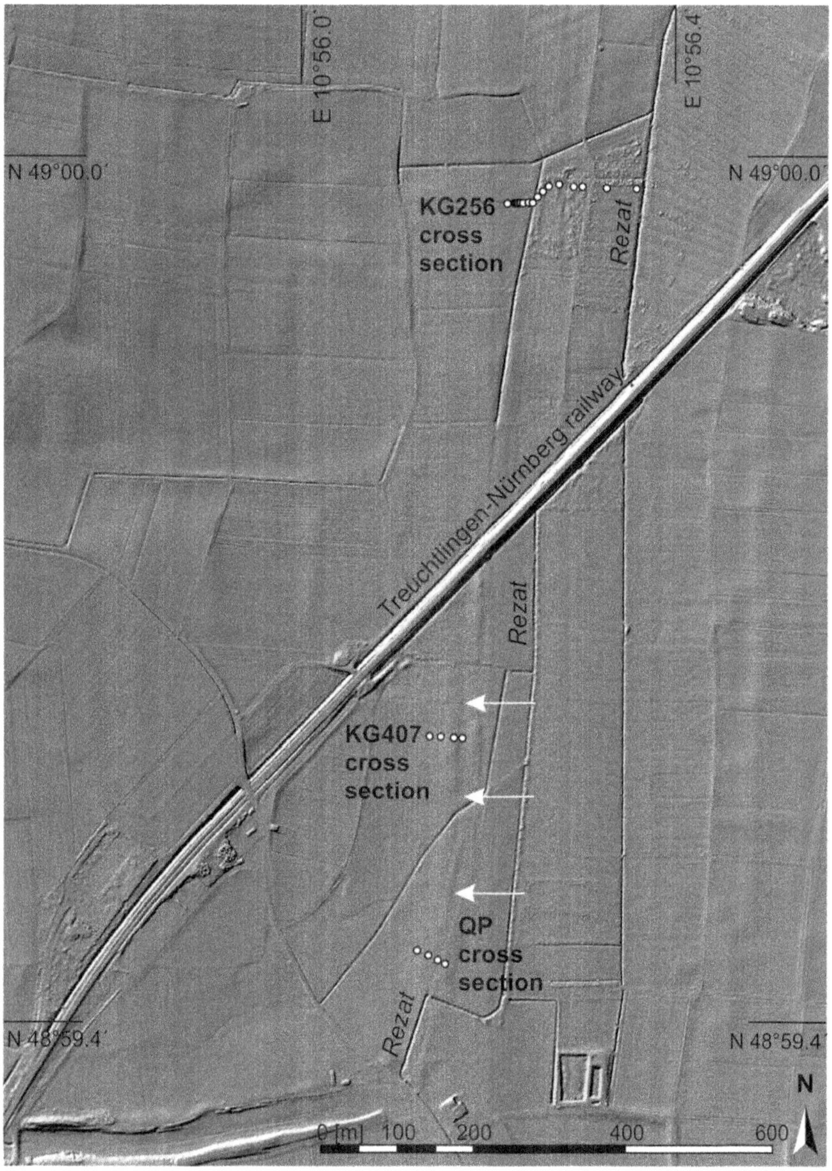

Figure 10. Airborne LIDAR scan with drilling positions in the Northern and North-Eastern Sections. White arrows indicate the assumed course of the canal. The cross sections are documented in Figs. 14, 15, and 16. Raw data source: LIDAR data are cordially provided by Bavarian Land Surveying Office for illustrative purposes.

specific magnetic susceptibilities (40,000 to 53,000 10^{-6} SI units). The formation of these layers remains unclear so far. As the black and red clayey to silty layers and the capping reddish-brownish colluvial layer (0 to 130 cm) seem to be incised in the half-bog A horizon as well as in the bank remnants we assume a Carolingian to post-Carolingian age of the magnetic anomaly.

KG-407 cross section (Fig. 10 and 13) consists of four cores (KG-407-1 to KG-407-4). KG 407-1 is exactly positioned in the assumed canal course. The lower sequence reveals sandy fluvial fills (Fig. 13). The greyish layer is poor in organic matter and fines up to a covering greyish clay layer. The middle sequence is characterized by peat and sapropel layers which are interpreted as fills of the Carolingian trench. The layers indicate a semi-terrestrial to limnic environment. A ^{14}C sample from a seed of the peat layer (137 cm below surface) features a calibrated age of 1444–1464 AD (Table 2). Hence, the canal exhibited an open body of water for at least 650 years, assuming a Carolingian age of the excavated trench at this point. KG 407-2 core represents a sequence of the undisturbed fluvial fills. The lower part of the sequence features organic-poor fine-sandy deposits. The medium sequence shows clayey layers with shallow organic-rich interruptions. Similar to the QP cross section the fluvial sequence ends with an *in situ* half-bog A horizon due to the adjacent Rezat fen. The half-bog A horizon is buried by a sandy bank featuring the former excavated trench material. The half-bog A horizon and the covering bank material can also be detected in the cores KG 407-3 and KG 407-4 (Fig. 13).

North-Eastern Section

Within the North-Eastern Section the course of the canal is unknown. Whereas in the south of the Nuremberg-Treuchtlingen railway the topography slightly mirrors the remnants of two

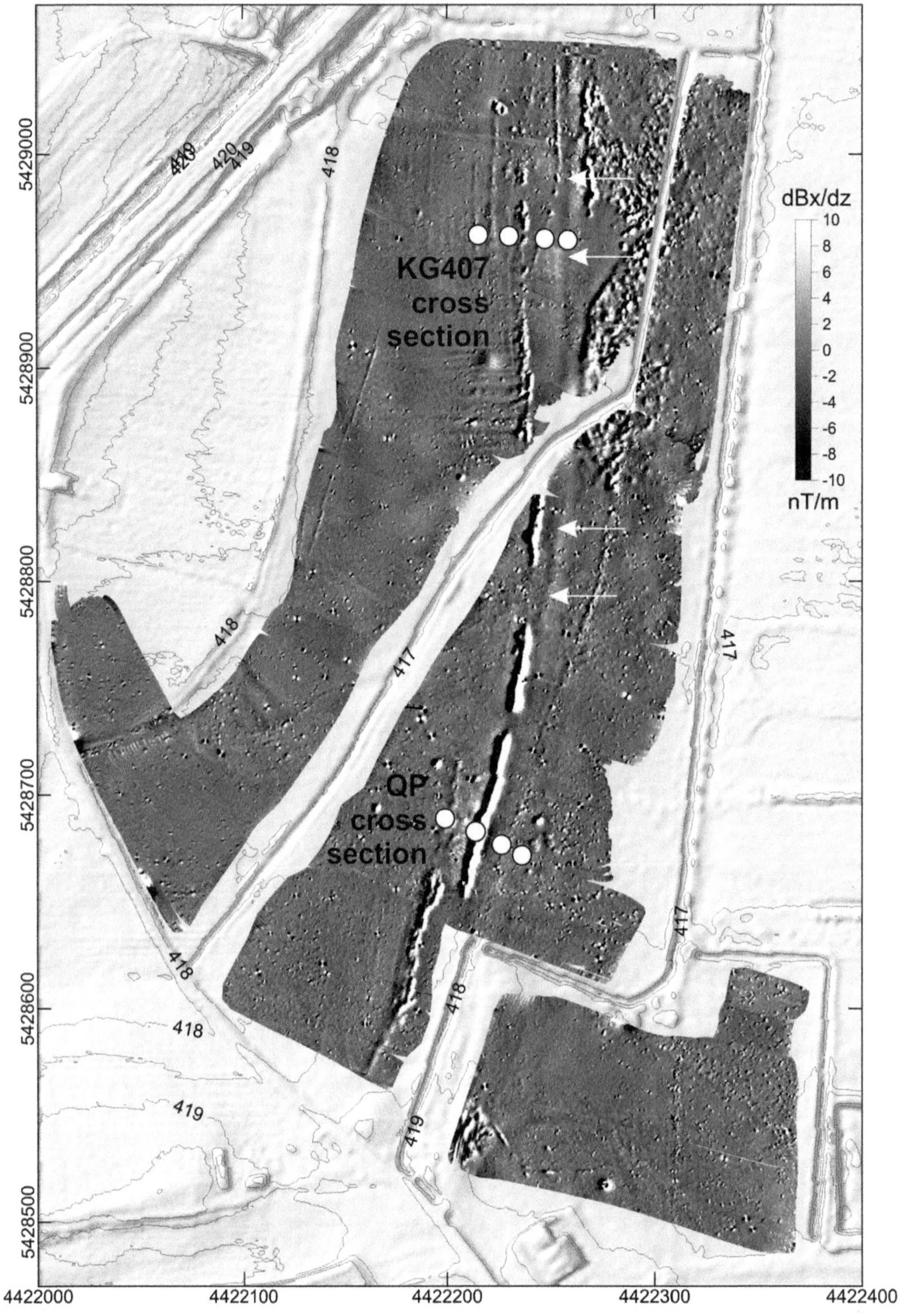

Figure 11. SQUID magnetogram from the Northern Section. White dots mark the drilling positions. White arrows feature the course of canal. The image represents an area of 600 m×400 m. The magnetogram is embedded in a LIDAR illustration, which raw data are cordially provided by Bavarian Land Surveying Office.

parallel banks (Fig. 3), these banks are not any more visible in the North-Eastern Section. Additionally, no *in situ* dating from a possible trench bottom is available yet which might prove the presence of a Carolingian canal here. However, linear North-South straightened magnetic anomalies are also detectable in the North-Eastern Section close to the railway where no topographic features are detectable by LIDAR. Thus, we conclude that the assumed Carolingian canal course is revealed and precisely

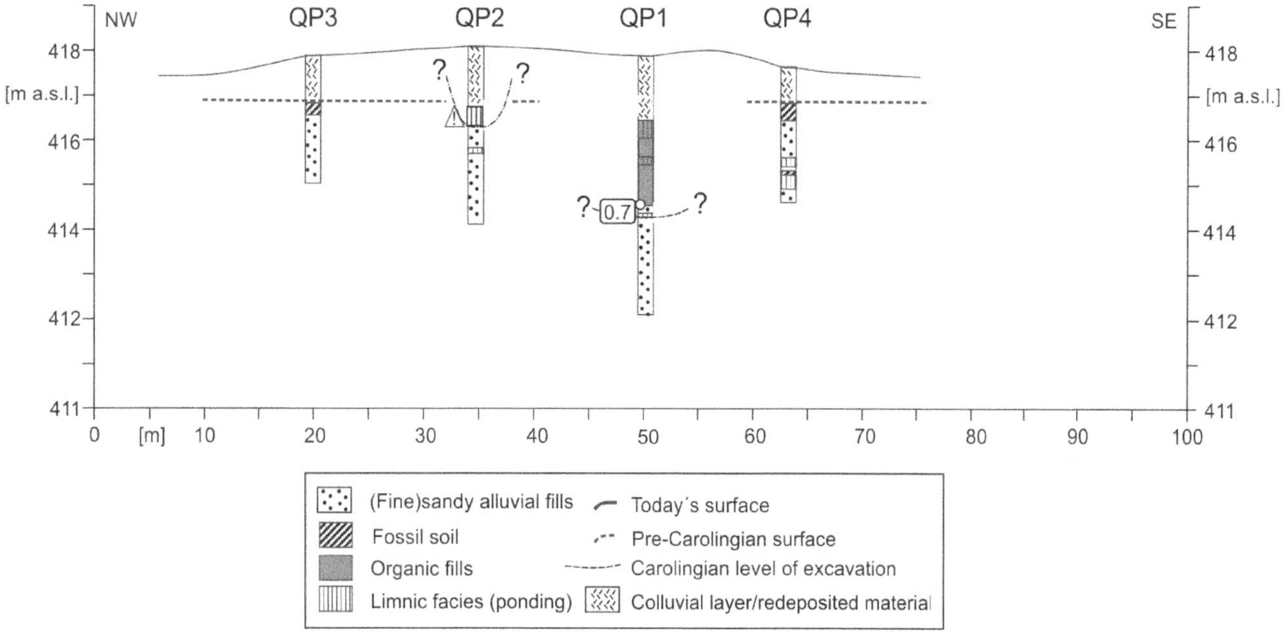

Figure 12. Stratigraphical documentation of the QP cross section within the Northern Section of the canal. The drawing shows a 42 m long cross section with the different facies types and the Carolingian level of excavation.

marked by these magnetic signatures. Fig. 10 shows the drilling positions of our examined KG 256 cross section through the Rezat fen in the North-Eastern Section. At the bottom of the cross section medium to fine-sandy greyish fluvial deposits fines up to a greyish sandy loam poor in organic matter. The sandy loam is covered by a 50 to 100 cm thick organic layer featuring slightly decomposed Rezat fen peat deposits. The peat layer terminates with a half-bog *in situ* A horizon, which indicates high decomposition of the upper peat layer. Between a modern drainage trench and the present-day straightened course of the Rezat River in the East (Fig. 14) the half-bog A horizon is covered by a colluvial-like brownish sandy loam.

In the west of the modern drainage trench the slightly decomposed peat layer is absent. In contrast, the organic-rich material in the cores 256–3, –6, –7, –8 and –9 (Fig. 14) reveals no clear structures of botanical remnants. In core 256–3 the organic-rich layer attains an extraordinary thickness and depth, which does not match with adjacent cores 256–7 and 256–8. Furthermore, the ubiquitous basal organic-poor sandy loamy layer is absent here. In conclusion, we postulate purely from the stratigraphical findings

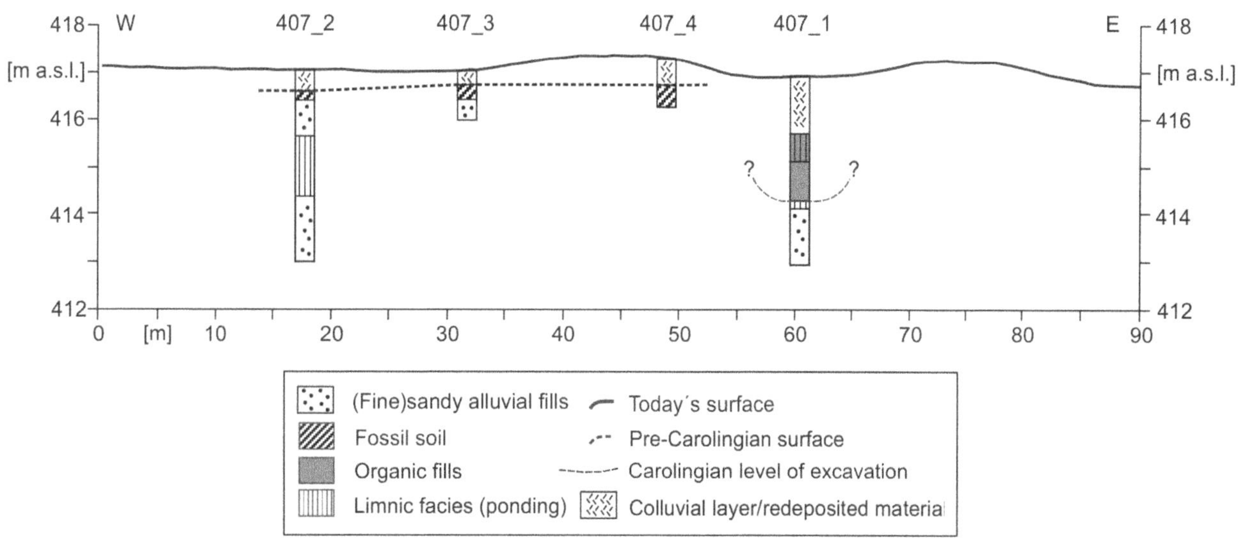

Figure 13. Stratigraphical documentation of the KG 407 cross section within the Northern Section of the canal. The drawing shows a 90 m long cross section with the different facies types and the Carolingian level of excavation.

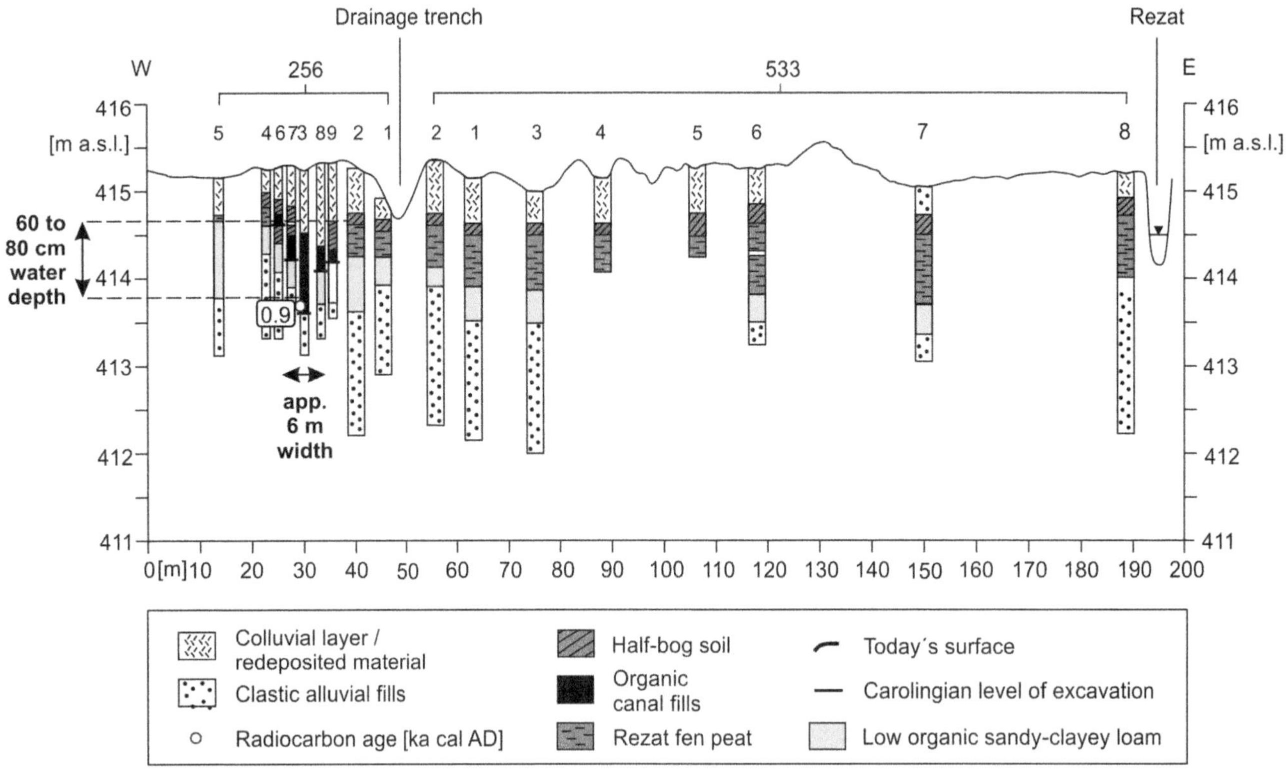

Figure 14. Stratigraphical documentation of the KG 256 cross section within the North-Eastern Section. The cross section shows the in situ layers of the Rezat fen in the East and the trench structure in the West. For the first time there is in situ evidence for a Carolingian trench in the North-Eastern Section. The trench bottom is shown by horizontal bars.

an excavated basal sandy loamy layer which was subsequently filled up with organic-rich trench deposits. Thus, we are sure that we detected an artificial sedimentological sequence here which might correspond to a former canal construction. We have clear evidence for a Carolingian trench based on the result of a radiocarbon dating. A wood macro-remnant from the trench bottom reveals a calibrated ^{14}C age of 829–927 AD.

Discussion

Evidence for Early and High Medieval ponding

Grain-size analyses, the classification of the organic layers, and the chronostratigraphical documentation of the trench fills of the Central Section and of the West-East Section clearly show that the fills can be divided into separate depositional units. The banks and the colluvial deposits at the base of the trench are made of the same material that constitutes the underlying basal Quaternary fluvial deposits of the region. In contrast, there is strong evidence for clayey to silty layers in the trench fills which reveal remnants of suspension limited stillwater deposition and, therefore, the evidence of Carolingian and post-Carolingian ponding (Table 3). These findings are strongly supported by numerous sapropel layers in the Central Section as well as in the West-East and Northern Sections. Hence, there is no doubt about the existence of ponds at the Fossa Carolina during the Early and High Medieval periods (Table 3). Chronological data of previous Fossa Carolina archaeological studies were limited to historical records and not supported by radiocarbon dating. Thus, in previous surveys all organic-rich layers were classified as a signal for a Carolingian canal [33]. For the first time, our data indicate enduring ponding phases from the Carolingian period up to the High Medieval period. We have evidence that at least isolated zones of the Fossa Carolina exhibit open water bodies for hundreds of years. We have to consider the enduring use of the ponds for fisheries as well.

Regarding the ^{14}C data from the Central Section and the West-East Section, major peat growing phases and periods of sapropel accumulation were predominant during the 9th to the 13th centuries. Here, the Early Medieval ponding took place mainly during the 9th century, and the High Medieval sapropels are dated between the 11th and 13th centuries.

Evidence for abrupt refill processes and timber use

Numerous basal trench fills from the Central Section point to an abrupt refill process, starting immediately after the Carolingian excavation phases. We have evidence for at least 1 to 1.5 metres of Carolingian re-deposited material in cores A and K. Especially in core K the Early Medieval age inversion (Fig. 6 and Table 1) at the lowermost sequence points to an abrupt refill process shortly after the excavation of the trench. These findings correspond with the historical sources which describe a collapse of the Carolingian canal project due to unconsolidated bank materials slipping back in the trench after heavy rainfalls [24]. However, the reconstructed Early Medieval open water bodies in numerous core stratigraphies do not support a definite collapse theory.

It is striking that the analyses of the wood macro-remains indicate numerous *Quercus sp.* at the bottom of the trench fills (e.g. cores A, Q, S, and QP1). All ^{14}C ages of these *Quercus* samples reveal ages slightly before 793 AD (Tables 1 and 2). We conclude that the dated *Quercus* samples from the trench bottom reveal

Table 3. Georchaeological findings of the different canal sections: summarizing overview.

Canal section	Early Medieval trench	Early Medieval ponds	High Medieval ponds	Timber use	Chute	Weir	Abrupt Carol. refills	Stepped ponds	Summit canal concept	Number of drillings
(I) Central	+	+	+	+	o	o	+	+	+	10
(II) West-East	+	+	+	+	-	(+)	(+)	+	+	23
(III) Northern	+	+	+	+	o	o	o	(+)	+	8
(IV) North-Eastern	+	(+)	(+)	o	o	o	o	(+)	+	17
(V) Rezat Headwater [31]	o	-	-	o	o	o	o	o	(+)	2
(VI) Altmühl floodplain	o	o	o	o	o	o	o	o	o	Ongoing study

A plus sign (+) indicates clear *in situ* stratigraphical and chronological evidence, a circle (o) feature additional research needs, and a minus sign (-) indicates no *in situ* evidence.

evidence for timber use and mirror the advanced construction level of the canal. This corresponds with archaeological findings from the Northern Section. Here, a recent archaeological excavation suggests the use of *Quercus* boards for the stabilisation of the Carolingian canal edges [53].

Early Medieval inland navigation requirements

Two geoarchaeological cross sections from the West-East Section and from the North-Eastern Section set limits for the required width of the planned Carolingian canal. Both cross sections reveal evidence for a conceptual width of the open water body between 5 to 6 metres. In the North-Eastern Section the maximum stage of the Carolingian water body can be limited by the level of the adjacent half-bog A horizon which features at least seasonal terrestrial conditions. Keeping this level in mind, the maximum water depth of the canal in the North-Eastern Section is limited to 60 to 80 cm. A width of 5 to 6 metres and a depth of 60 to 80 cm allow a crossing way passage of boats with a maximum width of about 2.50 m and a maximum draft of about 50 cm which corresponds to excavated Carolingian cargo scows with a payload of several tons [13], [54]–[57]. Although the final development of the entire waterway between the Altmühl and Rezat rivers is not proven yet, the reconstructed canal widths provide helpful information for the improvement of knowledge about Early Medieval inland navigation requirements in Central Europe. So far, there has been only few data available about the navigability of small inland waterways in Central Europe [10], [19], [58]. A couple of years ago, an Early Medieval hythe was discovered and excavated at the adjacent southern Franconian Schwarzach River, a small tributary of the Altmühl River [11], [59]. The headwater of the Schwarzach River is also limited by a valley divide of the Central European Watershed and hydrologically comparable with the Rezat River. Hence, there is evidence for the general integration of the small Franconian rivers and streams in the regional inland shipping networks across the Central European Watershed.

Summit canal: advanced construction level of the known Carolingian canal course

On the basis of our drilling findings we are able to reconstruct a Fossa Carolina longitudinal profile with the levels of Carolingian excavation depths from Graben village to the Rezat River (Fig. 15). It is considerable that the excavation depths decrease along the Central and West-East course of the canal. In the zone of the Central European Watershed (reddish line in Fig. 15) the level of the trench bottom does not follow the level of the natural watershed. Here, the canal features maximum Carolingian excavation depths which correspond to maximum size of the adjacent banks. This clearly shows that a summit of the canal was not planned in the zone of the natural valley watershed. This required an additional Carolingian work load of enormous extent and we need a hydro-engineering explanation here. When we prolong the longitudinal profile into the Northern and North-Eastern Section, we have evidence for a summit canal concept and that the summit of the Carolingian canal is located around 1000 metres further to the East compared with the position of the natural watershed (Fig. 15). Here, we postulate a planned hydro-engineering concept. The artificial displacement of the watershed allows the opportunity to supply the summit of the canal with water from the Rezat headwater (Fig. 3 and Fig. 15). The Rezat spring reveals strong seasonal amplitudes in discharge due to the weak water yield of the shallow karst aquifers in this zone of the Franconian Jura [60], [61]. However, considering recent hydro-climatic conditions [62], [63] and summer discharge gauging the

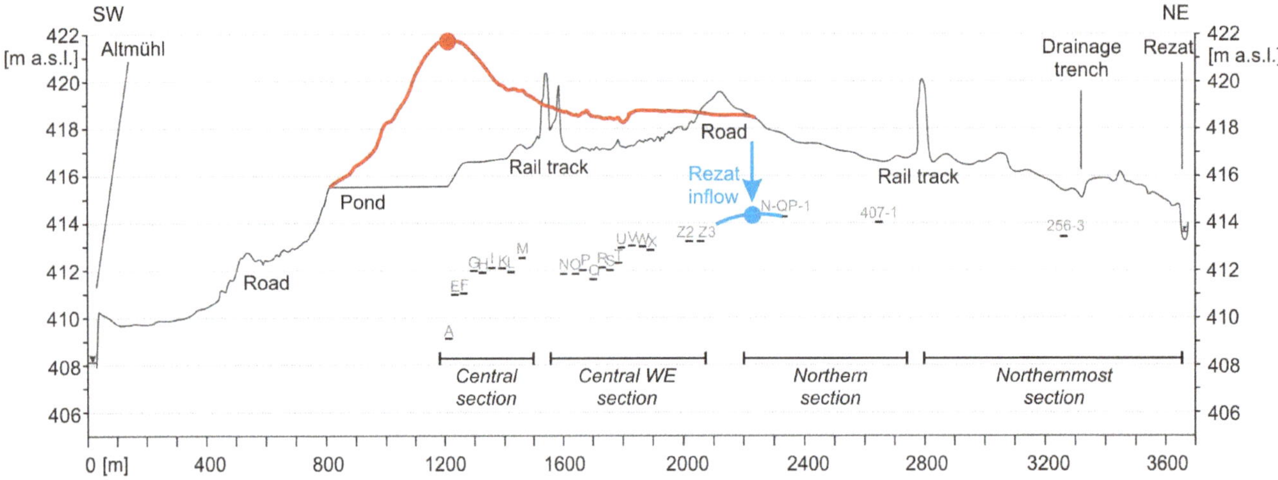

Figure 15. Synthetic longitudinal profile of the Fossa Carolina from Graben to Rezat River. a) The red line shows the Pre-Carolingian surface and watershed, the red dot indicates the summit the natural watershed. b) The short horizontal bars feature the Carolingian levels of excavation deduced from the drilling stratigraphies, and the blue dot indicates the artificial summit of the Carolingian trench bottom. c) The solid black line represents the level of the recent trench surface following the course of the canal.

Rezat spring features a base flow of 6 l/s in the zone of the summit of the canal. This corresponds with a minimum daily water supply of around 500 cubic meters.

Although recent geoarchaeological research discards Koch's [15] reservoir theory for the Rezat headwater [35] the discovery of the trench bases clearly shows how the Carolingian breaching of the watershed took place and provides new evidence for the outstanding Early Medieval knowledge in hydro-engineering. Apart from the Fossa Carolina summit canal, there is only poor evidence for shipping canals in Early Medieval Europe. The Early Medieval Kanhave canal on Samsø Island is an exception [64]. However, historical sources generally document Early Medieval knowledge in the construction of mill canals and aqueducts in Central Europe [25], [65]–[68]. Even in the adjacent Paar floodplain and in tributaries of the Altmühl River we have archaeological evidence for mill canal construction and Early Medieval competence in hydro-engineering [11], [69].

European dimension of the project: extension of the canal course to the north

Due to its extraordinary geomorphological favourableness the valley watershed between the Swabian Rezat and Altmühl Rivers played an important role in the Early Medieval trade and transport system. Here, the comparatively easy crossing over the European Watershed is characterised by a channelisation of supra-regional transport axes [29]. During the Carolingian era the Main-Rednitz-Regnitz-Swabian Rezat headwater system was equipped with a chain of royal courts and monasteries reflecting a central role for supra-regional trading and transport [16], [70]. Initiating with the Carolingian royal court at Hallstatt the succeeding courts upstream at Forchheim, Fürth, Roth and Weissenburg represent daily stages of an inland shipping or trading route towards the European watershed [19]. The supra-regional importance of crossing the land passage between the Swabian Rezat River and the Altmühl River justifies the enormous effort put in the construction of a Carolingian canal [1], [25], [71]. The geoarchaeological results from the Central and West-East Sections indicate an extraordinarily advanced Carolingian construction level of the canal (Table 3). Additional drilling campaigns in the Altmühl floodplain and archaeological excavations are required to clarify the location of the connection between the canal and the Altmühl River. However, our new SQUID geomagnetic data and geoarchaeological *in situ* findings from the Northern and North-Eastern Section prove for the first time the extension of the canal course in a northerly direction close to the confluence with the Rezat River. This leads to a total length of the constructed Carolingian canal of at least 2300 metres.

Conclusions

The trench fills of the Fossa Carolina indicate numerous ponding phases during the Early and High Medieval periods (Table 3). We have evidence that at least isolated zones of the Fossa Carolina exhibit an open body of water for hundreds of years. The basal sequences of the trench fills feature occasionally sandy layers rich in isolated plant and wood remnants. These layers feature ^{14}C age inversions and might be the result of abrupt and unintentional refilling processes shortly after the Carolingian construction phases. However, these abrupt refills are often covered by sediment sequences which indicate the later evidence for open water bodies. Hence, there is no clear evidence for an abrupt and final collapse of the Carolingian canal construction.

The reconstruction of two canal cross sections indicates a width of approximately 5 to 6 metres and a depth of 60 to 80 cm for the open water bodies. This would have allowed a crossing way passage in the West-East and Northern Section of the canal for Carolingian cargo scows with a payload of several tons. The synthesized longitudinal profile of the trench fills indicate clear evidence for a summit canal as the final hydro-engineering concept. We conclude that the artificial and labour intensive displacement of the watershed allows the opportunity to supply the summit of the canal with Rezat spring water. We detected an extraordinarily advanced construction level for a navigable waterway which mirrors the enormous Carolingian effort in crossing the Central European Watershed and in improving supra-regional inland navigation networks.

Supporting Information

Figure S1 Geological sketch of the study area. Redrawn [39], [74].

Figure S2 Chronostratigraphical documentation of core K. Anthracological findings from the Central Section are presented in a synthetic form and normalized to the levels of core K. The black arrow marks the level of the Carolingian excavation depth.

Figure S3 Stratigraphical documentation of core O. Anthracological findings from the West-East Section are presented in a synthetic form and normalized to the levels of core O. The black arrow marks the level of the Carolingian excavation depth.

Table S1 Technical and historical terminology related medieval inland navigation and canalisation. The table shows specific terms which are used in the manuscript in a stringent and standardised form. The High Medieval Period from the late 10th century to early 14th century corresponds to the Central Middle Ages following the Oxford terminology [75].

File S1 Core K and O: examples of our detailed chronostratigraphical approach. We show two representative cores from the Central Section and the West-East Section to document our stratigraphical approach in a more detailed form.

Acknowledgments

We would like to thank Birgit Schneider and Katja Pöhlmann for their cooperation in conducting grain size and CNS analyses. For methodological support in classifying the organic layers, we are most thankful to Annett Krüger. We would like to thank Stefan Dunkel as part of the SQUID prospection team. We thank André Kirchner for providing a geological sketch of the study area. We would like to thank John P. Hart, Balázs Bradák and two anonymous reviewers for their detailed and very helpful comments and suggestions.

Author Contributions

Conceived and designed the experiments: CZ SBH PE. Performed the experiments: EL CZ AS JB MS CM. Analyzed the data: EL CZ AS SL MS CM. Wrote the paper: CZ EL LW.

References

1. McCormick M (2002) The origins of the European economy. Communications and commerce A.D. 300–900. New York: Cambridge University Press. 1130 p.
2. Carver M, Loveluck C (2013) Maritime and Marine Historic Environment Research Framework. Appendix 5: Early Medieval Chapter. Original Early Medieval Period Working Group Resource Assessment and Research Agenda Document. Available: http://archaeologydataservice.ac.uk. Accessed 3 September 2013.
3. Theuws F (2012) River-based trade centres in early medieval northwestern Europe. Some 'reactionary' thoughts. In Gelichi S, Hodges R (Eds.) From one sea to another. Trading places in the European and Mediterranean early Middle ages. Proceedings of the International Conference Comacchio 27th-29th March 2009. Seminari internazionali del Centro Interuniversitario per la Storia e l'Archeologia dell'Alto Medioevo 3. Turnhout: Brepols, pp. 25–45.
4. Sindbaek SM (2007) The Small World of the Vikings: Networks in Early Medieval Communication and Exchange. Norwegian Archaeological Review 2007: 59–74.
5. Johanek P (1986) Der fränkische Handel der Karolingerzeit im Spiegel der Schriftquellen *(The Franconian trade during the Carolingian period in the mirror of historical sources)*. In Düwel K, Jankuhn H, Siems H, Timpe D (Eds.) Untersuchungen zu Handel und Verkehr der vor- und frühgeschichtlichen Zeit in Mittel- und Nordeuropa, Part IV. Der Handel der Karolinger- und Wikingerzeit. Abhandlungen der Akademie für Wissenschaften Göttingen. Phil.-Hist. Kl. 3 Nr. 156, Göttingen: Vandenhoeck & Ruprecht. pp. 7–68.
6. Bernhardt JW (1993) Itinerant Kingship and Royal Monasteries in Early Medieval Germany. Cambridge Studies in Medieval Life and Thought 21. Cambridge: Cambridge University Press. 376 p.
7. Elmshäuser K (2002) Facit Navigium. Schiffahrt auf Seine, Marne, Mosel und Rhein in Quellen zur frühmittelalterlichen Grundherrschaft *(Facit Navigium. Navigation on Seine, Marne, Mosel and Rhine Rivers detected in sources of Early Medieval seigniory)*. In Elmshäuser K, Ellmers D (Eds.). Häfen, Schiffe, Wasserwege. Zur Schiffahrt des Mittelalters. Schriften des Deutschen Schiffahrtsmuseums 58, Hamburg: Convent. pp. 22–53.
8. McKitterick R (2008) Charlemagne. The formation of a European identity. Cambridge: Cambridge University Press. 476 p.
9. Freund S (2007) Flüsse und Flusstäler als Wirtschafts- und Kommunikationswege *(Rivers and valleys as corridors for economies and communication)*. Bonn: Arbeitskreis für Historische Kulturlandschaftsforschung in Mitteleuropa (ARKUM). 457 p.
10. Eckholdt M (1980) Schifffahrt auf kleinen Flüssen Mitteleuropas in Römerzeit und Mittelalter *(Roman and Medieval navigation on small streams in Central Europe)*. Schriften des Deutschen Schiffahrtsmuseums 14. Oldenburg: Stalling. 119 p.
11. Liebert T (2013) Siedlungskomplex Großhöbing - Mühlen und Bootsländen als Einrichtungen ländlicher Zentralorte *(Großhöbing settlement – mills and hythes as facilities of rural major centres)*. In Ettel P, Werther L (Eds.) Zentrale Orte und zentrale Räume des Frühmittelalters in Süddeutschland. RGZM Tagungen 18, Mainz: RGZM. pp. 141–159.
12. Westerdahl C (Ed.) (2006) The significance of Portages. Proceedings of the first international conference on the significance of Portages. BAR international series 1499. Oxford: Archaeopress. 273 p.
13. Ellmers D (1972) Frühmittelalterliche Handelsschiffahrt in Mittel- und Nordeuropa *(Early Medieval merchant shipping in Central and Northern Europe)*. Neumünster: Wachholtz. 358 p.
14. Kalmring S (2007) Schiff – Hafen – Stadt. Mittelalterliche Hafenanlagen in Nordeuropa *(Ship – Harbour – City. Medieval harbours in Northern Europe)*. In v. Carnap-Bornheim C, Radtke C (Eds.) Es war einmal ein Schiff. Archäologische Expeditionen zum Meer. Hamburg: Mare. pp. 171–204.
15. Koch R (2002) Fossa Carolina. Neue Erkenntnisse zum Schifffahrtskanal Karls des Großen *(Fossa Carolina. New findings about Charlemagne's navigation canal)*, In Elmshäuser K (Ed.) Häfen – Schiffe – Wasserwege. Zur Schifffahrt des Mittelalters. Schriften des Deutschen Schifffahrtsmuseums 58. Hamburg: Convent. pp. 54–70.
16. Ettel P (2011) Der Main als Kommunikations- und Handelsweg im Frühmittelalter - Fossa Carolina, Burgen, Königshöfe und der überregionale Handelsplatz Karlburg *(The Main River as corrodor for communication and trade during the Early Medieval – Fossa Carolina, castles, royal courts and the supra-regional Karlburg trading centre)*. In Niedersächsisches Institut für historische Küstenforschung (Ed.). Flüsse als Kommunikations- und Handelswege. Siedlungs- und Küstenforschung im südlichen Nordseegebiet. Rahden/Westf. pp.201–226.
17. Leitholdt E, Zielhofer C, Berg-Hobohm S, Schnabl K, Kopecky-Hermanns B, et al. (2012) Fossa Carolina: The First Attempt to Bridge the Central European Watershed. – A Review, New Findings, and Geoarchaeological Challenges. Geoarchaeology 27: 88–104.
18. Grewe K (2008) Tunnels and Canals. In Oleson JP (Ed.) The Oxford handbook of engineering and technology in the classical world. Oxford: Oxford University Press, pp. 319–336.
19. Ellmers D (1993) Die Verkehrssituation zwischen Obermain und Altmühl in der Zeit Karls des Großen *(Traffic between Upper Main and Altmühl Rivers during Charlemagne's era)*. In Koch R, Leininger G (Ed.) Der Karlsgraben – Ergebnisse neuer Erkundungen. Bau intern Special Issue. München: Lipp. pp. 4–7.
20. Renard E (2009) La politique militaire de Charlemagne et la paysannerie franque. Francia 36, 1–33.
21. Hammer CI (2007) From Ducatus to Regnum. Ruling Bavaria under the Merovingians and early Carolingians. Collection Haut Moyen Âge 2. Turnhout: Brepols. 384 p.
22. Rau R (1993) Quellen zur karolingischen Reichsgeschichte Teil 1. *(Sources of the history of the Carolingian realm, part 1)* Die Reichsannalen. Einhards Leben Karls des Großen. Zwei Leben Ludwigs. Nithards Geschichte. Ausgewählte Quellen zur deutschen Geschichte des Mittelalters 5. Darmstadt: Wissenschaftliche Buchgesellschaft. 484 p.
23. Beck F (1911) Der Karlsgraben: Eine historische, topographische und kritische Abhandlung *(The Fossa Carolina: a historical, topographical and critical essay)*. Nürnberg: Korn. 91 p.
24. Hack A (2014) Der Bau des Karlsgrabens nach den Schriftquellen *(The construction of the Fossa Carolina – a historical review)*. In Ettel P, Daim F, Berg-Hobohm S, Werther L, Zielhofer C (Eds.). Großbaustelle 793 - Das Kanalprojekt Karls des Großen zwischen Rhein und Donau. Mosaiksteine – Forschungen am Römisch-Germanischen Zentralmuseum 11. Mainz: RGZM. pp. 53–62.

25. Squatriti P (2002) Digging Ditches in Early Medieval Europe. Past and Present 2002: 11–65.
26. Hofmann HH (1969) Kaiser Karls Kanalbau *(Charlemagne's canal)*. Sigmaringen: Thorbecke. 68 p.
27. Koch R, Kerscher H, Küster H (1993) Fossa Carolina – 1200 Jahre Karlsgraben *(1200-year-old Fossa Carolina)*. Denkmalpflege Informationen D 19.
28. Spindler K (1998) Der Kanalbau Karls des Großen. Seine Reflexion in den mittelalterlichen Quellen und der aktuelle archäologische Forschungsstand *(Charlemagne's canal in the mirror of historical sources and current state of archaeological research)*. In Spindler K (Ed.) Mensch und Natur im mittelalterlichen Europa – Archäologische, historische und naturwissenschaftliche Befunde. Klagenfurt: Wieser. pp. 47–99.
29. Eigler F (2000) Die früh- und hochmittelalterliche Besiedlung des Altmühl-Rezat-Rednitz-Raums *(Early and High Medieval colonisation of the Altmühl, Rezat and Rednitz catchments)*. Eichstätter Geographische Arbeiten 11. München: Profil, 488 p.
30. Irsigler F (2000) Zur wirtschaftlichen Bedeutung der frühen Grundherrschaft *(The economic role of early feudality)*. In Dilcher G, Violante C (Eds.). Strukturen und Wandlungen der ländlichen Herrschaftsformen vom 10. zum 13. Jahrhundert. Schriften des Italienisch-Deutschen Historischen Instituts in Trient 14. Berlin: Duncker & Humblot. pp. 165–187.
31. Lebecq S (2000) The role of the monasteries in the system of production and exchange of the Frankish world between the seventh and the beginning of the ninth centuries. In Hansen IL, Wickham CJ (Eds.) The long eighth century: production, distribution and demand, Leiden: Brill. pp. 121–148.
32. Schwarz K (1962) Der Main-Donau-Kanal Karls des Großen *(Charlemagne's Main-Danube canal)*. In Werner J (Ed.). Aus Bayerns Frühzeit. Friedrich Wagner zum 75. Geburtstag. München: C.H. Beck'sche Verlagsbuchhandlung. pp. 321–328.
33. Koch R, Leininger G (1993) Der Karlsgraben. Ergebnisse neuer Erkundungen *(Fossa Carolina – new findings of current surveys)*. Bau intern Special Issue. München: Lipp. pp. 11–15.
34. Küster H (1994) Pollenanalytische Untersuchungen im Bereich des Karlsgrabens *(Palynological research in the vicinity of the Fossa Carolina)*. Das Archäologische Jahr in Bayern 1993: 135–138.
35. Berg-Hobohm S, Kopecky-Hermanns B (2012) Naturwissenschaftliche Untersuchungen in der Umgebung des Karlsgrabens (Fossa Carolina) *(Natural scientific research in the vicinity of the Fossa Carolina)*. Bericht der Bayerischen Bodendenkmalpflege 52: 403–418.
36. Leitholdt E, Krüger A, Zielhofer C (2014) The medieval Peat Layer of the Fossa Carolina – Evidence for bridging the Central European Watershed or climate control? Z Geomorph 58 Suppl 1: 189–209.
37. Keller WE (1993) Der Karlsgraben, 1200 Jahre, 793–1993 *(Fossa Carolina, 1200-year-old, 793–1993)*. Treuchtlingen: Keller. 95 p.
38. Schmidt-Kaler H (1993) Geologie und Landschaftsentwicklung im Rezat-Altmühl-Bereich *(Geological structure and landscape dynamics in the Rezat and Altmühl valleys)*. (Bau intern Special Issue. München: Lipp. pp.8–10.
39. Schmidt-Kaler H (1976) Erläuterungen zur Geologischen Karte von Bayern 1:25.000. Blatt 7031 Treuchtlingen *(Elaboration of the Bavarian geological map 1:25.000. Treuchtlingen 7031 map sheet)*. München: Bayerisches Geologisches Landesamt.
40. Pohl J, Stöffler D, Gall H, Ernstson K (2010) The Ries impact crater. In Roddy DJ, Pepin RO, Merrill RB (Eds.) Impact and Explosion Cratering. New York : Pergamon Press, 343–404.
41. Bartington G, Chapman CE (2004) A high-stability fluxgate magnetic gradiometer for shallow geophysical survey applications. Archaeol Prospect 11: 19–34.
42. Bartington G (2013) Operation Manual for Grad601 Magnetic Gradiometer. Bartington Instruments Ltd. Oxford.
43. Geoscan Research (2005) Geoplot Version 3.00v for Windows - Instruction Manual Version 1.97. Bradford.
44. Linzen S, Schultze V, Chwala A, Schüler T, Schulz M et al. (2009) Quantum Detection meets Archaeology - Magnetic Prospection with SQUIDs, highly sensitive and fast. In Reindel M, Wagner GA (Eds.) New Technologies for Archaeology. Berlin, Heidelberg: Springer. pp. 71–85.
45. Schneider M, Stolz R, Linzen S, Schiffler M, Chwala A et al. (2013) Inversion of geo-magnetic full-tensor gradiometer data. J Appl Geophysics 92: 57–67.
46. Zielhofer C, Lee C, Rollefson G, Wächter S, Hoffmeister D et al. (2012) The decline of the early Neolithic population center of 'Ain Ghazal and corresponding earth-surface processes, Jordan Rift Valley. Quat Research 78: 427–441.
47. Eckelmann W (Ed.) (2005) Bodenkundliche Kartieranleitung *(Instruction for pedological mapping)*. Stuttgart: Schweizbart. 438 p.
48. Danzeglocke U, Jöris O, Weninger B (2013) Calpal – the calibration program of the Cologne radiocarbon laboratory. Available: http://www.calpal-online.de/ Accessed 5 March 2014.
49. Ullrich B, Günther T, Rücker C (2008) Electrical resistivity tomography methods for archaeological prospection. In Posluschny A, Lambers K, Herzog I (Eds.) Layers of Perception. Proceedings of the 35th International Conference on Computer Applications and Quantitative Methods in Archaeology (CAA), Berlin, April 2–6, 2007. Kolloquium für Vor- und Frühgeschichte 10. Bonn: Habelt. p. 25.
50. Günther T (2004) Inversion Methods and Resolution Analysis for the 2D/3D Reconstruction of Resistivity Structures from DC Measurements, PhD Thesis. Freiberg: TU Bergakademie.
51. Tucker ME (2001) Sedimentary Petrology: an Introduction to the Origin of Sedimentary Rocks. Oxford, London: Blackwell. 272 p.
52. Röder J (1974/1975) Fossatum Magnum – Der Kanal Karls des Großen *(Fossatum Manum – Charlemagne's canal)*. Jahresbericht der Bayerischen Bodendenkmalpflege 15/16: 121–130.
53. Werther L, Feiner D (2014) Der Kanal im Focus der Archäologie *(Fossa Carolina – archaeological research foci)*. In Ettel P, Daim F, Berg-Hobohm S, Werther L, Zielhofer C (Eds.). Großbaustelle 793 - Das Kanalprojekt Karls des Großen zwischen Rhein und Donau. Mosaiksteine – Forschungen am Römisch-Germanischen Zentralmuseum 11. Mainz: RGZM. pp. 33–40.
54. McGrail S (1987) Ancient boats in Northwest Europe. The archaeology of water transport to AD 1500. London, New York: Longman. 321 p.
55. Ellmers D, Hoffmann P (1990) Ein Frachter aus der Zeit Karls des Großen *(A cargo ship of Charlemagne's era)*. Bremer Archäologische Blätter N.F. 1, 33–37.
56. Obladen-Kauder J (1994) Ein karolingischer Flußkahn aus Kalkar-Niedermörmter *(The Carolingian scow of Kalkar-Niedermörmter)*. Archäologie im Rheinland 1993, 98–99.
57. Flatman J (2010) Maritime and Marine Historic Environment Research Framework. Technical Appendix 3. Archaeological Remains of Vessels from Great Britain, Northern Ireland and Ireland 6th - 16th Centuries AD. Available: http://archaeologydataservice.ac.uk. Accessed 3 September 2013.
58. Molkenthin R (2006) Straßen aus Wasser: Technische, wirtschaftliche und militärische Aspekte der Binnenschifffahrt im Westeuropa des frühen und hohen Mittelalters *(Inland waterways: technical, economical and military aspects of inland navigation in Early and High Medieval Western Europe)*. Berlin: Lit. 187 p.
59. Herzig F (2009) Dendroarchäologie. Mensch und Umwelt - eine Wechselwirkung, eingraviert in Holz *(Dendroarchaeology: man and environment – an interaction engraved in wood)*. Bericht der Bayerischen Bodendenkmalpflege 50: 225–236.
60. Zielhofer C (2004) Hydrographical and hydrochemical characteristics of karst water components (southern Franconian Jura) – a contribution to fresh water protection and karst morphogenesis. Z Geomorph Suppl 136: 113–134.
61. Zielhofer C (2009) The protection potential of soils and post-Jurassic covers against nitrate pollution in the Southern Franconian Jura, Germany. Z Geomorph 53: 371–386.
62. Kottek M, Grieser J, Beck C, Rudolf B, Rubel F (2006) World Map of the Köppen-Geiger climate classification updated. Meteorol Z 15: 259–263.
63. Mühr B (2007) Available: http://www.klimadiagramme.de/Deutschland/weissenburg2.html. Accessed 5 March 2014.
64. Jørgensen NA (2000) Kanhave-Kanal *(Kanhave canal)*. Reallexikon der Germanischen Altertumskunde 16. Berlin, New York: de Gruyter. pp. 221–222.
65. Elmshäuser K (1992) Kanalbau und technische Wasserführung im Mittelalter *(Medieval canal construction and hydrological engineering)*. Technikgeschichte 59, 1–26.
66. Grewe K (2000) Water Technology in Medieval Germany. In Squatriti 2000 (Ed.) Working with water in medieval Europe. Technology and resource use. Technology and change in history 3. Leiden, Boston, Köln: Brill. pp. 129–160.
67. Squatriti P (Ed.) (2000) Working with water in medieval Europe. Technology and resource use. Technology and change in history 3. Leiden, Boston, Cologne: Brill. 448 p.
68. Henigfeld Y, Schwien JJ, Werlé M (2010) L'apport de l'archéologie à la connaissance de la ville médiévale: le cas de Strasbourg *(The archaeological retrieval of knowledge of the Medieval city: the case of Strasbourg)*. In Chapelot J (Ed.) Trente ans d'archéologie médiévale en France. Un bilan pour un avenir, IXe congrès international de la Société d'Archéologie Médiévale (Vincennes, 16–18 juin 2006). Paris. pp. 351–368.
69. Czysz W (1998) Die ältesten Wassermühlen: Archäologische Entdeckungen im Paartal bei Dasing *(The oldest water mills: archaeological excavations in the Paar valley near Dasing)*. Thierhaupten: Klostermühlenmuseum. 55 p.
70. Störmer W (Ed.) (2008) Mittelalterliche Klöster und Stifte in Bayern und Franken *(Medieval monasteries and convents in Bavaria and Franconia)*. St. Ottilien: EOS. 526 p.
71. Hardt M (2007) Die Donau als Verkehrs- und Kommunikationsweg zwischen der ostfränkischen Residenz Regensburg und den Zentren an der mittleren Donau im 9. Jahrhundert *(The Danube as corridor for traffic and communication between Regensburg city of residence in Eastern Franconia and the major centres of the Danube middle course during the 9th century)*. Siedlungsforschung: Archäologie - Geschichte - Geographie 25 : 103–120.
72. Greenbelt MD (2009) EOSDIS, Goddard Space Flight Center (GSFC) National Aeronautics and Space Administration (NASA). Available: https://wist.echo.nasa.gov/api/. Accessed 6 September 2013.
73. OpenStreetMap (2014) http://www.openstreetmap.org. Accessed 18th June 2014.
74. Cohen KM, Finney SC, Gibbard PL, Fan JX (2013) The ICS International Chronostratigraphic Chart. Episodes 36: 199–204.
75. Power D (2006) The Central Middle Ages - 950–1320. Short Oxford History of Europe. Oxford: Oxford University Press. 320 p.

14

The Role of the Sedimentary Regime in Shaping the Distribution of Subtidal Sandbank Environments and the Associated Meiofaunal Nematode Communities: An Example from the Southern North Sea

Michaela Schratzberger[1]*, Piers Larcombe[2]

1 Centre for Environment, Fisheries and Aquaculture Science, Lowestoft Laboratory, Lowestoft, Suffolk, United Kingdom, 2 RPS MetOcean, Subiaco, Western Australia, Australia

Abstract

We combined sediment and faunal data to explore the role of the sedimentary regime in shaping the distribution of subtidal sandbank environments and the associated meiofaunal nematode communities at Broken Bank and Swarte Bank, in the southern North Sea. A variety of sediment transport processes occur in the area, differing in the frequency and magnitude of sediment mobility, and the continuum between erosion, translation and sediment accumulation. The seabed contained a variety of bedforms, including longitudinal furrows, and small to very large sandwaves. The bed sediments were dominated by fine and medium sands, with admixtures of silt and gravel. Based on sedimentary bedforms and grain size analysis, a total of 11 sedimentary facies were delineated, of which 8 were analysed in detail for their relationships with the meiofauna. The sedimentary facies fell clearly into groups of facies, respectively representing high, high-moderate and moderate, and episodic sediment mobility. For those sedimentary facies where daily movement of sediments and bedforms occurred ('high' sediment mobility), the resulting spatially homogeneous environments were dominated by an impoverished nematode community comprising small deposit feeders and large predators. Resistance to sediment movement and the ability to exploit alternative food sources were prominent functional features of the successful colonisers. Those facies characterised by relatively infrequent sediment mobility ('episodic' and 'high-moderate and moderate' sediment mobility) comprised a heterogeneous suite of benthic habitats, containing taxonomically and functionally diverse assemblages of nematodes of various sizes, feeding types and reproductive potential. Faunal distribution patterns here indicated trade-offs between the resistance to sediment movement, environmental tolerance and competitive abilities. Our focus on diverse assemblages of organisms with high turnover times, inhabiting highly dynamic sedimentary environments, has revealed new animal-sediment relationships of relevance to pure and applied science.

Editor: David William Pond, Scottish Association for Marine Science, United Kingdom

Funding: The work was supported by the Department for Environment, Food and Rural Affairs (Defra, research project AE1148), by EDF Energy and by RPS MetOcean. Early work on the manuscript was supported by the Marine Environment Protection Fund (MEPF), a delivery partner of Defra's Aggregate Levy Sustainability Fund (ALSF). The funders had no role in study design, data collection and analysis, decision to publish, or preparation of the manuscript.

* Email: michaela.schratzberger@cefas.co.uk

Introduction

Scientists, socio-economists, environmental managers, policy-makers and the public all have interests in marine environmental studies and their outcomes. A common motivation of marine studies is the implicit or explicit search for indicators of natural and anthropogenic change, with an aim of identifying and assessing human impacts and the effects of management actions. As an example policy driver, the EU Habitats Directive for the protection of species and habitats [1] requires that management plans be created in order to detect and respond to human-induced change. In particular, such plans are to be prepared for the candidate Special Areas of Conservation (SAC), including subtidal mobile sandbanks, on which we expand below. Thus, it is a requirement that plans are able to detect human-induced change as distinct from the various types and timescales of natural variability, and that actions are identified which are able to be enacted to counter such (assumed) detrimental change. In most environments this is not a simple task, but in the dynamic marine sedimentary environment of much of the western European shelf it appears to be a major challenge indeed.

Effective conservation of marine habitats in European seas requires specific conservation objectives. These need to be achieved by closely linked management measures that regulate the activities of those marine industries which may threaten the integrity of the habitat, or the viability of the associated populations [2]. Conservation objectives for subtidal sandbanks generally demand that, subject to natural processes, the extent and distribution of sandbank habitats, their faunal structure and function, their gross morphology (e.g. depth, distribution and profile) and the supporting processes on which sandbank habitats rely, are maintained or recovered [3]. In order to be able to measure the success or otherwise of such management goals, detailed investigations of the links between the sedimentary habitat and the biota are essential.

Reviewing the literature on animal-sediment relationships, [4] found little evidence that animal distributions are determined by any of the sediment variables derived from grain size alone. Indeed, [5] summarised the results of nine recent studies, which used abiotic variables such as percentage mud, percentage calcium carbonate, percentage gravel, total organic carbon content, a bed disturbance factor (named seabed exposure by [6]), water depth, latitude, longitude, slope, turbidity and distance to ocean. In these studies, between 25 and 75% of variability in faunal distribution patterns was accounted for; most of the stronger relationships involved using physical measures of the sea floor and sediments. Hence, meaningful and predictive explanations for faunal distributions are likely to emerge if these are evaluated relative to the suite of hydrodynamic and sediment transport processes that are responsible for sediment distributions. It is also important to understand the physical form of the sediments (bathymetry and the nature of sedimentary bedforms) at a range of spatio-temporal scales, and the details of the grain size distribution.

A general appreciation of the highly heterogeneous nature of continental shelf sediments has progressively emerged in recent decades. Multi-beam bathymetric surveys have revealed numerous types of sedimentary and geomorphological features on the sea floor that occur at a variety of spatial scales, ranging from a few meters to several kilometres [7,8,9,10]. Ground-truthing acoustic data to confirm the nature of the seabed is an important process, but the next step of improving understanding of the relationships with the fauna is hampered by gaps in our knowledge. These gaps include the difficulty of identifying organisms to fine taxonomic levels [11] and the comparative shortage of information on those organisms with a range of life-history characteristics which tie them closely to the sediment in which they live, such as meiofauna (comprising invertebrates that are retained on a 500–1000 μm mesh; [12]).

There are obvious possible dependencies of meiofauna upon the distribution and mobility of sediments [13,14,15]. When patches of suitable habitat are spatially or temporally discontinuous, an organism's dispersal capability will greatly affect its ability to colonise available space. Where the dispersal mechanism is closely related to the transport mechanisms of the associated sediment (e.g. meiofaunal nematodes) we would expect that the distribution and location of the sediment to be a key factor in controlling faunal distribution patterns. Mobile sediments at subtidal sandbanks could affect meiofaunal assemblages by altering the composition of sedimentary environments, thereby changing their suitability for settling organisms and/or by increasing the resuspension of organisms as a result of sediment deposition and/or sediment instability.

A detailed description of the benthic fauna associated with two linear sandbanks off the North Norfolk coast in the southern North Sea by [16] provided a general overview of the benthic ecology of the area. Here, we further explore these data, taking account of the topographic complexity of the surveyed sandbank habitats, to improve our understanding of the dynamics of the sandbank ecosystem. As such, this scoping study has two objectives: 1) to qualitatively describe the nature and origins of several kinds of sedimentological features and processes at two subtidal sandbanks in the southern North Sea, explicitly taking account of hydrodynamics and sediment transport processes, the bathymetry and nature of sedimentary bedforms, and grain size distributions and 2) to depict the taxonomic and functional characteristics of meiofaunal nematode communities which occupy these environments. We use our findings to better understand the way in which the physical dynamics of the seabed and its spatial heterogeneity might combine to determine the number and type of available niches and hence influence the diversity and composition of meiofaunal nematode communities.

Field Surveys and Sample Processing

All data reported in this article were collected by the Centre for Environment, Fisheries and Aquaculture Science (Cefas), as part of a government-funded research project. As an executive agency of the UK Government, Cefas has an exemption from The Crown Estate (i.e. land owner of our study site) to collect sediment and other scientific material (except vertebrates) in UK waters. The activities involved in the survey of the seabed and the collection of benthic invertebrates were not subject to specific regulation and did not involve endangered or protected species. Throughout the survey, scientists and ship crew acted in a responsible and environmentally friendly manner.

We analysed the bathymetry, sedimentary bedforms, sediments and meiofaunal nematodes at and near Broken Bank (BB) and Swarte Bank (SB), in the southern North Sea, to examine the association of faunal assemblages and their sedimentary habitat. The sampling strategy was not designed to fully characterise nematode-sediment relationships at the two sandbanks, but rather to enable us, with minimal cost and effort, to make comparisons between nematode communities inhabiting a diverse range of sedimentary environments.

Study site

Broken Bank and Swarte Bank are the outermost of the Norfolk Banks in the southern North Sea. Broken Bank lies 67 km offshore, and has dimensions typical of the tidally-controlled open-shelf linear sandbanks [17] of the outer Norfolk Banks area, at about 30 km long, 1 km wide and with relief of around 30 m above the surrounding sea floor (Fig. 1). Both banks are asymmetric, with a gently sloping SW face and a relatively steep NE face. For the Norfolk Banks in general, based on analyses of past charts and repeat bathymetric surveys, bank migration over decadal and century timescales is generally to the NE, at rates up to 11–15 m yr^{-1} [18,19], although rates of 40 m yr^{-1} have been recorded [20]. Banks may repeatedly shorten and lengthen, and may migrate SW then NE. For all banks, the bathymetric changes at the head and tail appear high compared to the sideways migration rates [19]. The surface tidal currents in the area of Broken Bank and Swarte Bank are strong, with speeds of 1.6 m s^{-1} and 1.0 m s^{-1} on spring and neap tides, respectively [21]. The bed sediments on and around the banks are loosely-consolidated, well-sorted, fine-grained sands which are transported in an overall clockwise rotation [22]. This transport is thought to take place

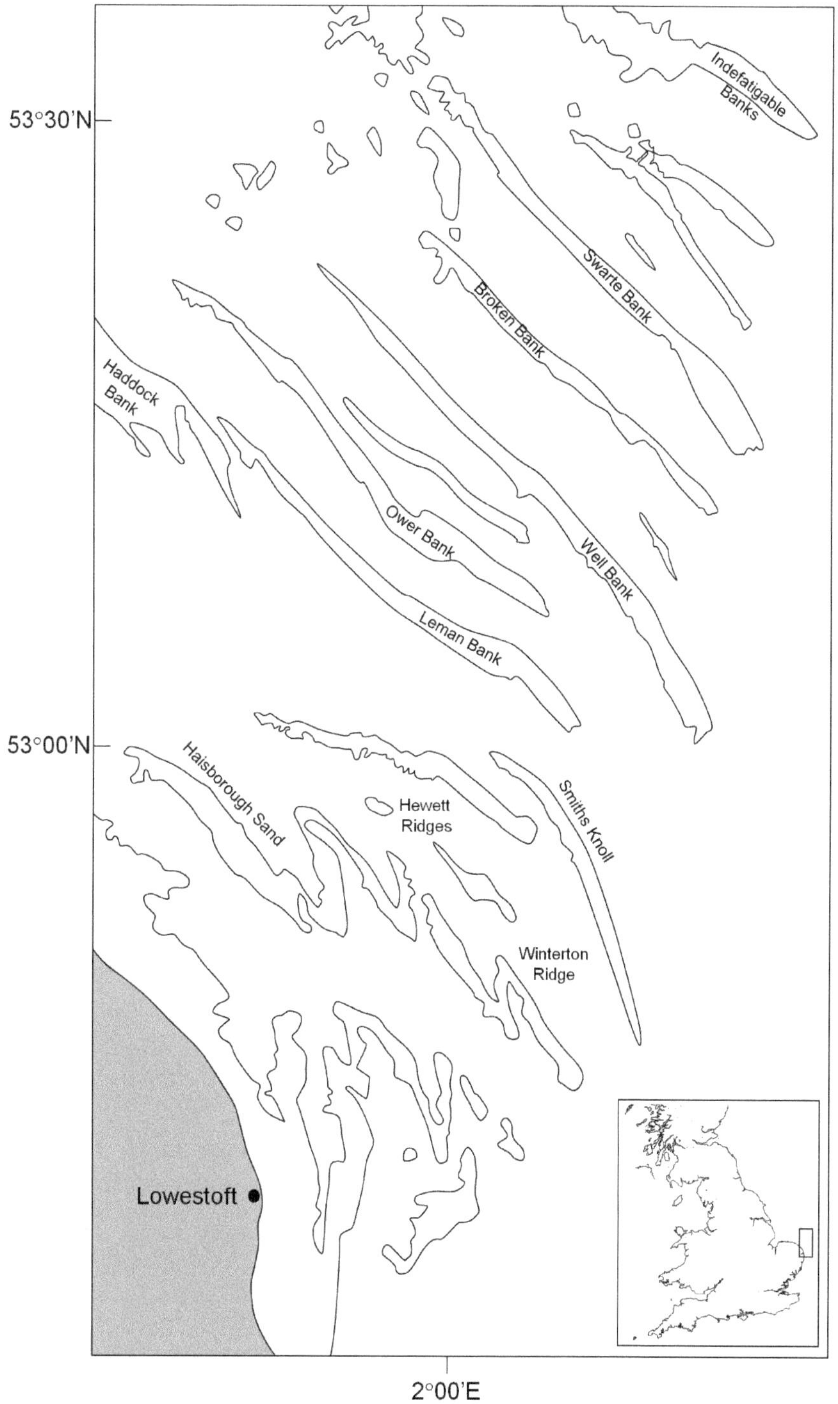

Figure 1. Location of Broken Bank and Swarte Bank in the southern North Sea. Figure adapted from [16].

nominally as bedload, under the influence of tidal currents alone (see also [23]). The addition of wave-induced bed stresses or storm surges can act to increase the transport towards the bank crests and lead to some sand being moved into suspension [22].

Geophysics

In April 2006, the southern portions of Broken Bank and Swarte Bank were surveyed using RV Cefas Endeavour to run a Kongsberg EM3000D multi-beam echo-sounder to provide high-resolution bathymetric data, and a Benthos 1500 digital side-scan sonar to measure seabed acoustic reflectance, which can indicate substrate hardness [11]. The acoustic survey was carried out in parallel to a multi-disciplinary sampling programme designed to provide a general description of the benthic fauna (including meiofauna, macrofauna, epifauna and fish) associated with Broken Bank and Swarte Bank [16]. Individual swath widths covered approximately 150 m on the seabed, and survey lines were designed to run parallel and perpendicular to the overall NW-SE alignment of the bathymetric features. More complete coverage was performed in selected areas of sedimentary interest, including a N-S strip 3×1 km across the SW face, crest and NE face of Broken Bank, and an area 500–800 m wide extending for 7 km along the NE flank of Swarte Bank (Fig. 2 and 3). The survey included coverage of the swale between the banks and an extended swath line off Broken Bank to the SSE. The data were processed using Fledermaus to drape the side-scan images onto the measured bathymetry, with tidal corrections applied. The morphology of the banks and their associated sedimentary bedforms were measured on screen using the Fledermaus viewing software ('iview4D'), leading to the delineation of a series of areas defined by their sedimentary morphology (bedform groups).

Sediments

Sediments from both Broken Bank and Swarte Bank were sampled by means of a 1 m^2 Day grab (Fig. 2; Table 1). Four samples each were collected at five stations along the crest of each sandbank (X, A, B, C, D) and at single sites either side of the sandbanks (F, I). Single samples were collected at four other stations located on the deeper surrounding seabed of each sandbank (E, G, H, J). From each grab sample, two sub-samples were collected with a Perspex corer (3.6 cm diameter, 10 cm^2 surface area) to a depth of 5 cm, one for the analysis of grain size distribution and one for the study of meiofaunal nematodes. The location of sampling stations was largely dictated by the seabed topography because sufficient draught was necessary under the vessel to allow the collection of sediment samples.

Sub-samples for grain size analysis were combined by station and frozen at −20°C prior to analysis. After thawing, the 22 sediment samples were wet sieved through a 500 μm sieve, and the fraction> 500 μm was oven dried at 90°C for 24 h. This fraction was then dry sieved at 0.5 phi intervals, down to 1 phi (500 μm). The fraction <500 μm was freeze dried and analysed on a Malvern Mastersizer S laser particle sizer. The sieve and laser sizer data were merged to form one uniform distribution spanning the size range 0.49 μm −22 mm. It is recognised that freezing samples can alter the grain size distribution, especially of clay-sized grains, but microscope and size analysis both indicated that no clay-sized grains were present in any samples, consistent with previous findings [22], and all samples were treated identically.

Grain size data were grouped using the computer programme ENTROPY [24]. Entropy analysis has been widely used to analyse grain size distributions, avoiding the use of simplified and often inaccurate measures of the shape of the distribution (such as the mean, skewness and standard deviation). Details of the data processing technique have been described in numerous previous studies [24,25,26,27,28,29] and are not repeated here. The analysis was used to group the data into a number of grain size groups, ranging in number from two (over-simplified) to 12 (over-complicated) inclusive, the spatial distributions of which were assessed visually. The choice of the number of groups to use in subsequent interpretations is a compromise between achieving a high R(s) statistic (i.e. statistically explaining more of the data; [30]) and accommodating empirical observations within a sedimentologically reasonable model of sediment distribution (e.g. [24]).

Distribution patterns of the various sedimentary bedforms and the grain size data were used together to delineate a series of sedimentary facies for the study area. In the geological literature, there are a number of ways that the term facies has been used [31], including regarding a rock's measured distinctive lithological and/ or biological characteristics, as well as the interpreted sedimentary processes and/or sedimentary environment. As used here, the term sedimentary facies means those distinctive physical characteristics of the sediments that occur under particular conditions of sedimentation, reflecting the sum of sedimentary and depositional processes and the environment (e.g. [31]) and which act to distinguish them from adjacent deposits. Geologists have long used fossil evidence of the associated biology as an integral part of defining sedimentary facies.

Meiofauna

All faunal samples were fixed in 5% formaldehyde in filtered (63 μm) sea water. After washing the meiofauna samples onto a 63 μm sieve, the fauna from all replicate samples were extracted by floatation in Ludox 40 [32]. The extraction was repeated three times before the extracted material was evaporated slowly in anhydrous glycerol and mounted on slides for identification and counting.

Nematodes were identified to genus or species level, using taxonomic keys based on a phylogenetic classification scheme [33,34,35,36]. Species and genera were then assigned to a number of biological traits, i.e. a number of morphological and life-history characteristics, thought or known to represent an important ecological function [37]. Here, six traits were included: mouth morphology, tail shape, pattern of the body wall (cuticle), body size, body shape and life-history strategy (Table 2). There were a total of 25 categories of these traits. A matrix of these biological traits was constructed by assigning to each nematode species/ genus its affinity to each trait category.

Species and trait diversity and dominance in each sample were expressed as Hill numbers N_1 and N_2 [50], respectively. These evenness measures describe different aspects of the community and differ only in their tendency to include or ignore the relatively rare species or traits. Bartlett's and Cochran's tests were used to test for homogeneity of variance before facies-specific differences in species and trait diversity, as well as the relative abundance of functional groups (Table 2) were explored using a one-way analysis of variance (ANOVA). Following the detection of significant differences ($P \leq 0.05$) between facies, the Tukey HSD multiple comparisons test was used.

Non-metric multi-dimensional scaling (MDS) ordination using the Bray-Curtis similarity measure was applied to relative abundance data to compare spatial patterns in the taxonomic (i.e. species) and functional (i.e. traits) composition of nematode communities amongst the putative sedimentary facies. One-way analysis of similarities (ANOSIM), the multivariate analogue of analysis of variance, was used to test for significant taxonomic and functional differences between sedimentary facies. Similarity percentages (SIMPER) were calculated to identify the species and traits which both characterised *a priori* sedimentary facies and distinguished between them. The dissimilarity between faunal distribution patterns recorded in the sedimentary facies was computed and an average calculated. We then calculated the contribution of each species and trait to the average similarity within each sedimentary facies and the average dissimilarity between facies. All statistical analyses were performed using the software packages Statgraphics Plus version 3.3 and Primer version 6.1.5 [51].

Disturbance versus mobility

A variety of publications have used the term disturbance to cover a wide range of anthropogenic and natural sediment transport and mixing processes including dredging, dredge material emplacement, trawling and natural sediment mobility [52,53,54]. In our study, we are dealing specifically and purely with the natural movement of sediment, and hence use sediment mobility for maximum clarity. Where we use the term disturbance,

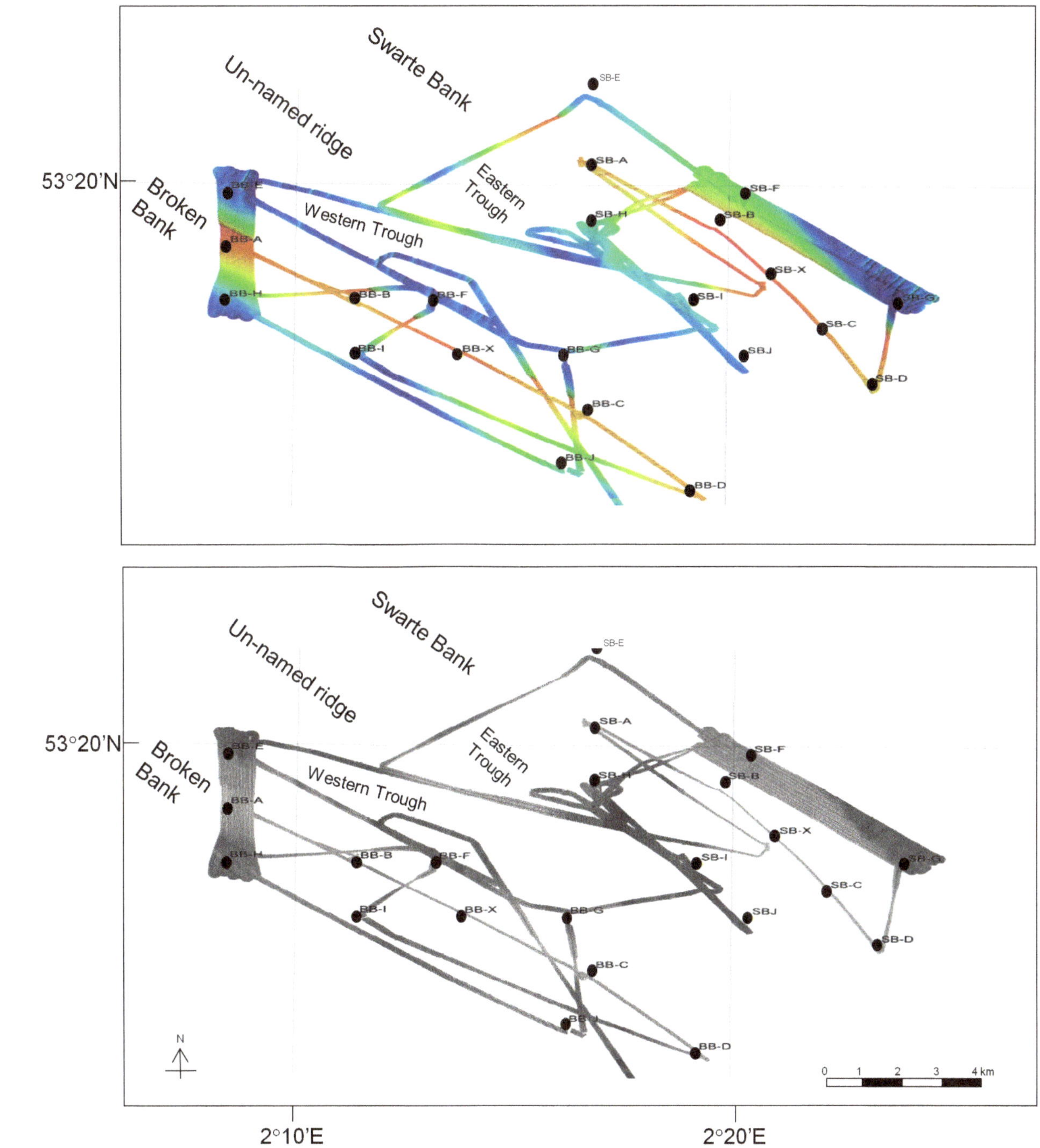

Figure 2. Acoustic survey coverage. The acoustic survey coverage, showing (upper) recorded depth (red = shallow, dark blue = deep) and (lower) acoustic backscatter, where darker areas indicate greater bed roughness. Sampling stations are also shown. BB = Broken Bank, SB = Swarte Bank.

it generally refers to the disturbance experienced by organisms during a period of seabed mobility.

Results

Sandbank form and superposed bedforms

The banks' gross morphologies include measured lee slopes of nearly 5° (Table 3). Superposed (flow-normal) bedforms include a wide variety of megaripples and sandwaves, of different planform, full-beddedness, size and symmetry (orientation). Some bedforms occur in extensive bedform fields and others in isolation (see [55] for explanation of terminology). Superposed in many areas on both sandbanks are smaller megaripples, which, where resolved, have a steep morphology, indicating that they are probably active. In contrast, some areas between Broken Bank and Swarte Bank and areas associated with their lower stoss flanks contain fields of longitudinal furrows, whose formative processes [56] indicate that these will only be active episodically.

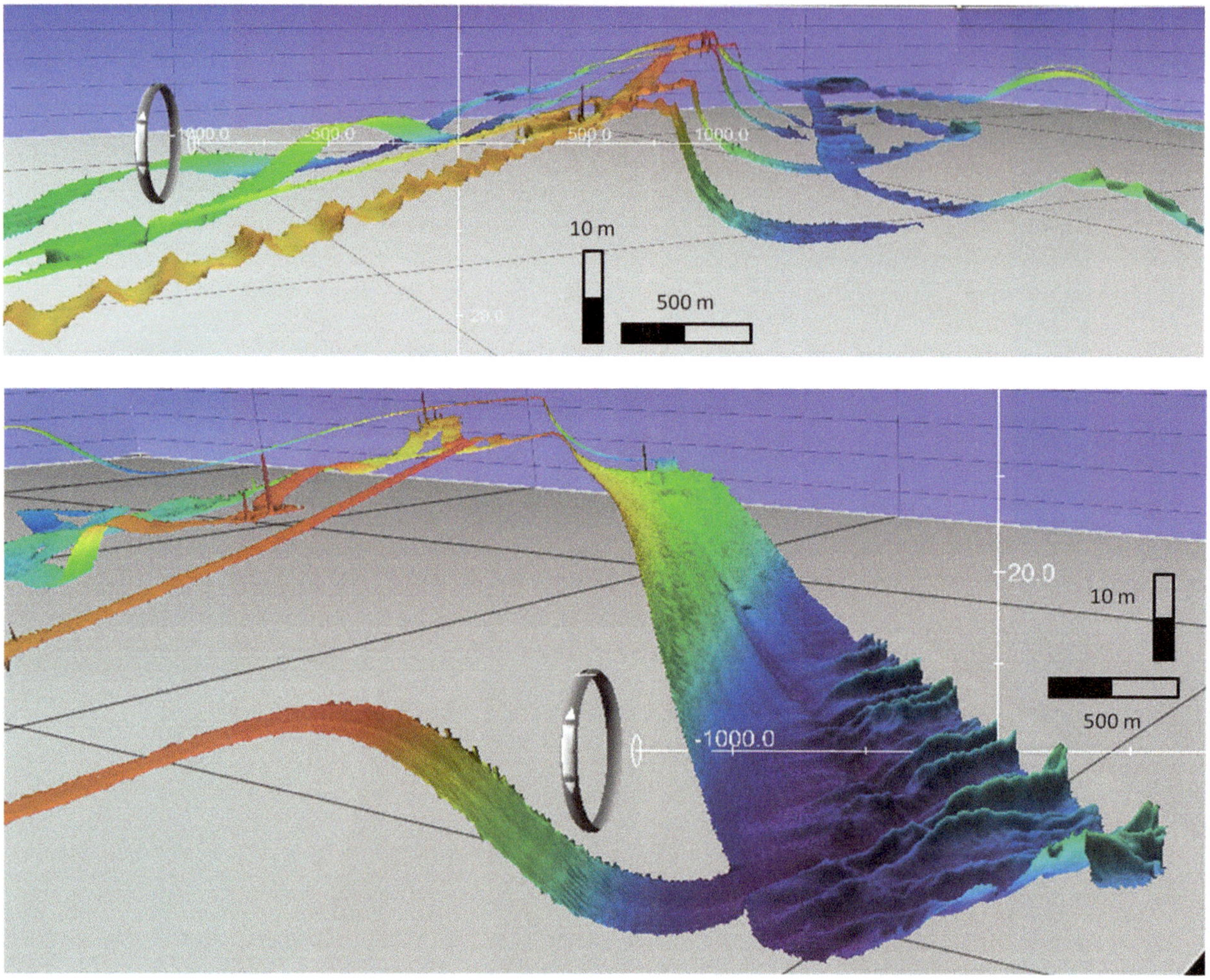

Figure 3. Bathymetry of the surveyed area. Oblique colour-contoured bathymetry of the surveyed area. Vertical exaggeration ×35. Black grid squares 5 km apart. Scale bars refer to the intersection point of the white axes. Upper - View towards NW of the asymmetrical cross-section of Broken Bank (sharp crestline in centre) and the low rounded crest of the Un-named ridge (right). Note large asymmetric sandwaves migrating up the SE stoss of Broken Bank (bottom left; details in Table 4). Lower - View towards the NNW of the megarippled lee slope (bedform group v; centre) of Swarte Bank, with field of large asymmetric sandwaves (bedform group i) in the trough migrating to the SE (right; details in Table 4).

Table 1. Positions of the 22 stations sampled in the southern North Sea.

Swarte Bank			**Broken Bank**		
Station	**Latitude**	**Longitude**	**Station**	**Latitude**	**Longitude**
SB-X	53.30 N	2.35 E	BB-X	53.27 N	2.23 E
SB-A	53.34 N	2.28 E	BB-A	53.31 N	2.14 E
SB-B	53.32 N	2.33 E	BB-B	53.29 N	2.19 E
SB-C	53.28 N	2.37 E	BB-C	53.25 N	2.28 E
SB-D	53.26 N	2.39 E	BB-D	53.22 N	2.32 E
SB-E	53.37 N	2.28 E	BB-E	53.33 N	2.14 E
SB-F	53.33 N	2.34 E	BB-F	53.29 N	2.22 E
SB-G	53.29 N	2.40 E	BB-G	53.27 N	2.27 E
SB-H	53.32 N	2.28 E	BB-H	53.29 N	2.14 E
SB-I	53.29 N	2.32 E	BB-I	53.27 N	2.19 E
SBJ	53.27 N	2.34 E	BB-J	53.23 N	2.27 E

Table 2. Morphological characteristics and life-history of nematode species and genera.

Functional group	Functional importance	Trait categories
Mouth morphology [38]	The mouth cavity shows great diversity in form and reflects the food ingested by the nematodes. It can thus be a good indicator of a species' feeding strategy. Feeding type groupings are fundamental to carbon and energy fluxes through ecosystems, and are linked to nutrient cycling.	Selective deposit feeder, non-selective deposit feeder, epistrate feeder, predator
Tail shape [39,40,41]	The tail shape is important in locomotion and reproduction. Long, filiform tails, for example, are considered a special adaptation to fine sand and muddy sediments where only an incomplete interstitial system exists. In these sediments, the tail enables animals to retract from dead-end interstitial passageways that are too narrow to allow the worm to turn around and escape.	Short/round, conical, clavate, elongated/ filiform
Pattern of the body wall [42,43]	The body wall (cuticle) is used as a support and leverage point for movement. It protects the body from drying out and other harsh environments. Nematodes with elaborate cuticular ornamentation tend to be associated with coarser, silt-free sediments, which may be correlated both with their mode of locomotion and with the need for mechanical protection in unstable substrata.	Smooth, annulated, striated, rows of dots/ punctuated, rows of structures, with desmen
Body size and shape [44,45,46]	Body size and shape affect physiological and ecological features of populations, including metabolic rates, tolerance to chemicals and anoxia, ability to move or migrate, vulnerability to predation etc. Slender nematodes are able to move swiftly through the sediment, but are vulnerable to predation. Predation pressure on stout species may be reduced but so is mobility.	Body size: <1 mm, 1 - 2 mm, 2 - 4 mm,> 4 mm. Body shape: Long/thin, slender, stout
Life-history strategy [47,48,49]	Many small species have short generation times of usually about one month or less with high reproduction rates. This largely r-selected life-history strategy is in contrast to the longer life-cycles and fewer offspring of more K-selected species. Some of these often larger species have an annual reproductive cycle.	Coloniser (largely r-selected), intermediate (2 categories), persister (largely K-selected)

Functional groups were chosen because they have been shown or inferred to be functionally important and are thus potentially ecologically meaningful (detailed in [37]). Terminology adapted from [34].

The bedforms (numbered i – viii) superposed on and between the sandbanks vary across the survey area in a relatively ordered fashion (Fig. 4; detail in Table 4). On the upper stoss flanks (i.e. the SW side), the banks themselves have few clear sedimentary bedforms and show only low relief (<0.5 m; bedform group iii), with, in places, trains of solitary or isolated large sandwaves (bedform group iv). The sandbank lee slopes exhibit very extensive fields of megaripples (bedform group v), with bedform asymmetries indicating bedload sediment transport to the SE. There are also smaller patches of large sandwaves, which are either symmetrical or indicative of net bedload transport to the SE (bedform groups i, vi and vii). The swales between the banks, and areas low on the stoss flanks have a relatively rough topography, which includes areas of longitudinal furrows (bedform group ii) and fields of undulating megaripples (bedform group viii) which occur superposed on the furrows in places. Here, the net bedload transport inferred from the megaripples is generally to the NW.

The key physical features are therefore that (Fig. 4; Table 4):

(a) For the positive relief features of Broken Bank and Swarte Bank, bedform transport is relatively active, with transport of sand to the NW across the stoss slope and along the crest, and to the SE along the lee slope, i.e. in an apparent clockwise circulation (agreeing with [22]).

(b) Between the two sandbanks, including the Western Trough, the Un-named ridge and the Eastern Trough, the bed contains a variety of bedforms and a complex distribution, closely associated with the major bathymetric relief.

(c) The flat seabed away from the sandbanks is relatively starved of sand, and here fast storm-associated shelf flows have caused seabed erosion in the form of longitudinal furrows. Where superposed by megaripples, the furrows have not recently been active.

(d) Overall, the troughs and lower stoss slopes of Broken Bank and Swarte Bank contain coarser sediments and have a rougher seabed, together with bedforms (longitudinal furrows) which indicate episodically fast flows. This is in contrast to the upper stoss where the sediment is finer and where the mostly smoother seabed is inferred to be more continuously mobile.

Sediment granulometry

The seabed sediments are mostly fine and medium sands, with a modal grain size of 180 – 250 μm, and with a minor gravel component in some samples. Based on their entire size distributions, the samples fall into four broad grain size groups (numbered a–d; Fig. 5A, B). In relative terms, the slightly gravelly sediments of groups a and b are poorly sorted and tend to occur off the main banks in areas with uneven topography and high acoustic roughness, including in areas close to the northern termini of longitudinal furrows. The groups' characteristics are relatively simple and, in combination with the bathymetric and side-scan datasets, relationships between them are readily interpretable (Table 5). The data are consistent with the finding of [22] that sand transport on the banks themselves will generally occur as bedload, but with suspension of sand likely at times of superposed waves; this may help explain the relatively poorly defined bedforms on the upper stoss of the banks. During severe storms, mobilisation of all the sampled grain sizes is possible.

Sedimentary facies

The survey area comprised a total of 11 sedimentary facies, eight of which (numbered 1 – 8) were included in our statistical analyses (Fig. 6). Facies are primarily based on the physical evidence for the frequency, magnitude and inferred direction of sediment movement, based mostly on bedforms, and the tendency

Table 3. Morphological characteristics of the positive relief features forming Broken Bank and Swarte Bank.

	Length	Max. width	Max. relief above surrounding seafloor	Water depth at sandbank crest	Max. stoss[a] slope (in surveyed area)		Max. lee slope (in surveyed area)	
	km	km	m	m	degrees	m/100m	degrees	m/100 m
Broken Bank	33	1.1	30	13–16	1.3	2.2	4.9	8.5
Swarte Bank	37	1.4	28	12–15	0.6	1.1	4.0	7.0

[a]Many mobile sandy sedimentary bedforms are asymmetrical in vertical cross-section. For tidal sandbanks, sand may be transported over the bedforms in different directions at different times, and may circulate around it in the long term. Nonetheless, there is often a long-term net direction of movement of the bedform, indicated by its asymmetry. Sand is generally transported (obliquely) up the longer shallower stoss slope, over the crest and down the shorter steeper lee slope, which faces the overall direction of movement of the sandbank.

for long-term erosion or accumulation, based on the evidence of exposed coarse substrates. Facies with less than three observations were omitted from our statistical analyses, including 'Active large full-bedded sandwaves' at stations BB-G and SB-G (total of two observations), 'Washed-out sandwaves' at the lower lee to trough of Broken Bank (single observation at station BB-E) and the 'Washed-out megarippled stoss' of Broken Bank (single observation at station BB-H).

The facies names highlight the major features which define each facies and which distinguish them from adjacent deposits (Table 6; Fig. 6). The sedimentary facies delineated for the study area differed with regard to 1) their location across the study area, 2) their observed grain size distribution and seabed topography, and 3) their inferred history, intensity and frequency of sediment mobility. The concept of mobility, as used here, can thus include factors such as duration, magnitude, period since previous mobilisation, time of year, depth, and physical nature (i.e. erosion, transport, accumulation, mixing and unmixing, covering and uncovering of lag surfaces etc.) as well as associated changes in oxidation state and other biogeochemical parameters. With a view towards timescales of around a month, in line with meiofauna turnover times, our eight facies fall into four groups with respect to their relative level of sediment mobility. Facies 1 is termed high, facies 2, 3 and 4 are high-moderate, facies 5 is moderate and facies 6, 7 and 8 are episodic (Table 6).

Distribution of nematode species and functional traits relative to sedimentary facies

At all stations, nematodes consistently comprised> 81% of the total meiofauna abundance. A total of 9171 nematodes was identified, with abundances per sample ranging from 47 in washed-out megaripples at the stoss off Broken Bank (station BB-D) to 434 in active megaripples at Swarte Bank (station SB-F). Nematode assemblages in the study area were taxonomically and functionally diverse. A total of 118 species was recorded which fell into 77 trait groups.

Results from the one-way ANOVA revealed significant differences between sedimentary facies in terms of taxonomic and trait diversity ($N_{1(species)}$: $F = 6.71$, $P < 0.01$; $N_{1(traits)}$: $F = 4.66$, $P < 0.01$) and dominance ($N_{2(species)}$: $F = 6.96$, $P < 0.01$; $N_{2(traits)}$: $F = 3.72$, $P = 0.02$). Taxonomic and trait diversity (N_1) were significantly lower in facies 1 and 8 ('Active megaripples at Swarte Bank' and 'Sandwaves over longitudinal furrows off the banks') than in most of the others. High numbers of some dominant species and traits in facies 1 resulted in a low N_2 value, whereas the low N_2 value observed in facies 8 was a consequence of the distribution of nematodes across a limited number of species and traits (Fig. 7).

Functionally, nematode assemblages were dominated by epistrate-feeding species that combined fast turnover times and an ornamented, relatively small, slender body (Table 7). A total of five chromadorid and cyatholaimid species fell into this category. Another multi-species trait group, differing from the dominant trait group only in the type of body ornamentation, as well as two single-species trait groups (comprising *Xyala striata* and *Metadesmolaimus pandus*, respectively) were also highly influential in determining taxonomic and functional distribution patterns across sedimentary facies.

Results from SIMPER and ANOSIM analyses (Tables 6 and 8) indicate that there were differences in nematode community structure between facies. Nematode communities assigned to facies 1 ('Active megaripples at Swarte Bank') exhibited relatively high average similarity in taxonomic and functional composition, probably reflecting the relative homogeneity of the sedimentary

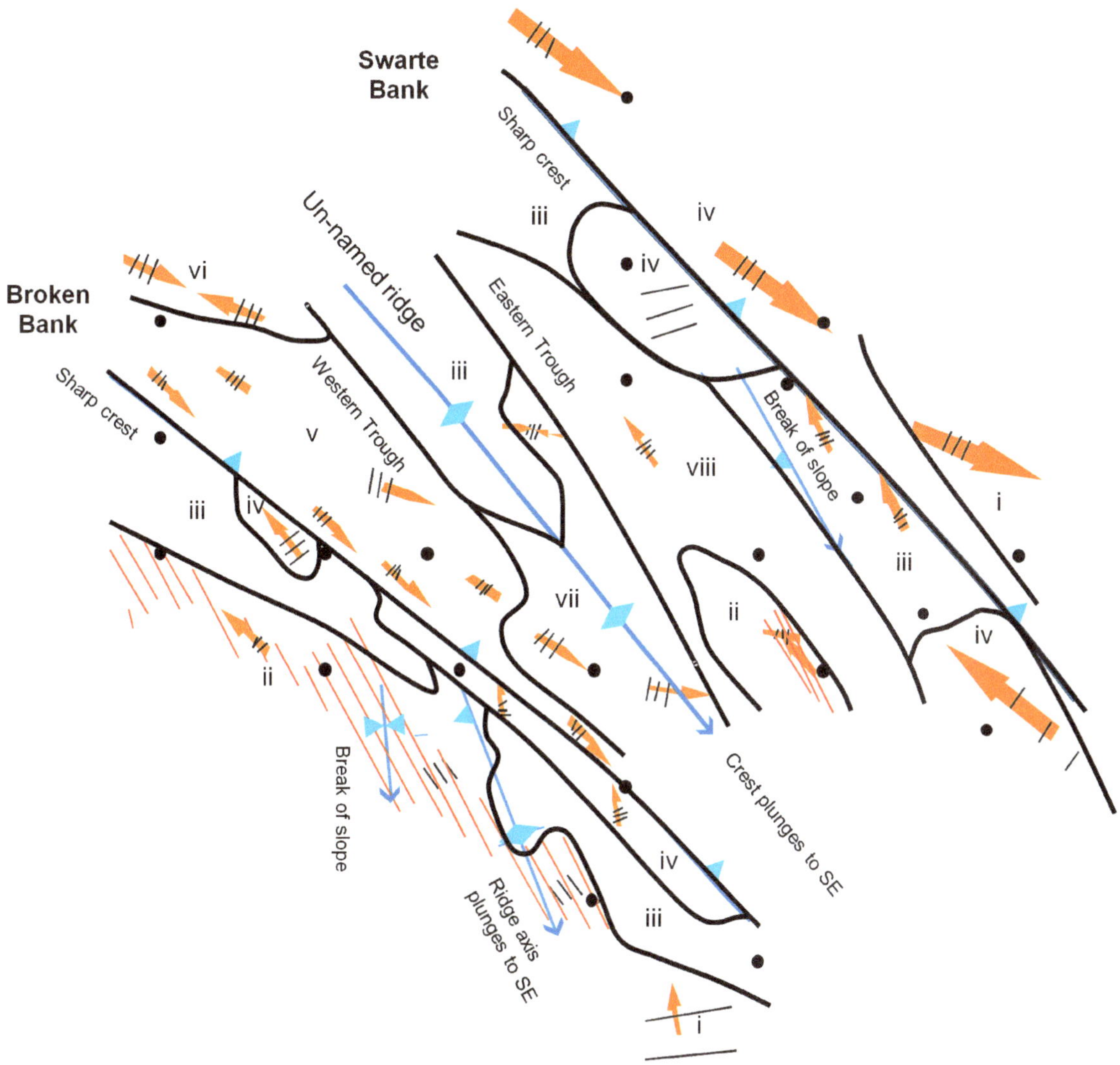

Figure 4. Bedforms in the study area. The 22 stations (black dots) fall into eight broad bedform groups (numbered i – viii; see Table 4), differing in the nature of contained grain size groups (see Table 5). Note - all blue arrows point downhill, indicating the orientation of a significant bed slope. Orange arrows indicate inferred bedform transport direction, with superposed black lines indicating orientation of sandwave crests and sandwave spacing: close lines = <100 m; medium = 100 – 300 m; far apart => 300 m. Narrow parallel orange lines indicate areas and orientation of longitudinal furrows.

environment. In contrast to other facies, the communities of facies 1 were taxonomically more similar than functionally, indicating that those species that were unique to individual samples were mostly single-species trait groups. The samples were highly dominated by contrasting trait groups, comprising *Metadesmolaimus pandus* and large predators, respectively. The trait groups differed not only with regard to feeding habit, but also in cuticle pattern (striated versus smooth) and body size (<1 mm versus> 2 mm). The relative abundance of non-selective deposit feeders was significantly higher ($F = 4.66$, $P < 0.01$) in this facies and that of epistrate feeders significantly lower ($F = 7.07$, $P < 0.01$) compared to other sedimentary environments (Fig. 8). This was the only sedimentary facies where larger predatory nematodes were present in notable numbers ($F = 12.14$, $P < 0.01$; bottom of Fig. 9).

A high proportion of nematodes with a heavily cuticularised body wall was a common and significant functional feature of facies 2 ('Active full-bedded sandwaves at the crest of Broken Bank'; top of Fig. 9) as well as other hydrodynamically active facies containing either megaripples (e.g. facies 3 and 4) or sandwaves (e.g. facies 5). In this group of facies, the proportion of epistrate feeders tended to be higher than the relative abundance of non-selective deposit feeders (Fig. 8; Table 6). The species- and trait-rich facies 5 ('Active large solitary sandwaves at the stoss/apron of Swarte Bank') was associated with the most taxonomically and functionally varied assemblages. The dominant trait groups differed little from those encountered in facies 8, but their relative abundance was notably different. In facies 5, the proportion of *Metadesmolaimus pandus* was lower, with associated higher relative abundance of heavily ornamented nematodes such as *Xyala striata* (top of Fig. 9). This facies comprises a range of sedimentary environments, which are correspondingly colonised by functionally diverse assemblages (Table 6). These assemblages encompassed deposit and epistrate feeders of various sizes (<1 – 2 mm), body ornamentations, ranging from moderately (i.e. striated or punctuated) to heavily (i.e. annulated or structured)

Table 4. Summary characteristics of main bedform groups.

Bedform group	Location	Principal bedforms	Orientation/Asymmetry (degrees)	Size			Description
				Length or spacing	Height or depth	Width or crest length	
Broken Bank							
i	South of Broken Bank	Very large sandwaves	080 – 090, lee to N	~ 600 m	<5 m	> 250 m	Two full-bedded, straight-crested very large sandwaves, with sandwaves in their troughs.
ii	At and beyond the western edge of Broken Bank	Longitudinal furrows overprinted by full-bedded sandwaves	Furrows – 325–330 Sandwaves – 040, lee to NW	Spacing 80 – 180 m	0.3 – 0.5 m	> 2 km long	Longitudinal furrows with superposed sandwaves.
iii	Upper stoss and crest						Bedforms poorly defined, low relief (<0.5 m).
iv	Central and southern crest	Sandwaves	060 – 090, lee to NW	80 – 100 m	2 – 3 m	> 500 m	Highly asymmetric, full-bedded straight crested sandwaves.
v	Lee slope	Sandwaves	Lee to SE	<40 m	<1.5 m on upper slope		Extensive and uniform field of full-bedded straight-wavy crested sandwaves.
Western Trough							
vi	Lee slope of Broken Bank	Large sandwaves	020 (convergence in the north)	100 – 300 m	> 5 m	Unclear	Complex large relief, including asymmetric straight-wavy crested sandwaves.
v	Western Trough	Small sandwaves	020 – 050	10 – 20 m	<0.3 m	Unclear	Symmetry unclear.
vii	Northern part of Western Trough	Sandwaves	Lee to SE	120 – 150 m	<0.5 m	> 200 m	Wavy-crested.
Un-named ridge							
vii	Trough and southern lee slope	Large sandwaves	350 – 020, lee to E	150 m	<2 m	> 250 m	Asymmetric straight-crested sandwaves.
vii	Lower E slope of un-named bank	Sandwaves	005 – 015 (convergence)	80 – 110 m	<1 m	> 300 m	Asymmetric straight-wavy crested sandwaves.
Eastern Trough							
viii	Broad swale west of Swarte Bank and central stoss slope of Broken Bank	Small sandwaves	050 – 070, some have lee to N	~ 20 m	<0.5 m	> 80> 150 m (some straight crested> 400 m)	Undulating to rough relief, patchy. Centre and S has topography parallel to longitudinal furrows. Discrete fields of straight-wavy crested sandwaves.
Swarte Bank							
iv	NW part of crest (and southern apron)	Large sandwaves	060 – 090, lee to NW	200 – 600 m	2 – 3 m (0.5 – 2 m in E)	> 500 m	Train of five solitary highly asymmetric sandwaves.

Table 4. Cont.

Bedform group	Location	Principal bedforms	Orientation/Asymmetry (degrees)	Size			Description
				Length or spacing	Height or depth	Width or crest length	
v	Lee slope	Sandwaves	Crests 030 – 050, lee to SE	30 – 40 m (smaller at boundary with i)	<1 m on upper slope, <0.4 m on lower slope	> 200 m	Very extensive and uniform field of full-bedded straight-crested sandwaves along the bank's lee slope.
i	South-eastern edge of lee slope	Large sandwaves	030 – 040, lee to SE	50 – 200 m	2 – 6 m	> 800 m	Asymmetric full-bedded, wavy-straight crested (some double-crested). Height increases to the E.
iii	Upper stoss of Swarte Bank				0.3 – 0.5 m		Bedforms poorly defined, low relief (<0.5 m).

cuticularised, and varied reproductive potential (coloniser to intermediate).

Metadesmolaimus pandus (23% of all nematodes) and the dominant multi-species trait group (24% of all nematodes) were abundant in facies 8, ('Sandwaves over longitudinal furrows') as was a limited range of other trait groups which combined a deposit- or epistrate-feeding habit with a striated, medium-sized (1 – 2 mm), slender body. Nematode assemblages in this facies exhibited a relatively high similarity with those recorded in facies 6 and 7 (Tables 6 and 8).

From the above, the key faunal features are that:

(a) Taxonomically and functionally distinct nematode communities were present in facies 1 and 5, characterised by low and high faunal diversity, respectively (Table 6).

(b) There was a considerable overlap in the species and trait composition of nematode assemblages collected in facies 2, 3 and 4. Heavily ornamented nematodes were common here and numbers of epistrate feeders generally exceeded that of other feeding types.

(c) Some sedimentary environments located in the Western Trough and Eastern Trough and off the banks (i.e. facies 6, 7 and 8) differed little with regard to their taxonomic and functional structure. Small and medium-sized deposit- or epistrate-feeding species dominated nematode assemblages.

These key features tend to indicate a range of faunal similarities between sets of facies, which led to readily interpretable patterns in the MDS ordination (Fig. 10). Distinct nematode communities were present in active megaripples at Swarte Bank (facies 1, termed high mobility), where spatially homogeneous bedforms are mobilised daily by tides. These communities form a tight cluster at the bottom right hand side of the ordination plots. Nematode assemblages from environments where sediment mobility was more intense yet infrequent or absent on timescales of a month (facies 6 – 8, termed episodic mobility) form a broad cluster across the centre and upper part of the ordination plots. This is consistent with the variable suite of sedimentary environments. Data from facies 8 are particularly widely spaced on all plots. Sampling stations where nematodes experience high to moderate intensities and frequencies of sediment mobility (facies 2 – 5, termed high-moderate and moderate mobility) form a medium-sized cluster, within which most facies overlap except for facies 3, representing relatively active megaripples, which forms a separate cluster.

Discussion

Interstitial habitats such as those occupied by meiofauna are broadly distributed across the planet, and the complexity of the associated physical sedimentary environments has been present throughout the evolutionary history of meiofauna (e.g. [57]). Hence, it is clear that processes operating in the geological past have shaped what we sample and see today. A key relevant timescale is the last few thousand years, when the southern North Sea was inundated by the rising Post-Glacial sea [58] and became subject to marine processes [59]. At shorter timescales, the periods between major storms and the various tidal cycles became relevant to the benthic biology. Put simply, the benthic fauna we sample depends in part on the day-to-day modern processes but also on the episodic storm-associated processes, and the history of both.

A variety of sediment transport processes in our survey area, operating at different magnitudes and frequencies, created a wide range of sedimentary environments for meiofaunal nematodes (Table 6). The presence of active sedimentary bedforms, such as

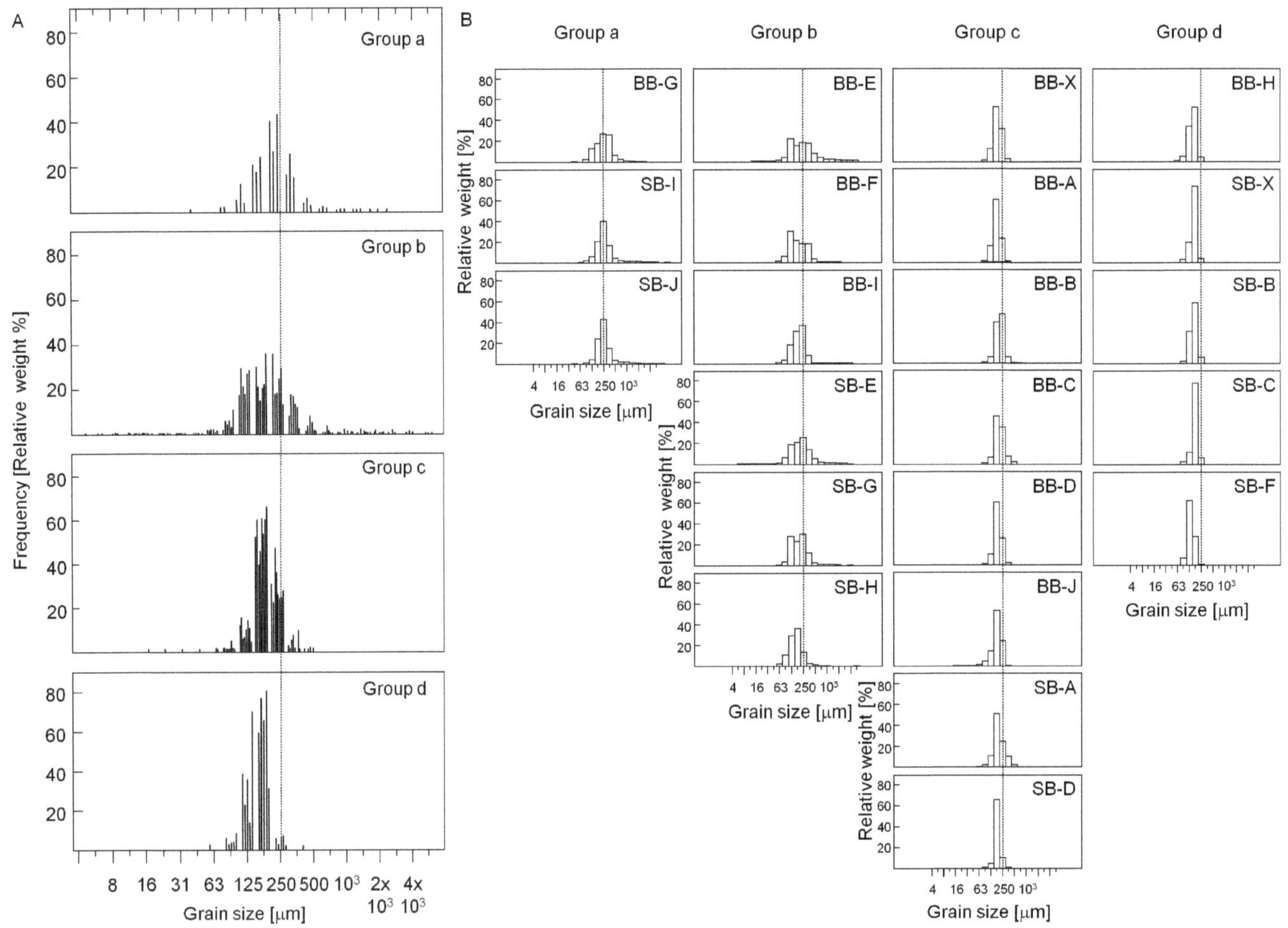

Figure 5. Grain size distribution of all 22 sediment samples. A. All samples of each group, plotted together to illustrate the key differences between the groups. B. All samples within each group. The vertical dotted line at 250 μm helps illustrate the between-group shift in the peak of the grain size distribution. Note that grain size groups a and b have a minor gravel component.

Table 5. Summary characteristics of grain size groups.

Grain size group	Sediment	Main characteristics and water depth	Spatial distribution (i.e. samples forming the group)	Comparison between groups	Interpreted sedimentological processes
a	Slightly gravelly unimodal medium sand	Modal size ~ 250 μm, with up to 25% fine sand. 34 - 39 m	Samples in the Western Trough (BB-G) and the southern part of the Eastern Trough (SB-I, SB-J).	Coarsest sediments, with less sand than other groups. Sand is medium-grained.	Sand fraction probably mobile on regular basis, over a gravel lag and/or medium sand is being eroded from the seabed.
b	Slightly gravelly fine and medium sand	Poorly sorted fine-medium sand with a minor gravel component. 34 - 40 m	In the troughs between the main banks and low on the main banks' stoss slopes (BB-E, BB-F, BB-I, SB-E, SB-G, SB-H).	Equivalent to group a plus fine sand. All sizes present across all groups.	Mixture of other groups; likely to represent a thin layer of mobile fine sands over surface of group a material.
c	Unimodal fine to medium sand	Modal size ~ 180 μm, with up to 30% medium sand. 17 - 33 m	Along the crest of Broken Bank (BB-A, BB-B, BB-X, BB-C, BB-D) and Swarte Bank (SB-A), and on the stoss slope of SE Broken Bank (BB-J) and Swarte Bank (SB-D).	Relatively well sorted.	Regular movement of sand with, in calm seas, perhaps only a few tides per fortnight with no movement. Long-term accumulation.
d	Unimodal fine sand	15 - 34 m	On the crest and NE flank of Swarte Bank (SB-F, SB-B, SB-X, SB-C) and one sample on the northern stoss of Broken Bank (BB-H).	Relatively well sorted.	Daily movement of sand within megaripple fields. Long-term accumulation.

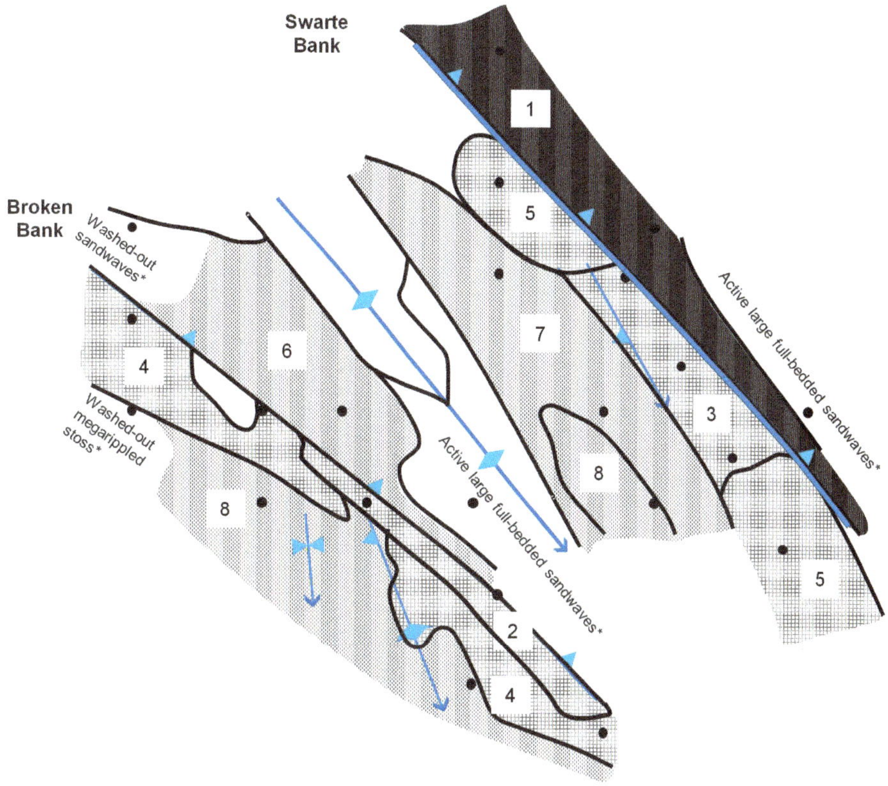

Figure 6. Distribution of sedimentary facies groups in the study area. Facies shadings are grouped according to their relative level of sediment mobility (Table 6). The 22 stations (black dots) fall across a range of sedimentary facies, numbered 1 – 8 (details in the text and in Table 6). Facies with less than three observations (*) were excluded from statistical analyses. For explanation of annotations see Fig. 4.

sandwaves (i.e. crest spacing of 10s – 100s m) and megaripples (i.e. crest spacing of 1s – 10s m), for example, have important consequences for the benthic ecology of sandbanks. Sandwave and megaripple crests tend to have different sediments, sometimes subtly, than their troughs [60,61,62,63], providing a variety of different niches for colonising meiofauna. Previously published review articles on the effects of physical seabed disturbance on benthic ecosystems [64,65] suggest that species-specific properties such as population growth rates and tolerance to environmental change affect a species' susceptibility to disturbance as well as its recovery from it. However, only recently have scientists begun to consider the key role of the sedimentary regime in determining the pre-disturbance faunal distribution patterns and as a factor in the post-disturbance phase of 'recovery' [66,67].

Benefits of the sedimentary facies approach

Delineating and using sedimentary facies can help to better understand the way in which the local complexity of the seabed might determine the number and type of available niches for meiofaunal nematode communities. In our context, sedimentary facies inherently include the key features of the bedforms and sediments themselves (e.g. [31]), thus incorporating many aspects of sediment mobility and the tendency for erosion or accumulation, which vary across environments. Therefore, sedimentary facies tend to group various combinations of physical and time-variable factors, improving the likelihood of forming meaningful and readily interpretable relationships with the infaunal data.

With an emphasis on the sandy facies appropriate to our study area, the general relationships between the groups of sedimentary facies representing high, high-moderate and moderate, and episodic sediment mobility can be illustrated by using three key parameters: the frequency of sediment mobility, its magnitude, and the continuum between erosion, translation (i.e. sediment throughput) and sediment accumulation (Fig. 11). Our facies groups fall into different areas of this diagram. This diagram helps visualise some of the concepts of our data and its interpretation, and we hope helps others in designing measurement and monitoring programmes, and interpreting collected data.

High mobility

A common assumption of ecological theories addressing the causes of faunal variability in space and time is that species or functional groups with different morphology, physiology or behaviour dominate under different environmental regimes [68,69,70,71] and this is consistent with our findings. In facies 1 ('Active megaripples at Swarte Bank'), for example, daily movement of sediments and bedforms will most likely have led to the resuspension, burial and/or mechanical damage of microalgal cells [72]. It is thus logical to find that epistrate feeders, whilst abundant elsewhere on the sandbanks, were less numerous here. We hypothesise that organic material accumulates preferentially in megaripple troughs at slack tide, some of which is subsequently buried by migrating ripple crests during succeeding tidal flows, a recognised sedimentary process [60] which has been postulated as important for meiofauna on intertidal sandflats (e.g. [73]). This subsequently attracts deposit-feeding nematodes from surrounding areas such as the small-sized *Metadesmolaimus pandus* who, together with a contrasting trait group comprising large predatory nematodes (including *Viscosia*), dominated colonist

Table 6. Physical features and key meiofaunal characteristics of sedimentary facies.

	Active megaripples at Swarte Bank (1)	Active full-bedded sandwaves at the crest of Broken Bank (2)	Megarippled stoss of Swarte Bank (3)	Washed-out megaripples at the stoss of Broken Bank (4)	Active large solitary sandwaves at the stoss/apron of Swarte Bank (5)	Active large solitary sandwaves in the Western Trough (6)	Washed-out undulating bed with sand-wave fields in the Eastern Trough (7)	Sandwaves over longitudinal furrows off the banks (8)
Station(s)	SB-E, SB-F	BB-X, BB-C	SB-X, SB-B, SB-C	BB-A, BB-B, BB-D	SB-A, SB-D	BB-F	SB-H, SB-I	BB-I, BB-J, SB-J
No. of obs.	5	8	12	12	8	4	5	6
GS groups	b, d	c	d	c	c	b	a, b	a, b, c
Bedf. groups	v	iv	iii	iii	iv	v	viii	ii
Rel. level of sediment mobility	High	High-moderate	High-moderate	High-moderate	Moderate	Episodic	Episodic	Episodic
Description (partly interpretative)	Sediments and megaripples mobile on a daily basis, except possibly at neap tides.	Large sandwaves migrate frequently.	Sediments mobile on a daily basis, but megaripples less frequently so.	Sediments mobile on a daily basis, but megaripples less frequently so.	Large sandwaves migrate frequently.	Large sandwaves migrate, superposed by small sandwaves and megaripples. Gravel grains widespread.	Sandy grains regularly mobile across and formed into sandwaves and megaripples at times, perhaps during storms, when widespread gravel grains also mobile.	Furrows formed and reactivated during severe storms, then become inactive for long periods, during which sandwaves become superposed and finer sediment accumulates in furrows.
Sedimentary environment (partly interpretative)	Little spatial variation in surface sediment, and probably also homogenous to a depth of a few decimetres.	Minor gradients in the nature of benthic habitats across the sandwaves. Sediments probably homogenous to a depth of a few decimetres.	Little spatial variation in surface sediment, and homogenous to a depth of a few decimetres.	Little spatial variation in surface sediment, and probably also homogenous to a depth of a few decimetres.	Moderate gradients in the nature of benthic habitats across the sand-waves, although sediments relatively similar; different ha-bitats in the troughs. Variation with depth, especially across the troughs.	Moderate gradients in the nature of benthic habitats across sandwaves; different habitats in the troughs. Patchy degree of variation with depth, especi-ally across the troughs between sandwaves.	Moderate gradients in the nature of benthic habitats. Potential rapid accumulation after storms.	Range of benthic habitats in and between furrows and sandwaves, affected by the time since creation or re-activation and rate of infilling. Significant spatial and depth-related differences in sediment type.
Nematode assemblages	Highly similar nematode communities dominated by contrasting trait groups, comprising small deposit feeders and large predators.	High proportions of nematodes with a heavily cuticularised body wall; a ratio of epistrate feeders vs. non- selective deposit feeders> 1.	Considerable faunal similarity to facies 2; subtle differences in the relative abundance of dominant species and traits.		Varied assemblages encompassing deposit and epistrate feeders of various sizes, body ornamentations and reproductive potential.	Considerable overlap in the species and trait composition with facies 8. Trait similarity between facies 6 and 7 exceeds similarity within the individual facies.		High abundance of the dominant multi-species trait group and a limited range of other trait groups combining a deposit or epistrate-feeding habit with a medium-sized, slender body.

Facies number in brackets. GS = grain size.

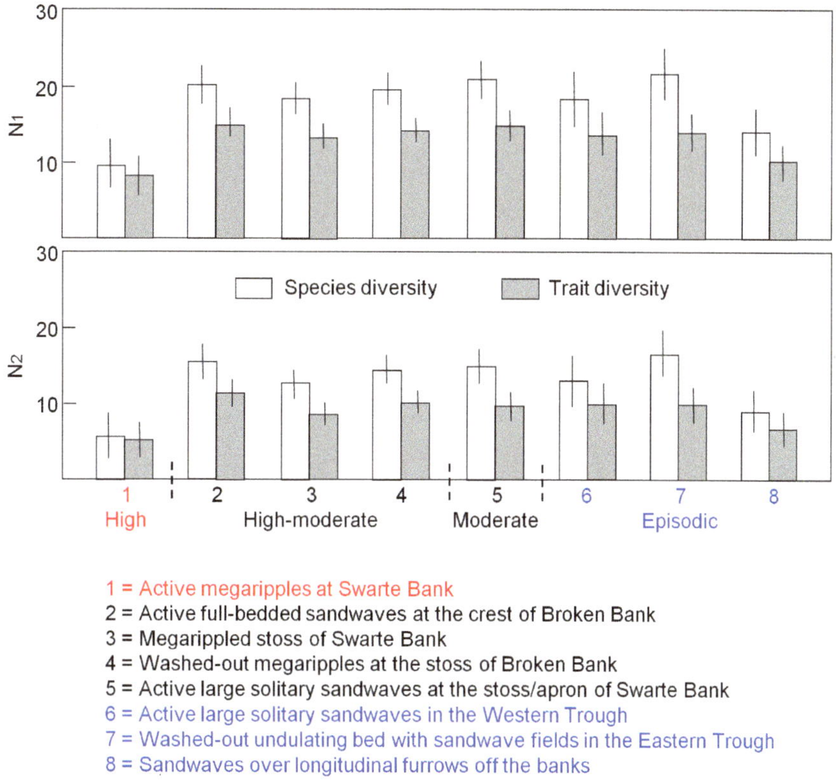

Figure 7. Species and trait diversity of nematode assemblages. Mean (± 95% pooled confidence interval) species and trait diversity of nematode assemblages recorded in eight sedimentary facies coded by relative level of sediment mobility. Diversity is expressed as Hill numbers N_1 and N_2.

communities. These disturbance-resistant nematodes [74] were most efficient in exploiting available food sources.

High-moderate mobility

Larger sedimentary bedforms comprising small sandwaves and washed-out megaripples are located along the crests of both banks (facies 2 – 4) and indicate relatively high magnitudes and frequencies of bed sediment transport. However, in contrast to facies 1, the bedforms may persist for longer (i.e. weeks to years as opposed to hours to days) and be more spatially varied. Accordingly, we recorded a range of coexisting epistrate feeders and sub-dominant deposit feeders. As has been shown for other dynamic sedimentary environments [75,76], much of the available space was filled by competitors most resistant to physical damage, in our case species with a heavily cuticularised body wall.

Moderate mobility

Solitary sandwaves on the lower stoss and apron of Swarte Bank (facies 5) were characterised by moderate spatial gradients in the benthic habitat across and along the bedform (Tables 4 and 6). Compared to smaller, more frequently active bedforms (i.e. megaripples and smaller sandwaves, see above), the physical

Table 7. Taxonomic affiliation and morphological/functional characteristics of nematode trait groups which discriminate between sedimentary facies.

	Dominant multi-species trait group	Sub-dominant multi-species trait group	Single-species trait group	Single-species trait group
No. of species	5	3	1 (*Xyala striata*)	1 (*Metadesmolaimus pandus*)
Taxonomic affiliation	Chromadoridae, Cyatholaimidae	Microlaimidae	Xyalidae	Xyalidae
Buccal morphology	Epistrate feeder	Epistrate feeder	Non-selective deposit feeder	Non-selective deposit feeder
Tail shape	Conical	Conical	Conical	Clavate
Cuticle pattern	Rows of dots/punctuated	Striated	Structures	Striated
Body size	1-2 mm	1-2 mm	1 - 2 mm	<1 mm
Body shape	Slender	Slender	Slender	Slender
Life-history strategy	Coloniser	Coloniser	Intermediate	Coloniser

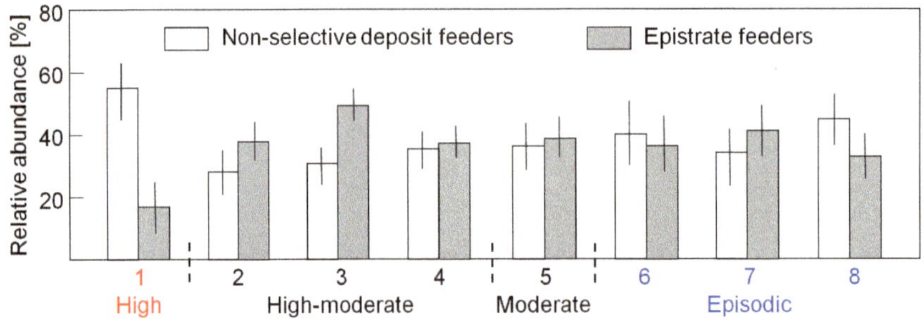

Figure 8. Relative abundance of non-selective deposit feeders and epistrate feeders. Mean (± 95% pooled confidence interval) relative abundance of non-selective deposit feeders and epistrate feeders recorded in eight sedimentary facies coded by relative level of sediment mobility. Facies numbers as in Fig. 7.

conditions here appear less critical in controlling the success or failure of species and functional groups. Biological stress such as intense competition for food and space is likely to be gradually mediated through biological interactions (see review by [77], [78]). Because different functional types use different resources, use the same resources in different ways or use the environment in a way that allows them to reduce or avoid competition with other functional types [79], the relatively high environmental heterogeneity increased the number of microhabitats for nematodes. The functional types recorded in our samples showed responses indicative of the trade-offs between resistance to disturbance, environmental tolerance and competitive abilities. Deposit and epistrate feeders of various sizes, body ornamentations and reproductive potential coexisted in taxonomically and functionally diverse colonist communities.

Episodic mobility

The relatively low elevation seabed between Broken Bank and Swarte Bank comprised sandwave fields of varying topography and extent (e.g. facies 6 and 7). In common with facies 8, at the steep lee slope of Swarte Bank, these areas of seabed probably undergo repeated and complex ecological transitions, associated with the irregular timing and nature of seabed mobilisation (e.g. as a result of storms in the study area). The initiation and episodic regeneration of bedforms (including longitudinal furrows) by strong storm-associated flows is related to very high rates of sediment transport, involving largely passive transport of meiofauna, and at these times, much of the study area would constitute a solely physically driven benthic system (sensu [80]). However, between such events, the area occupied by facies 6 – 8 would experience longer periods of being a more biologically influenced system, geologically characterised by the inter-storm formation of superposed sandwaves and infilling of the furrows, and where nematode communities respond to both the varied sedimentary environment but also to various biological interactions.

Such an episodically influenced change in seabed character probably also extends to those facies located North and West of the furrowed areas, away from the SE tips of the banks. It is likely that, especially during periods of decreasing sediment transport associated with the decaying phase of storms, a phase of relatively high net sediment accumulation occurs here over a few hours, and perhaps up to a day or so, depending upon the particular storm's characteristics. This phase of accumulation would be followed by an extended period where colonisation and assemblage development will take place, associated with the ambient, strongly tidal,

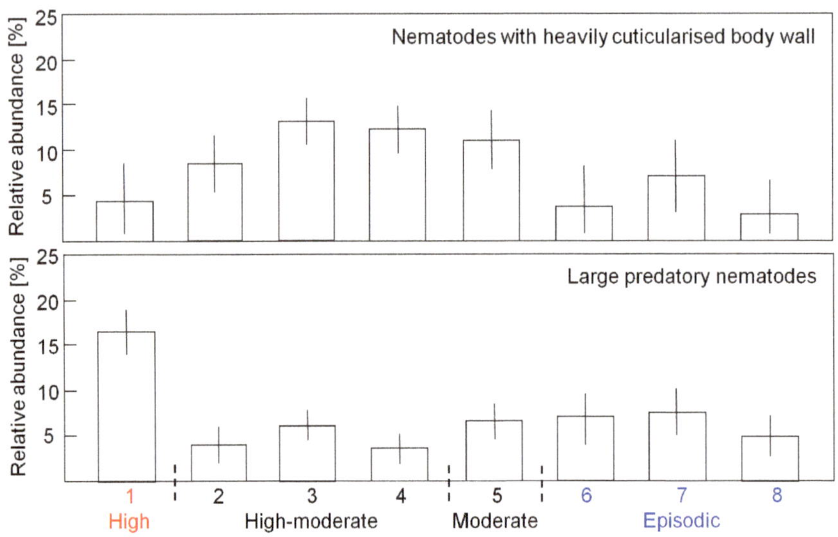

Figure 9. Relative abundance of nematodes exhibiting specific functional traits. Mean (± 95% pooled confidence interval) relative abundance of nematodes exhibiting a heavily cuticularised body wall (top) and mean (± 95% pooled confidence interval) relative abundance of large predators (bottom) recorded in eight sedimentary facies coded by relative level of sediment mobility. Facies numbers as in Fig. 7.

Table 8. Faunal differences (R-values) between sedimentary facies.

	Active megaripples at Swarte Bank (1)	Active full-bedded sandwaves at the crest of Broken Bank (2)	Megarippled stoss of Swarte Bank (3)	Washed-out mega-ripples at the stoss of Broken Bank (4)	Active large solitary sandwaves at the stoss/apron of Swarte Bank (5)	Active large solitary sandwaves in the Western Trough (6)	Washed-out undulating bed with sandwave fields in the Eastern Trough (7)
Active full-bedded sandwaves at the crest of Broken Bank (2)	**1/0.99**						
Megarippled stoss of Swarte Bank (3)	**1/0.99**	**0.83/0.80**					
Washed-out megaripples at the stoss of Broken Bank (4)	**0.99/0.94**	**1**	**0.54/0.43**				
Active large solitary sandwaves at the stoss/apron of Swarte Bank (5)	**0.99/0.90**	**0.51/0.35**	**0.40/0.47**	0.22/0.03			
Active large solitary sandwaves in the Western Trough (6)	**0.89/0.73**	**0.99/0.90**	**1/0.93**	**0.97/0.75**	**0.97/0.69**		
Washed-out undulating bed with sandwave fields in the Eastern Trough (7)	**0.97/0.88**	**0.78/0.77**	**0.93/0.85**	**0.65/0.51**	**0.70/0.46**	0.68/0.28	
Sandwaves over longitudinal furrows off the banks (8)	**0.50/0.34**	**0.74/0.66**	**0.92/0.87**	**0.78/0.65**	**0.73/0.57**	0.14/−0.18	0.44/0.34

R-values derived from pair-wise analysis of similarities (ANOSIM) tests for differences between sedimentary facies based on relative abundance data of species/functional traits. Bold indicates significant difference at $P < 0.01$. Facies number in brackets.

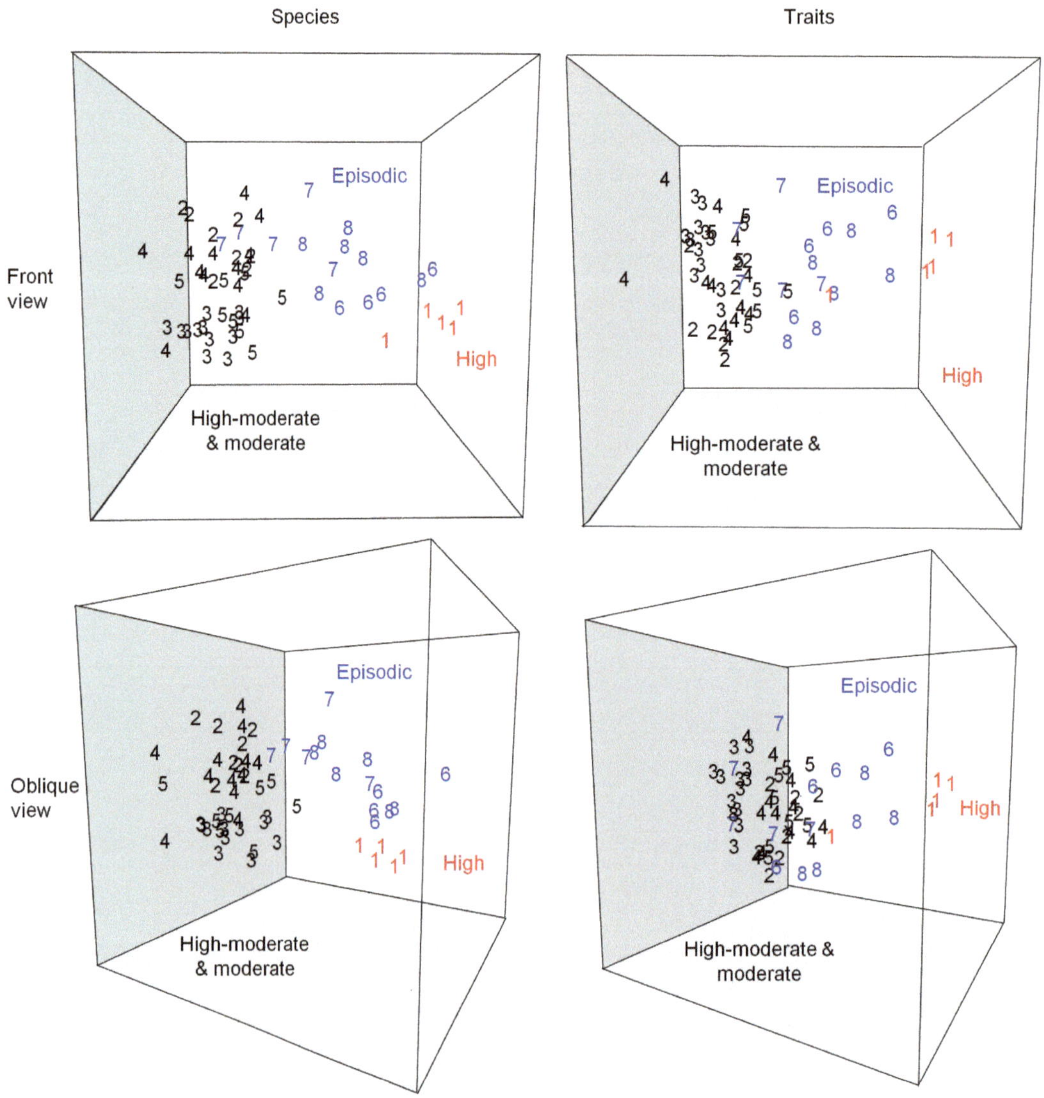

Figure 10. Non-parametric multi-dimensional scaling (MDS) ordination of nematode assemblages. 3D-view of non-parametric MDS ordination of nematode assemblages coded by facies (numbered as in Table 6 and Fig. 7) and relative level of sediment mobility (see text for details). The plots are derived from a similarity matrix of relative abundance of species (left) and traits (right).

oceanographic regime. Resident nematode communities appear to reflect such sedimentary complexity. In our particular case, the dominance of small- to medium-sized deposit- and epistrate-feeders of high reproductive potential most likely represents communities at a relatively early post-mobility stage.

Implications for future studies

Taxonomically and functionally disparate communities inhabiting the sedimentary facies defined in our survey area can be considered a complex product of the various sedimentary contrasts inherent within different sedimentary facies and the general sedimentary nature of the area being considered. Regarding variation within facies, the sedimentary environments experienced by meiofauna will be distinctly different across the variety of sub-environments across a sandwave-dominated facies, where troughs, stoss slopes, crests and lee slopes are dynamically and texturally different. There will also be great differences in those facies where longitudinal furrows occur with differing levels of infill. Further, sandwaves are much larger physical features than megaripples, and therefore their rates of horizontal migration and morphological change tend to be much slower. In our study area, sandwaves also tend to be much more patchy in their distribution than megaripples. Therefore, we might expect that 1) a similar number of sediment grabs in sandwave facies is more likely to include a greater variety of seabed types (and meiofauna) than in megarippled facies and that 2) the more frequent and regular mobility of megaripples tends to favour the presence of fewer meiofauna species and functional groups.

The long-term (i.e. annual to decadal) tendencies for seabed erosion, sediment throughput or sediment accumulation are important because they exert key controls upon habitat type and sediment mobility (Fig. 11). For example, in an environment where sandwaves occur, a long-term tendency towards erosion might increase habitat complexity by producing a series of solitary sandwaves [55] with layers of less mobile sediment between individual bedforms. In contrast, a long-term tendency towards accumulation might be associated with full-bedded sandwaves and lower habitat complexity.

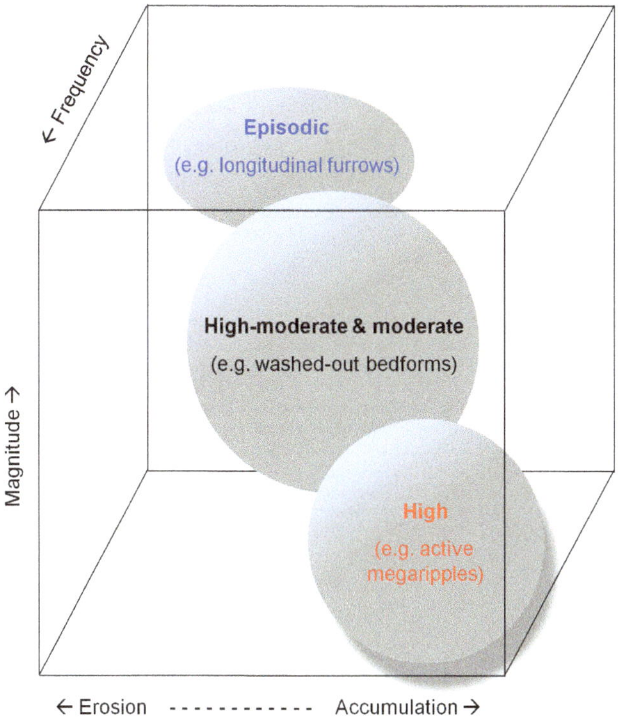

Figure 11. Conceptual diagram of three key controls on sedimentation in the study area. General relative locations of groups of sedimentary facies representing high, high-moderate and moderate, and episodic sediment mobility, plotted against the frequency of sediment mobility, its magnitude, and the continuum between erosion, translation (i.e. sediment throughput) and sediment accumulation.

Two main points emerge from our data. Firstly, a broad-scale characterisation of sedimentary environments is unable to account for the microhabitat heterogeneity to which the associated meiofauna will tend to respond. A key aim of future studies should be to incorporate fine-scale spatial information on sedimentary environments into the analysis of broader-scale animal-sediment relationships. For deeper insights, such studies should build on a good conceptual understanding of sedimentary processes operating at a range of spatial and temporal scales, including the movement of sediment across the full range of marine sedimentary bedforms [56], each of which will have important ecological associations (e.g. [81] and our study).

Secondly, small-scale faunal distribution patterns do not always reflect only those marine processes occurring concurrently with sample collection. Whilst this is not a new finding, our results emphasise how important past events can be in affecting animal distributions. Caution must be exercised when attempts are made to explain spatial distributions of animals from physical or biological conditions measured at a single point in time. For example, analyses conducted of pre-storm samples from our facies 8 ('Sandwaves over longitudinal furrows') would inevitably generate different faunal results, and hence differing expectations regarding state, than analyses of samples collected shortly after major storms. Further, for those areas where longitudinal furrows occur, there would be relatively high variation in sediment types after storm-associated currents which activate the furrows, but the degree of variation would change through time. Within the furrows themselves, sediments would accumulate relatively rapidly and become finer through time, favouring mud-dwelling meiofauna whereas immediately outside the furrows, the sediments would change in completely different fashion in response to the re-development of ripples and megaripples, with associated different meiofaunal responses.

Conclusions

The variety of sedimentary regimes at and around Broken Bank and Swarte Bank clearly shapes the meiofaunal nematode communities, and the interaction of local hydrodynamics and sediment transport is clearly a key driver of the processes involved. Their influence occurs in a variety of different ways, and, importantly, in ways that would not have been predictable from using 1) existing generalised theories of seabed disturbance based on frequency or intensity alone [64,65,82], as opposed to including duration, period since previous event, etc., 2) sediment summary descriptors alone, as opposed to sedimentary facies, nor using 3) faunal analyses based on taxonomic relationships alone, as opposed to including functional traits.

Our focus on diverse assemblages of organisms with high turnover times (i.e. meiofaunal nematodes), inhabiting highly dynamic sedimentary environments (i.e. subtidal sandbanks and inter-bank swales), has revealed interesting animal-sediment relationships worthy of investigation by future studies of these and a broader range of marine environments. Results from such studies should facilitate a new generation of hypotheses concerning the relative importance of different environmental drivers in structuring benthic communities, and movement towards their rigorous testing. This will provide better indications of the stability of the sediments that benthic communities inhabit and allow better insights into how such communities may respond to natural sediment mobility and to anthropogenic disturbance. Taking account of the past evolution and modern processes of the sedimentary environment is paramount. Contemporary faunal distribution patterns reflect the integrated response of species and functional groups to a wide suite of sedimentary processes, so that effective descriptors of those sedimentary processes will help reveal meaningful animal-sediment relationships.

As noted in the introduction, the EU Habitats Directive requires that "management plans be created in order to detect human-induced change"; such words are easy to write but are likely to be very difficult to achieve. In the case of the UK's candidate Special Areas of Conservation (SAC), which include subtidal mobile sandbanks, there is much to be learnt before such plans can realistically be formed. Our scoping study has indicated some physically and ecologically meaningful measures of seabed physical texture that will help improve seabed characterisation and associations with benthic assemblages.

Acknowledgments

We thank the scientists, officers and crew of RV Cefas Endeavour for practical assistance in the collection of samples and the conduct of surveys. We are grateful to colleagues Claire Mason for carrying out grain size analyses, Karema Warr for processing meiofauna samples and Annie Brown, Jon Rees and Christian Wilson for processing side-scan and multi-beam data.

Author Contributions

Conceived and designed the experiments: MS. Performed the experiments: MS. Analyzed the data: MS PL. Contributed reagents/materials/analysis tools: MS PL. Wrote the paper: MS PL.

References

1. European Commission (1992) Council Directive 92/43/EEC of 21 May 1992 on the conservation of natural habitats and of wild fauna and flora. Available: http://ec.europa.eu/environment/nature/legislation/habitatsdirective/index_en.htm.
2. Pedersen SA, Fock H, Krause J, Pusch C, Sell AL, et al. (2009) Natura 2000 sites and fisheries in German offshore waters. ICES Journal of Marine Science 66: 155–169.
3. Joint Nature Conservation Committee (2012) Offshore Special Area of Conservation: North Norfolk sandbanks and Saturn Reef. Conservation objectives and advice on operations. Available: http://jncc.defra.gov.uk/pdf/NNSandbanksandSaturnReef_ConservationObjectives_AdviceonOperations_6.0.pdf)
4. Snelgrove PVR, Butman CA (1994) Animal-sediment relationships revisited: cause versus effect. Oceanography and Marine Biology: an Annual Review 32: 111–177.
5. Harris PT (2012) Surrogacy. In: Harris PT, Baker EK, editors. Seafloor geomorphology as benthic habitat. GeoHAB atlas of seafloor geomorphic features and benthic habitats. UK: Elsevier. pp.93–108.
6. Post AL, Wassenberg TJ, Passlow V (2006) Physical surrogates for macrofaunal distributions and abundance in a tropical gulf. Marine and Freshwater Research 57: 469–483.
7. Belderson RH, Kenyon NH, Stride AH, Stubb AR (1972) Sonographs of the sea floor - a picture atlas. Amsterdam: Elsevier. 185 p.
8. Baptist MJ, van Dalfsen J, Weber A, Passchier S, van Heteren S (2006) The distribution of macrozoobenthos in the southern North Sea in relation to meso-scale bedforms. Estuarine, Coastal and Shelf Science 68: 538–546.
9. Clements AJ, Strong JA, Flanagan C, Service M (2009) Objective stratification and sampling-effort allocation of ground-truthing in benthic-mapping surveys. ICES Journal of Marine Science 67: 628–637.
10. Galparsoro I, Borja Á, Legorburu I, Hernández C, Chust G, et al. (2010) Morphological characteristics of the Basque continental shelf (Bay of Biscay, northern Spain); their implications for Integrated Coastal Zone Management. Geomorphology 118: 314–329.
11. McArthur M, Brooke B, Przeslawski R, Ryan DA, Lucieer V, et al. (2010) On the use of abiotic surrogates to describe marine benthic biodiversity. Estuarine, Coastal and Shelf Science 88: 21–32.
12. Giere O (2009) Meiobenthology. The microscopic motile fauna of aquatic sediments. Second edition. Berlin: Springer Verlag. 527 p.
13. Palmer MA, Gust G (1985) Dispersal of meiofauna in a turbulent tidal creek. Journal of Marine Research 43: 179–210.
14. Palmer MA (1988) Dispersal of marine meiofauna: a review and conceptual model explaining passive transport and active emergence with implications for recruitment. Marine Ecology Progress Series 48: 81–91.
15. Boeckner MJ, Sharma J, Proctor HC (2009) Revisiting the meiofauna paradox: dispersal and colonisation of nematodes and other meiofaunal organisms in low- and high-energy environments. Hydrobiologia 624: 91–106.
16. Ellis JR, Maxwell TAD, Schratzberger M, Rogers SI (2011) The benthos and fish of offshore sandbank habitats. Journal of the Marine Biological Association of the United Kingdom 91: 1319–1335.
17. Dyer KR, Huntley DA (1999) The origin, classification and modelling of sand banks and ridges. Continental Shelf Research 19: 1285–1330.
18. Caston VND (1972) Linear sand banks in the southern North Sea. Sedimentology 18: 63–78.
19. Pravotorov IA (1983) Application of the cartographic method in the study of submarine sand ridges for navigational purposes. Geomorphologiya 2: 39–43.
20. Johnson MA, Caston GF (1984) Assessment of Ower Bank stability. Institute of Oceanographic Sciences Internal Report. 51pp.
21. Balson PS, Harrison DJ (1988) Marine Aggregate Survey Phase 1: Southern North Sea. British Geological Society Marine Report No. 86/38. 20 pp. and maps.
22. Collins MB, Shimwell SJ, Gao S, Powell H, Hewitson C, et al. (1995) Water and sediment movement in the vicinity of linear sandbanks: the Norfolk Banks, southern North Sea. Marine Geology 123: 125–142.
23. Huntley DA, Amos CL, Williams JJ, Humphery JD (1991) Estimating bedload transport on continental shelves by observations of ripple migration - an assessment. In: Soulsby RL, Bettess R, editors. Proceedings of EUROMECH 262 - Sand transport in rivers, estuaries and the sea, Balkema, Rotterdam, Netherlands, pp.17–22.
24. Woolfe KJ, Michibayashi K (1995) 'Basic' entropy grouping of laser-derived grain size data: an example from the Great Barrier Reef. Computers and Geosciences 21: 447–462.
25. Forrest J, Clark NR (1989) Characterizing grain size distribution: evaluation of a new approach using a multivariate extension of entropy analysis. Sedimentology 36: 711–722.
26. Woolfe KJ, Larcombe P, Stewart LK (2000) Shelf sediments adjacent to the Herbert River delta, Great Barrier Reef, Australia. Australian Journal of Earth Sciences 47: 301–308.
27. Orpin AR, Kostylev VE (2006) Towards a statistically valid method of textural sea floor characterization of benthic habitats. Marine Geology 225: 209–222.
28. Mikkelsen OA, Curran KJ, Hill PS, Milligan TG (2007) Entropy analysis of in-situ particle size spectra. Estuarine, Coastal and Shelf Science 72: 615–625.
29. Okada T, Larcombe P, Mason C (2009) Estimation of the spatial distribution of dredged material disposed of at sea by using particle-size distributions and metal concentrations. Marine Pollution Bulletin 58: 1164–1177.
30. Semple RK, Youngman CE, Zeller RE (1972) Economic regionalisation and information theory: an example. Discussion paper 28, Department of Geography, Ohio State University, Columbus, USA. 64 p.
31. Reading HG, Levell BK (1996) Controls on the sedimentary rock record, in: Reading HG, editor. Sedimentary environments: processes, facies and stratigraphy (Third edition). Cambridge: Cambridge University Press. pp.5–35.
32. Somerfield PJ, Warwick RM (1996) Meiofauna in marine pollution monitoring programmes. A laboratory manual. Lowestoft: Ministry of Agriculture, Fisheries and Food, Directorate of Fisheries Research. 71 p.
33. Platt HM, Warwick RM (1983) Free-living marine nematodes. Part I. British enoploids. Cambridge: Cambridge University Press. 307 p.
34. Platt HM, Warwick RM (1988) Free-living marine nematodes. Part II. British chromadorids. Cambridge: Cambridge University Press. 502 p.
35. Warwick RM, Platt HM, Somerfield PJ (1998) Free-living marine nematodes. Part III. British monhysterids. Shrewsbury: Field Studies Council. 296 p.
36. Deprez T (2005) The NeMys database. World Wide Web electronic publication, Available: www.nemys.ugent.be, version (2/2008).
37. Schratzberger M, Warr K, Rogers SI (2007) Functional diversity of nematode communities in the southwestern North Sea. Marine Environmental Research 63: 368–389.
38. Wieser W (1953) Die Beziehung zwischen Mundhöhlengestalt, Ernährungsweise und Vorkommen bei freilebenden marinen Nematoden. Arkiv för Zoologi 4: 439–484.
39. Riemann F (1974) On hemisessile nematodes with flagilliform tails living in marine soft bottoms and on microtubes found in deep sea sediments. Mikrofauna des Meeresbodens 40: 1–15.
40. Thistle D, Sherman KM (1985) The nematode fauna of a deep-sea site exposed to strong near-bottom currents. Deep-Sea Research 32: 1077–1088.
41. Thistle D, Lambshead PJD, Sherman KM (1995) Nematode tail-shape groups respond to environmental differences in the deep sea. Vie et Milieu 45: 107–115.
42. Ward AR (1975) Studies on the sublittoral free-living nematodes of Liverpool Bay. II. Influence of sediment composition on the distribution of marine nematodes. Marine Biology 30: 217–225.
43. Wright KA (1987) The nematode's cuticle: its surface and the epidermis: function, homology, analogy: a current consensus. Journal of Parasitology 73: 1077–1083.
44. Jensen P (1987) Differences in microhabitat, abundance, biomass and body size between oxybiotic and thiobiotic free-living marine nematodes. Oecologia 71: 564–567.
45. Soetaert K, Muthumbi A, Heip C (2002) Size and shape of ocean margin nematodes: morphological diversity and depth-related patterns. Marine Ecology Progress Series 242: 179–193.
46. Vanaverbeke J, Soetaert K, Vincx M (2004) Changes in morphometric characteristics of nematode communities during a spring phytoplankton bloom deposition. Marine Ecology Progress Series 273: 139–146.
47. Bongers T (1990) The maturity index: an ecological measure of environmental disturbance based on nematode species composition. Oecologia 83: 14–19.
48. Bongers T, Alkemade R, Yeates GW (1991) Interpretation of disturbance-induced maturity decrease in marine nematode assemblages by means of the Maturity Index. Marine Ecology Progress Series 76: 135–142.
49. Bongers T, de Goede RGM, Korthals GW, Yeates GW (1995) Proposed changes of c-p classification for nematodes. Russian Journal of Nematology 3: 61–62.
50. Hill MO (1973) Diversity and evenness: a unifying notation and its consequences. Ecology 54: 427–432
51. Clarke KR, Gorley RN (2006) PRIMER v6. User manual. Plymouth: Primer-E Ltd. 190 p.
52. Jennings S, Pinnegar JK, Polunin NVC, Warr KJ (2001) Impacts of trawling disturbance on the trophic structure of benthic invertebrate communities. Marine Ecology Progress Series 213: 127–142.
53. Wilber DH, Ray GL, Clarke DG, Diaz RJ (2008) Responses of benthic infauna to large-scale sediment disturbance in Corpus Christi Bay, Texas. Journal of Experimental Marine Biology and Ecology 365: 13–22.
54. Foden J, Rogers SI, Jones AP (2009) Recovery rates of UK seabed habitats after cessation of aggregate extraction. Marine Ecology Progress Series, 390: 15–26

55. Ashley GM (1990) Classification of large-scale bedforms: a new look at an old problem. Journal of Sedimentary Petrology 60: 160–172.
56. Belderson RH, Johnson MA, Kenyon NH (1982) Bedforms, in: Stride AH, editor. Offshore tidal sands - processes and deposits. London: Chapman and Hall. pp.27–57.
57. Rundell RJ, Leander BS (2010) Masters of miniaturisation: convergent evolution among interstitial eukaryotes. Bioassays 32: 430–437.
58. Shennan I, Lambeck K, Flather F, Horton B, McArthur J (2000) Modelling western North Sea palaeogeographies and tidal changes during the Holocene. In: Shennan I, Andrews J, editors. Holocene Land-Ocean Interaction and Environmental Change around the North Sea (Geological Society Special Publications 166). London: Geological Society. pp.299–319.
59. Ward I, Larcombe P (2008) Determining the preservation rating of submerged archaeology in the post-glacial southern North Sea: a first-order geomorphological approach. Environmental Archaeology 13: 59–83.
60. Reineck H-E, Singh IB (1980) Depositional sedimentary environments. Berlin: Springer-Verlag. 549 p.
61. Visser MJ (1980) Neap-spring cycles reflected in Holocene subtidal large-scale bedform deposits: a preliminary note. Geology 8: 543–546.
62. Langhorne DN (1982) A study of the dynamics of a marine sandwave. Sedimentology 29: 571–594
63. Nio SD, Yang C (1991) Diagnostic attributes of clastic tidal deposits: A review. In: Smith DG, Zaitlin BA, Reinson GE, Rahmani RA, editors. Clastic tidal sedimentology, Canadian Society of Petroleum Geologists Memoir 16: 3–28.
64. Hall SJ (1994) Physical disturbance and marine benthic communities: life in unconsolidated sediments. Oceanography and Marine Biology: an Annual Review 32: 179–239
65. Jennings S, Kaiser MJ (1998) The effects of fishing on marine ecosystems. Advances in Marine Biology 34: 201–352
66. Harris P, Hughes M (2012) Predicted benthic disturbance regimes on the Australian continental shelf: A modelling approach. Marine Ecology Progress Series 449: 13–25
67. Harris PT (2014) Shelf and deep-sea sedimentary environments and physical benthic disturbance regimes: A review and synthesis. Marine Geology 353: 169–184
68. Petraitis PS, Latham RE, Niesenbaum RA (1989) The maintenance of species diversity by disturbance. Quarterly Review of Biology 64: 393–418.
69. Tilman D (1990) Constraints and trade-offs: toward a predictive theory of competition and succession. Oikos 58: 3–15.
70. Steneck RS, Dethier MN (1994) A functional group approach to the structure of algal-dominated communities. Oikos 69: 476–498.
71. Wellborn GA, Skelly DK, Werner EE (1996) Mechanisms creating community structure across a freshwater habitat gradient. Annual Review of Ecology and Systematics 27: 337–363.
72. Delgado M, de Jonge VN, Peletier H (1991) Effect of sand movement on the growth of benthic diatoms. Journal of Experimental Marine Biology and Ecology 145: 221– 231.
73. Hogue EW, Miller CB (1981) Effects of sediment microtopography on small-scale spatial distributions of meiobenthic nematodes. Journal of Experimental Marine Biology and Ecology 53: 181–191.
74. Vanaverbeke J, Vincx M (2008) Short-term changes in nematode communities from an abandoned intense sand extraction site on the Kwintebank (Belgian Continental Shelf) two years post-cessation. Marine Environmental Research 66: 240–248.
75. Connell JH (1978) Diversity in tropical rain forests and coral reefs. Science 199: 1302–1310.
76. Raes M, Vanreusel A (2006) Microhabitat type determines the composition of nematode communities associated with sediment clogged cold-water coral framework in the Porcupine Seabight (NE Atlantic). Deep-Sea Research I 53: 1880–1894.
77. Branch GM (1984) Competition between marine organisms: ecological and evolutionary implications. Oceanography and Marine Biology: an Annual Review 22: 429–593.
78. Wilson WH (1990) Competition and predation in marine soft-sediment communities. Annual Review in Ecology and Systematic 21: 221–241.
79. Hooper DU, Chapin FS, Ewel JJ, Hector A, Inchausti P, et al. (2005) Effects of biodiversity on ecosystem functioning: a consensus of current knowledge. Ecological Monographs 75: 3–35.
80. Sanders HL (1968) Marine benthic diversity: a comparative study. American Naturalist 102: 243–282.
81. Wilson JB (1982) Shelly faunas associated with temperate offshore tidal deposits. In: Stride AH, editor. Offshore tidal sands - processes and deposits. London: Chapman and Hall. pp.126–171.
82. Huston MA (1996) Biological diversity. Cambridge: Cambridge University Press. 681 p.

Burrows of the Semi-Terrestrial Crab *Ucides cordatus* Enhance CO_2 Release in a North Brazilian Mangrove Forest

Nathalie Pülmanns[1]*, Karen Diele[2], Ulf Mehlig[3], Inga Nordhaus[1]

1 Department of Ecology, Leibniz Center for Tropical Marine Ecology (ZMT), Bremen, Germany, **2** School of Life, Sport and Social Sciences, Edinburgh Napier University, Edinburgh, United Kingdom, **3** Instituto de Estudos Costeiros, Universidade Federal do Pará - Campus de Bragança, Bragança, PA, Brazil

Abstract

Ucides cordatus is an abundant mangrove crab in Brazil constructing burrows of up to 2 m depth. Sediment around burrows may oxidize during low tides. This increase in sediment-air contact area may enhance carbon degradation processes. We hypothesized that 1) the sediment CO_2 efflux rate is greater with burrows than without and 2) the reduction potential in radial profiles in the sediment surrounding the burrows decreases gradually, until approximating non-bioturbated conditions. Sampling was conducted during the North Brazilian wet season at neap tides. CO_2 efflux rates of inhabited burrows and plain sediment were measured with a CO_2/H_2O gas analyzer connected to a respiration chamber. Sediment redox potential, pH and temperature were measured in the sediment surrounding the burrows at horizontal distances of 2, 5, 8 and 15 cm at four sediment depths (1, 10, 30 and 50 cm) and rH values were calculated. Sediment cores (50 cm length) were taken to measure the same parameters for plain sediment. CO_2 efflux rates of plain sediment and individual crab burrows with entrance diameters of 7 cm were 0.7–1.3 $\mu mol\ m^{-2}\ s^{-1}$ and 0.2–0.4 $\mu mol\ burrows^{-1}\ s^{-1}$, respectively. CO_2 released from a *Rhizophora mangle* dominated forest with an average of 1.7 *U. cordatus* $burrows^{-1}\ m^{-2}$ yielded 1.0–1.7 $\mu mol\ m^{-2}\ s^{-1}$, depending on the month and burrow entrance diameter. Laboratory experiments revealed that 20–60% of the CO_2 released by burrows originated from crab respiration. Temporal changes in the reduction potential in the sediment surrounding the burrows did not influence the CO_2 release from burrows. More oxidized conditions of plain sediment over time may explain the increase in CO_2 release until the end of the wet season. CO_2 released by *U. cordatus* and their burrows may be a significant pathway of CO_2 export from mangrove sediments and should be considered in mangrove carbon budget estimates.

Editor: Navnith K.P. Kumaran, Agharkar Research Institute, India

Funding: N.P. and I.N. were funded by the Leibniz Center for Tropical Marine Ecology (project 6201, http://www.zmt-bremen.de). K.D. received funding from the MASTS pooling initiative (The Marine Alliance for Science and Technology for Scotland, http://www.masts.ac.uk/) and its support is gratefully acknowledged. MASTS is funded by the Scottish Funding Council (grant reference HR09011) and contributing institutions. The European Commission funded the contributions by U.M. within this study via the international cooperation through the "Coastal Research Network on Environmental Changes" (CREC, http://www.forst.tu-dresden.de/CREC/) as part of its 7th 825 Framework Programme. The CREC project is classified as a Marie Curie Action (FP7-PEOPLE-2009-IRSES). The funders had no role in study design, data collection and analysis, decision to publish, or preparation of the manuscript.

Competing Interests: The authors have declared that no competing interests exist.

* Email: nathalie.puelmanns@gmail.com

Introduction

In mangrove ecosystems, semi-terrestrial burrowing crabs are important bioturbators affecting the biogeochemistry of the sediment and carbon cycling [1–5]. Their bioturbation activity includes constructing and maintaining of burrows leading to the reworking of sediment particles and organic material. Organic material, such as litter, is also reworked by their feeding, ingestion and defecation and can become buried in the sediment leading to carbon storage [1,6–8]. Crabs accumulate carbon as body biomass by feeding on mangrove litter which otherwise would have been washed away by the tide [9–14]. However, crabs may also account for carbon depletion in the sediment. Their burrows present an extension of the sediment surface area [1,15], thus, at low tide sediment surrounding the burrows can become oxidized by atmospheric oxygen. Atmospheric oxygen enters only a few millimeters into the mangrove sediment [8,16,17] where oxic reduction can take place. During burrow maintenance, crabs may mix more oxidized sediment inside their burrows with surrounding less oxidized sediment. The resulting increase in reduction potential affects the sediment surrounding the burrows up to a distance of several centimeters [18,19]. This may alter the type of microbial carbon oxidation from mainly anaerobic (e.g. sulfate reduction or methanogenesis) to partially aerobic decomposition processes (e.g. aerobic respiration or iron reduction) [1,8,20,21], which can result in higher CO_2 efflux rates [22] and consequently to carbon loss from the sediment [6]. Thus, bioturbation by crabs and associated changes in the microbial decomposition process possibly influence CO_2 release from the sediment.

To better evaluate the role of burrowing crabs in carbon cycling, it is important to quantify the amount of carbon stored and released by the crabs and their burrows. Bouillon et al. [23]

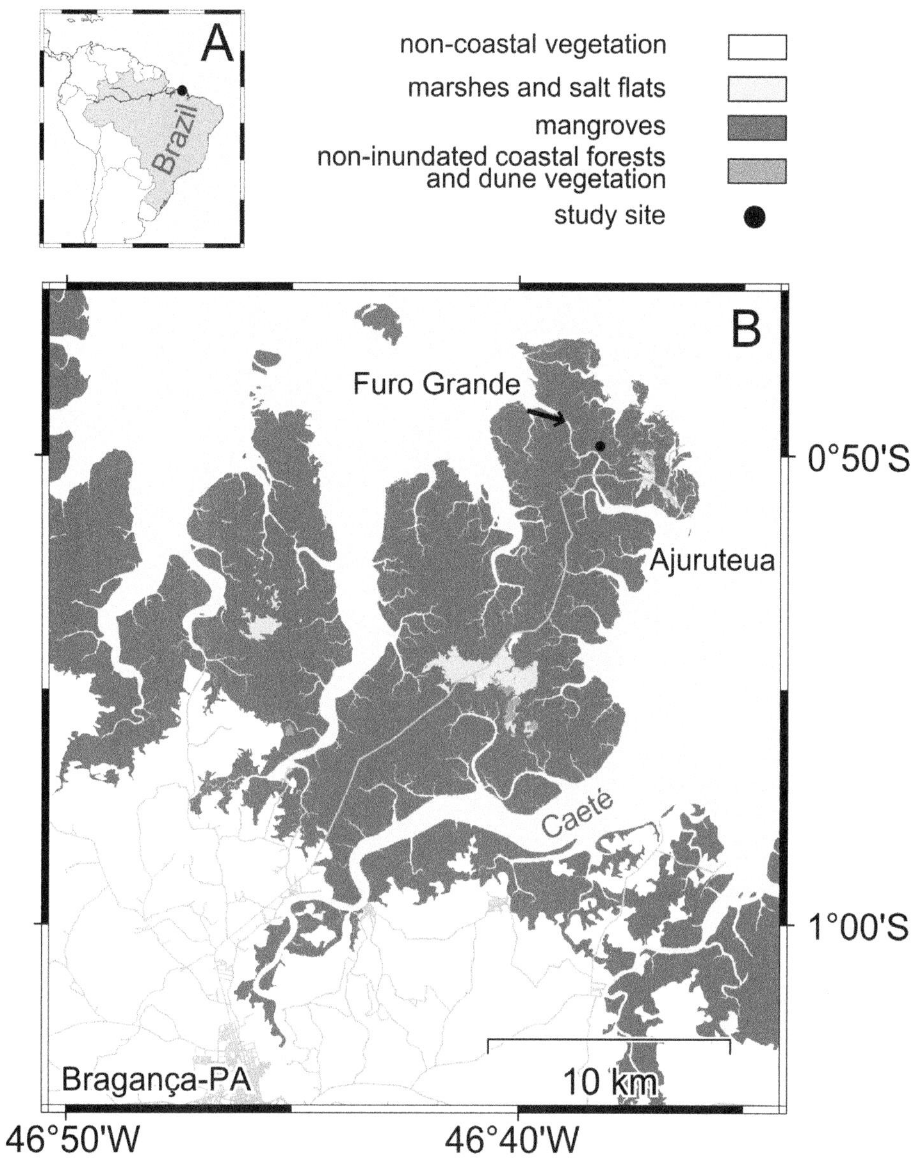

Figure 1. Study area on the Ajuruteua peninsula, Northern Brazil. The study was performed in Brazil (A) near the city of Bragança (46°38'W 0°50'S, B). The study site is marked by a black dot and is located at the tidal channel Furo Grande (B).

assume that the CO_2 released by burrows, animal respiration and roots is currently underestimated and may be an uncertain factor within the global carbon budget estimates for mangrove forests. Studies investigating the CO_2 release by crab burrows or the reduction potential in the sediment surrounding the burrows have only been conducted for the relatively small fiddler crabs [18,19,24]. Inhabited fiddler crab burrows increased the CO_2 release from mangrove sediment approximately 2 to 5 times compared to plain sediment [19,24]. 58% of the released CO_2 originated from the sediment surrounding the burrows and the remainder from the crab inside the burrow [24]. Studies characterizing the reduction potential in sediment surrounding fiddler crab burrows in salt marsh and mangrove sediments found a 1.5–4 cm thick oxidized zone, where iron reduction replaced the usually predominant sulfate reduction [18,19]. This zone also presented the highest CO_2 production rates. The rates decreased with increasing distance from the burrow, indicating that the change in reduction potential can influence carbon decomposition processes [18]. The availability and quality of carbon as well as the temperature can further influence carbon decomposition processes and thereby CO_2 efflux from the sediments [8,21].

Araújo Jr. et al. [25] investigated the bioturbation activity of the large and abundant crab *Ucides cordatus* (Ucididae), which, together with much smaller fiddler crabs, influence with their burrows the microtopography of the sediment in Brazilian mangrove forests. Araújo Jr. et al. [25] concluded from measurements of the reduction potential in sediment cores from bioturbated and non-bioturbated sites that *U. cordatus* burrows oxidized the sediment. However, the study did not provide information on the supposed gradient and the horizontal range of the reduction potential in the sediment surrounding the burrows, likewise, no CO_2 efflux rates were measured.

In the present study the effects of bioturbation by *U. cordatus* on carbon dynamics and sediment reduction potentials were evaluated by (i) quantifying the CO_2 efflux rate of individual burrows compared to plain sediment and (ii) measuring the reduction potential in radial profiles in the sediment around burrows at different depths in a *Rhizophora mangle* forest, respectively. We hypothesized that (i) sediment with *U. cordatus* burrows has higher CO_2 efflux rates than sediment without burrows and (ii) reduction potentials in radial profiles of the sediment surrounding the burrows down to 50 cm depth decrease gradually from the burrow wall, approximating conditions of non-bioturbated sediment.

Materials and Methods

The artisanally harvested study species *Ucides cordatus* is listed as endangered in the Brazilian legislation due to the risk of overfishing. All research activities were allowed by a research permit granted by SISBIO (Sistema de Autorização e Informação em Biodiversidade, authorization number 30007-1). Field work was performed in the extractive reserve "Reserva Extrativista Marinha de Caeté-Taperaçu".

Study area

The study area is located on the Ajuruteua peninsula, the western bank of the Caeté river estuary in North Brazil (Figure 1). Field work was performed in the high intertidal zone near the tidal channel Furo Grande (46°38′W 0°50′S) (Figure 1, B). The region has semidiurnal tides with amplitudes of 2 to 5 m. During spring tide days the study site is flooded twice a day, while during neap tide days tidal heights are too low to inundate the whole mangrove forest [26,27]. The forest is dominated by the mangrove tree *Rhizophora mangle* (Rhizophoraceae). Other mangrove tree species are also present including *Avicennia germinans* (Acanthaceae) and *Laguncularia racemosa* (Combretaceae) [26].

Mean annual temperature in 2012 was 26.3±1°C (mean ± standard deviation, [28]). Total precipitation recorded at the weather station in Tracuateua (50 km from the study area) was 1552 mm (Figure 2). In Northern Brazil, the wet season generally occurs from January to August and the dry season (monthly precipitation <100 mm) from September to December [28]. Surface water salinity of the tidal channel Furo Grande varied between 22 and 37 throughout the year 2012 (unpublished data).

In the study area *U. cordatus* is the only large burrowing crab (max. carapace width of 9.9 cm in the Caeté estuary, [29]). An average density of 1.7 ind. m^{-2} was reported for the Furo Grande area in *R. mangle* dominated forests [30]. *U. cordatus* prefers to construct its burrow around the root system of *R. mangle*, because large roots structure the sediment and thereby increase sediment stability and provide shelter against predators [31]. Burrows of *U. cordatus* can have various forms. They can be simple burrows with one opening and one corridor, U-shaped with more than one opening, or even more complex structures with a maximum of three openings [32]. In northeast Brazil simple burrows are the most common type (more than 85% of all investigated burrows, n = 735) [32]. In the Furo Grande area simple burrows are also the most common burrow type (unpublished observation). Such burrows have the following morphology: Their corridors initially descend with a slight slope and then bend vertically downwards until forming a terminal burrow chamber ([14], Figure 3). At our study site at the Furo Grande total burrow corridor lengths (initial horizontal and vertical part) ranged from 73 to 219 cm with an average of 123±32 cm (mean ± standard deviation, n = 30, own data). The inner walls of the burrows of *U. cordatus* have no conspicuous inner mucus layer such as known from Polychaetes (e.g. reviewed by Kristensen and Kostka [33]). The burrows of the other crab species, for example sympatric fiddler crabs, also seem to be unlined (own observation, [34]).

CO_2 efflux rates

CO_2 efflux rates of both burrows and plain sediment were measured in April, July, September and October 2012 around neap tides during the daytime. Due to technical reasons, measurements of sediment with burrows and plain sediment were performed during two succeeding neap tide cycles. In October 2012, the sampling of plain sediment was performed two neap tide cycles later.

U. cordatus burrows with only few *R. mangle* stilt roots nearby were chosen for the measurements to minimize the effect of root respiration. Furthermore, freshly excavated and blackish sediment had to be in front of their entrances indicating the presence of a live crab and ensuring sampling of recently bioturbated sediment. The diameter of the burrow entrances and if possible the current

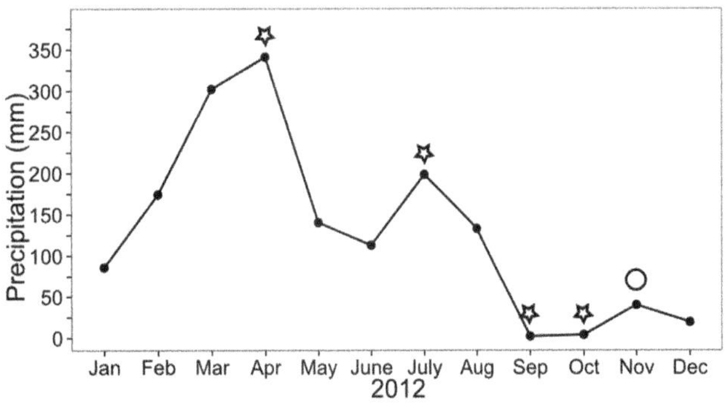

Figure 2. Total monthly precipitation in 2012. Precipitation (mm) was recorded by the weather station in Tracuateua (50 km southwest from the study site), Pará, Brazil [28]. The asterisks mark the sampling months (CO_2 efflux rate and rH of burrows and control sediment). The circle marks the only control sampling (CO_2 efflux rate and rH), which had to be conducted two neap tide cycles later than the corresponding sampling in October 2012 due to technical reasons.

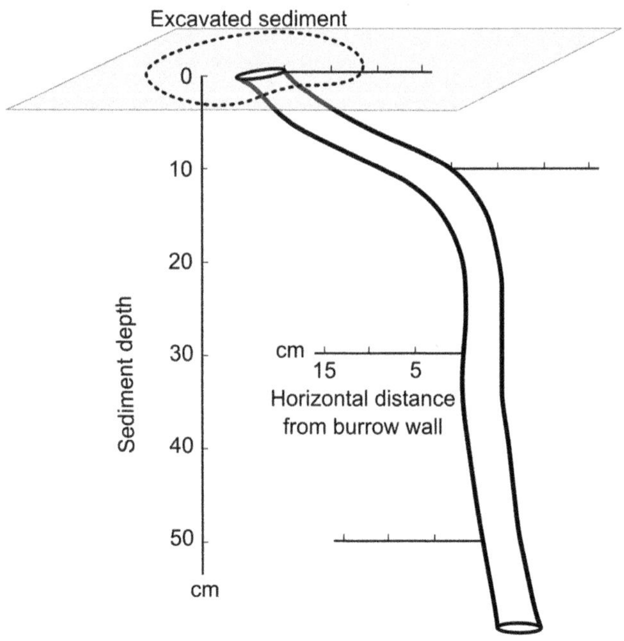

Figure 3. Vertical and horizontal sampling scheme of the sediment surrounding *U. cordatus* burrows for the rH measurements.

depths from the sediment surface to the water surface inside the burrow were recorded with a measuring tape before taking measurements. If the water surface inside the burrows was too low to measure, measurement was delayed to the rH measurements, when parts of the burrow walls were removed to facilitate access. The area of the burrow wall exposed to air was estimated based on the entrance diameter and the distance to the water surface.

For measuring the CO_2 efflux rate of a burrow, a PVC collar with 20 cm diameter was positioned around the entrance and slightly pushed several centimeters into the sediment. To avoid any influence of CO_2 released during and after the installation of the collar, the measurements were delayed for 1 h. A respiration chamber connected to a CO_2/H_2O infrared gas analyzer (LI-8100A, LI-COR, Biosciences) was then placed on top of the PVC collar. Since the respiration chamber was opaque no CO_2 was consumed through photosynthesis during measurements. The CO_2 efflux was recorded over 2 min and the measurement repeated five times. Between replicates, the chamber was opened for 25 s to release the accumulated CO_2. Most probably, these measurements included the respiration of the crab inside the burrow. The contribution of crab respiration to the measured CO_2 efflux rate was estimated by a separate experiment (see below). Sediment temperature was measured outside the collar at 2 cm sediment depth with a thermocouple (OMEGA Engineering).

As a control, the CO_2 efflux rate of plain sediment without *U. cordatus* burrows and at the most with very few fiddler crab burrows was performed, following the same measurement protocol as for the sediment with burrows. However, four replicates were taken, because preliminary tests had shown that variance of measured values for plain sediment was smaller than for sediment with *U. cordatus* burrows.

The CO_2 efflux rates were calculated for a linear flux rate [35] taking into account the surface sediment temperature and surface area inside the collar. For the measurement with *U. cordatus* burrows, the surface area inside the collar also included the burrow wall surface area. It also had to be considered that the relationship between burrow opening area and surface sediment area inside the collar did not represent the average relationship of burrow opening area to plain sediment area at the study site due to specific selection of single burrows and the small collar area. Under natural conditions, the proportion of burrow to plain area is lower (1.7 ind. m^{-2}, [30]). Therefore, the calculated CO_2 efflux rates inside collars with a crab burrow could not be directly compared with rates of plain sediment. To nevertheless allow for a comparison between the CO_2 efflux rates of plain sediment with sediment containing burrows at a given burrow density, the following calculations were performed: First, the CO_2 efflux rates of crab burrows without the surface sediment were estimated. To do this, the corresponding monthly mean CO_2 efflux rates of the plain sediment measurements were subtracted from the calculated CO_2 efflux rate of the burrows, considering the areal proportions of each term as follows:

$$F_b = (F_t \times A_t - F_s \times A_s)/A_b$$

F is the CO_2 efflux rate and A the area, which are both linked to plain sediment s, burrow b, total measured area t or total CO_2 efflux t. The total measured area refers to the surface area inside the PVC collar. Mean control CO_2 efflux rates of each month were used for F_s. From this calculation values for F_b (µmol $m^{-2}\ s^{-1}$) were obtained. From these values we derived CO_2 efflux rate of individual burrows (µmol $burrows^{-1}\ s^{-1}$) with entrance diameters of 5, 6 and 7 cm corresponding to the range of burrow diameters sampled in the field. CO_2 efflux rates for mangrove sediment based on the CO_2 efflux rate of plain sediment and individual burrows were then obtained with the help of size-specific crab burrow density data. Two burrow densities were chosen, one represents the mean *U. cordatus* burrow density in a *R. mangle* dominated forests with 1.7 ind. m^{-2} for the Furo Grande area [30]. To simplify the calculation, this density of crabs can be related to 1.7 $burrows^{-1}\ m^{-2}$ since most *U. cordatus* burrows have only one opening. The other chosen burrow density was 11.9 $burrows^{-1}\ m^{-2}$ (unpublished data) representing the increased burrow density around dense *R. mangle* stilt roots.

Respiration of *U. cordatus*

Measuring the CO_2 efflux rate of a burrow without a live crab was not feasible as removal of the crab would have strongly disturbed the sediment. To nevertheless allow an estimation of the proportion of the CO_2 efflux rate contributed by the crabs, their respiration rates were determined in an additional experiment. Adult male *U. cordatus* were caught nearby the study site. The carapace width (cm) of each animal was measured and the wet mass (g) was calculated by the regression formula for male *U. cordatus* provided by Diele [36]:

$$\text{Wet mass} = 0.4489 \times \text{carapace width}^{2.9533}$$

In the laboratory, crabs were placed in a PVC basin (20 cm diameter, 6 cm height) with the respiration chamber mounted on top. Under natural conditions crabs rest in air and water (own observation). Therefore, measurements were performed with different crabs in a dry basin (1. treatment: respiration in air, n = 14) and in a basin containing seawater (2. treatment: respiration in seawater, water level 4 cm, salinity 32, n = 14). CO_2 released by each crab was measured five times for 10 min with breaks of 25 s in-between to open the chamber and release the accumulated CO_2. Crabs were not acclimated inside the chamber prior to measurements. Therefore, the five consecutive measurements presented a gradient of crab respiration from a least acclimated state to a more acclimated one. CO_2 respiration rates of crabs (µmol CO_2 $kg^{-1}\ s^{-1}$) were calculated as described above, but related to wet mass instead of surface area in the calculation.

In the next step the contribution of crab respiration to the CO_2 efflux rate was estimated for individual burrows with a burrow entrance diameter (BED) of 5, 6 and 7 cm. Since the inhabitants of the burrows used for CO_2 measurements were not captured, their carapace width was estimated from the BED by a least squared regression formula based on data collected by Piou et al. [31]:

$$\text{Carapace width} = (\text{BED} - 0.32) \div 0.97\ (R^2 = 0.5, n = 310)$$

The wet mass was then estimated from the calculated carapace width with the formula for male *U. cordatus* by Diele [36] as indicated above. With the estimated wet mass of crabs inhabiting burrows with entrance diameters of 5, 6, and 7 cm, respectively, the respiration rate of individual crabs (µmol CO_2 $crab^{-1}\ s^{-1}$, in air and in seawater) was calculated based on the first and the last measurement of the five consecutive respiration measurements. These two measurements represent a "worst" (crab least acclimated, resting in air) and a "best" case scenario (crab with the longest acclimatization time and resting in water), since preliminary tests had shown that crabs resting in air had higher CO_2 efflux rates compared to crabs resting in water.

Reduction potential

After the CO_2 efflux rate measurements in the field were performed, reduction potentials of both burrow walls and of non-bioturbated sediment were measured at the selected sites on the same day. Redox potential (± 1.0 mV), pH (± 0.1) and temperature ($\pm 0.1°C$) were measured in the sediment excavated by the crabs outside the burrow and in the upper 2 cm of the burrow water with a Sartorius ORP (redox) combination electrode and a WTW Sentix 41 pH-electrode connected to a WTW portable meter (Multi 340i), respectively. The above mentioned parameters were also measured in the sediment surrounding the burrow (from here on referred to as "burrow wall sediment" - BWS) at sediment depths of 1, 10, 30 and 50 cm. To be able to measure the reduction potential in deeper sediment depths, one half of the burrow was carefully opened. When the desired measuring depth inside the burrows was below the water table, the water was manually removed with a small vial before taking the measurement. Measurements beyond 50 cm sediment depth were technically too difficult due to the rapid refilling of the burrow with water from below. At each sediment depth, the electrodes were inserted perpendicularly to the burrow wall, thereby subsequent readings at horizontal distances of 2, 5, 8 and 15 cm from the BWS were obtained (Figure 3). As reduction potentials in the sediment around fiddler crab burrows have been observed as being influenced at a range of 1–4 cm [18,19] and *U. cordatus* burrows are larger, we choose a larger maximal horizontal distance of 15 cm to insure that the change in reduction potential were within the measured radius.

In addition to the burrows, non-bioturbated sediment (referred to as the control) was also measured for comparison. Sediment cores were taken at least 15 cm away from burrows and stilt roots to sample areas where the influence of burrows and roots on the surrounding sediment was most likely low. A peat sampler (Eijkelkamp) of 50 cm length and 6 cm diameter was used to obtain the cores. Redox potential, pH and temperature of the sediment core were measured at depths of 1, 10, 30 and 50 cm.

Redox potential, temperature and pH values of each measurement were used to calculate the rH values [37]. These values are an indicator of the reduction force of a redox system and range between 0 (strongly reducing conditions) and 42 (strongly oxidizing conditions).

Statistical analyses

The statistical analyses were carried out following protocols for data exploration and analysis of Zuur et al. [38,39] using the statistical programming environment R [40] with the "nlme" [41] and the "ggplot2" [42] package. Data were checked for outliers (Cook distance) and these were, if necessary, removed. For information on data sets used for the analyses see Data S1.

Burrow entrance diameters, depth from the surface sediment to the water surface inside burrows, and rH values of burrow water and excavated sediment were each separately analyzed with a one way ANOVA for differences over time. When the trend over time was not linear, time was considered as a factor and a Tukey post-hoc test was applied to detect differences throughout the wet season.

Gaussian linear mixed-effects (LME) models were used for the following analyses, as they can account for inner variation between sediment cores or burrows [38,43,44]. The CO_2 efflux rates of burrows (F_b) and controls were analyzed with a LME model to test for differences over time. The random intercept of the model allowed for heterogeneity among burrows and between the sampling sites of plain sediment. To find the optimal set of fixed terms, a backwards model selection was used based on the maximum likelihood ratio test (ML) and/or on the Akaike Information Criterion (AIC). The validity of the model was checked by examination of diagnostic plots of model residuals. Final models were checked for homogeneity between residuals versus fitted values and covariates. Independence was examined by plotting residuals versus time. The final model was generated and evaluated with the restricted maximum likelihood estimation (REML) and these models are presented in Tables S1, S2, S3, S4, S5 and S6. All following LME models were analyzed as described above. Estimated values from statistical analyses are presented as mean $\pm$ standard error (se). Respiration rates of crabs were analyzed with a LME model to test for difference between treatments, number of measurements (from least acclimated to acclimated) and their respective interaction terms. The random intercept accounted for heterogeneity between individual crabs. No function for the temporal correlation structure was considered in the repeated measurements of the same crab, to be able to detect those differences between repeated measurements.

The rH data for the BWS were analyzed with a LME model to test for differences among sediment depths, horizontal distances from the burrow wall, time and their interaction terms. The random intercept accounted for heterogeneity among individual burrows. Control rH values were tested with a LME model for differences between sediment depths, time and their interaction term. The random term allowed for heterogeneity between sediment cores. Since general trends over time for the control and burrow rH data had already been analyzed in the two LME models and to keep the following models simple, control and burrow rH data were analyzed separately for each month. Monthly data sets of rH values from burrows and controls were analyzed in four LME models to test for differences among horizontal sampling levels, sediment depths and their interaction terms. BWS values of 2 (BWS2), 5 (BWS5), 8 (BWS8) and 15 cm (BWS15) and the control (>15 cm) represented the five horizontal sampling levels. BWS values and the control value were set as categorical covariates, because the exact horizontal distances of the control cores to the neighboring below ground burrow corridors was unknown. The random term allowed for heterogeneity among sampling locations (sediment core or burrow).

A Pearson correlation was applied to measure the relationship between surface rH values and CO_2 efflux rates of the plain sediment. Because the CO_2 efflux rate of a burrow could be associated to several rH values in the burrow wall sediment (from horizontal distances of 2 cm in several sediment depths) no correlation was performed.

Results

CO_2 efflux rates

Burrow entrance diameters did not significantly differ over time ($F = 3.3$, $df = 1$, $p = 0.07$, Table 1) insuring that burrows of different samplings were comparable to each other. Depths from the sediment surface to the water surface inside burrows changed over time ($F = 9.9$, $df = 1$, $p < 0.001$) and were largest in October (Tukey post-hoc: $p \leq 0.002$, Table 1). The mean burrow wall area exposed to air was 0.04 ± 0.02 m^2 (mean $\pm$ standard deviation, $n = 78$) and ranged from 0 (burrow completely filled with water) to 0.1 m^2.

Examining the CO_2 efflux rates of burrows (F_b) for differences over time showed that the LME model which included the term time was not better than the model without this term (Likelihood Ratio/L. Ratio $= 1.1$, $df = 1$, $p = 0.3$, Table S1) indicating that measured values did not differ over time. By contrast, control sediment CO_2 efflux rates increased over time (Table 1); removing

Table 1. Field data of measured and estimated parameters.

	2012							
	April		**July**		**September**		**October**	
	Burrow entrance diameter (cm)							
Mean ± se (n)	6.6±0.3	(24)	5.9±0.2	(24)	6.2±0.1	(19)	5.8±0.3	(11)
Minimum	4.1		3.4		5.3		4.6	
Maximum	9.6		7.4		7.4		7.4	
	Burrow water level from the surface (cm)							
Mean ± se (n)	16.2±0.9	(24)	21.7±2.9	(24)	18.3±1.3	(19)	33.8±2.2	(11)
Minimum	7		5		10		27	
Maximum	26		50		29		50	
	CO_2 efflux rate of plain sediment (μmol m^{-2} s^{-1})							
Mean ± se (n)	0.7±0.07	(88)	0.9±0.07	(96)	1.3±0.07	(93)	1.3±0.06	(96)
	CO_2 efflux rate of individual burrows (μmol $burrow^{-1}$ s^{-1})							
BED: 5 cm, mean ± se (n)	0.1±0.01	(116)	0.2±0.02	(119)	0.1±0.01	(95)	0.1±0.01	(55)
BED: 6 cm, mean ± se (n)	0.2±0.01	(116)	0.3±0.03	(119)	0.1±0.01	(95)	0.2±0.02	(55)
BED: 7 cm, mean ± se (n)	0.3±0.02	(116)	0.4±0.04	(119)	0.2±0.01	(95)	0.2±0.03	(55)
	Reduction potential of burrow water							
Mean ± se (n)	16.2±0.9	(22)	16.8±0.5	(22)	18.0±0.9	(16)	16.2±0.7	(9)
	Reduction potential of excavated sediment							
Mean ± se (n)	15.5±0.8	(25)	16.1±0.7	(24)	16.0±0.7	(17)	17.5±0.9	(12)

Burrow entrance diameter, burrow water level from the surface, CO_2 efflux rate of plain sediment and inhabited burrows with specific diameter, reduction potential of burrow water and excavated sediment.
Abbreviation: BED = burrow entrance diameter (in cm), n = sample size, se = standard error.

the term time from the corresponding LME model led to a less suitable model, so the factor was retained (L. Ratio = 15.3, df = 1, $p<0.001$, Table S2). Calculated CO_2 efflux rates of individual burrows ranged from 0.1 µmol $burrows^{-1}$ s^{-1} for burrows with a BED of 5 cm to 0.4 µmol $burrows^{-1}$ s^{-1} for burrows with a BED of 7 cm (Table 1).

Respiration of *U. cordatus*

Mean carapace width and derived wet masses of crabs are given in Table 2. A LME model of the crab respiration data with the interaction term treatment×number of observations was significantly better than a model without this interaction (L. Ratio = 12.5, df = 1, $p<0.001$, Table S3). This indicates that crab respiration rates varied considerably between individuals of one treatment, but decreased during the five consecutive measurements. Crabs resting in seawater had the lowest respiration rates (Table 2). During the measurements, the least acclimated crabs produced foam in front of their mouthparts, whereas more acclimated crabs did not foam and stopped respiring for a few seconds up to several minutes. Respiration rates of individual crabs with a wet mass corresponding to burrow entrance diameters of 5, 6 and 7 cm, respectively, are presented for the "worst" (crab least acclimated, resting in air) and "best" case scenarios (crab with the longest acclimatization time and resting in water) in Table 2. The crabs' contribution to the CO_2 efflux rates of individual burrows was 55–100% in the "worst" case and 20–60% in the "best" case scenario (Table 1, Table 2).

Estimates for CO_2 release of mangrove sediment

Calculated CO_2 efflux rate estimates of mangrove sediment with 1.7 and 11.9 $burrows^{-1}$ m^{-2} increased over time. The reason for this was an increase in CO_2 efflux rates of plain sediment over time due to seasonality, while individual burrow CO_2 efflux rates remained constant over time (Figure 4). Estimates of CO_2 efflux rate of sediment with *U. cordatus* burrows (Table 3, Figure 4) are 15 to 500% (considering 1.7 and 11.9 $burrows^{-1}$ m^{-2}, respectively) higher than for plain sediment (Table 3). Despite the small contribution of burrow openings to total sediment surface area (< 5%), *U. cordatus* burrows, especially in densely rooted sediment, can increase the CO_2 efflux rate of plain sediment by 92 to 500% (Table 3, Figure 4).

Reduction potential

The rH values of both burrow water (F = 0.6, df = 1, $p = 0.4$, Table 1) and excavated sediment (F = 2.3, df = 1, $p = 0.1$, Table 1) did not differ over time. The final model of the BWS rH data indicates that rH values decreased with sediment depth and horizontal distance as well as over time (Figure 5); a LME model of the BWS rH data with a three way interaction term was not significantly better than a model without this interaction (L. Ratio = 4.4, df = 9, $p = 0.9$, Table S4). Furthermore, the interaction terms horizontal depth×time (L. Ratio = 1.0, df = 3, $p = 0.8$, Table S4) and sediment depth×time (L. Ratio = 3.2, df = 3, $p = 0.4$, Table S4) were eliminated. The model with the interaction term horizontal distance×sediment depth was better than without interaction (L. Ratio = 17.2, df = 9, $p = 0.045$, Table S4). However, we continued with the simpler (using fewer degrees of freedom) and slightly better model (AIC of model with the interaction term: 4389.1 and without interaction term: 4388.4) by excluding the interaction term. Finally, comparing models where each main term was excluded once with the original model containing all three main terms showed that the latter was best (sediment depth: L. Ratio = 555.0, df = 3, $p<0.001$; horizontal distance: L. Ratio = 35.6, df = 3, $p<0.001$; time: L. Ratio = 20.7, df = 1, $p<0.001$, Table S4).

The model of the control rH data with the interaction term sediment depth×time was significantly better (L. Ratio = 10.1, df = 1, $p = 0.0015$, Table S5) than the model without this term indicating that control rH values decreased with sediment depth, but had different depth-rH curves over time (Figure 5). Further, the surface rH values positively correlated with the CO_2 efflux rates of plain sediment (correlation coefficient = 0.18, $p<0.001$).

In the next step, BWS and control rH data were compared separately for each sampling campaign. The comparison of the four models including and excluding the interaction term sediment depth×time showed that the interaction term significantly improved the four models (April: L. Ratio = 18.0, df = 4, $p = 0.001$, July: L. Ratio = 33.2, df = 4, $p<0.001$, September: L. Ratio = 38.3, df = 4, $p<0.001$, October: L. Ratio = 44.0, df = 4, $p<0.001$, Table S6). Hence, at any point of time rH values decreased with increasing sediment depth, and depth-rH curves differed between horizontal sampling levels, in the way that rH values decreased mostly with increasing horizontal distances (Figure 5). Measured pH values are not shown in detail; they ranged from 3.5 to 7.1 for plain sediment and from 4.3 to 7.8 for burrow wall sediment.

Discussion

Measured CO_2 efflux rates of individual burrows are composed of three elements: The respiration of the inhabiting crab itself, the CO_2 released by the burrow wall sediment and the CO_2 released by the water inside the burrow.

The crabs' contribution to the CO_2 efflux rate of burrows

In our study measurements of CO_2 exhaled by isolated crabs showed that high CO_2 production by respiration was related to stress. Crabs in our study, particularly those kept without seawater in the PVC basin, were probably not fully acclimated during the 50 min respiration measurements, because CO_2 respiration of crabs resting in air were always higher in the five measurements compared to crabs resting in water (Table 2). Some crabs exhaled four or fewer times in 10 min (see minimum values in Table 2) suggesting reasonable acclimatization.

During field measurements, crabs were sometimes seen inside their burrows when the respiration chamber was opened between measurements. The crabs did not appear to be stressed by the chamber, since they moved slowly and did not produce foam, as they do when they are handled or otherwise stressed [45]. Therefore, respiration rates of *U. cordatus* measured in the laboratory for the "worst" case scenario (crabs with no acclimatization time and resting in air) are probably overestimating the crabs' respiration under field conditions. Respiration rates for the "best" case scenario (crabs with longest acclimatization time resting in water) on the other hand, probably resemble field conditions more closely. Our data thus suggest that crabs contributed about 20–60% to the CO_2 efflux rates of individual burrows, which leaves the burrow CO_2 efflux rates without crabs to range between 0.04–0.32 µmol $burrows^{-1}$ s^{-1} for burrows of 5 to 7 cm entrance diameter (Table 1). Respiration rates of an acclimated *U. cordatus* specimen (250 g wet mass, in seawater) are 0.19±0.04 µmol CO_2 $crab^{-1}$ s^{-1} and are similar to rates recorded for a large sesarmid crab (*Neoepisesarma versicolor*, 250 g wet mass, in seawater) of 0.24 µmol CO_2 $crab^{-1}$ s^{-1} [46].

Table 2. Data of *U. cordatus* respiration measurements.

	Crab in air		Crab in seawater	
	Carapace width (cm)			
Mean ± se (n)	7.3±0.1	(14)	7.0±0.1	(14)
Median	7.4		7.0	
Minimum	6.5		6.2	
Maximum	7.8		7.5	
	Wet mass (g)			
Mean ± se (n)	161.6±6.3	(14)	140.4±5.9	(14)
Median	162.4		137.7	
Minimum	113		98.2	
Maximum	193.5		172.4	
	Crab respiration (μmol CO_2 kg^{-1} s^{-1})			
Mean ± se (n) of measurement 1	2.3±0.4	(13)	1.4±0.2	(13)
Mean ± se (n) of measurement 2	1.4±0.2	(13)	1.0±0.2	(12)
Mean ± se (n) of measurement 3	1.1±0.1	(13)	0.9±0.2	(12)
Mean ± se (n) of measurement 4	0.9±0.1	(12)	0.7±0.1	(12)
Mean ± se (n) of measurement 5	0.9±0.1	(12)	0.8±0.2	(12)
Minimum of all measurement (n)	0.4	(63)	0.04	(61)
Maximum of all measurement (n)	4.1	(63)	2.4	(61)
	Crab respiration - worst case (μmol CO_2 $crab^{-1}$ s^{-1})			
BED: 5 cm, WM: 28 g, mean ± se (n)	0.11±0.02	(13)	0.07±0.01	(13)
BED: 6 cm, WM 55 g, mean ± se (n)	0.19±0.03	(13)	0.12±0.02	(13)
BED: 7 cm, WM 93 g, mean ± se (n)	0.30±0.05	(13)	0.19±0.02	(13)
	Crab respiration - best case (μmol CO_2 $crab^{-1}$ s^{-1})			
BED: 5 cm, WM: 28 g, mean ± se (n)	0.04±0.004	(12)	0.04±0.01	(12)
BED: 6 cm, WM 55 g, mean ± se (n)	0.07±0.01	(12)	0.06±0.01	(12)
BED: 7 cm, WM 93 g, mean ± se (n)	0.12±0.01	(12)	0.10±0.02	(12)

Carapace width, estimated wet mass after Diele [36], average respiration rates during the five measurements and respiration rate of individual crabs of a specific wet mass inhabiting burrows with different burrow entrance diameters for the "worst" case scenario (crab not acclimated, values calculated from measurement 1) and for the "best" case scenario (crab with longest acclimatization time, values calculated from measurement 5).
Abbreviation: BED = burrow entrance diameter (in cm), n = sample size, se = standard error, WM = wet mass.

CO_2 efflux rates of burrows and sediment

CO_2 efflux rates of inhabited *U. cordatus* burrows were at least 20 times higher (Table 1) than estimated rates from inhabited fiddler crab burrows (0.005 μmol $burrows^{-1}$ s^{-1}) measured in a Tanzanian mangrove forest [24]. Higher CO_2 efflux rates of *U. cordatus* burrows are probably caused by the larger body size and higher respiration of the crab and the corresponding larger burrow size (and depth) compared with *Uca* spp. and their burrows (average wet mass of *Uca* spp.: 3.2 g, 0.002 μmol CO_2 $crab^{-1}$ s^{-1}, average burrow entrance diameter not reported, [24]). Mean CO_2 efflux rates of plain sediment in our study ranged from 0.7 to 1.3 μmol m^{-2} s^{-1} and were similar to other mangrove sediments [20,22], e.g. to values measured by Lovelock [47] for sites in the Caribbean, Australia and New Zealand, which ranged from −0.25 to 2.97 μmol m^{-2} s^{-1}. In line with our first hypothesis, CO_2 efflux rates of mangrove sediment containing burrows inhabited by *U. cordatus* were 15 to 500% higher than for plain sediment depending on burrow density and month (Table 3). In a Tanzanian mangrove forest, Kristensen et al. [24] recorded CO_2 efflux rates of sediment with inhabited *Uca* spp. burrows in areas without obvious above ground roots. Measured rates ranged between 0.68 μmol m^{-2} s^{-1} (68±17 $burrows^{-1}$ m^{-2}) and 1.38 μmol m^{-2} s^{-1} (237±53 $burrows^{-1}$ m^{-2}) and were similar to ours with 1.0±0.02 until 1.7±0.05 μmol m^{-2} s^{-1} based on an average *U. cordatus* burrow density of 1.7 $burrows^{-1}$ m^{-2} (Table 3). Although CO_2 release of individual fiddler crab burrows is much lower than that of *U. cordatus*, the burrow density of fiddler crabs at the Tanzanian study site was much higher (68–626 $burrows^{-1}$ m^{-2}, [24]) than burrow densities of *U. cordatus* recorded near our study site (1.7 $burrows^{-1}$ m^{-2} at an average, Diele et al. [30]). Thus, a high density of burrows may compensate for a lower per-burrow CO_2 efflux rate.

In addition to *U. cordatus* burrows, *Uca* spp. burrows were also present at our study site (28 to 105 $burrows^{-1}$ m^{-2} with an average of 57±18 $burrows^{-1}$ m^{-2}, mean±standard deviation, n = 42, with each sample area 1 m^2, unpublished data). Future investigations should focus on the entire crab community of a mangrove site, involving all burrowing crabs, to obtain more precise estimates of overall CO_2 efflux rates from the mangrove sediment.

The amount of CO_2 released by sediment or burrows depends on the prevalence of organic carbon in the sediment and its oxidation pathways by microorganisms [48–50]. In a study on the Ajuruteua peninsula the organic matter content and mean cell

Table 3. CO_2 efflux rates of mangrove sediment.

	2012							
	April		**July**		**September**		**October**	
Density: 1.7 burrows^{-1} m^{-2}	**CO_2 efflux rate of mangrove sediment (μmol m^{-2} s^{-1})**							
BED: 5 cm, mean ± se (n)	1.0±0.02	(116)	1.2±0.03	(119)	1.5±0.01	(95)	1.5±0.02	(55)
BED: 6 cm, mean ± se (n)	1.1±0.02	(116)	1.3±0.05	(119)	1.6±0.02	(95)	1.6±0.03	(55)
BED: 7 cm, mean ± se (n)	1.2±0.03	(116)	1.5±0.06	(119)	1.7±0.02	(95)	1.7±0.05	(55)
Density: 11.9 burrows^{-1} m^{-2}	**CO_2 efflux rate of mangrove sediment (μmol m^{-2} s^{-1})**							
BED: 5 cm, mean ± se (n)	2.4±0.12	(116)	3.2±0.22	(119)	2.5±0.08	(95)	2.7±0.17	(55)
BED: 6 cm, mean ± se (n)	3.2±0.17	(116)	4.2±0.32	(119)	3.0±0.12	(95)	3.4±0.24	(55)
BED: 7 cm, mean ± se (n)	4.1±0.23	(116)	5.4±0.44	(119)	3.6±0.16	(95)	4.1±0.33	(55)
Density: 1.7 burrows^{-1} m^{-2}	**Increase of CO_2 efflux rate to plain sediment (%)**							
BED: 5 cm (AP: 0.3%)	43		33		15		15	
BED: 6 cm (AP: 0.5%)	57		44		23		23	
BED: 7 cm (AP: 0.7%)	71		67		31		31	
Density: 11.9 burrows^{-1} m^{-2}	**Increase of CO_2 efflux rate to plain sediment (%)**							
BED: 5 cm (AP: 2.3%)	243		256		92		108	
BED: 6 cm (AP: 3.4%)	357		367		131		162	
BED: 7 cm (AP: 4.6%)	486		500		177		215	

Rates and the increase in % compared to plain sediment for average burrow densities of 1.7 burrows^{-1} m^{-2} for *Rhizophora mangle* dominated areas at the Furo Grande [30] and 11.9 burrows^{-1} m^{-2} for densely rooted areas around *R. mangle* trees.
Abbreviation: AP = areal proportion of the burrow opening per m^2, BED = burrow entrance diameter (in cm), n = sample size, se = standard error, WM = wet mass.

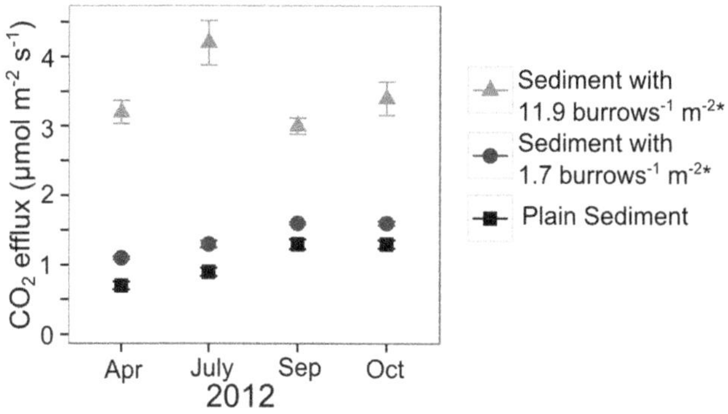

Figure 4. CO_2 efflux rates (µmol m^{-2} s^{-1}) of plain sediment and sediment with a burrow density of 1.7 and 11.9 burrows^{-1} m^{-2}. *For these estimates all burrow entrance diameters were set at 6 cm.

numbers of bacteria were measured in the BWS of *U. cordatus* burrows (70 cm depth) and were similar to corresponding values of the surface sediment [51]. Therefore, microbial activity in BWS may be more similar to surface sediment than to non-bioturbated sediment at 70 cm sediment depth. However, CO_2 released from burrows may not exclusively originate from carbon oxidation in the BWS, but also from the carbon stock in deeper sediment layers which may be released via the burrows [20,24]. Furthermore, the amount of CO_2 released from the water accumulated inside the burrow is unknown. The water may contain CO_2 originating from bacterial, meio- and macrofaunal respiration and from tidal input. How the bicarbonate system of the burrow water is constituted is also an unknown, e.g. how such systems react to increasing/decreasing CO_2 concentrations and whether changing water levels inside burrows influence the burrow water CO_2 gas exchange. CO_2 dissolved in water may also be partly exported by the groundwater during the ebb tide [52].

Apart from microbial decomposition in the sediment, CO_2 is produced by root respiration. Information about root respiration of mangrove tree species is scarce; root respiration rates are reported to be low for *R. mangle* trees. Kristensen et al. [24] estimated the amount of CO_2 released by pneumatophores of *Sonneratia alba* and *Avicennia marina* in a Tanzanian mangrove forest, yielding an equivalent or higher efflux rate than sediment with burrows. Due to technical reasons we had to avoid measuring near aboveground aerial roots. However, underground root respiration may have contributed to the CO_2 released by burrows and, equally, by surface sediment without burrows. As *U. cordatus* prefers areas with *R. mangle* roots when constructing its burrows [31], their burrows possibly function as an exhalant channel for CO_2 produced by root respiration. To test this hypothesis, further work on CO_2 release must overcome technical restrictions and focus on areas where crabs and mangrove tree roots occur closely together.

Temporal and spatial patterns of reduction potentials

Reduction potentials of the burrow water and the excavated sediment were similar during the wet season and relatively oxidized compared to the BWS and control rH values in deeper sediment depths. This indicates that burrow water is regularly mixed with more oxidized water or oxygen by, for example, tidal flushing or groundwater flow [53–55]. Further, excavated sediment oxidizes at the surface when in contact with atmospheric oxygen. Thus, bioturbation by crabs significantly oxidizes previously reduced sediment.

In contrast to the stable values in the burrow water and the excavated sediment, rH values in the BWS changed over time. In April, BWS had the lowest rH values. At 1 and 10 cm sediment depth rH values were even lower compared to plain sediment. The rising water column inside the burrows (Table 1) and waterlogging of the sediment caused by the high precipitation rates in this month may have led to anoxic conditions along the burrow wall. In contrast, during the last sampling at the end of the wet season rain showers, temporally filling up burrows became less frequent and water levels sank to 50 cm below the surface (Table 1). The longer exposure of BWS to air at the end of the wet season led to more oxidized conditions at 30–50 cm sediment depths compared with plain sediment. Additionally, from July onwards BWS rH and control rH values increased with decreasing precipitation rates (Figure 2, Figure 5).

The spatial pattern of the *U. cordatus* BWS rH values indicated a wider oxidized zone, reaching up to 15 cm away from the burrow, compared to the *Uca sp.* burrows, where the oxidized zone reached only 1.5–4 cm into the surrounding sediment [18,19]. This difference is most likely due to the much larger size and depth of *U. cordatus* burrows, leading to improved solute exchange conditions between the BWS and tidal water, groundwater and atmospheric air compared to the smaller fiddler crab burrows. Why rH values for the up to 15 cm horizontal distance from the burrow wall did not approximate control values remains unclear. However, an approximation may have been masked by the high variance of the rH values due to nearby below ground burrow galleries or roots [56], which could not be seen during the measurements. Since rH values in the BWS were not always more oxidized compared to plain sediment we reject our second hypothesis. The reduction potential of the sediment is further affected by different environmental and biotic factors such as water logging, tree roots and age of the forest stand [1,18,19,48,49,57].

Temporal pattern in CO_2 release

CO_2 efflux rates of plain surface sediment increased over time. Possibly surface sediment became more oxidized towards the end of the wet season and a change in the microbial decomposition process may thus have resulted in the different CO_2 efflux rates [22]. Another possible reason could be that CO_2, which is normally released over the sediment surface, may have been dissolved after rainfalls and stored in rain water filling sediment pore space. This would have increased the sediment CO_2 concentrations, while less CO_2 would have been released from the sediment, as observed in our study at the beginning of the wet season. Corresponding results have been obtained by studies performed in a temperate pinewood forest [58,59]. In contrast to the CO_2 efflux rates of plain surface sediment, the CO_2 efflux rates of individual burrows did not change over time, despite simultaneously varying rH values. The reasons for this are not clear.

Conclusions

Until now, the ecological role of *Ucides cordatus* has been mostly discussed in terms of carbon retention through litter feeding. The results of the present study emphasize the crabs' importance for carbon export via CO_2 release. Respiration by *U. cordatus* and CO_2 release by its burrows increase the CO_2 efflux rate compared to plain sediment by 15–71% assuming a conservative average density of 1.7 burrows^{-1} m^{-2} for *Rhizophora mangle* dominated forests at the Caeté estuary in North Brazil. This increase in CO_2 efflux rate is particularly high considering

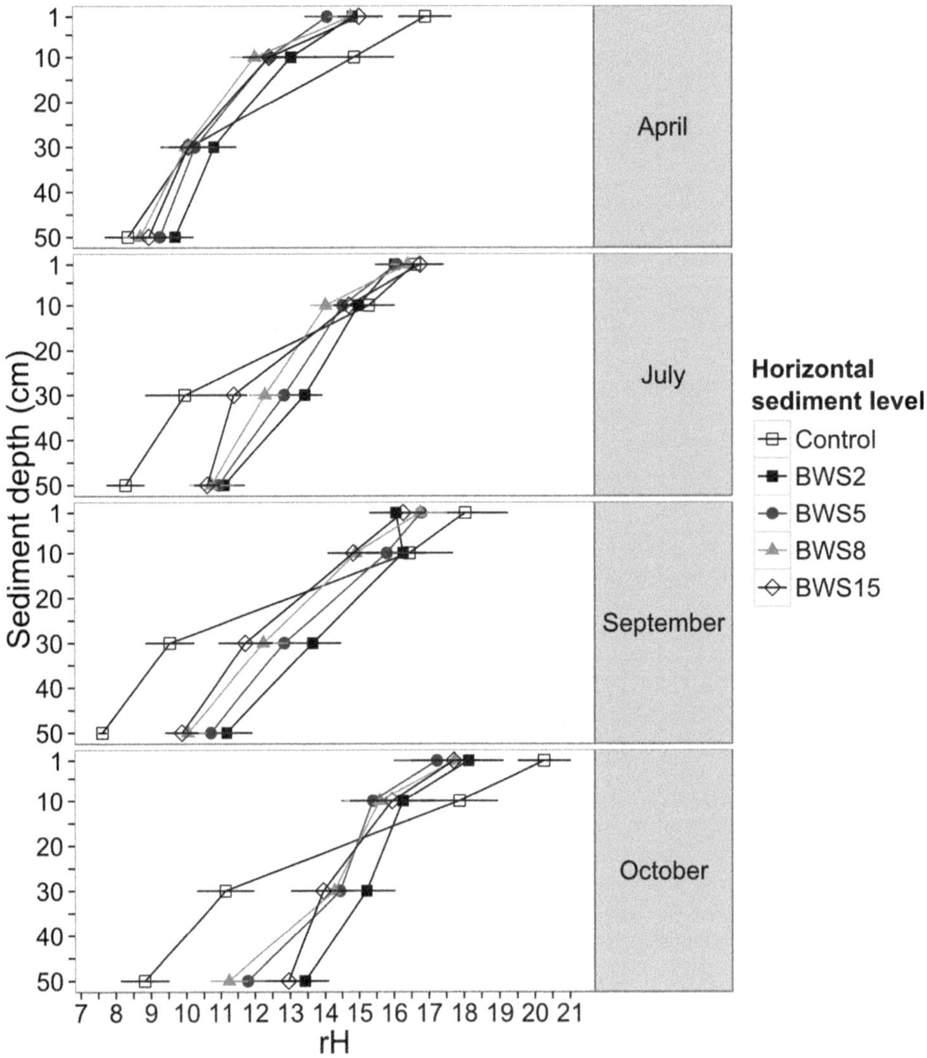

Figure 5. Burrow and plain sediment rH values for each sampling campaign in the year 2012. Mean sediment rH values ± standard error (se) plotted against sediment depth (cm) (sample size (n) for each month: April: n = 308, July: n = 363, September: n = 282 and October: n = 240). Different grey shades and symbols represent the different horizontal sampling distances from the burrow wall sediment (BWS, BWS2 = 2 cm, BWS5 = 5 cm, BWS8 = 8 cm, BWS15 = 15 cm and control >15 cm).

the relatively low proportion (0.3–0.7%) for which crab burrow entrances of *U. cordatus* account for per sediment area. Furthermore, burrows lead to changes in the reduction potential in the sediment and may thereby indirectly influence microbial decomposition pathways and CO_2 release. This study further demonstrates that the reduction potential around the burrow walls underlies spatio-temporal variations which need to be considered to improve estimates of CO_2 release from mangrove sediments. Future studies should disentangle carbon release and carbon storage by crabs and determine whether these two processes are in balance. Moreover, we underline the importance of including CO_2 efflux rates of burrows and respiration rates of crabs into carbon budget estimates of mangrove ecosystems.

Supporting Information

Table S1 Final linear mixed-effects model of burrow CO_2 efflux rate data.

Table S2 Final linear mixed-effects model of control CO_2 efflux rate data.

Table S3 Final linear mixed-effects model of crab respiration data.

Table S4 Final linear mixed-effects model of burrow rH data.

Table S5 Final linear mixed-effects model of control rH data.

Table S6 Final linear mixed-effects model of control-burrow rH data.

Data S1 Complete data set used for statistical analyses.

Acknowledgments

We thank the Instituto do Estudos Costeiros (IECOS) of the Federal University of Pará (UFPa), Brazil for the support during field and laboratory work. Further we would like to express our thanks to the fishermen and students helping us during this time. We are also grateful to Cyril Piou for providing additional data.

Author Contributions

Conceived and designed the experiments: NP KD UM IN. Performed the experiments: NP. Analyzed the data: NP. Contributed reagents/materials/analysis tools: NP UM. Contributed to the writing of the manuscript: NP KD UM IN.

References

1. Kristensen E (2008) Mangrove crabs as ecosystem engineers; with emphasis on sediment processes. J Sea Res 59: 30–43.
2. Smith III TJ, Boto KG, Frusher SD, Giddins RL (1991) Keystone species and mangrove forest dynamics: the influence of burrowing by crabs on soil nutrient status and forest productivity. Estuar Coast Shelf Sci 33: 419–432.
3. Smith NF, Wilcox C, Lessmann JM (2009) Fiddler crab burrowing affects growth and production of the white mangrove (*Laguncularia racemosa*) in a restored Florida coastal marsh. Mar Biol 156: 2255–2266.
4. Lee SY (1998) Ecological role of grapsid crabs in mangrove ecosystems: a review. Mar Freshw Res 49: 335–343.
5. Lee SY (2008) Mangrove macrobenthos: Assemblages, services, and linkages. J Sea Res 59: 16–29.
6. Andreetta A, Fusi M, Cameldi I, Cimò F, Carnicelli S, et al. (2014) Mangrove carbon sink. Do burrowing crabs contribute to sediment carbon storage? Evidence from a Kenyan mangrove system. J Sea Res 85: 524–533.
7. Kristensen E, Penha-Lopes G, Delefosse M, Valdemarsen T, Quintana CO, et al. (2012) What is bioturbation? The need for a precise definition for fauna in aquatic sciences. Mar Ecol Prog Ser 446: 285–302.
8. Kristensen E (2000) Organic matter diagenesis at the oxic/anoxic interface in coastal marine sediments, with emphasis on the role of burrowing animals. Hydrobiologia 426: 1–24.
9. Robertson AI (1986) Leaf-burying crabs: Their influence on energy flow and export from mixed mangrove forests (*Rhizophora* spp.) in northeastern Australia. J Exp Mar Bio Ecol 102: 237–248.
10. Robertson AI, Daniel PA (1989) The influence of crabs on litter processing in high intertidal mangrove forests in tropical Australia. Oecologia 78: 191–198.
11. Emmerson WD, McGwynne LE (1992) Feeding and assimilation of mangrove leaves by the crab *Sesarma meinerti* de Man in relation to leaf-litter production in Mgazana, a warm-temperate southern African mangrove swamp. J Exp Mar Bio Ecol 157: 41–53.
12. Twilley RR, Pozo M, Garcia VH, Rivera-Monroy VH, Zambrano R, et al. (1997) Litter dynamics in riverine mangrove forests in the Guayas River estuary, Ecuador. Oecologia 111: 109–122.
13. Nordhaus I, Wolff M, Diele K (2006) Litter processing and population food intake of the mangrove crab *Ucides cordatus* in a high intertidal forest in northern Brazil. Estuar Coast Shelf Sci 67: 239–250.
14. Schories D, Barletta-Bergan A, Barletta M, Krumme U, Mehlig U, et al. (2003) The keystone role of leaf-removing crabs in mangrove forests of North Brazil. Wetl Ecol Manag 11: 243–255.
15. Katz LC (1980) Effects of burrowing by the fiddler crab, *Uca pugnax* (Smith). Estuar Coast Mar Sci 11: 233–237.
16. Revsbech NP, Sørensen J, Blackburn TH, Lomholt JP (1980) Distribution of oxygen in marine sediments measured with microelectrodes. Limnol Oceanogr 25: 403–411.
17. Michaels RE, Zieman JC (2013) Fiddler crab (*Uca* spp.) burrows have little effect on surrounding sediment oxygen concentrations. J Exp Mar Bio Ecol 448: 104–113.
18. Gribsholt B, Kostka JE, Kristensen E (2003) Impact of fiddler crabs and plant roots on sediment biogeochemistry in a Georgia saltmarsh. Mar Ecol Prog Ser 259: 237–251.
19. Nielsen OI, Kristensen E, Macintosh DJ (2003) Impact of fiddler crabs (*Uca* spp.) on rates and pathways of benthic mineralization in deposited mangrove shrimp pond waste. J Exp Mar Bio Ecol 289: 59–81.
20. Kristensen E, Bouillon S, Dittmar T, Marchand C (2008) Organic carbon dynamics in mangrove ecosystems: A review. Aquat Bot 89: 201–219.
21. Canfield DE (1994) Factors influencing organic carbon preservation in marine sediments. Chem Geol 114: 315–329.
22. Leopold A, Marchand C, Deborde J, Chaduteau C, Allenbach M (2013) Influence of mangrove zonation on CO_2 fluxes at the sediment–air interface (New Caledonia). Geoderma 202–203: 62–70.
23. Bouillon S, Borges A V., Castañeda-Moya E, Diele K, Dittmar T, et al. (2008) Mangrove production and carbon sinks: A revision of global budget estimates. Global Biogeochem Cycles 22: 1–12.
24. Kristensen E, Flindt MR, Ulomi S, Borges AV, Abril G, et al. (2008) Emission of CO_2 and CH_4 to the atmosphere by sediments and open waters in two Tanzanian mangrove forests. Mar Ecol Prog Ser 370: 53–67.
25. Araújo Jr JMC, Otero XL, Marques AGB, Nóbrega GN, Silva JRF, et al. (2012) Selective geochemistry of iron in mangrove soils in a semiarid tropical climate: effects of the burrowing activity of the crabs *Ucides cordatus* and *Uca maracoani*. Geo-Marine Lett 32: 289–300.
26. Krause G, Schories D, Glaser M, Diele K (2001) Spatial patterns of mangrove ecosystems: the bragantinian mangroves of northern Brazil (Bragança, Pará). Ecotropica 7: 93–107.
27. Krause G (2010) The geography of the Bragança coastal region. In: Saint-Paul U, Schneider H, editors. Mangrove Dynamics and Management in North Brazil. Springer Berlin Heidelberg, Vol. 211. 19–34.
28. INMET (2013) Available: http://www.inmet.gov.br/portal/index.php?r=estacoes/estacoesConvencionais. Accessed 2013 Jan 15.
29. Diele K, Koch V (2010) Growth and mortality of the exploited mangrove crab *Ucides cordatus* (Ucididae) in N-Brazil. J Exp Mar Bio Ecol 395: 171–180.
30. Diele K, Koch V, Saint-Paul U (2005) Population structure, catch composition and CPUE of the artisanally harvested mangrove crab *Ucides cordatus* (Ocypodidae) in the Caeté estuary, North Brazil: Indications for overfishing? Aquat Living Resour 18: 169–178.
31. Piou C, Berger U, Feller IC (2009) Spatial structure of a leaf-removing crab population in a mangrove of North-Brazil. Wetl Ecol Manag 17: 93–106.
32. Araújo MSLC de, Calado TCS (2011) Burrows architecture of the crab *Ucides cordatus* (LINNAEUS, 1763) (Crustacea, Decapoda, Ucididae) in a mangrove swamp of Brazil. Trop Oceanogr 39: 155–165.
33. Kristensen E, Kostka JE (2005) Macrofaunal Burrows and Irrigation in Marine Sediment: Microbiological and Biogeochemical. In: Kristensen E, Haese RR, Kostka JE, editors. Interactions between Macro- and Microorganisms in Marine Sediments. Washington, DC: American Geophysical Union. 125–157.
34. Curran HA, Martin AJ (2003) Complex decapod burrows and ecological relationships in modern and Pleistocene intertidal carbonate environments, San Salvador Island, Bahamas. Palaeogeogr Palaeoclimatol Palaeoecol 192: 229–245.
35. LI-COR Biosciences (2010) LI-8100 Instruction Manual. Rev. 1.0.
36. Diele K (2000) Life history and population structure of the exploited mangrove crab *Ucides cordatus* (Linnaeus, 1763) (Decapoda: Brachyura) in the Caeté estuary, North ZMT Contribution 9, University Bremen, Germany.
37. Pöpel F (2000) Lehrbuch für Abwassertechnik und Gewässerschutz. Müller Jur.Vlg.C.F.
38. Zuur AF, Ieno EN, Walker N, Saveliev AA, Smith GM (2009) Mixed Effects Models and Extensions in Ecology with R. New York: Springer.
39. Zuur AF, Ieno EN, Elphick CS (2010) A protocol for data exploration to avoid common statistical problems. Methods Ecol Evol 1: 3–14.
40. R Core Team (2012) R: A Language and Environment for Statistical Computing. Vienna, Austria.
41. Pinheiro JC, Bates DM, DebRoy S, Sarkar D, R Core Team (2012) nlme: Linear and nonlinear mixed effects models: 1–102.
42. Wickham H (2009) ggplot2: Elegant graphics for data analysis.
43. Pinheiro JC, Bates DM (2000) Mixed effects models in S and S-Plus. New York: Springer.
44. Zuur AF, Ieno EN, Smith GM (2007) Analysing Ecological Data. New York: Springer.
45. Diele K, Koch V (2010) Comparative population dynamics and life histories of North Brazilian mangrove ccabs, genera *Uca* and *Ucides* (Ocypodoidea). In: Saint-Paul U, Schneider H, editors. Mangrove Dynamics and Management in North Brazil. Springer Berlin Heidelberg, Vol. 211. 275–285.
46. Thongtham N, Kristensen E (2005) Carbon and nitrogen balance of leaf-eating sesarmid crabs (*Neoepisesarma versicolor*) offered different food sources. Estuar Coast Shelf Sci 65: 213–222.
47. Lovelock CE (2008) Soil respiration and belowground carbon allocation in mangrove forests. Ecosystems 11: 342–354.
48. Alongi DM, Tirendi F, Trott LA, Xuan TT (2000) Benthic decomposition rates and pathways in plantations of the mangrove *Rhizophora apiculata* in the Mekong delta, Vietnam. Mar Ecol Prog Ser 194: 87–101.
49. Alongi DM, Sasekumar A, Tirendi F, Dixon P (1998) The influence of stand age on benthic decomposition and recycling of organic matter in managed mangrove forests of Malaysia. J Exp Mar Bio Ecol 225: 197–218.
50. Kristensen E, Andersen FØ, Holmboe N, Holmer M, Thongtham N (2000) Carbon and nitrogen mineralization in sediments of the Bangrong mangrove area, Phuket, Thailand. Aquat Microb Ecol 22: 199–213.
51. Nordhaus I, Wolff M (2007) Feeding ecology of the mangrove crab *Ucides cordatus* (Ocypodidae): food choice, food quality and assimilation efficiency. Mar Biol 151: 1665–1681.
52. Alongi DM, Carvalho NA, Amaral AL, da Costa A, Trott L, et al. (2012) Uncoupled surface and below-ground soil respiration in mangroves: implications for estimates of dissolved inorganic carbon export. Biogeochemistry 109: 151–162.
53. Xin P, Jin G, Li L, Barry DA (2009) Effects of crab burrows on pore water flows in salt marshes. Adv Water Resour 32: 439–449.
54. Susilo A, Ridd P V., Thomas S (2005) Comparison between tidally driven groundwater flow and flushing of animal burrows in tropical mangrove swamps. Wetl Ecol Manag 13: 377–388.

55. Heron SF, Ridd PV (2008) The tidal flushing of multiple-loop animal burrows. Estuar Coast Shelf Sci 78: 135–144.
56. Ferreira TO, Otero XL, de Souza Junior VS, Vidal-Torrado P, Macías F, et al. (2010) Spatial patterns of soil attributes and components in a mangrove system in Southeast Brazil (São Paulo). J Soils Sediments 10: 995–1006.
57. Marchand C, Baltzer F, Lallier-Vergès E, Albéric P (2004) Pore-water chemistry in mangrove sediments: relationship with species composition and developmental stages (French Guiana). Mar Geol 208: 361–381.
58. Maier M, Schack-Kirchner H, Hildebrand EE, Schindler D (2011) Soil CO_2 efflux vs. soil respiration: Implications for flux models. Agric For Meteorol 151: 1723–1730.
59. Maier M, Schack-Kirchner H, Hildebrand EE, Holst J (2010) Pore-space CO_2 dynamics in a deep, well-aerated soil. Eur J Soil Sci 61: 877–887.

16

Effects of Atrazine, Metolachlor, Carbaryl and Chlorothalonil on Benthic Microbes and Their Nutrient Dynamics

Daniel Elias*, Melody J. Bernot

Ball State University, Department of Biology, Muncie, Indiana, United States of America

Abstract

Atrazine, metolachlor, carbaryl, and chlorothalonil are detected in streams throughout the U.S. at concentrations that may have adverse effects on benthic microbes. Sediment samples were exposed to these pesticides to quantify responses of ammonium, nitrate, and phosphate uptake by the benthic microbial community. Control uptake rates of sediments had net remineralization of nitrate ($-1.58\ NO_3\ \mu g\ gdm^{-1}\ h^{-1}$), and net assimilation of phosphate ($1.34\ PO_4\ \mu g\ gdm^{-1}\ h^{-1}$) and ammonium ($0.03\ NH_4\ \mu g\ gdm^{-1}\ h^{-1}$). Metolachlor decreased ammonium and phosphate uptake. Chlorothalonil decreased nitrate remineralization and phosphate uptake. Nitrate, ammonium, and phosphate uptake rates are more pronounced in the presence of these pesticides due to microbial adaptations to toxicants. Our interpretation of pesticide availability based on their water/solid affinities supports no effects for atrazine and carbaryl, decreasing nitrate remineralization, and phosphate assimilation in response to chlorothalonil. Further, decreased ammonium and phosphate uptake in response to metolachlor is likely due to affinity. Because atrazine target autotrophs, and carbaryl synaptic activity, effects on benthic microbes were not hypothesized, consistent with results. Metolachlor and chlorothalonil (non-specific modes of action) had significant effects on sediment microbial nutrient dynamics. Thus, pesticides with a higher affinity to sediments and/or broad modes of action are likely to affect sediment microbes' nutrient dynamics than pesticides dissolved in water or specific modes of action. Predicted nutrient uptake rates were calculated at mean and peak concentrations of metolachlor and chlorothalonil in freshwaters using polynomial equations generated in this experiment. We concluded that in natural ecosystems, peak chlorothalonil and metolachlor concentrations could affect phosphate and ammonium by decreasing net assimilation, and nitrate uptake rates by decreasing remineralization, relative to mean concentrations of metolachlor and chlorothalonil. Our regression equations can complement models of nitrogen and phosphorus availability in streams to predict potential changes in nutrient dynamics in response to pesticides in freshwaters.

Editor: Christopher Joseph Salice, Texas Tech University, United States of America

Funding: Funding provided by Indiana Water Resources Research Consortium subaward from the United States Department of Interior, United States Geological Survey as part of the federal Water Resources Research Act of 1984 (Program 104B)"(MJB, DE), http://www.iwrrc.org/. The funders had no role in study design, data collection and analysis, decision to publish, or preparation of the manuscript.

Competing Interests: The authors have declared that no competing interests exist.

* Email: delias@bsu.edu

Introduction

Agricultural activities, such as crop protection via pesticides, are increasing in response to global human population growth (projected to reach 9 billion by 2050; [1]). In the last decade, U.S. pesticide sales have increased ~10% with 80% of these pesticides used for agricultural activities [1]. The continued growth of the human population, coupled with the need for more efficient agricultural practices, will undoubtedly yield future increases in the occurrence of pesticides in freshwaters. Further, despite decades of research on agricultural pesticides, recent calls to action have highlighted the need to fill critical knowledge gaps in our understanding of how pesticides may adversely affect aquatic ecosystems [2], [3].

Once in the aquatic ecosystem, pesticides may have adverse effects on organisms ranging from direct toxicity to indirect effects such as changes in growth or behavior [4]. At higher concentrations, such as those following spring runoff, direct mortality results across diverse organisms including tadpoles [4], bluegill [5], and benthic organisms (e.g., amphipods and chironomids, [6]). However, at lower concentrations, sub-lethal effects can result in altered respiration rates [7], organismal growth [4] and fecundity [8]. In streams, benthic microbes are an important component of aquatic ecosystems and are integral in nutrient and energy dynamics [9]. For example, nitrate, ammonium and phosphorus are cycled by benthic microbes through assimilation and remineralization [10]. These processes are influenced by N concentration in freshwaters [11], [12]. At saturated conditions, such as those in agricultural streams with high input of N from fertilizer runoff, more N is available, due to microbial uptake saturation (i.e. biota have reach their N demand; [13]) or an increase in heterotrophic mineralization [11]. Further, at increasing N concentrations, PO_4 often becomes a secondary limiting nutrient [12].

In the Midwestern U.S., two herbicides (atrazine and metolachlor), one insecticide (carbaryl) and one fungicide (chlorothalonil) have both high usage rates and prevalence in receiving waters [14–17]. Atrazine is a triazine herbicide used predominantly in corn production for control of broadleaf and grassy weeds [18] with a half-life in water at pH seven of 86 days [19]. Metolachlor is a chloroacetanilide herbicide that inhibits mitosis and cell division [19]. Metolachlor is stable in water at pH seven [19]. Atrazine and metolachlor were detected in U.S. freshwaters at peak concentrations of 201 µg/L and 77.6 µg/L (Table 1). Carbaryl is a carbamate family insecticide that inhibits the enzyme acetylcholinesterase [19], with a half-life in water of 12 days at pH seven [19]. Chlorothalonil is a fungicide used in U.S. agriculture; it is stable in water at pH seven [19]. Its mode of action is by binding to glutathione and negatively affecting cellular respiration [7]. Carbaryl and chlorothalonil were detected in U.S. freshwaters at peak concentrations of 4.8 µg/L and 0.3 µg/L (Table 1).

These pesticides are detected in freshwater ecosystems at concentrations that adversely affect biota and human health (Table 2). However, research has focused primarily on the impacts of agricultural pesticides on fish and invertebrates; little is known about how exposure to pesticides may directly influence benthic nutrient dynamics and overall ecosystem function [20]. For example, ecotoxicology studies addressing the effects of pesticides focus primarily on non-benthic vertebrates (e.g., bluegill, [5]), and benthic invertebrates (e.g., amphipods and chironomids, [6]) with few studies conducted on sediment microbial dynamics [21]. Benthic microbial communities influence nutrient cycling (e.g. uptake, remineralization) by affecting fluxes as consumers or sources. Thus, benthic microbial communities are an important component of the freshwater ecosystem [22]. These nutrient dynamics are affected by the presence of pesticides [23]. Specifically, pesticides can reduce microbial activity that contributes to nutrient cycling (e.g. *Volvox* spp., *Botryococcus* spp., *Synedra* spp.) [24], and change species composition by favoring microbes with enhanced pesticide degradation capacities. Also, pesticides can become nutrient sources by providing carbon, nitrogen or phosphorus to some microorganisms [25], and alter nitrogen and/or phosphorus cycles [26]. Thus, there is a need to understand the direct effect of pesticides on sediment nutrient dynamics, and how these changes can affect whole-ecosystem pools and fluxes of nutrients [27].

We measured the effects of atrazine, metolachlor, carbaryl and chlorothalonil on benthic microbial nutrient dynamics by quantifying net assimilation and remineralization rates of ammonium, nitrate and phosphate in laboratory mesocosms. We hypothesized that pesticides with a broad mode of action and higher affinity to organic matter such as chlorothalonil (disruption of cellular respiration, log K_{ow}: 2.9) and metolachlor (inhibitor of mitosis and cellular division, log K_{ow}: 3.4) would decrease microbial nitrate and phosphate uptake rates. In contrast, pesticides with more specific modes of action and higher affinity to the aqueous phase (atrazine: blocks photosynthesis, log K_{ow}: 2.7 and carbaryl: inhibitor of synaptic activity, log K_{ow}: 2.4) were predicted to have no effect on nutrient cycling. Further, natural resources managers and stakeholders would be able to make general predictions of agricultural pesticides effects on the microbial community based on the mode of action and water/solid affinities of these pesticides.

Materials and Methods

Experimental mesocosms

Sediment and water collection was conducted in May 2012 at Ball State University field station property of Jakes Creek - Cooper farm/Skinner field (40.234493, −85.45235) and approval for these experiments was received following the appropriate procedures. This sample collection did not involve endangered or protected species. Jakes Creek is a 3rd order agriculturally-influenced stream with adjacent row crops (i.e. corn and soybean) in Muncie, Indiana within the Upper White River Watershed (UWRW). During the sampling time, Jakes Creek water temperature was 17°C, pH 7.92, depth 5 cm (at sampling location), and discharge was 37 L/s. -While stream samples were not collected for pesticides analysis at the time of this experiment; stream water and sediment samples were collected one week prior to this study at the same site that shows atrazine and metolachlor concentrations were below detection limits. Filtered water samples (ten 1000 mL and one 200 mL) were collected from the stream thalweg using a 60 mm syringe and subsequently filtered (Whatman© glass fiber filter; 0.7 µm nominal pore size) into acid-washed Nalgene© bottles. The 200 mL Nalgene plastic bottle was used to determine initial concentrations of nitrate, ammonium, and phosphate. A composite sediment sample (~2000 cm^3) was randomly collected from the top 5 cm of the stream benthos and placed into three Nalgene plastic bottles. Sediment samples were transported on ice and subsequently combined and homogenized using a USGS no. 5 sieve in the laboratory. Homogenized sediment (20 cm^3) and 60 ml filtered stream water were placed into each of 160 laboratory mesocosms (Fisherbrand sterile urine cup, 120 ml).

Stock solutions were prepared for atrazine (Atrazine 4L, 42.2% purity, Loveland, CO, US), metolachlor (Me-too-lachlor II, 84.4% purity, Drexel Chemical Company, TN, US), carbaryl (Sevin XLR Plus, 44.1% purity, Bayer, NC, US), and chlorothalonil (Bravo, 54% purity, Syngenta, NC, US) to achieve final stock concentrations of 10,000 µg/L for atrazine and metolachlor, 5,000 µg/L for carbaryl, and 8,000 µg/L for chlorothalonil. Aliquots from each stock solution were added to mesocosms to

Table 1. Detection frequency and concentrations of atrazine, metolachlor, carbaryl, and chlorothalonil in U.S. freshwaters.

Compound	Detection frequency (%)	Mean concentration (µg/L)	Maximum concentration (µg/L)	References
Atrazine (Herbicide)	78.1	2.4	201	[15–17]
Metolachlor (Herbicide)	71.1	1.2	77.6	[15–17]
Carbaryl (Insecticide)	18.1	0.013	4.78	[15–17]
Chlorothalonil (Fungicide)	0.033	<0.07	0.29 (290*)	[15–17]

Detection frequency and concentrations of atrazine, metolachlor, carbaryl, and chlorothalonil in U.S. freshwaters. Detection frequency was estimated throughout the U.S. across 50 basins (33 agricultural, 10 urban and 7 mixed); mean and maximum concentrations correspond to 83 agricultural streams. Annual mean detection frequencies for each compound at each site provide the proportion of water samples that have detectable levels of pesticides for a year period. *Chlorothalonil was detected at concentrations of 290 µg/L in run-off near golf courses.

Table 2. Toxicity and Octanol-water partition coefficient of atrazine, metolachlor, carbaryl and chlorothalonil to daphnids, green algae, and humans.

Compound	Daphnids	Green algae	Humans	Octanol-Water partition coefficient
	(mg/L)	(mg/L)	(mg/kg/bw/d)	
	NOEC	NOEC	ADI	Log K_{ow}
Atrazine	0.25	0.1	0.02	2.7
Metolachlor	0.7	57.1*	0.1	3.4
Carbaryl	0.25	-	0.0075	2.4
Chlorothalonil	0.009	0.033	0.015	2.9

Toxicity of atrazine, metolachlor, carbaryl and chlorothalonil in mg/L to daphnids and green algae, and mg/kg of body weight (bw) per day (d) to humans. No observed effect concentrations (NOEC) for daphnids and green algae were calculated by chronic tests of 21 days and 96 hours, respectively. Acceptable daily intake (ADI).* Half maximal effective concentration (EC50) of metolachlor on growth after 72 hours (19).

reach ten target treatment concentrations for each pesticide with four replicates for each treatment.

Mean and peak atrazine (2 µg/L, 201 µg/L) and metolachlor (1 µg/L, 78 µg/L) are detected at concentrations ~2–3 orders of magnitude higher than carbaryl (0.01 µg/L, 5 µg/L) or chlorothalonil (0.07 µg/L, 0.3 µg/L) throughout the U.S. (14–17). Thus, the treatment concentrations used were selected to include these environmentally relevant concentrations and appropriately represent pesticide occurrence in streams. Treatment solutions used in this study ranged from 0 µg/L (control) to the maximum concentrations detected in U.S. freshwaters for atrazine (200 µg/L), metolachlor (80 µg/L), carbaryl (4 µg/L) and chlorothalonil (0.5 µg/L) (Table 2).

Water from each mesocosm was removed after 24 h using a 10 mL syringe, subsequently filtered as above and placed into vials (two analytical replicates ~5 ml) for analysis of nitrate and phosphate via ion chromatograph (DIONEX, ICS-3000). The colorimetric phenol-hypochlorite technique [28], [29] was used to quantify ammonium concentrations. Initial concentrations (background) of nitrate, ammonium, and phosphate were also analyzed following the analytical methods above. Sediment dry mass in each mesocosm was quantified using an analytical balance (OHAUS, Adventurer SL AS64).

Data analysis

Nutrient uptake rates were calculated for phosphate, ammonium and nitrate as changes in concentration over time (24 h) per g of dry mass (sediment) in response to treatments as (22):

$$Nutrient\,uptake\,rate = \frac{(C_f - C_i)\cdot V}{T\cdot gdm}$$

Where: Cf = Final concentration (mg/L); Ci = Initial concentration (mg/L); V = Volume (L) in the jar$T$ = time (h); gdm = g dry mass (g). Negative nutrient uptake rates indicated net remineralization of nutrients and positive nutrient uptake rates indicated net assimilation of nutrients (Table 3). Uptake rates were divided by the average of the control treatment for each pesticide (N = 16) to assess the effects of different treatments for a particular pesticides. Thus, in this study, a response ratio relative to controls was used as the response variable. Further, data were log-transformed to meet normality assumptions for statistical analyses. SigmaPlot© 12.0 software was used for linear and nonlinear regression analyses of response to pesticide concentration. The Akaike Information criterion (AIC) was used to select the best fit model among the different polynomial candidate models. Further, to develop predictive models of microbial response to pesticides in agricultural waters mean and peak concentrations of metolachlor and chlorothalonil throughout U.S. freshwaters (Table 1) were used to calculate nitrogen and phosphorus uptake rates.

Results

Nitrate, phosphate, and ammonium uptake rates varied less than one order of magnitude across pesticide treatments (Table 3). Phosphate uptake rates were three orders of magnitude greater than ammonium uptake rates across treatments, though both phosphate (mean = 180.56 µg gdm^{-1} h^{-1}) and ammonium (mean = 0.34 µg gdm^{-1} h^{-1}) uptake rates were net assimilative (Table 3). In contrast, we observed net remineralization of nitrate (mean = −32.22 µg gdm^{-1} h^{-1}) across pesticides (Table 3). Overall, these nutrient dynamics are expected in nitrogen-saturated agricultural ecosystems (i.e., microbial nitrate assimilation is saturated and PO_4 becomes a limiting nutrient).

Nitrate dynamics

Control nitrate uptake rate was ~20x higher on average than nitrate uptake influenced by pesticides (i.e., increasing remineralization in presence of pesticides). Nitrate uptake rates in response to atrazine, metolachlor, and carbaryl treatments exposure ranged from net assimilation (consumption/removal) to remineralization (source/addition) in the water column of our mesocosms (Table 3). In contrast, increasing concentrations of chlorothalonil yielded increasing nitrate remineralization ($p = 0.02$, $r^2 = 0.83$, Fig. 1A). No other pesticides had significant effects on nitrate uptake rates ($p > 0.05$).

Phosphate dynamics

Phosphate uptake rates in response to pesticides exposure was ~100x higher than control phosphate uptake rates. Thus, phosphate assimilation increased in the presence of atrazine, metolachlor, carbaryl, and chlorothalonil relative to the control (Table 3). Further, phosphate uptake was negatively related to increasing chlorothalonil concentration ($p < 0.001$, $r^2 = 0.87$, Fig. 1B). Similarly, the metolachlor effect on phosphate uptake rate ($p = 0.005$, $r^2 = 0.91$, Fig. 1C) followed a cubic relationship with increasing phosphate uptake at lower concentrations (0 to 10 µg/L), and decreasing rates at higher concentrations (10 to 80 µg/L). No other pesticides resulted in significant effects on phosphate uptake rates ($p > 0.05$).

Table 3. Uptake rates for nitrate, phosphate, and ammonium in response to pesticides exposure.

Pesticide	Nitrate (μg gdm^{-1} h^{-1})	Phosphate (μg gdm^{-1} h^{-1})	Ammonium (μg gdm^{-1} h^{-1})
Atrazine	−25.88 (−45.2-4.30)	126.93 (83.1-169)	0.36 (0.23-0.45)
Metolachlor	−21.34 (−40.9-2.10)	169.54 (120-214)	0.39 (0.03-0.53)
Carbaryl	−25.13 (−56.6-8.80)	192.11 (37.9-260)	0.31 (0.13-0.53)
Chlorothalonil	−56.54 (−83.1-−3.7))	233.67 (39.8-601)	0.28 (0.13-0.47)
Mean uptake rate	**−32.22**	**180.56**	**0.34**
Control	−1.58 (−1.62-−1.49)	1.34 (7.03E-05-4.4)	0.03 (0.01-0.04)

Uptake rates for nitrate, phosphate, and ammonium (μg gdm^{-1} h^{-1}) across pesticide and control (no pesticide) treatments. Range noted in parentheses. Mean uptake rate was calculated for each nutrient across pesticides.

Ammonium dynamics

Ammonium uptake rates in response to pesticides exposure was ~10x higher than control ammonium uptake rate. Thus ammonium assimilation increased in the presence of atrazine, metolachlor, carbaryl, and chlorothalonil relative to the control (Table 3). The ammonium uptake rate varied in response to metolachlor treatments ($p = 0.023$, $r^2 = 0.83$, Fig. 1D), increasing with lower concentrations of metolachlor (0 to 10 μg/L), followed by a decline at higher concentrations (10 to 80 μg/L). No other pesticides had significant effects on ammonium uptake rates ($p > 0.05$).

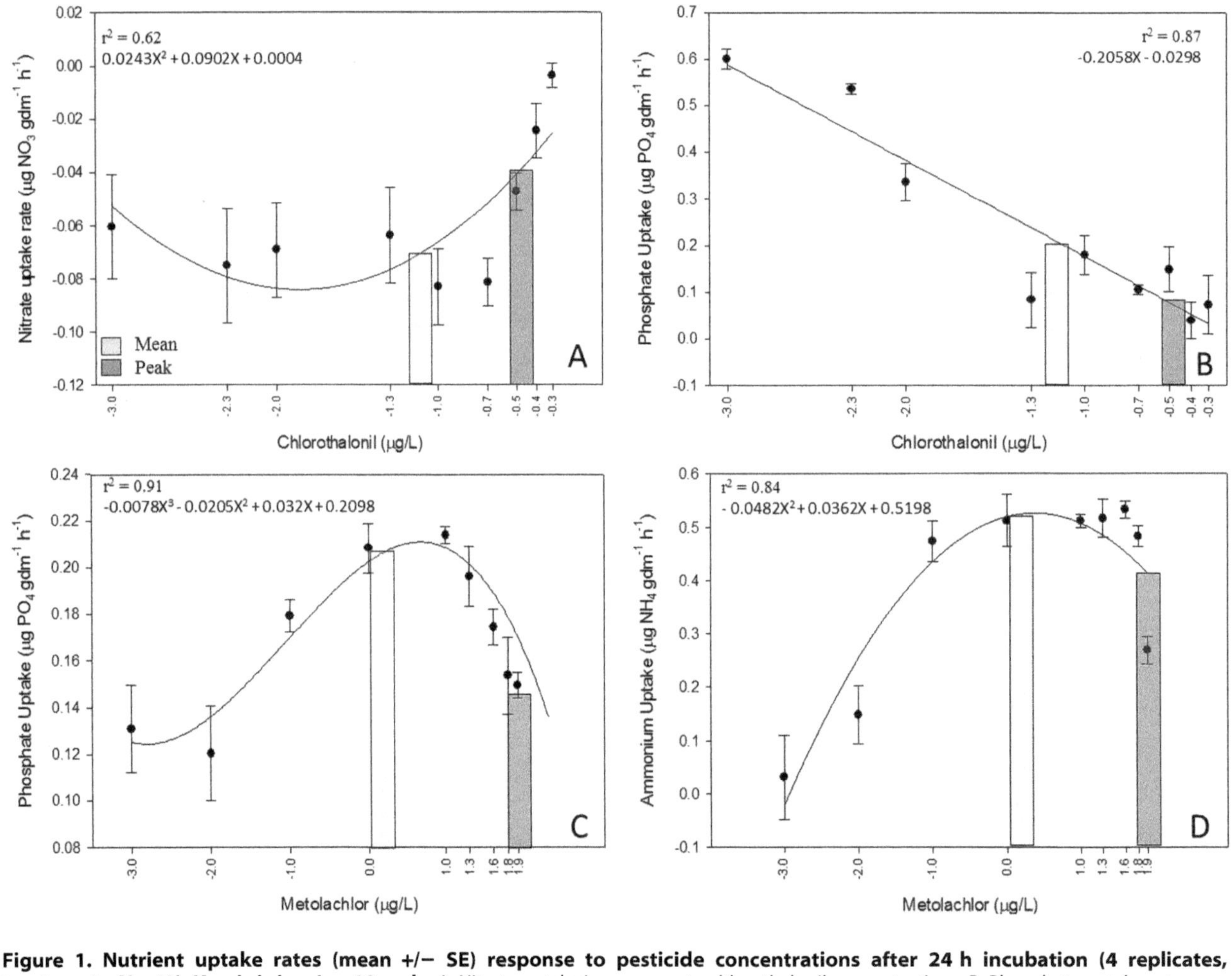

Figure 1. Nutrient uptake rates (mean +/− SE) response to pesticide concentrations after 24 h incubation (4 replicates, 10 treatments, N = 40). X-axis is in a Log10 scale. A: Nitrate uptake in response to chlorothalonil concentrations. B: Phosphate uptake rate response to chlorothalonil concentrations. C: Phosphate uptake rate response to metolachlor concentrations. D: Ammonium uptake rate response to metolachlor concentrations. Columns represent predicted uptake rates for each nutrient calculated at mean and peak concentrations of metolachlor and chlorothalonil measure in U.S. freshwaters (Table 2).

Predicting nutrient response to pesticides

Changes in stream ecosystem nitrogen and phosphorus uptake rates were modeled across metolachlor and chlorothalonil concentrations detected throughout the U.S. (Fig. 1). These changes were generated with polynomial regressions from this experiment ($p<0.05$).

The predicted ammonium uptake (0.52 NH_4 μg gdm^{-1} h^{-1}) in response to mean metolachlor concentrations was ~20% higher than ammonium uptake (0.42 NH_4 μg gdm^{-1} h^{-1}) at peak metolachlor concentrations. At mean concentrations of metolachlor, the predicted phosphate uptake (212.2 PO_4 μg gdm^{-1} h^{-1}) was ~40% lower than phosphate uptake (372.41 PO_4 μg gdm^{-1} h^{-1}) at peak metolachlor concentrations. At mean and peak chlorothalonil concentrations, there is net remineralization of nitrate (−71.99 NO_3 μg gdm^{-1} h^{-1} and −41.87 NO_3 μg gdm^{-1} h^{-1}, respectively). Further, there was net assimilation of phosphate in response to mean and peak concentrations of chlorothalonil (207.88 and 77.81 PO_4 μg gdm^{-1} h^{-1}, respectively).

Overall, there was decreased ammonium assimilation and remineralization of nitrate in the presence of peak concentrations of metolachlor and chlorothalonil (Fig. 1). Phosphate assimilation increased at peak concentrations of metolachlor and decreased at peak concentrations of chlorothalonil (Fig. 1). Peak chlorothalonil is predicted to have the greatest effect on phosphate and nitrate uptake rates, by decreasing net assimilation and remineralization, respectively, over 50% relative to mean concentrations of these pesticides.

Discussion

Research focuses primarily on the impacts of agricultural pesticides on fish and invertebrates [5], [6]; however, little is known about effects on microbial communities [20]. Our ecotoxicological research showed that benthic microbes' nutrient uptake rate response is likely a function of pesticide chemical characteristics, and how these changes can affect nutrient dynamics, due to nutrient availability.

Variation in nutrient uptake rates is likely a result of differences in baseline nutrient concentrations and the biotic community. For example, control ammonium uptake rates in this study (0.026 μg NH_4 gdm^{-1} h^{-1}) were ~3x lower than control ammonium uptake rates in Bunch and Bernot [30]. These results suggest dissimilar nutrient demand due to unique biotic communities and nutrient conditions across sampling sites and temporal variability.

In our study, mean ammonium uptake rate was ~1.5x higher than ammonium uptake under enriched nitrate conditions and ~2.5x lower under enriched ammonium conditions reported in Bunch and Bernot [30]. Thus, ammonium uptake is higher in the presence of pesticides than under enriched nitrate conditions and lower under enriched ammonium conditions, due to microbial adaptations to nutrient availability and metabolic responses in presence of pesticides. Our nitrate uptake results were similar to yields with net remineralization under enriched ammonium conditions reported in Bunch and Bernot [30], though remineralization was 10 times lower than rates measured in response to pesticides in our study.

The effects of metolachlor and chlorothalonil on nutrient uptake rates suggest a unique biotic community at this site, represented by mostly heterotrophic benthic microbes. Our results suggest a toxic effect of metolachlor and chlorothalonil on the benthic microbial community that is reflected by the increased nitrate remineralization, and reduced assimilation of ammonium and phosphate. This increase in remineralization rates could be an outcome of cellular lysis or a stress mechanism [31]. Further, a biotic community characterized by autotrophs [22] had ammonium and nitrate uptake rates three orders of magnitude higher than benthic uptake in our study; possibly, under these study conditions, primary producers have a higher assimilation rate than heterotrophic benthic microbes in response to the available forms of nitrogen [22]. Further, control phosphate uptake rates in our study were ~5x lower than rates previously measured with phosphorus enrichment (0.1–2 mg/L; 34). Thus, phosphate is likely a limiting nutrient in our system. However, in presence of pesticides, phosphate uptake was two orders of magnitude higher than the rates measured under phosphorus limiting conditions [32] possibly, in addition to available phosphorus in the water column, microbes were potentially degrading pesticides as a source of phosphorus [33].

Nutrient dynamics in the presence of pesticides are dependent on the physicochemical characteristics (e.g. sorption kinetics, modes of action) of each pesticide. Sorption kinetics of pesticides and their corresponding index (Octanol-water partition coefficient, K_{ow}) determine the affinity of organic contaminants to either the water column or sediments [34]. Atrazine and carbaryl have a higher affinity to the aqueous phase (Log K_{ow}: 2.7 and 2.4, respectively), relative to metolachlor and chlorothalonil (Log K_{ow}: 3.4 and 2.9, respectively). Thus, atrazine and carbaryl are likely more prevalent in the water column and less available to the sediment microbial communities, with minimal effect on benthic microbial activity. In contrast, metolachlor and chlorothalonil have a higher affinity to solids and higher prevalence in sediment, potentially affecting benthic microbial activity. Our interpretations of pesticide availability based on their water/solid affinities supports our results of no effects for atrazine and carbaryl (Log $K_{ow}<2.7$, $p>0.05$) within the tested range, decreasing nitrate remineralization and decreasing phosphate uptake in response to chlorothalonil, and decreasing ammonium and phosphate uptake in response to metolachlor (Log $K_{ow}>2.9$, $p<0.05$).

Nutrient dynamics are also affected by the pesticide mode of action. Atrazine and carbaryl have specific modes of action; atrazine blocks photosynthesis, and carbaryl inhibits the activity of acetylcholinesterase, an enzyme of insects, fish, mammals [35]. Thus, the specificity of atrazine and carbaryl, and a sediment microbial community dominated by heterotrophs with no synaptic activity [36] may explain the lack of significant effects of these pesticides on benthic microbes within the tested range (Table 3). In contrast, metolachlor and chlorothalonil are broad spectrum pesticides [19]. In our study, ammonium and phosphate uptake rates decreased with increasing metolachlor concentrations. These decreasing ammonium and phosphate uptake rates could be due to metolachlor inhibition of mitosis and cell division. Similarly, phosphate and nitrate uptake rates were affected by chlorothalonil; possibly this fungicide affects benthic microbes by disrupting cellular respiration. Thus, pesticides with non-specific modes of action (e.g. metolachlor, chlorothalonil) are more likely to have a significant effect on nutrient dynamics of sediment microbes, consistent with our study results.

Metolachlor and chlorothalonil not only affect nutrient dynamics of the sediment microbial community; they can also affect other processes. In our study, at increasing concentrations of chlorothalonil, there is a decrease in phosphate uptake, thus there is more phosphate available for organismal consumption. In agricultural streams, where phosphorus is the limiting nutrient [37], increasing availability of phosphate can lead to algal blooms, eutrophication, hypoxia, loss of biodiversity (e.g. fish kills), and loss of aesthetic value of these habitats [38]. In contrast, in the presence of peak concentrations of metolachlor, there is an increase in phosphate assimilation, which could mitigate the excess phosphate in streams [39]. Similarly, there is an increase of ammonium availability due to inhibited uptake rates in response to metolachlor, which in turn could potentially increase biological activity of pesticide resistant microbes [21], [25], [26], [33]. Nitrate remineralization also decreases in response to chlorothalonil, reducing nitrate availability, and potentially further mitigating excess nitrogen in these habitats [40].

Mesocosms have a critical role in understanding the mechanisms driving ecological processes [42], [43] and provide a bridge between smaller, better controlled experiments and the larger freshwater ecosystems [44]. For example, [43] revealed that the effects of nutrients on primary producers are similar in artificial habitats across five orders of magnitude in size. Also, [44] mentioned that mesocosms help disentangle direct from indirect effects over scale. Thus, our conclusions try to bridge what we observed in the laboratory level and what could potentially occur at an agricultural influenced stream level.

Pesticide occurrence and concentrations in streams are dictated primarily by land-use. Streams receiving run-off from agricultural landscapes frequently have the highest concentrations of pesticides, compared to forested, mixed-use and urban lands [41]. These high concentrations occur as pulses that are coupled by the seasonality of agricultural practices [16]. Peak concentrations of metolachlor and chlorothalonil are detected from April to August [16]. Thus, the effects of these pesticides on nutrient dynamics are highest during this critical time. At peak metolachlor and chlorothalonil concentrations there is a decrease of predicted ammonium assimilation and nitrate remineralization. Further, at peak concentrations of metolachlor there is increased phosphate assimilation and at peak concentrations of chlorothalonil there is decreased phosphate assimilation (Fig. 1).

Our findings demonstrate individual effects of these pesticides on sediment nutrient dynamics that are likely driven by a pesticides' mode of action and water/sediment affinities. More studies are required to understand the net effect on ecosystems and to address their synergistic or antagonistic effects as mixtures. The regression equations we generated can complement models of nitrogen and phosphorus availability in streams to predict the potential changes in nutrient dynamics in response to increasing presence of pesticides in lotic ecosystems.

Acknowledgments

We thank James Justice, Mark Pyron and two anonymous reviewers for helpful comments.

Author Contributions

Conceived and designed the experiments: DE MJB. Performed the experiments: DE. Analyzed the data: DE. Contributed reagents/materials/ analysis tools: DE MJB. Wrote the paper: DE MJB.

References

1. Enserink M, Hines PJ, Vignieri SN, Wigginton NS, Yeston JS (2013) Smarter pest control. The pesticide paradox. Introduction. Science 341: 728–729.
2. Köhler H-R, Triebskorn R (2013) Wildlife ecotoxicology of pesticides: can we track effects to the population level and beyond? Science 341: 759–765.
3. Murray KE, Thomas SM, Bodour AA (2010) Prioritizing research for trace pollutants and emerging contaminants in the freshwater environment. Environmental Pollution 158: 3462–3471.
4. Relyea RA (2006) The effects of pesticides, pH, and predatory stress on amphibians under mesocosm conditions. Ecotoxicology 15: 503–511.
5. Munn MD, Gilliom RJ, Moran PW, Nowell LH (2001) Pesticide toxicity index for freshwater aquatic organisms. Water Resources Investigations Report United States Geological Survey: 66.
6. Liess M, Ohe PC (2005) Analyzing effects of pesticides on invertebrate communities in streams. Environmental Toxicology and Chemistry 24: 954–965.
7. McMahon TA, Halstead NT, Johnson S, Raffel TR, Romansic JM, et al. (2012) Fungicide-induced declines of freshwater biodiversity modify ecosystem functions and services. Ecology Letters 15: 714–722.
8. Kashian DR, Dodson SI (2002) Effects of common-use pesticides on developmental and reproductive processes in Daphnia. Toxicology and Industrial Health 18: 225–235.
9. Covich AP, Palmer MA, Crowl TA (1999) The role of benthic invertebrate species in freshwater ecosystems: zoobenthic species influence energy flows and nutrient cycling. BioScience 49: 119–127.
10. Dodds WK, Lopez AJ, Bowden WB, Gregory S, Grimm NB, et al. (2002) N uptake as a function of concentration in streams. Journal of the North American Benthological Society 21: 206–220.
11. Bernot MJ, Dodds WK (2005) Nitrogen Retention, Removal, and Saturation in Lotic Ecosystems. Ecosystems 8: 442–453.
12. Earl SR, Valett HM, Webster JR (2006) Nitrogen saturation in stream ecosystems. Ecology 87: 3140–3151.
13. Davis JC, Wayne MG (1999) Nitrogen and phosphorus uptake in two Idaho (USA) headwater wilderness streams. Oecologia 119: 247–255.
14. United States Environmental Protection Agency (1987) Chlorothalonil: health advisory. U.S. Environmental Protection Agency, Washington, D.C.
15. Kolpin DW, Barbash JE, Gilliom RJ (1998) Occurrence of pesticides in shallow groundwater of the United States: Initial results from the National Water-Quality Assessment Program. Environ. Sci. Technol. 32: 558–566.
16. Larson SJ, Gilliom RJ, Capel PD (1999) Pesticides in Streams of the United States—Initial Results from the National Water-Quality Assessment Program. U.S. Department of the Interior, US Geological Survey.
17. Gilliom RJ, Barbash JE, Crawford CG, Hamilton PA, Martin JD, et al. (2006) Pesticides in the nation's streams and ground water, 1992–2001. U.S. Geological Survey Circular 1291: 172.
18. Kruger EL, Coats JR, Zhu BE (1996) Relative mobilities of atrazine, five atrazine degradates, metolachlor, and simazine in soils of Iowa. Environmental Toxicology and Chemistry 15: 691–695.
19. University of Hertfordshire (2013). The Pesticide Properties DataBase (PPDB) developed by the Agriculture & Environment Research Unit (AERU), University of Hertfordshire, 2006–2013.
20. Beketov MA, Kefford BJ, Schäfer RB, Liess M (2013) Pesticides reduce regional biodiversity of stream invertebrates. Proceedings of the National Academy of Sciences.
21. Satsuma K (2006) Characterisation of new strains of atrazine-degrading Nocardioides sp. isolated from Japanese riverbed sediment using naturally derived river ecosystem. Pest Management Science 62: 340–349.
22. Kemp MJ, Dodds WK (2002) The influence of ammonium, nitrate, and dissolved oxygen concentrations on uptake, nitrification, and denitrification rates associated with prairie stream substrata. Limnology and Oceanography 47: 1380–1393.
23. Gramlich J, Davis D (1967) Effect of atrazine on nitrogen metabolism of resistant species. Weeds: 157–160.
24. Neumann M, Dudgeon D (2002) The impact of agricultural runoff on stream benthos in Hong Kong, China. Water Research 36: 3103–3109.
25. Tappin AD, Loughnane JP, McCarthy AJ, Fitzsimons MF (2012) Removal of atrazine from river waters by indigenous microorganisms. Environmental Chemistry Letters 10: 89–96.
26. Doddamani HP, Ninnekar HZ (2001) Biodegradation of carbaryl by a Micrococcus species. Current Microbiology 43: 69–73.
27. Brown JH, Gillooly JF, Allen AP, Savage VM, West GB (2004) Toward a metabolic theory of ecology. Ecology 85: 1771–1789.
28. American Public Health Association AWWAWEF (1995) Standard methods for the examination of water and wastewater. Washington, D.C.: American Public Health Association.
29. Aminot A, Kirkwood DS, Kerouel R (1997) Determination of ammonia in seawater by the indophenol-blue method: Evaluation of the ICES NUTS I/C 5 questionnaire. Marine Chemistry 56: 59–75.
30. Bunch ND, Bernot MJ (2012) Nitrate and ammonium uptake by natural stream sediment microbial communities in response to nutrient enrichment. Research in Microbiology 163: 137–141.
31. Bronk D, Ward B (1999) Gross and net nitrogen uptake and DON release in the euphotic zone of Monterey Bay, California. Limnology and Oceanography 44: 573–585.
32. Klotz R (1985) Factors controlling phosphorus limitation in stream sediments. Limnology and Oceanography 30: 543–553.
33. Cook AM, Daughton CG, Alexander M (1978) Phosphorus-containing pesticide breakdown products: quantitative utilization as phosphorus sources by bacteria. Applied and Environmental Microbiology 36: 668–672.
34. Wauchope RD, Yeh S, Linders JBHJ, Kloskowski R, Tanaka K, et al. (2002) Pesticide soil sorption parameters: theory, measurement, uses, limitations and reliability. Pest Management Science 58: 419–445.
35. Durieux ED, Farver TB, Fitzgerald PS, Eder KJ, Ostrach DJ (2011) Natural factors to consider when using acetylcholinesterase activity as neurotoxicity biomarker in Young-Of-Year striped bass (Morone saxatilis). Fish Physiology and Biochemistry 37: 21–29.

36. Pester M, Friedrich MW, Schink B, Brune A (2004) pmoA-based analysis of methanotrophs in a littoral lake sediment reveals a diverse and stable community in a dynamic environment. Applied and Environmental Microbiology 70: 3138–3142.
37. Correll D (1999) Phosphorus: a rate limiting nutrient in surface waters. Poultry Science 78: 674–682.
38. Conley DJ, Paerl HW, Howarth RW, Boesch DF, Seitzinger SP, et al. (2009) Controlling eutrophication: nitrogen and phosphorus. Science 323: 1014–1015.
39. Gächter R, Meyer J (1993) The role of microorganisms in mobilization and fixation of phosphorus in sediments. In: Boers PCM, Cappenberg TE, van Raaphorst W, editors. Proceedings of the Third International Workshop on Phosphorus in Sediments: Springer Netherlands. pp. 103–121.
40. Zogg GP, Zak DR, Pregitzer KS, Burton AJ (2000) Microbial immobilization and the retention of anthropogenic nitrate in a northern hardwood forest. Ecology 81: 1858–1866.
41. Rinella FA, Janet ML (1998) Seasonal and spatial variability of nutrients and pesticides in streams of the Willamette Basin, Oregon, 1993–95: U.S. Department of the Interior, U.S. Geological Survey.
42. Frost T, Ulanowicz R, Blumenshine S, Allen T, Taub F, et al. (2001) Scaling issues in experimental ecology: freshwater ecosystems. Scaling relations in experimental ecology Columbia University Press, New York: 253–280.
43. Spivak AC, Vanni MJ, Mette EM (2011) Moving on up: can results from simple aquatic mesocosm experiments be applied across broad spatial scales? Freshwater Biology 56: 279–291.
44. Stewart RIA, Dossena M, Bohan DA, Jeppesen E, Kordas RL, et al. (2013) Mesocosm Experiments as a Tool for Ecological Climate-Change Research. Advances in Ecological Research: Global Change in Multispecies Systems, Pt 3 48: 71–181.

17

Galvanic Manufacturing in the Cities of Russia: Potential Source of Ambient Nanoparticles

Kirill S. Golokhvast[1]*, Anna A. Shvedova[2,3]

1 Scientific Educational Center of Nanotechnology, Far Eastern Federal University, Vladivostok, Russian Federation, **2** Pathology and Physiology Research Branch/NIOSH/CDC, Morgantown, West Virginia, United States of America, **3** Department Physiology and Pharmacology, School of Medicine, West Virginia University, Morgantown, West Virginia, United States of America

Abstract

Galvanic manufacturing is widely employed and can be found in nearly every average city in Russia. The release and accumulation of different metals (Me), depending on the technology used can be found in the vicinities of galvanic plants. Under the environmental protection act in Russia, the regulations for galvanic manufacturing do not include the regulations and safety standards for ambient ultrafine and nanosized particulate matter (PM). To assess whether Me nanoparticles (NP) are among environmental pollutants caused by galvanic manufacturing, the level of Me NP were tested in urban snow samples collected around galvanic enterprises in two cities. Employing transmission electronic microscopy, energy-dispersive X-ray spectroscopy, and a laser diffraction particle size analyzer, we found that the size distribution of tested Me NP was within 10–120 nm range. This is the first study to report that Me NP of Fe, Cr, Pb, Al, Ni, Cu, and Zn were detected around galvanic shop settings.

Editor: Vipul Bansal, RMIT University, Australia

Funding: The work is financially supported by the Scientific Fund of the Far Eastern Federal University (No. 13-06-0318-m_a), the Presidential Grant for supporting young scientists (MK-1547.2013.5) and the Grant of the Ministry of Education and Science of the Russian Federation (No. 14.594.21.0006). The funders had no role in study design, data collection and analysis, decision to publish, or preparation of the manuscript.

Competing Interests: The authors have declared that no competing interests exist.

* Email: droopy@mail.ru

Introduction

Galvanics is a technology involving electrolytic precipitation of a thin layer of metal (Me) on a surface of pure Me or alloys. The products of galvanics can be used for protection from corrosion, to increase wear resistance, to protect from cementation, and for decorative purposes [1].

Many industries, including automotive, electronics, aerospace, hardware, jewelry, heavy equipment, appliances, tires, and telecommunications, use Me finishing for manufacturing goods [2]. The electroplating, plating, polishing, anodizing, and coloring industry is classified under the Standard Industrial Classification (SIC) code 3471 and includes establishments primarily engaged in all types of Me finishing [3]. Metallic coatings change the surface properties of the workpiece forming a composite material with new properties that could not be achieved by the material alone. The coating's function is usually as a durable, corrosion-resistant protective layer, while the core material provides a load-bearing function [4]. Common coating materials used include aluminum, lead, tin, zinc, and combinations of these metals.

It has been previously shown that airborne particulate matter (PM) collected from various urban and industrial sites contained a number of hard and soluble Me of Cd, Cu, Mn, Ni, Pb, Zn [5]. The potential adverse effects of hard and soluble metals on the environment and on habitat species are of major concern [6,7]. Studies conducted at a waste water treatment plant, which is in close proximity to a Ni/Cr plating plant, showed accumulation of different hard Me including Cu, Ni, Cr and Zn [8,9].

Recent epidemiological studies have shown a strong association between daily morbidity and mortality with increased air pollution [10–14]. The hazard posed by galvanic productions to an urban environment is an emerging issue that recently drew public attention and concern [15–18]. Chemical analysis of washing and sewage effluents coming to adjacent aquatic sources from locally producing small shops revealed the release of a number of harmful impurities including a high level of heavy Me that exceeded the environmental standard in Russia [19]. A high source of residual hazardous Me (Fe, Cr, Zn, Cu, Pb, Cd, Sn, Ni) and an elevated level of strong acids (e.g., hydrochloric, sulfuric, nitric, phosphoric, hydrocyanic and fluoric), used for electro-plating and/or galvanic fabrications, were seen in water outside of producing areas [19].

No published data is currently available regarding Me NP found around galvanic shops. Therefore, the goal of the current study was to assess whether Me NP could be traceable in snow sediments near industrial galvanic plants.

Methods

Sample collections

Snow sediments were collected for analysis of Me contaminants within several industrial areas of Blagoveshchensk and Ussuriisk (Russian Federation). The plans of cities were taken from OpenStreetMap (http://www.openstreetmap.org/copyright). The samples were gathered near large highways, repair plants, heat power plants and galvanic shops after snowfalls between

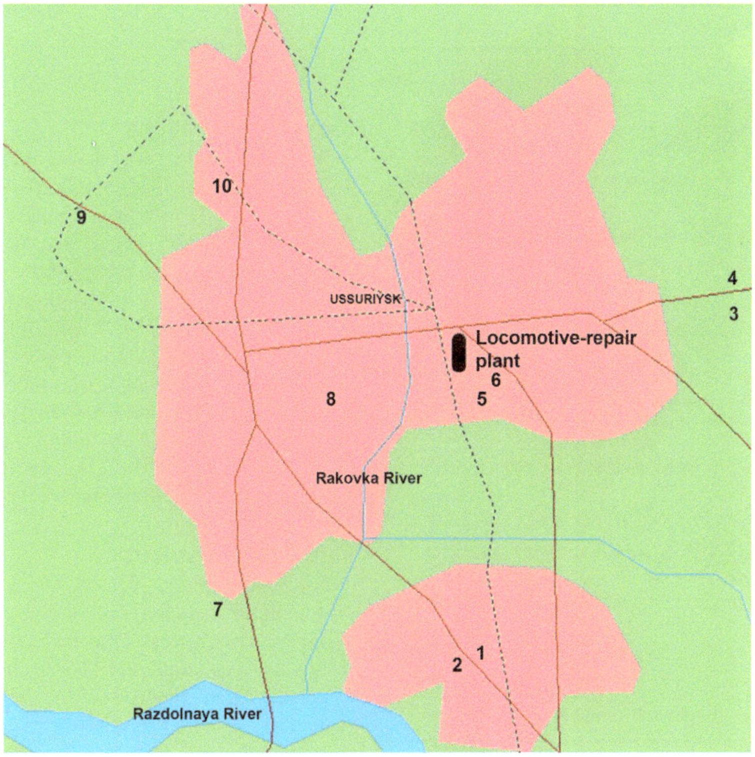

Figure 1. Schematic map of the locations of the sampling of snow in the territory of Ussuriisk (stations of sampling are described in Table 1. The black rectangle in the map denotes the locomotive-repair plant. (Earth Science and Remote Sensing Unit, NASA-Johnson Space Center. "The Gateway to Astronaut Photography of Earth." <http://eol.jsc.nasa.gov/Info/use.htm>09/16/2014 14:13:21).

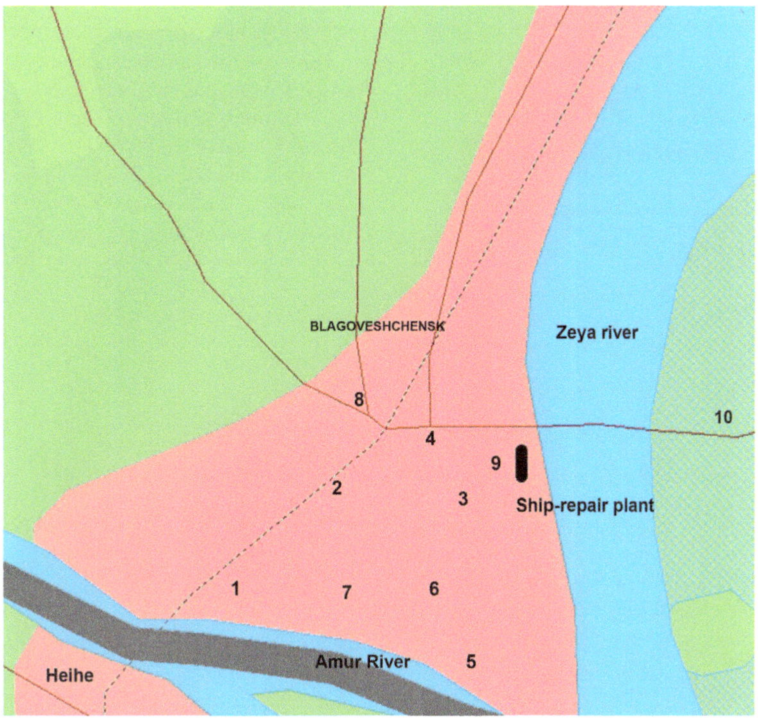

Figure 2. Schematic map of the locations of the sampling of snow in the territory of Blagoveshchensk (the stations of sampling are described in Table 2). The black rectangle in the map designates the ship-repair plant. (Earth Science and Remote Sensing Unit, NASA-Johnson Space Center. "The Gateway to Astronaut Photography of Earth." <http://eol.jsc.nasa.gov/Info/use.htm>09/16/2014 14:13:21).

Table 1. Sampling areas in Ussuriysk.

Station	Sample station location
1	District "Sugar Factory." Major road.
2	District "Sugar Factory." Major road. Large company - JSC "Primorsky sugar."
3	Bypass route on the leeward side. Road and park zone.
4	Bypass route - the opposite site. Road and park zone.
5	Industrial zone. Near Ussuri locomotive-repair plant (Blucher Avenue, 19), near railway station "Ussurijsk" 5 m from major road in public zone.
6	Industrial zone. The sample is taken by the house Blucher Avenue, 36, in front of the Ussuri locomotive-repair plant.
7	Utesnoe Village, park zone.
8	Down Town of the Ussuriysk with traffic 1500 car per hour and small boilers.
9	Large boiler, regularly on railroad tracks with carbon come.
10	Northern district of the city. There are several small boilers, and substantial flow of goods.

Table 2. Sampling areas in Blagoveshchensk.

Stations	Description of the sampling sites
1	First City Hospital. Parks.
2	Thermal power station, a coal-fired plant.
3	Exhibition center and highway.
4	Railway station.
5	Lenin Square. Shore of the Amur River. Major road.
6	Vicinity of shopping center "Mega." Large motor isolation.
7	Crossroad of Gorky and Kalinin streets.
8	Kalinin street. A major ring road.
9	Theatralnaya street. A major ring road. Ship-repair plant (Pushkin street, 189).
10	Traffic police post on the outskirts of the city. Major road.

2010–2012 (Fig. 1–2 and Tables 1–2). Samples were collected from public areas; therefore, no specific permissions were required.

Sample preparations

Snow sediments were chosen as sorbents for the assessment of airborne PM [20]. This method was employed to measure the level of Me NP in water samples derived from thawed snow sediments gathered from park zones (control) and the industrial sites. The snow samples were gathered from a 1 m^2 area within 200–500 m around the galvanic shop. The top layer of freshly fallen snow was collected into 3-liter plastic polyvinylchloride containers. The snow samples were thawed, evaporated to 60 ml and sterilized. All samples collected from the clean areas (control, n = 3) and industrial zones (exposed, n = 3) were stored and kept in the dark at 4°C until processed. Two independent experiments were done to assess Me NP content in snow sediment collected from two cities (see Materials S1 for the details).

Particle size distribution

For measurements of particle size distribution, a laser particle size analyzer supplied with Fritch MaS software (Analysette 22 NanoTec Fritsch, Idar-Oberstein, Germany) was employed. This technology can analyze PM size distribution of the wet or dry dispersion units separately or simultaneously with automated switching features. Optimal dispersion was accomplished in the NanoTec integral wet dispersion unit by using a combination of a robust, variable speed centrifugal pump with powerful ultrasonification, according to the manufacturer's manual. The software controls all of the functions of the NanoTec wet dispersion unit. The sample was added in the open dispersion chamber on the top of the instrument. The samples (60 ml) were diluted in water (150 ml) and then tested.

Scanning electron microscopy (SEM)

For scanning electron microscopy, the water from the samples was evaporated and the dry PM was covered with platinum using a turbo-pumped sputter coater Q150T (Quorum Technologies, Lewes, United Kingdom). The collected PM were characterized by X-ray diffraction (XRD) analysis. Morphology of PM was evaluated by SEM using Zeiss EVO 40XVP (Zeiss, Oberkochen, Germany) with an energy dispersive spectrometer INCA Energy (Oxford Instruments, Abingdon, United Kingdom) and Hitachi S-

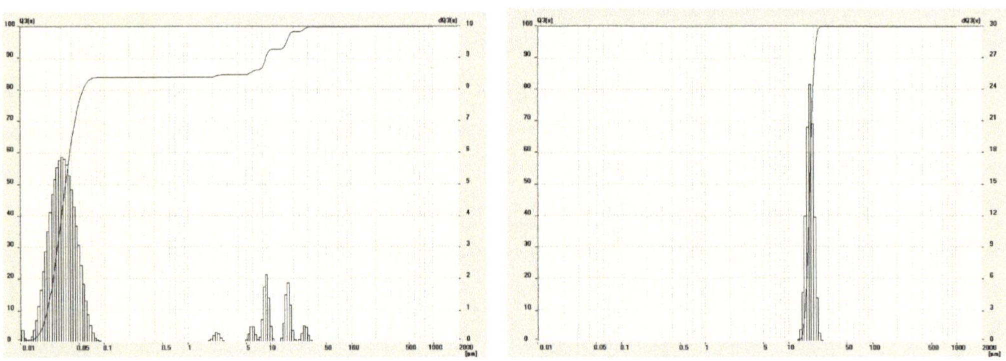

a b

Figure 3. The profile of particle size distribution in snow samples collected from the western section, **Ussiriisk (point 6) (a).** The size of the particles from the samples collection ranges between 10–70 nm and between 2–3, 5–7, 7–10, 7–12, 20–30 and 30–40 μm. The profile of particle size distribution in snow samples collected from the bypass road, Ussuriisk (point 3) **(b).** The size of the particles from the samples collection ranges between 14–23 μm.

Table 3. Morphometric parameters of the particles in thawed snow samples collected from points 3 and 6 in Ussuriisk.

	point 6						point 3
Size, μm	0,01–0,07	2–3	5–7	7–12	20–30	30–40	14–23
Proportion, %	84	1	3	5	5	2	100
Arithmetic mean diameter, μm	2.06±0.12						18.26±1.21
Specific surface area, m^2/sm^3	171.15						0.13

Table 4. Morphometric data of the particles in thawed snow samples collected from points 1–5 and 7–10 in Ussuriisk.

Points	1	2	4	5	7	8	9	10
Arithmetic mean diameter, μm	519.41±33.2	104.7±9.0	146.99±8.8	386.19±22.1	432.22±38.9	15.60±1.2	120.85±9.4	7.48±1.0
Specific surface area, sm^2/sm^3	5554.4	3324.8	8331.6	2346.7	1436.9	5286.4	3264.4	9303.1

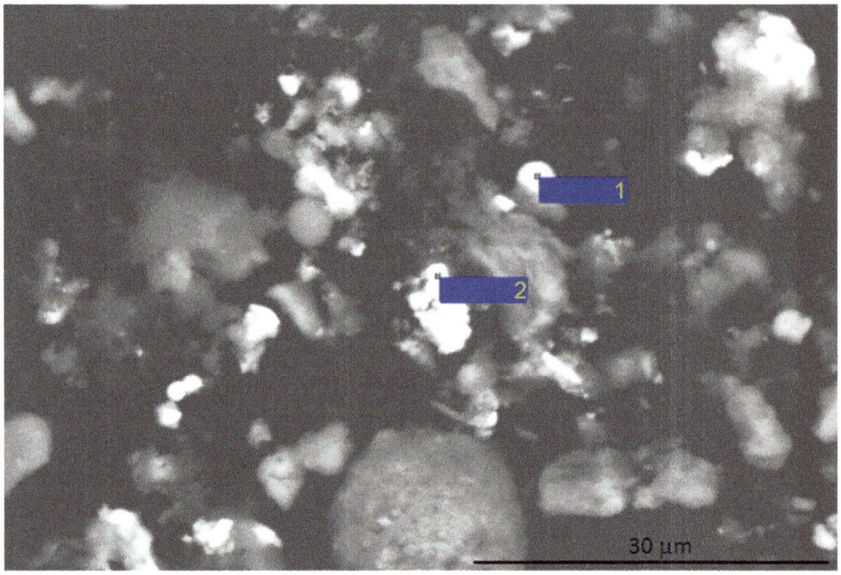

Figure 4. Images of the survey micrograph of particles from a snow sample in the western section of Ussuriisk (point 6), executed in reflection mode. White - metal particles. 1 and 2 indicating particles tested by EDAX method.

3400N (Hitachi, Tokio, Japan) with an energy dispersive spectrometer Ultra Dry (Thermo Fisher Scientific, Waltham, USA). Two microscopes were used to differentiate the size of collected Me NP.

Measurements of Me in collected samples

A high resolution inductively coupled plasma mass spectrometer (HR-ICP-MS) employing Element XR (Thermo Fisher Scientific, Waltham, USA) was used for assessment of Me in collected snow PM samples. Prior to evaporation, the liquid portion of the samples was stored at 4°C and analyzed according to technique CV 3.18.05-2005 (standard by JSC "Water Research and Control Center;" http://aqua-analyt.com).

Statistical Analysis

Results were compared by one-way ANOVA using the all pairwise multiple comparison procedures (Holm-Sidak method) or Dunnett's multiple comparisons to control, and Student's unpaired t test with Welch's correction for unequal variances. All results are presented as means ± SE. P values of <0.05 were considered to indicate statistical significance.

Results

The snow sediments were collected from Blagoveshchensk and Ussuriisk during 2010–2012. Snow samples were collected from the area of 200–500 m around galvanic shops, large highways and heat power stations. The maps of the areas and collection spots are presented in Fig. 1/Table 1, and Fig. 2/Table 2. Me NP were only found in snow sediments collected from areas in both cities near enterprises employing galvanic fabrication. In Ussuriisk, for example, Me NP were present in snow nearby the locomotive-repair plant and the carriage refrigerator depot (Fig. 3a, point 6). No Me NP were detected in control samples collected from clean/green region of a bypass road at the western part of the city (Fig. 3b, point 3) or other areas of Ussuriisk (Table 3, 4).

Apparently, an average particle size distribution of Me NP collected from clean areas fluctuates from 7.48±0.97 to 519.41±33.21 µm (Fig. 3b, Tables 3–4) while size distribution of Me NP gathered from industrial settings ranged 2.06±0.12 µm (Fig. 4, Tables 3).

The data showing surface area of Me NP collected from all areas in Ussuriisk are presented in Table 3–4. The Me NP collected from industrial point 6 (Table 3) had a surface area 1400-fold higher compared to those collected from clean area sources.

Table 5. Elemental composition of two Me microparticles tested in the snow samples collected from Ussuriisk (point 6) by EDAX method.

Element	Spectrum 1		Spectrum 2	
	mass. %	atom. %	mass. %	atom. %
O	7.11	20.52	24.20	50.44
Al			2.93	3.62
Si	3.18	5.24	4.12	4.89
Fe	89.71	74.24	68.75	41.05

Table 6. Results of the mass spectrometry of samples of the snow collected from 1–10 districts in Ussuriisk (the data presented in ppb).

district/element	^{111}Cd	^{207}Pb	^{27}Al	^{52}Cr	^{56}Fe	^{59}Ni	^{63}Cu	^{66}Zn
1	-	-	2.47±1.15	-	0.91±0.32	0.77±0.04*	1.32±0.25	4.73±1.02*
2	-	-	0.63±0.05*	-	0.29±0.03*	3.61±0.27*	2.74±0.76*	4.32±0.41*
3	0.13±0.02	0.02±0.004	3.46±1.04	-	1.22±0.11	1.46±0.09	0.77±0.8	12.2±1.54
4	0.03±0.01*	0.02±0.004	6.77±1.02*	0.09±0.01*	1.71±0.26	1.57±0.24	0.82±0.12	6.97±1.04
5	1.3±0.03*	5.52±0.36*	9.2±1.29*	2.71±0.18*	22.10±2.19*	4.71±0.62*	1.72±0.03	31±4.59*
6	0.6±0.04*	8.75±0.88*	29.1±3.79*	3.02±0.21*	120.02±13.89*	6.6±0.91*	2.33±0.12*	223±31.45*
7	0.03±0.01*	0.77±0.12	3.62±0.54	0.09±0.04	9.66±1.45*	0.35±0.05*	0.61±0.09	276±41.34*
8	0.07±0.01	1.57±0.23	3.95±0.59	0.25±0.04*	4.12±0.62*	0.39±0.06*	0.81±0.12	234±35.14
9	0.05±0.01	0.26±0.04*	6.56±0.98*	0.32±0.05*	3.54±0.53*	1.52±0.23	3.93±0.59*	3.39±0.51*
10	0.03±0.01*	0.82±0.12*	7.29±1.09*	0.1±0.01	8.36±1.25*	0.39±0.06*	0.41±0.06	273±41.01*

"-" below detection limit, *$p<0.05$, vs control (point 3).

To confirm that these NPs contained metals, scanning electron microscopy and the energy dispersive X-ray spectral analysis (EDX-analysis) were used for assessment of a dry suspension of the collected snow sediments. The most characteristic images of the particles are shown in Fig. 4. The data presented in Table 5 indicates the results of EDX-analysis of two heavier particles shown in white in Fig. 4.

All studied particles were smaller or comparable in size to the diameter of the electron beam focus of the SEM. Therefore, Si and Al were also detected by EDX-analysis from nearby aluminosilicate impurities (Table 5). Chemical analysis of the data presented in Table 5 revealed that iron in these NPs was in oxidized form. The shapeless (xenomorphic) particles were oxidized more strongly (Fe_2O_3) (range 2) than the spheroidal smaller particles (FeO) (range 1) (Fig. 4).

Quantitative analysis by HR-ICP-MS of Me composition of the collected PM in snow samples from ten different Ussuriisk city areas is presented in Table 6. The content of metals present in snow from Ussuriisk did not differ from each other by more than ten times and did not exceed toxic levels. The elevated level of Fe and Cr were detected in the samples gathered from points 5 and 6. These Me were present in oxidized and salt form. We propose that the mechanism of NPs formed in the atmosphere was due to high temperature and the etching and fast evaporation of galvanic solutions. Formed galvanic steam satiated by salts stayed airborne and then precipitated in metallic and/or oxidized forms. Our observation demonstrates that at a distance greater than 250 m, no metals NPs were found. Therefore, airborne NP found in the snow sediments gathered within 200 to 250 m from galvanic shops were enriched with emitted Me. Additionally, the paradigm of wind flow mapping strictly corresponds to the NPs transferred from the plants.

In Blagoveshchensk's industrial areas, Me NP were present within a ship-repair plant (Fig. 5a, Table 7, point 9, Theatralnaya st.).

In contrast, no Me NP were detected in the samples gathered from First City Hospital park at point 1 (Figure 5b and Table 7).

Interestingly, the differences in average diameter of nanoparticles in dirty zones was 44 times less compared to clean ones, showing that nano-sized particles are predominately found within cites where galvanic plants are located. It also has to be mentioned that the specific surface area of the particles collected in Blagoveshchensks were 256 fold higher in dirty areas compared to clean areas (Figure 5a and 5b).

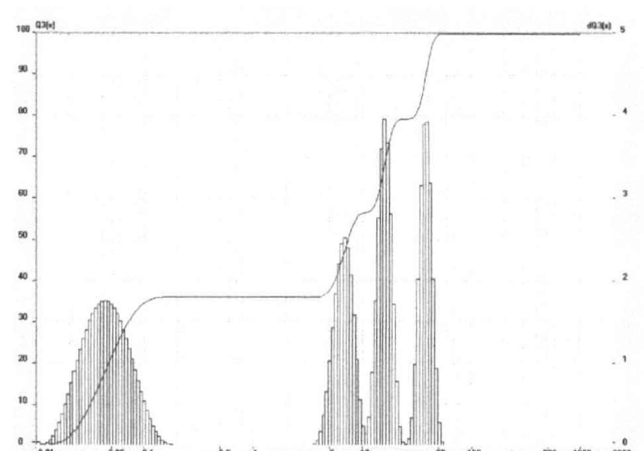

a

b

Figure 5. The profile of particle size distribution in snow samples collected from the Theatralnaya street, Blagoveshchensk (point 9) (a). The size of the particles from the samples collection ranges between 10–120 nm, 4–10, 10–30 and 40–50 µm. The profile of particle size distribution in snow samples collected from the First City Hospital, Blagoveshchensk (point 1) **(b).** The size of the particles from the samples collection ranges between 7–9 µm, 13–19, 170–350 and 500–900.

Table 7. Morphometric parameters of PM derived from snow samples collected at points 1–10 in Blagoveshchensk.

	point 9				point 1			
Size, μm	0.01–0.12	4–10	10–30	40–50	7–9	13–19	170–350	500–900
Proportion, %	36	20	24	20	1	3	23	73
Arithmetic mean diameter, μm	13.07±1.11				579.12±36.24			
Specific surface area, m^2/sm^3	56.41				0.022			

Table 8. Morphometric parameters of PM in collected snow samples (Blagoveshchensk) (points 2–8, 10).

Points	**2**	**3**	**4**	**5**	**6**	**7**	**6**	**8**	**10**
Arithmetic mean diameter	8.5±0,6	39.7±2.4	586.7±21.2	53.6±4.9	288.8±12.4	12.2±1.0	105.6±7.5	750.2±46.5	14.8±1.8
Specific surface area, sm^2/sm^3	7139.7	1532.1	686.4	3051.9	730.0	4956.0	587.5	582.8	4066.0

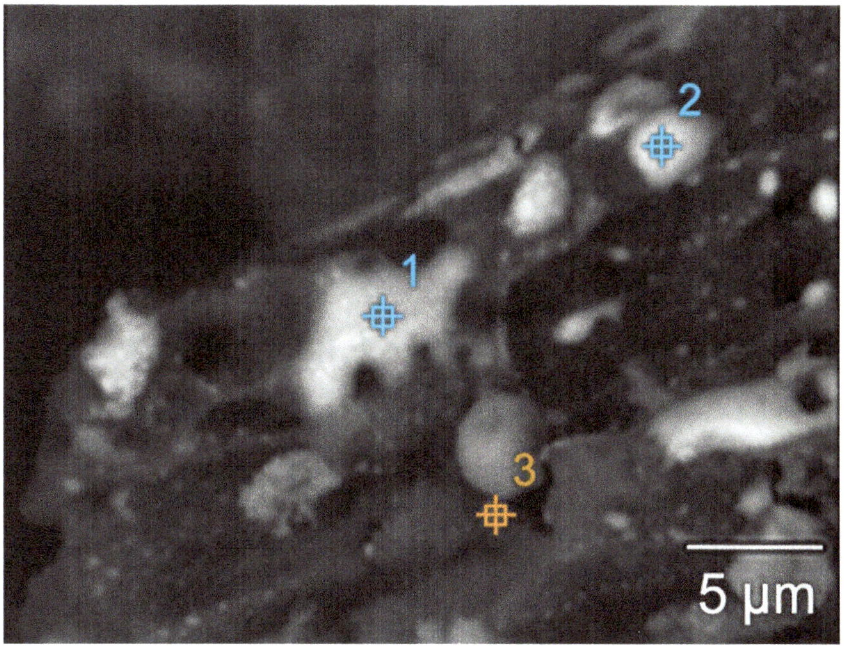

Figure 6. Image of the survey micrograph of particles from a snow sample around a ship-repair plant in Blagoveshchensk (point 9) imaged in reflection mode. White - metal particles. The large microsized "drops" consist of nanodimensional particles. Numbers 1–3 indicating particles tested by EDAX method.

The average arithmetic diameter of particles collected from other points (2–8, 10) in Blagoveshchensk fluctuated ranging from 8.47±0.61 to 288.83±12.36 µm (Table 8).

The chemical characterization of the particles seen in Figure 6 (marked as 1, 2 and 3) is presented in Table 9.

The large ferriferous particles detected by SEM are shown in Figure 4, 6 and 7. The surface of these PM was covered with nano grains, most likely formed due to aggregation of small iron particles (Figure 7).

The measurements by HR-ICP-MS indicate the low metallic concentrations of Fe and Cr in the samples from Blagoveshchensk (Table 10). Different Me composition was observed in the particles collected from ten different city areas. The content of metals collected from Blagoveshchensk did not differ from each other by more than ten times and did not exceed toxic levels. NP gathered from Ussuriisk areas had fewer metals than NP collected from Blagoveshchensk. This may be attributed to less numbers of airborne NP in areas of reduced traffic.

Table 9. Elemental composition of three Fe microparticles tested in the snow samples collected from Blagoveshchensk (point 9) by EDAX method.

Element	Spectrum 1	Spectrum 2	Spectrum 3
	Macc. %	Macc. %	Macc. %
C	8.52±0.18	8.54±0.16	26.72±0.32
O	40.18±0.61	37.58±0.57	18.30±0.46
Na			1.03±0.07
Al	8.49±0.18	8.79±0.16	7.21±0.16
Mg	1.85±0.15	1.64±0.14	0.82±0.07
Si	1.06±0.13	0.96±0.07	5.73±0.14
K			0.52±0.07
Cl			0.56±0.10
Ca	15.15±0.29	17.05±0.29	12.83±0.27
Mn		1.63±0.21	1.54±0.22
Fe	24.76±0.80	23.81±0.75	24.73±0.80
Sum	100.00	100.00	100.00

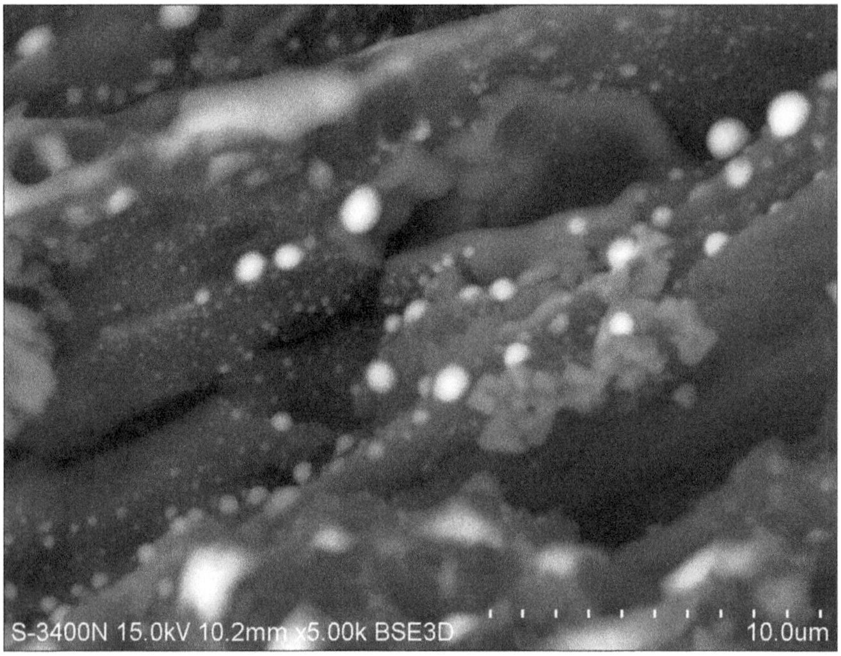

Figure 7. Image of the survey micrograph of particles derived from a snow sample collected from the area of a ship-repair plant in Blagoveshchensk (point 9). White - metal particles. NPs are observed to be absorbed by organic detrite (identity by morphology and C-peak in EDAX). Magnification: 5000×. Images are in reflection mode.

Discussion

A number of epidemiological studies have reported a strong correlation between the level of particulate air pollution and increased morbidity and mortality rates in both adults and children [21–24]. Particle count, composition, and surface properties are recognized as important for dosimetry and the assessment of adverse outcomes caused by air pollution [25]. The deposition of particles in the respiratory tract depends predominantly on their size; larger particles will accumulate in the upper and larger airways, while smaller particles will penetrate into the alveolar spaces. Thus 90% of inhaled 1 μm particles are deposited in the nasopharyngeal region while 20 nm particles have the highest deposition efficiency in the alveolar region (~50%) [26].

Me nanoparticles are very reactive and are known to cause wide range of toxic responses [27–29]. In welders exposed to a variety of fumes and gases (containing a mixture of oxides and salts of metals), the rate of pulmonary morbidity was relatively high [30–33]. An early meta-analysis among shipyard, mild steel and stainless steel welders revealed an increased lung cancer risk among all types of workers [34].

We have found elevated numbers of hard and soluble Me NP in snow sediments collected in close vicinities of galvanic manufacturers in two Russian cities. We are the first to report that hard and soluble Me NP consisting Pb, Al, Cr, Fe, Ni, Cu and Zn were detected around galvanic shop settings. Using different analytic techniques we observed that the size distribution of tested Me NP was within the 10–120 nm range.

Table 10. Metal composition in snow samples collected from 1–10 districts in Blagoveshchensk (Mass spectrometry, the data presented in ppb).

district/element	^{111}Cd	^{207}Pb	^{27}Al	^{52}Cr	^{56}Fe	^{59}Ni	^{63}Cu	^{66}Zn
1	0.03±0.01	0.02±0.04	6.77±1.02	0.09±0.01	1.71±0.26	1.57±0.24	0.82±0.12	6.97±1.04
2	0.21±0.01*	-	1.92±0.3*	0.12±0.01	-	2.2±0.11*	0.85±0.09	2.89±0.83*
3	0.14±0.02*	1.27±0.09*	4.34±0.87	0.91±0.04*	2.12±0.12	1.71±0.16	0.29±0.05*	6.78±0.93
4	0.49±0.03	1.42±0.07*	0.14±0.02*	-	1.42±0.11	2.42±0.32*	-	-
5	0.33±0.02*	0.29±0.02	4.66±0.92	-	1.13±0.09*	1.72±0.12	0.92±0.09	2.72±0.34*
6	-	1.11±0.08*	0.92±0.11*	0.27±0.03*	3.23±0.46*	0.88±0.08*	1.23±0.08*	-
7	0.39±0.01*	1.12±0.06	0.72±0.03*	1.09±0.03	0.93±0.08*	1.71±0.11	1.21±0.11	1.76±0.12*
8	0.52±0.05*	0.22±0.03*	2.72±0.94*	-	2.23±0.14	-	1.52±0.34*	0.57±0.02*
9	0.2±0.03*	2.56±0.38*	23.2±3.47*	0.25±0.04*	6.12±0.11*	1.12±0.17*	0.38±0.06*	310.8±46.45*
10	0.56±0.08*	-	2,64±0.33*	-	2.06±0.11	0.96±0.7*	-	2.69±0.31*

"-" - below the detection level; *$p<0.05$, vs control (point 1).

Potential exposure to Fe NP and Cr NP during production is an area of concern. Fe NP have unique magnetic properties which gives a high potential use in several biomedical applications, including magnetic drug targeting, magnetic detection, hyperthermia and magnetic resonance imaging [31]. Chromium is also an important industrial metal used in various processes, sometimes in the nanosized form.

A number of regulations have been implemented by the Department of Labor of Russian Federation targeting the safety of galvanic manufacture workers. There are two major regulatory documents "Sanitary standards designed for the industrial enterprises" and "The instruction on labor protection for the galvanizer man" which were designed to regulate the use of protective measures in the galvanic industry (CH 245-71, 1972; TOI R-31-205-97, 2011). In particular, the "Sanitary Standards Designed for the Industrial Enterprises" emphasizes the need to control exposure level around electrolytic plating workstations. Depending on the chemical galvanic technology used, certain areas were assigned for assessment of hazard identification providing industrial hygiene control in the area ranging from 100 m to 500 m. For good manufacturing practice, proper industrial control and a safe work environment, testing of airborne toxic fumes, Me and gases are recommended to be performed every six months (CH 245-71, 1972; TOI R-31-205-97, 2011). Additionally, when galvanic and plating technology is changed, all tests of airborne emitted galvanic contaminants are advised to be re-assessed. Furthermore, to capture harmful emissions, the use of a ventilation exhaust system has been strongly recommended to provide proper work safety ("The instruction on labor protection for the galvanizer man" TOI R-31-205-97, 2011) [35]. Unfortunately, the industrial ventilation devices currently used are not supplied with filters that provide efficient capturing of airborne nano-sized PM and fumes emitted during galvanic manufacturing.

Conclusions

Me nanoparticles of Fe, Cr and Fe/Cr-alloy are potentially released into the urban air during galvanic production. No Me NP were detected in snow sediments collected from clean areas in two industrial cities. Elevated levels of Me were detected in snow sediments in close proximity to galvanic shops. Future epidemiological studies are needed to determine whether the airborne Me NP that were found elicit adverse health effects.

Acknowledgments

Disclaimer: The findings and conclusions in this report are those of the authors and do not necessarily represent the views of the National Institute for Occupational Safety and Health. Mention of trade names or commercial products do not constitute endorsement or recommendation for use.

Author Contributions

Conceived and designed the experiments: KSG AAS. Performed the experiments: KSG. Analyzed the data: KSG AAS. Contributed reagents/materials/analysis tools: KSG. Contributed to the writing of the manuscript: KSG AAS.

References

1. Burkin AR (1969) Production of some metal powders by hydrometallurgical processes. Powder Met 12: 243–250.
2. Andrus ME (2000) A review of metal precipitation chemicals for metal-finishing applications. Metal Finishing 98(11): 20–23.
3. U.S. Imports, Exports, and Tariff Data, 1989–2001, John Romalis, Peter K. Schott NBER Working Paper No. 9387. Issued in December 2002 NBER Program(s).
4. Valli J (1986) A review of adhesion test methods for thin hard coatings. J Vac Sci Technol A 4: 3007–12.
5. Voutsa D, Samara C (2002) Labile and bioaccessible fractions of heavy metals in the airborne particulate matter from urban and industrial areas. Atmospheric Environment 36: 3583–3590.
6. Babich H, Stotzky G (1985) Heavy metal toxicity to microbe-mediated ecologic processes: A review and potential application to regulatory policies. Environ Res 36(1): 111–137.
7. Guilizzoni P (1991) The role of heavy metals and toxic materials in the physiological ecology of submersed macrophytes. Botany 42, no. 1–3: 87–109.
8. Silva JE, Paiva AP, Soares D, Labrincha A, Castro F (2005) Solvent extraction applied to the recovery of heavy metals from galvanic sludge. Journal of Hazardous Materials 120, no. 1–3: 113–118.
9. Rossini G, Bernardes A (2006) Galvanic sludge metals recovery by pyrometallurgical and hydrometallurgical treatment. Journal of Hazardous Materials 131, no. 1–3: 210–216.
10. Campen MJ, Nolan JP, Schladweiler MC, Kodavanti UP, et al. (2001) Cardiovascular and Thermoregulatory Effects of Inhaled PM-Associated Transition Metals: A Potential Interaction between Nickel and Vanadium Sulfate. Toxicological Sciences 64: 243–252.
11. Andersen ZJ, Wahlin P, Raaschou-Nielsen O, Ketzel M, Scheike T, et al. (2008) Size distribution and total number concentration of ultrafine and accumulation mode particles and hospital admissions in children and the elderly in Copenhagen, Denmark. Occupational and Environmental Medicine 65(7): 458–466.
12. Anderson JO, Thundiyil JG, Stolbach A (2012) Clearing the air: a review of the effects of particulate matter air pollution on human health. J Med Toxicol 8(2): 166–75.
13. Zhang Z, Chau PY, Lai HK, Wong CM (2009) A review of effects of particulate matter-associated nickel and vanadium species on cardiovascular and respiratory systems. Int J Environ Health Res 19(3): 175–85.
14. Chen LC, Lippmann M (2009) Effects of metals within ambient air particulate matter (PM) on human health. Inhal Toxicol 21(1): 1–31.
15. Vinogradov SS (2002) Ecologically safe galvanic production. 2nd edition. Moscow: Globus. 352 p.
16. Özverdi A, Erdem M (2010) Environmental risk assessment and stabilization/solidification of zinc extraction residue: I. Environmental risk assessment. Hydrometallurgy 100(3–4): 103–109.
17. Erdem M, Özverdi A (2011) Environmental risk assessment and stabilization/solidification of zinc extraction residue: II. Stabilization/solidification. Hydrometallurgy 105(3–4): 270–276.
18. Ozdemir OD, Piskin S (2012) Characterization and environment risk assessment of galvanic sludge. J Chem Soc Pakistan 34(4): 1032–1036.
19. Ryabukhin AG, Koshelev IV (2002) Waste-free technology of cleaning washing and galvanics and pickling of metals sewage. Reports of the Chelyabinsk scientific center Ural Branch RAS 4: 111–120.
20. Candelone JP, Hong S, Pellone C, Boutron CF (1995) Post-Industrial Revolution changes in large-scale atmospheric pollution of the northern hemisphere by heavy metals as documented in central Greenland snow and ice. J Geophysic Res: Atmospheres 100(8): 16605–16616.
21. Bates D (1995) The Effects of Air Pollution in Children. Environ Health Perspect. 103(6): 49–54.
22. Brook RD, Franklin B, Cascio W, Hong Y, Howard G, et al. (2004) Air pollution and cardiovascular disease: A statement for healthcare professionals from the expert panel on population and prevention science of the American Heart Association. Circulation 109: 2655–2671.
23. Brook RD, Rajagopalan S, Pope CA 3rd, Brook JR, Bhatnagar A, et al. (2010) Particulate Matter Air Pollution and Cardiovascular Disease: An Update to the Scientific Statement From the American Heart Association. Circulation 121: 2331–2378.
24. Cakmak S, Dales R, Kauri LM, Mahmud M, Ryswyk KV, et al. (2014) Metal composition of fine particulate air pollution and acute changes in cardiorespiratory physiology. Environ Pollut 189: 208–214.
25. Raza A, Bellander T, Bero-Bedada G, Dahlquist M, Hollenberg J, et al. (2014) Short-term effects of air pollution on out-of-hospital cardiac arrest in Stockholm. Eur. Heart J. 35(13): 861–867.
26. Kreyling WG, Semmler-Behnke M, Möller W (2006) Health implications of nanoparticles. J Nanoparticle Res 8: 543–562.
27. Murray AR, Kisin E, Inman A, Young SH, Kommineni C, et al. (2013) Oxidative stress and dermal toxicity of iron oxide NPs in vitro. Cell Biochem Biophys 67(2): 461–76.

28. Fadeel B, Pietroiusti A, Shvedova A (eds.) (2012) Adverse effects of engineered nanomaterials: Exposure, toxicology and impact on human health. Elsevier, Academic Press. 360 p.
29. Yanamala N, Kagan VE, Shvedova AA (2013) Molecular modeling in structural nano-toxicology: Interactions of nano-particles with nano-machinery of cells. Adv. Drug Deliv. Rev. 65(15): 2070–2077.
30. Soenen SJ, De Cuyper M, De Smedt S, Braeckmans K (2012) Investigating the toxic effects of iron oxide NPs. Meth Enzymol 509: 195–224.
31. Szalay B, Tatrai E, Nyiro G, Vezer T, Dura G (2012) Potential toxic effects of iron oxide NPs in in vivo and in vitro experiments. J Appl Toxicology 32(6): 446–453.
32. Zhu X, Tian S, Cai Z (2012) Toxicity assessment of iron oxide NPs in zebrafish (Danio rerio) early life stages. PLoS ONE 7(9): e46286.
33. Horie M, Nishio K, Endoh S, Kato H, Fujita K, et al. (2013) Chromium (III) oxide NPs induced remarkable oxidative stress and apoptosis on culture cells. Environ Toxicol 28(2): 61–75.
34. Moulin JJ, Wild P, Romazini S, Lasfargues G, Peltier A, et al. (1998) Lung cancer risk in hard-metal workers. Am J Epidemiol 148(3): 241–8.
35. The Standard instruction on labor protection for the galvanizer of TOI R-31-205-97 (2011) Regulations on Labor Protection. (available on www.mcfr.ru/journals/63/262/37335/).

The B Vitamins Nicotinamide (B_3) and Riboflavin (B_2) Stimulate Metamorphosis in Larvae of the Deposit-Feeding Polychaete *Capitella teleta*: Implications for a Sensory Ligand-Gated Ion Channel

Robert T. Burns[1]*, Jan A. Pechenik[1], William J. Biggers[2], Gia Scavo[2], Christopher Lehman[2]

1 Department of Biology, Tufts University, Medford, Massachusetts, United States of America, **2** Department of Biology, Wilkes University, Wilkes-Barre, Pennsylvania, United States of America

Abstract

Marine sediments can contain B vitamins, presumably incorporated from settled, decaying phytoplankton and microorganisms associated with decomposition. Because B vitamins may be advantageous for the energetically intensive processes of metamorphosis, post-metamorphic growth, and reproduction, we tested several B vitamins to determine if they would stimulate larvae of the deposit-feeding polychaete *Capitella teleta* to settle and metamorphose. Nicotinamide and riboflavin individually stimulated larvae of *C. teleta* to settle and metamorphose, generally within 1–2 hours at nicotinamide concentrations as low as 3 μM and riboflavin concentrations as low as 50 μM. More than 80% of the larvae metamorphosed within 30 minutes at a nicotinamide concentration of 7 μM. The pyridine channel agonist pyrazinecarboxamide also stimulated metamorphosis at very low concentrations. In contrast, neither lumichrome, thiamine HCl, pyridoxine HCl, nor vitamin B_{12} stimulated larvae of *C. teleta* to metamorphose at concentrations as high as 500 μM. Larvae also did not metamorphose in response to either nicotinamide or pyrazinecarboxamide in calcium-free seawater or with the addition of 4-acetylpyridine, a competitive inhibitor of the pyridine receptor. Together, these results suggest that larvae of *C. teleta* are responding to nicotinamide and riboflavin via a chemosensory pyridine receptor similar to that previously reported to be present on crayfish chela and involved with food recognition. Our data are the first to implicate B vitamins as possible natural chemical settlement cues for marine invertebrate larvae.

Editor: Tilmann Harder, University of New South Wales, Australia

Funding: This work was supported in part by funding from the Howard Hughes Medical Institute to the Biology Department of Wilkes University and Wilkes University Mentoring Task Force grant to WB, GS, and CL. Wilkes University and Tufts FRAC funded the publication of this manuscript. The funders had no role in study design, data collection and analysis, decision to publish, or preparation of the manuscript.

Competing Interests: The authors have declared that no competing interests exist.

* Email: robert.burns@tufts.edu

Introduction

The larvae of many benthic marine invertebrates are planktonic and are thus 'forced to wander' the sea until they become competent to metamorphose and locate a site suitable for settlement [1]. Particular chemical cues then promote larval settlement and subsequent metamorphosis by competent larvae of many species [2–4]. These chemical cues may indicate the presence of appropriate food, conspecific adults to mate with, or other environmental factors that signal the suitability of a site to live in for juveniles and adults. For example, larvae of the sea urchin *Holopneustes purpurascens* settle and metamorphose in response to histamine, which leaches away from *Delisea pulchra*, their algal food source [5–7]. Similarly, larvae of the tube-building polychaete worm *Phragmatopoma californica* settle and metamorphose in response to several different fatty acids that are produced by conspecific tube-dwelling adults, signaling the presence of potential mates [8]. Negative recruitment cues that prevent larvae from metamorphosing in specific areas have also been identified and can also be important in determining species distributions [9]. However, for the vast majority of marine invertebrates, the specific chemical settlement cues remain undefined.

Capitella teleta, previously known as *Capitella* sp. I, is a small (~20 mm long×1 mm wide), opportunistic, deposit-feeding marine polychaete found in salt marsh sediments and in disturbed and polluted areas such as busy harbors, sewage outflows, and some regions affected by oil spills [10,11]. Its planktonic, non-feeding, metatrochophore larvae will typically metamorphose in the presence of salt-marsh sediments, making it a convenient model for studies of substrate selection [12–14]. Larvae of *C. teleta* metamorphose rapidly when in the presence of salt-marsh sediment; in one study 90% of the tested larvae metamorphosed within 30 minutes of treatment [13]. The active component of the natural settlement cue contained within the salt-marsh sediment is currently unknown; however, larvae of *C. teleta* have been shown to settle and metamorphose preferentially in response to sediments with high organic content, and those with a low carbohydrate to protein ratio [15,16]. Sediments forced through a 0.45 μm filter can still stimulate larvae to metamorphose; however, filtering the sediment to 0.22 μm removed the cue, suggesting that the cue is bound to small particulates [15]. Also, combusting the sediment at 500°C for 6 h removed the cue from the sediment: the resulting ash did not stimulate metamorphosis, supporting the notion that the cue is organic [15].

Juvenile hormones and juvenile hormone-active chemicals are known to induce settlement and metamorphosis of *C. teleta* larvae [17,18]. Extracts prepared from marine sediments displayed juvenile hormone-activity in insect bioassays, suggesting that the chemicals in the sediments that induced settlement and metamorphosis may be similar in structure to juvenile hormones (JHs) or have juvenile hormone-activity. In addition, the induction of settlement and metamorphosis by JH was found to involve activation of protein kinase C and further activation of calcium channels [18]. Exposing *C. teleta* larvae to 400 nM calcium ionophore A23187, a membrane soluble chemical that shuttles calcium ions into cells, induced all of the larvae to metamorphose in less than one hour. Whether or not the larvae could metamorphose in calcium-free seawater however was not explored.

As has been found for some other invertebrate larvae, larvae of *C. teleta* settle and metamorphose in response to serotonin and the serotonin-reuptake inhibitor fluoxetine [19,20], indicating that stimulation of the larval nervous system is involved in mediating the chemosensory response to chemical cues. Furthermore, different inhibitors of nitric oxide synthase such as S-methylisothiourea and aminoguanidine hemisulfate were also found to induce larval settlement and metamorphosis [20], adding *C. teleta* to a growing list of marine invertebrates, including some gastropods [21–23], echinoderms [24], and ascidians [25,26] whose metamorphosis is inhibited by the presence of endogenous nitric oxide. While we currently know much about the intermediate steps of the signal transduction cascade leading to metamorphosis of *C. teleta*, we do not know which chemicals are acting as natural settlement cues for larvae of *C. teleta* or how these chemicals actually initiate this signal transduction cascade.

Larval metamorphosis, juvenile growth, and adult reproduction are all energy-requiring processes; not surprisingly the growth and reproduction of *Capitella* are affected by diet [27]. Plant nutrients have also been shown to be quickly assimilated into the growing oocytes of *Capitella* [28] and to supply adequate nutrition for larval development and subsequent metamorphosis. Dietary B vitamins are essential co-factors for biochemical energy production, and are known to be essential for the growth and development of some annelids [29]. In this respect it should be beneficial for larvae of *C. teleta* to settle and metamorphose in locations with adequate levels of these vitamins to support post-metamorphic growth. Because B vitamins are likely to be found within the marine sediments that the deposit-feeding adults of *C. teleta* consume and are required for growth and survival, we suspected that the larvae would respond to at least some B vitamins as a natural settlement cue. To test our hypothesis, we added several B vitamins individually to seawater and monitored the metamorphosis of newly-released larvae of *C. teleta*. In addition we conducted a series of experiments using various pharmacological agents to determine how these vitamins were stimulating metamorphosis.

Results

Effects of different B vitamins

Of the 7 B vitamins tested, only riboflavin (B_2), nicotinamide (B_3), and nicotinic acid (another form of B_3) stimulated metamorphosis in *C. teleta*. Thiamine HCl (B_1), pyridoxine HCl (B_6), biotin (B_7), and cobalamin (B_{12}) did not stimulate any metamorphosis within 24 h at the tested concentration of 500 μM (data not shown). In contrast, nicotinamide stimulated metamorphosis in a nearly dose-dependent manner at concentrations as low as 3–8 μM, with larvae initiating metamorphosis within 30 min (Figure 1). Nicotinic acid stimulated 70.8% +/−8.3 (s.e.m.) of the treated larvae to metamorphose within 24 h. Also, riboflavin concentrations of 50–200 μM stimulated nearly all larvae of *C. teleta* to metamorphose within 5 hours (Figure 2). However, the larvae did not respond to the riboflavin breakdown product lumichrome even when tested for 24 h at the maximum concentration of 1 mM (data not shown).

Effects of pyridine channel activators and inhibitors

Because concentrations of nicotinamide as low as 3–6 μM stimulated larvae of *C. teleta* to metamorphose quickly in our experiments, we hypothesized that larvae of *C. teleta* were sensing nicotinamide via a pyridine activated ion channel, something first characterized in the walking legs of the crayfish *Austrapotamobius torrentium* [30–32]. We therefore compared dose responses between nicotinamide and pyrazinecarboxamide, an agonist of the pyridine-activated ion channel, at 1 μM, 4 μM, and 8 μM. Another commonly occurring nutritive chemical, beta nicotinamide adenine dinucleotide (β-NAD), also stimulated the nicotinamide-activated ion-channel to open, with half-maximal rate of opening (K_M) at 1 mM [31]. We therefore treated larvae of *C. teleta* with β-NAD at the concentrations of 0.5, 1, and 5 mM to determine if β-NAD also stimulated metamorphosis. As shown in Figure 3, the pyridine-activated ion channel agonist pyrazinecarboxamide was found to also stimulate larvae of *C. teleta* to metamorphose, and did so at similar concentrations as nicotinamide. Although both chemicals stimulated larvae of *C. teleta* to metamorphose in a dose-dependent fashion, pyrazinecarboxamide stimulated more individuals (about 70%) to metamorphose at 4 μM than did nicotinamide at the same concentration (Figure 3). β-NAD, however, did not stimulate any metamorphosis at 0.5, 1, or 5 mM.

Effects of calcium free seawater

To determine if an influx of extracellular calcium is required for metamorphosis to proceed when pyridine-activated ion channels open, larvae of *C. teleta* were treated with nicotinamide or pyrazinecarboxamide while bathed in either normal ASW(control) or calcium-free ASW. Pre-treating the larvae in calcium-free ASW inhibited the metamorphic response of *C. teleta* larvae to both nicotinamide and pyrazinecarboxamide even at the high concentration of 40 μM (one-way ANOVA with Bonferroni post-hoc comparisons, $F(5,12) = 234.4$, $p<0.001$) (Figure 4). Larvae in calcium-free ASW settled to the bottom of the dish within a few minutes after the addition of nicotinamide or pyrazinecarboxamide; they then remained stationary, but did not metamorphose over the 24 h observation period. Control larvae in calcium-free ASW continued to swim normally and neither settled nor metamorphosed. At least 90% of larvae exposed to nicotinamide or pyrazinecarboxamide in ASW metamorphosed, indicating that most larvae were competent to metamorphose at the start of this experiment.

Effects of 4-acetylpyridine

4-acetylpyridine, a specific competitive antagonist of the pyridine-activated ion channel, was also tested on larvae of *C. teleta* in the presence of either nicotinamide or pyrazinecarboxamide [32]. Treating larvae with 200 μM 4-acetylpyridine

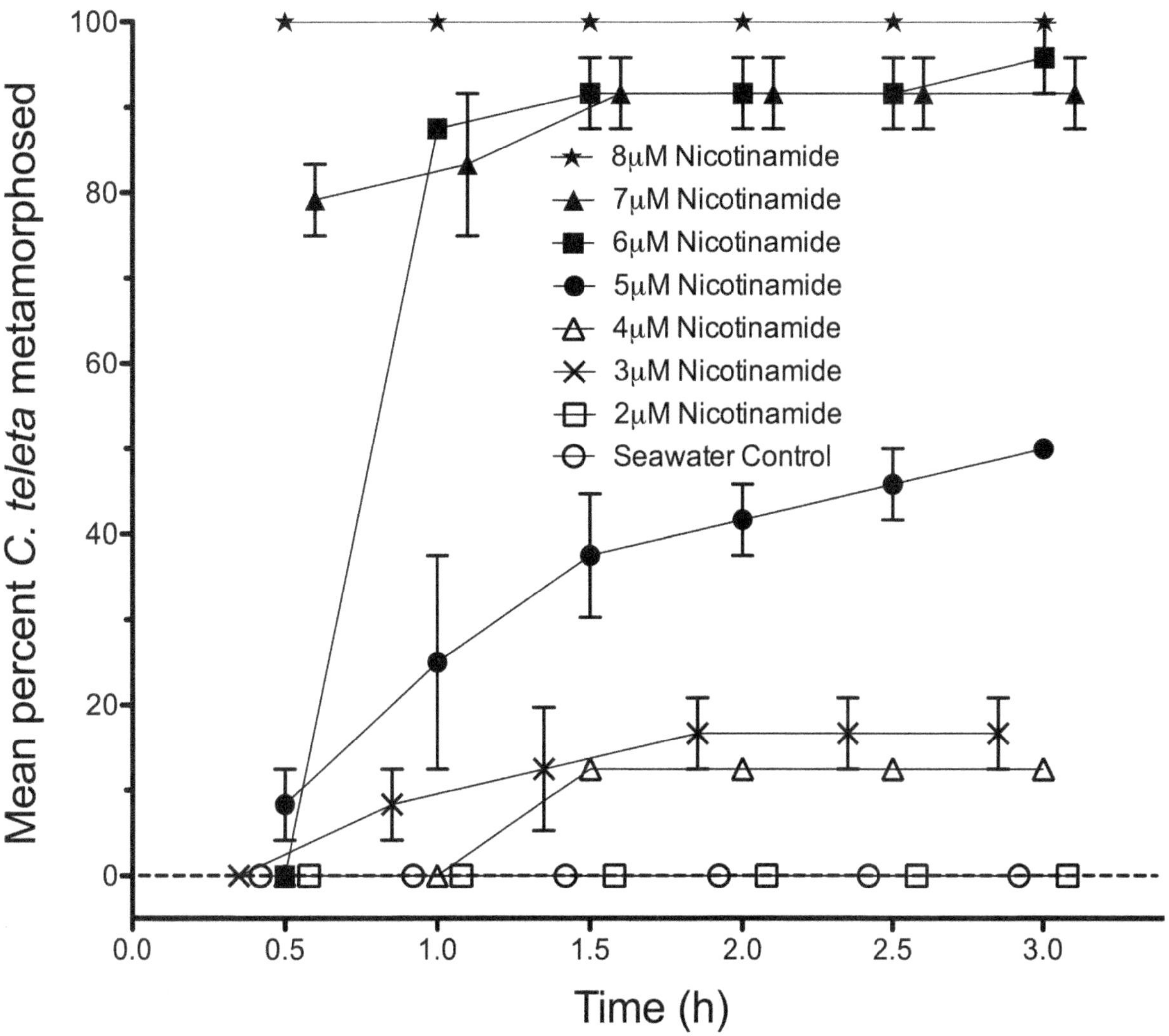

Figure 1. Promotion of metamorphosis by nicotinamide (vitamin B_3) in larvae of *Capitella teleta*. Each treatment consisted of 3 replicates with 8 larvae per replicate. Larvae were placed in 30 ppt artificial seawater (Instant Ocean) containing the indicated final concentration of nicotinamide. Artificial seawater acted as a negative control. Error bars represent +/−1 s.e.m.

inhibited them from metamorphosing in the presence of either 40 μM nicotinamide or 40 μM pyrazinecarboxamide (one-way ANOVA with Bonferroni post-hoc comparisons, $F(5,12) = 69.52$, $p<0.001$) (Figure 5). Only about half as many larvae metamorphosed within 24 h when exposed to both 4-acetylpyridine and either nicotinamide or pyrazinecarboxamide, indicating that the inhibitory concentration allowing for 50% settlement and metamorphosis (IC_{50}) is about 200 μM. No larvae exposed to both 4-acetylpyridine and either nicotinamide or pyrazinecarboxamide metamorphosed within the first 5 h of the experiment (data not shown) and no control larvae exposed to 200 μM 4-acetylpyridine alone settled or metamorphosed within 24 h. Again, the high response of larvae in the positive controls shows that most of the larvae were metamorphically competent when tested. Also, larvae were inhibited from metamorphosing in response to pyrazinecarboxamide in a dose-dependent fashion when tested with increasing doses of 4-acetylpyridine (one-way ANOVA with Bonferroni post-hoc comparisons, $F(7,16) = 53.92$, $p<0.001$) (Figure 6).

Effects of ketanserin

Sensory information relayed by sensory chemoreceptors such as olfactory receptor neurons usually involves sensory transduction through synapses within the central nervous system. In order to assess the involvement of the serotonergic nervous system in mediating the response of the larvae to nicotinamide and pyrazinecarboxamide, the larvae were first pre-exposed to ketanserin, an inhibitor of serotonin $5\text{-}HT_{2B}$ receptors, before induction by nicotinamide and pyrazinecarboxamide. Previous results have demonstrated that ketanserin blocks the response of *Capitella* larvae to natural mud sediment chemical cues, and also to the nitric oxide synthase inhibitors N-methyl-L-arginine and S-methylisothiourea [19,20], indicating the involvement of the serotonergic nervous system in mediating the settlement and metamorphosis of *C. teleta*. Pre-exposing *C. teleta* larvae to ketanserin inhibited most of the larvae of *C. teleta* from metamorphosing in response to both nicotinamide and pyrazinecarboxamide (one-way ANOVA with Bonferroni post-hoc comparisons, $F(5,12) = 122.5$, $p<0.001$) (Figure 7). No control larvae exposed to 2 μM ketanserin alone settled or metamorphosed within 24 h; however, all of the larvae exposed to the positive

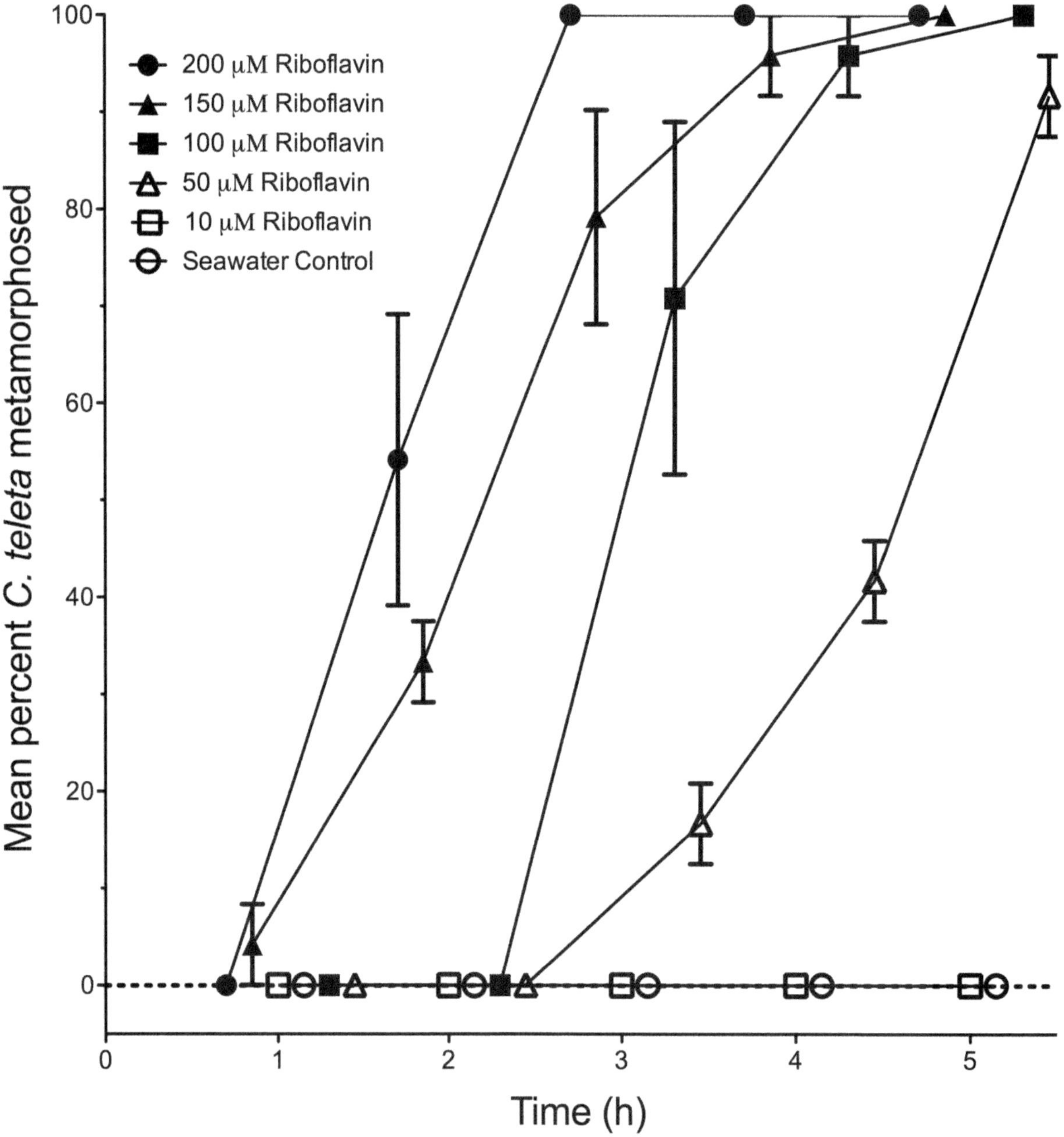

Figure 2. Promotion of metamorphosis by riboflavin (vitamin B_2) in larvae of *Capitella teleta*. Each treatment consisted of 3 replicates with 8 larvae per replicate. Larvae were placed in 30 ppt artificial seawater (Instant Ocean) containing the indicated final concentration of riboflavin. Artificial seawater acted as a negative control. Error bars represent +/−1 s.e.m.

controls of nicotinamide and pyrazinecarboxamide alone metamorphosed in less than an hour, indicating that most larvae were competent when tested.

Discussion

Overall, these data support our hypothesis that at least some B vitamins may act as chemical cues for habitat selection by the larvae of *C. teleta* and may stimulate the larvae of this species to metamorphose in the field. In particular, nicotinamide (B_3), nicotinic acid (B_3), and riboflavin (B_2) stimulated larvae of *C. teleta* to metamorphose, whereas thiamine HCl, pyridoxine HCl, biotin, vitamin B_{12}, and the riboflavin breakdown product lumichrome did not. It may be advantageous for marine larvae to settle and metamorphose in response to environmental riboflavin and nicotinamide, as riboflavin is required for the synthesis of FAD, and nicotinamide serves as a building block for the synthesis of NAD; both products are important electron carriers for the electron transport chain, ATP production, and other aspects of cellular metabolism. Substantial concentrations of these vitamins may serve as an adaptive signal that the local environment is suitable for the energetically intensive processes of juvenile development and adult reproduction. In this regard, B vitamins present in ocean waters have been demonstrated to be important for the growth of phytoplankton, zooplankton, and larger marine invertebrates [33–35]. However, even if B vitamins are contained

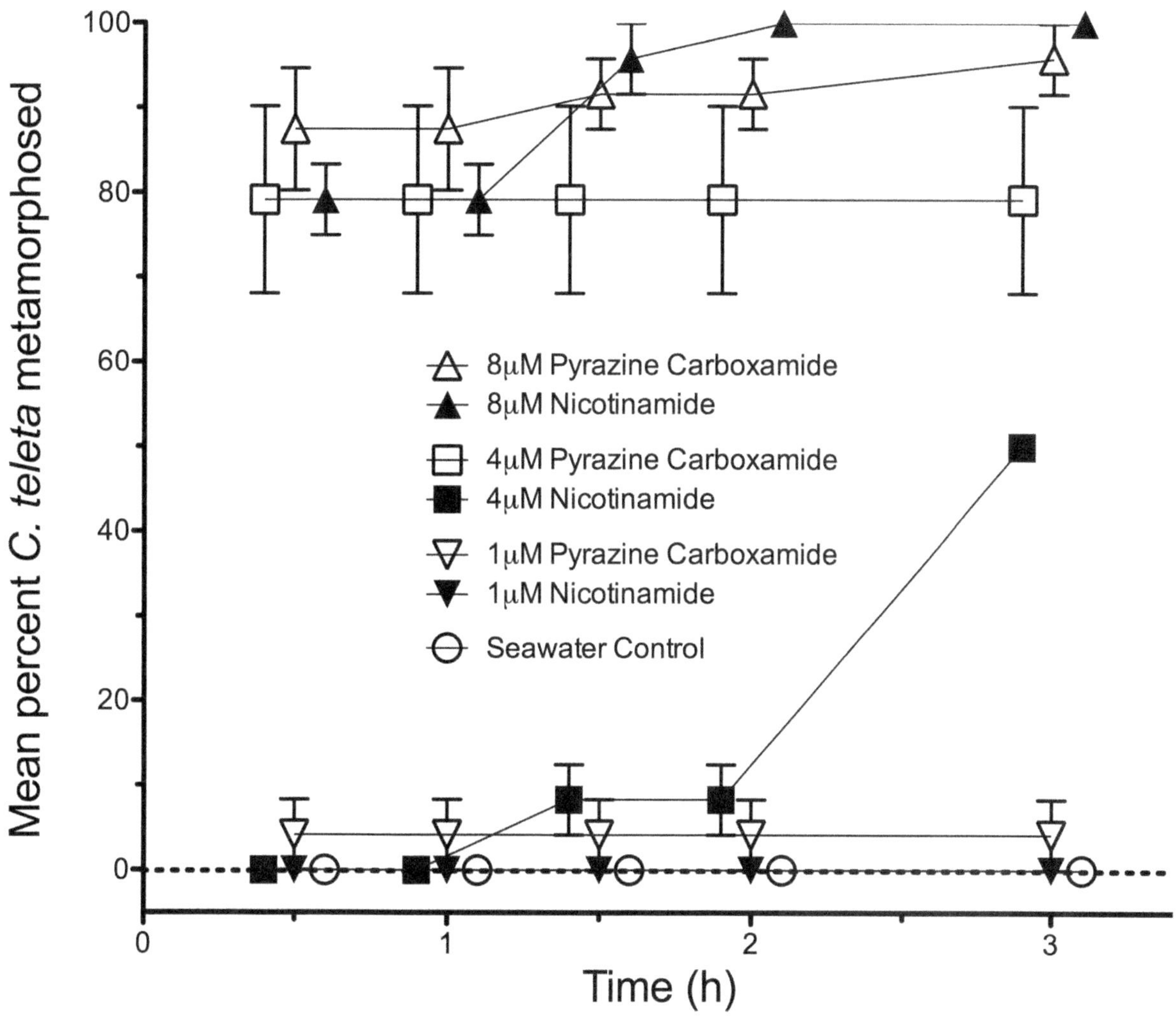

Figure 3. The effects of pyrazinecarboxamide and nicotinamide on metamorphosis of *C. teleta* when tested equal concentrations. Each treatment consisted of 3 replicates with 8 larvae per replicate. Artificial seawater (Instant Ocean) acted as a negative control. Error bars represent +/−1 s.e.m.

within marine sediments, they may not always be bioavailable to deposit feeders.

Although the stimulation of settlement and metamorphosis by vitamins seems not to have been investigated in other marine invertebrate species, the larvae of several other marine invertebrates have previously been demonstrated to settle and metamorphose in response to certain vitamin derivatives [2]. For example, larvae of the ascidian *Halocynthia roretzi* will settle and metamorphose in response to lumichrome, a photocatalytic breakdown product of riboflavin, although they did not respond to riboflavin itself [36]. This inducer was found to be naturally biosynthesized by the larvae and adults of *Halocynthia*. Lumichrome, however, did not induce settlement and metamorphosis in other ascidian species, and also did not stimulate metamorphosis in our study, although the larvae of *C. teleta* did respond to riboflavin. In other studies, larvae of the hydroid *Coryne uchidae* were found to settle and metamorphose in response to delta-tocopherol epoxides, compounds related in structure to the alpha-tocopherol (vitamin E) produced by the brown algae *Sargassum tortile* [37].

The vitamin requirements for the growth, reproduction, and development of *C. teleta* have so far been unexplored, to our knowledge; however, as previously mentioned, we know that growth, reproduction, and development of *C. teleta* depend strongly on diet and organic content of sediments [27,38], and that nutrients such as algal pigments, including carotenoids, are rapidly assimilated into developing oocytes [28]. As previously mentioned, larval settlement of *C. teleta* depends on the organic content of sediments, with larvae preferring sediments with a low carbohydrate to protein ratio [15,16]. Very little is known about marine annelid nutrition and vitamin requirements; however, an absolute requirement for riboflavin and nicotinamide, along with 6 other vitamins, was demonstrated using artificial culture media for the oligochaete annelid *Enchytraeus fragmentosus* [29].

We do not currently know the concentrations of riboflavin and nicotinamide present naturally in the environments that larvae of *C. teleta* settle and metamorphose in. It is likely, however, that recently deposited sediments, particularly those near sewage outflows, contain high levels of these compounds, sufficient to induce metamorphosis. Many phytoplankton species are known to contain large concentrations of B vitamins within their cells [39]. Several marine microorganisms, such as the bacteria *Shewanella spp.* [40] and *Micrococcus luteus* [41], and the yeast *Candida* [42], are also known to produce and release large amounts of riboflavin

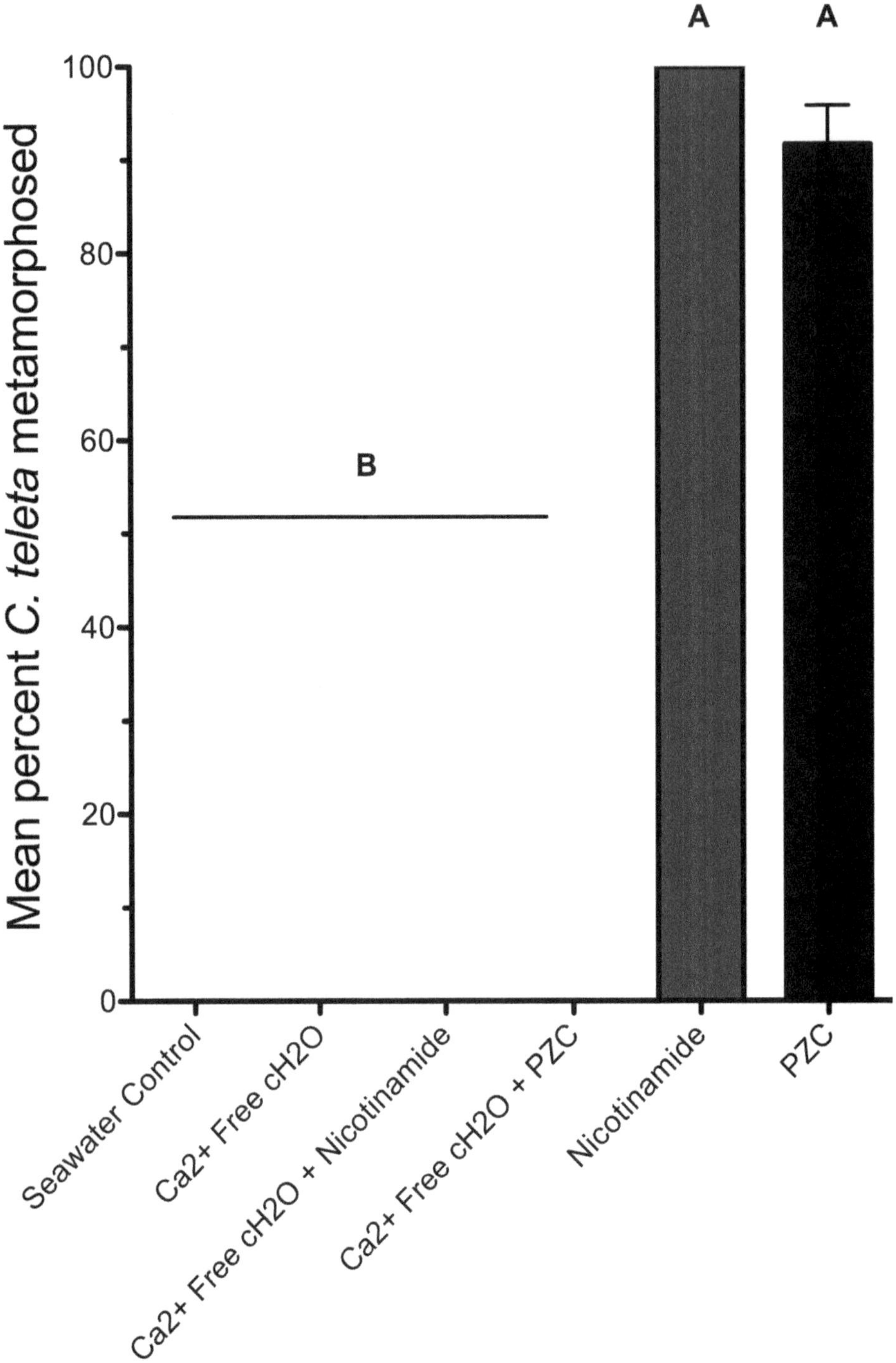

Figure 4. The effect of calcium-free seawater on metamorphosis of *C. teleta* treated with nicotinamide or pyrazinecarboxamide. Larvae were acclimated to calcium-free seawater for one hour before being transferred to a solution of calcium-free seawater and either nicotinamide or pyrazinecarboxamide (PZC) at 40 μM for 24 h. Each treatment consisted of 3 replicates with 8 larvae per replicate. Artificial seawater (Instant Ocean) acted as a negative control. Letters indicate significant ($P<0.05$) differences as determined by a Bonferroni post-hoc test. Error bars represent +1 s.e.m.

particularly when grown as biofilms. Remnants of settled, decomposing phytoplankton and the other microbes taking part in decomposition processes are likely to be a large component of marine sediments, which should therefore contain high concentrations of B vitamins. Indeed, biotin (B_7), thiamine (B_1), and vitamin B_{12} have been all identified in Pacific marine sediments [43]. Additional studies will be required to determine if riboflavin and/or nicotinamide are present in salt marsh sediments and sediments near sewage outfalls at the low concentrations that we have found to be inductive.

Although the biochemical and sensory mechanisms through which the larvae of *C. teleta* respond to nicotinamide and riboflavin still need more investigation, our results suggest that larvae may be sensing nicotinamide with chemosensory pyridine

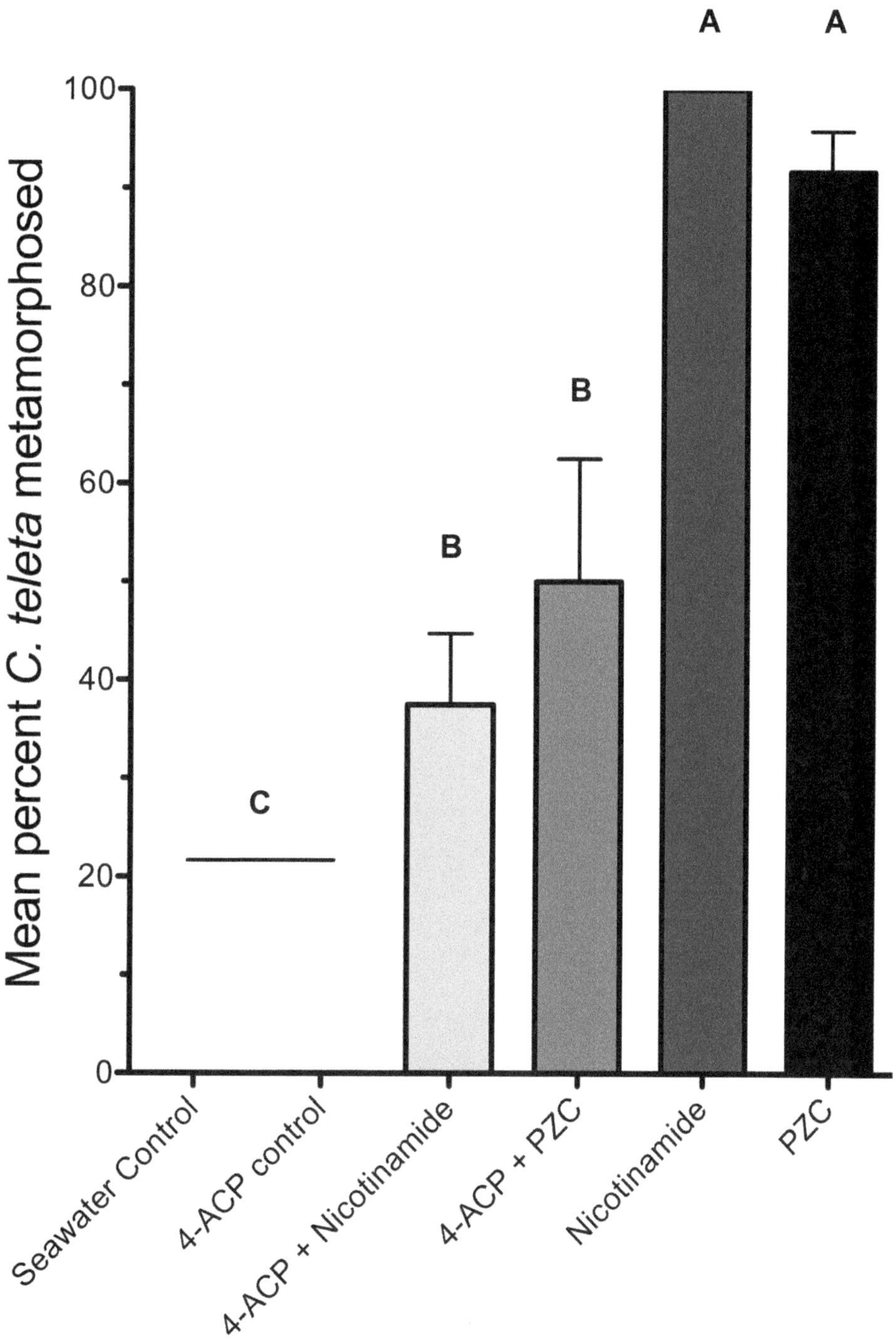

Figure 5. The effect of 4-acetylpyridine on metamorphosis of *C. teleta* in the presence of nicotinamide or pyrazinecarboxamide. Larvae were acclimated to 200 μM 4-acetylpyridine (4-ACP) for one hour before being transferred to a solution of 200 μM 4-acetylpyridine and either nicotinamide or pyrazinecarboxamide (PZC) at 40 μM for 24 h. Each treatment consisted of 3 replicates with 8 larvae per replicate. Artificial seawater (Instant Ocean) acted as a negative control. Letters indicate significant ($P<0.05$) differences as determined by a Bonferroni post-hoc test. Error bars represent + s.e.m.

receptors similar to those previously found in chemosensory sensilla on the walking legs of crayfish, and which are involved in the foraging behavior of those animals [30–32]. The inhibitory effects of 4-acetylpyridine on chemosensation of nicotinamide support this suggestion, as do the stimulatory effects of pyrazinecarboxamide on the larvae of *C. teleta*, a powerful agonist of these pyridine-activated ion channels [32]. In crayfish, pyrazinecarboxamide was found to be the most effective inducer, followed by nicotinamide; we have found that this is also the case for larvae of *C. teleta* (Table 1) [32]. Also, the IC_{50} for 4-acetylpyridine was found to be approximately 200 μM for *C. teleta* compared with a K_I of 70 μM for the crayfish *A. torrentium* [32]. This difference in effective concentrations for 4-ACP may at least be partly due to the fact that Hatt et al. [32] explored the effects of 4-ACP using isolated chemosensory neurons and electrophysiological techniques, whereas the effects on *C. teleta* were explored in

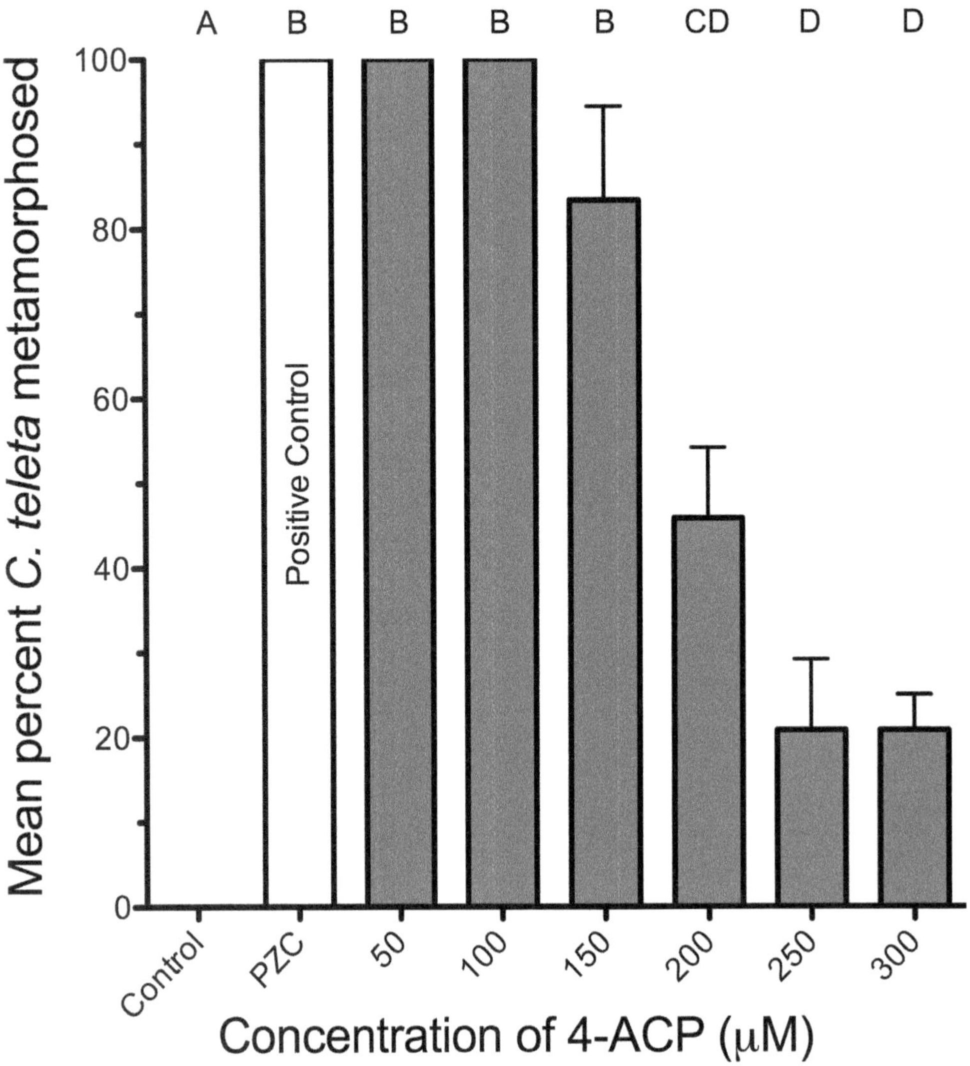

Figure 6. The inhibitory dose response of 4-acetylpyridine on metamorphosis of *C. teleta* after 24 h. Larvae were acclimated to the listed concentrations of 4-acetylpyridine (4-ACP) for one hour before being transferred to a solution of the same concentration of 4-acetylpyridine and pyrazinecarboxamide at 40 μM. Each treatment consisted of 3 replicates with 8 larvae per replicate. Artificial seawater (Instant Ocean) acted as a negative control. The PZC treatment acted as a positive control containing only 40 μM pyrazinecarboxamide and artificial seawater. Letters indicate significant ($P<0.05$) differences as determined by a Bonferroni post-hoc test. Error bars represent +1 s.e.m.

in vivo studies in which the chemical had to also penetrate the larvae. Interestingly, larvae did not metamorphose when exposed to β-NAD at concentrations as high as 5 mM. While nicotinamide, pyrazinecarboxamide, and nicotinic acid stimulated larvae to metamorphose at concentrations similar to those that stimulated action potentials in the crayfish, it may be possible that other pyridines such as β-NAD do not bind efficiently to this receptor at similar concentrations for both species. It would be interesting in this regard to also try a patch-clamp technique on the *C. teleta* larvae to demonstrate the presence of pyridine receptors.

Each year, more ligand-gated ion channels have been characterized as having roles in animal's taste reception [44], although the presence of pyridine receptors used for chemodetection in other animals besides crustaceans to our knowledge has not been reported. Signal transduction carried out by these ion channels may be calcium dependent [45]. Our results indicating that calcium-free seawater inhibited metamorphosis in response to nicotinamide also therefore agree with these data. Although larvae of *C. teleta* did not metamorphose in the presence of nicotinamide or pyrazinecarboxamide in calcium-free seawater in our study, they did settle to the bottom in those solutions within several minutes. Most likely, the nicotinamide or pyrazinecarboxamide ligand is binding to the channel and opening it; however, without calcium ions entering through the channel, the signal transduction pathway leading to metamorphosis is probably stalled shortly after initiation. These results parallel the results of Biggers and Laufer [18], showing that settlement and metamorphosis of *C. teleta* can be induced by very low concentrations of the calcium ionophore A23187.

Our results using the 5-HT_{2B} receptor inhibitor ketanserin suggest that the response to nicotinamide and riboflavin requires participation of the serotonergic nervous system as well, since ketanserin inhibited this response. These data suggest that the larvae detect nicotinamide and riboflavin in the environment via chemosensory neurons likely present in the chemosensory cilia noted by Eckelbarger and Grassle [46]. Sensory depolarization of these receptors may lead to the release of serotonin either directly by the chemosensory neurons or indirectly through synaptic

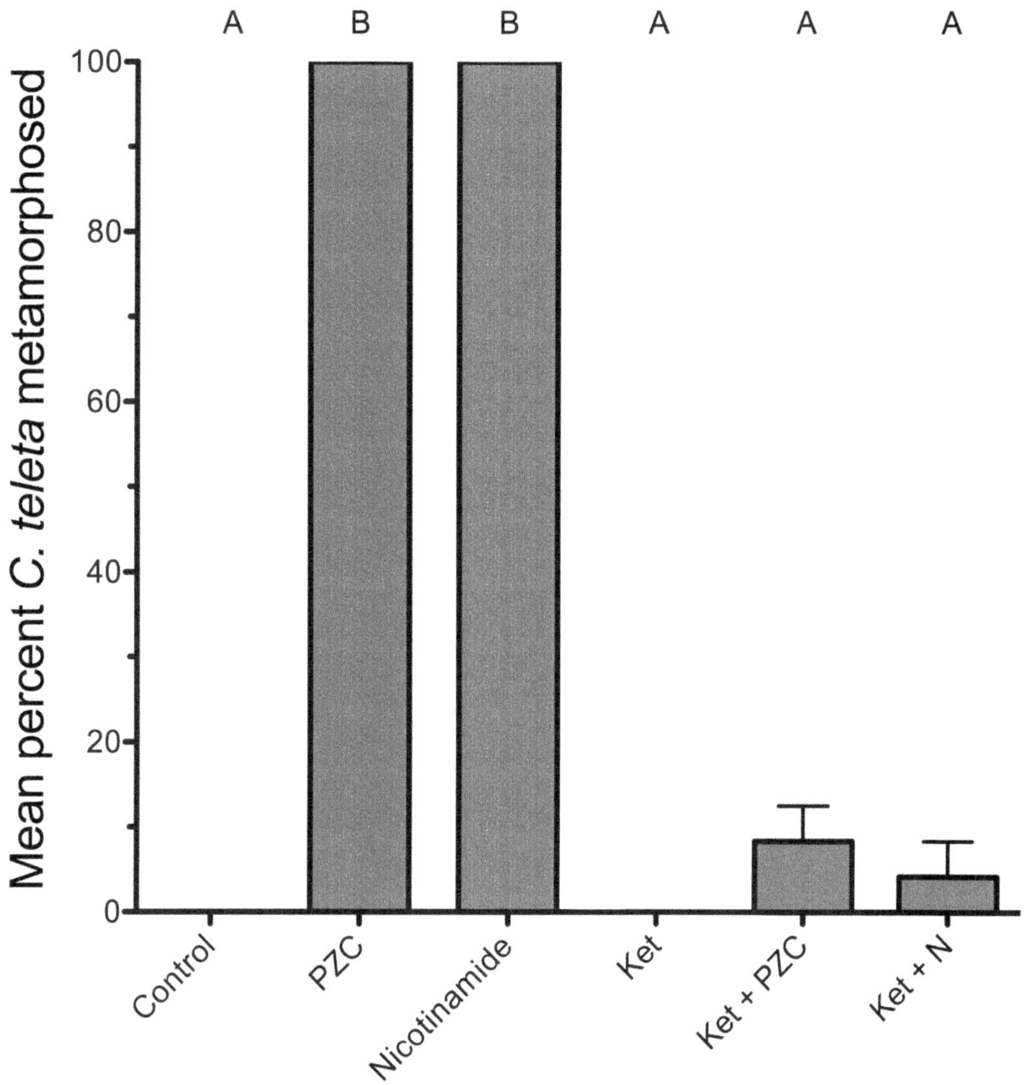

Figure 7. The effect of ketanserin on metamorphosis of *C. teleta* in the presence of nicotinamide or pyrazinecarboxamide. Larvae were acclimated to 2 μM ketanserin for three hours before being transferred to a solution of 2 μM ketanserin and either nicotinamide or pyrazinecarboxamide at 40 μM for 24 h. Each treatment consisted of 3 replicates with 8 larvae per replicate. Artificial seawater (Instant Ocean) acted as a negative control. Letters indicate significant ($P<0.05$) differences as determined by a Bonferroni post-hoc test. Error bars represent + s.e.m.

activation of other serotonergic neurons. Nitric oxide may play a role as a secondary messenger between the influx of ions caused by the opening of nicotinamide activated ion channels and serotonergic signaling. Larvae of *C. teleta* did not metamorphose in the presence NOS inhibitors when treated with ketanserin, a serotonin 5-HT_{2B} receptor antagonist [20]. Serotonin binding to receptors

Table 1. The lowest stimulatory concentrations of pyridines that induced larvae of *C. teleta* to metamorphose compared with the K_M of these same chemicals on the pyridine-activated ion channel of the crayfish *A. torrentium*.

	Species	
Chemical	***C. teleta***	***A. torrentium***
Pyrazinecarboxamide	1 μM	1.5 μM
Nicotinamide	3 μM	10 μM
Nicotinic acid	1 mM	>1 mM
4-acetylpyridine	200 μM	70 μM

The concentration of 4-acetylpyridine that inhibited 50% of the larvae of *C. teleta* is also compared with the IC_{50} of 4-acetylpyridine for the crayfish (adapted from Hatt and Schmiedel-Jacob, 1984).

then most likely leads to settlement and metamorphosis. In this regard, the serotonin 5-HT_{2B} receptor is known to be a metabotropic receptor involving G-protein activation, secondary messengers, calcium channel activation, and downstream kinase activation such as through mitogen activated protein kinases (MAPK) involved in mediating gene expression [47]. Other chemosensory structures as reviewed by Lindsay [48] may be also be involved in the detection of nicotinamide and riboflavin by these larvae.

Larvae of *C. teleta* may be foraging for an environment containing nutrient-rich sediment prior to metamorphosing and thereby may be exhibiting foraging behavior in choosing habitats for settlement and metamorphosis. By responding to the B vitamins nicotinamide and riboflavin as chemical settlement cues, they could be ensuring that a selected environment contains the proper vitamins to successfully grow and reproduce. As deposit feeders, *C. teleta* can utilize these vitamins only if they are contained within marine sediments. Here, larvae of *C. teleta* seem to be mirroring how adult crustaceans such as crayfish and lobsters sense nutrients within sediments via chemosensory hairs attached to their legs or antennules. The crayfish *A. torrentium* can sense nicotinamide and other pyridines while the spiny lobster *Panulirus argus* can sense ATP using their antennules [49]. Interestingly, the crayfish *Astacus astacus* has been demonstrated to have specific internal oesophageal receptors for nicotinamide [50], and the external receptors on the walking legs and antennules of crustaceans may represent primitive fore-runners of internal chemoreceptors such as the oesophageal receptors [49]. To our knowledge, the present paper is the first demonstration of the presence of nicotinamide or riboflavin receptors in an annelid or in invertebrate larvae.

Materials and Methods

Ethics statement

No permissions were required to collect sediment from the Little Sippewissett salt marsh (West Falmouth, MA) (41°34′34.4″N 70°38′08.6″W). This activity did not involve endangered or protected species.

Care and maintenance of larvae

Adults of *Capitella teleta* were provided by Dr. Judith Grassle (Rutgers University) and maintained in the lab since 2012 in 9 cm glass dishes containing 30 psu Instant Ocean artificial seawater (hereafter called ASW) at 18°C. Sediment collected from the Little Sippewissett salt marsh (West Falmouth, MA) was sieved through a 1 mm wire mesh screen, frozen for at least 24 h, aerated, and provided as food *ad libitum* [13,14]. Cultures were searched for brooding females every 2–3 days; brooding individuals were then transferred individually to 6 cm glass dishes containing ASW. Dishes containing brooding females were checked daily for swimming larvae; larvae were then pipetted into a separate, clean 6 cm glass dish containing ASW. All experiments used larvae released only within the previous 24 hours. *C. teleta* larvae are usually competent to settle and metamorphose within minutes after their escape from the brood tube [11,14], so that all larvae should have been competent at the start of our experiments.

General bioassay protocol

All experiments were conducted using either 12-well tissue-culture plates or 3 cm glass dishes, with each well or bowl containing 4 mL of a vitamin solution in ASW, or a negative control (ASW). Each experiment included 3 replicates per treatment, with 8 larvae per replicate. To avoid diluting the final treatment solution, larvae were first pipetted into an intermediate bath of treatment solution (~20 mL) before being pipetted into the final treatment vessel. Dishes were examined every 30 or 60 minutes under 20–50x magnification to count the number of newly metamorphosed juveniles. Larvae of *C. teleta* were considered to have metamorphosed when they elongated, lost their prototroch and telotroch cilia, and began crawling on the bottom of the vessel. All experiments were conducted at room temperature (20–23°C) under laboratory fluorescent lighting.

Testing the effects of B vitamins and other chemicals on larval metamorphosis

Thiamine HCl (B_1), riboflavin (B_2), nicotinamide (B_3), nicotinic acid (another form of B_3), pyridoxine HCl (B_6), biotin (B_7) and cobalamin (B_{12}) were obtained from Sigma Chemical Co. Thiamine HCl, nicotinic acid, pyridoxine HCl, biotin, and cobalamin were all tested on larvae of *C. teleta* at 500 μM. Nicotinic acid was also tested at 1 mM. Riboflavin was tested on larvae of *C. teleta* at 10, 50, 100, 150, and 200 μM. Nicotinamide was tested on larvae of *C. teleta* at 2, 3, 4, 5, 6, 7, and 8 μM. The concentrations tested for nicotinamide and riboflavin were based on preliminary pilot studies. The photo-degradation product of riboflavin, lumichrome, was obtained from Sigma Chemical and tested at concentrations of 100, 500, and 1000 μM. The pyridine receptor antagonists and agonists 4-acetylpyridine, pyrazinecarboxamide, and β-NAD were all purchased from Sigma Chemical Co. and tested in final concentrations as stated in results. The serotonin receptor antagonist ketanserin was purchased from Sigma Chemical Co. and prepared as a 10 mM stock solution in ethanol.

Testing the effects of 4-acetylpyridine, calcium-free ASW, and ketanserin on metamorphosis

Larvae were acclimated to ASW containing 200 μM 4-acetylpyridine for one hour and then transferred to a treatment containing either 200 μM 4-acetylpyridine + 40 μM nicotinamide, 200 μM 4-acetylpyridine + 40 μM pyrazinecarboxamide, or 200 μM 4-acetylpyridine as a negative control. Larvae that were not acclimated to 4-acetylpyradine were transferred to positive controls containing either 40 μM nicotinamide or 40 μM pyrazinecarboxamide.

Calcium-free ASW was prepared by omitting calcium chloride from the recipe given in [51]. Larvae that were in calcium-free ASW treatments were acclimated to calcium-free ASW for one hour and then were transferred to treatments containing either calcium-free ASW+40 μM nicotinamide, calcium-free ASW + 40 μM pyrazinecarboxamide, or calcium-free ASW only (negative control). Larvae that were not acclimated to calcium-free ASW were transferred to positive controls containing either 40 μM nicotinamide or 40 μM pyrazinecarboxamide to assess metamorphic competence.

Larvae were acclimated to ASW containing 2 μM ketanserin for three hours and then transferred to a treatment containing either 2 μM ketanserin + 40 μM nicotinamide, 2 μM ketanserin + 40 μM pyrazinecarboxamide, or 2 μM ketanserin as a negative control. Larvae that were not acclimated to ketanserin were transferred to positive controls containing either 40 μM nicotinamide or 40 μM pyrazinecarboxamide.

Data analysis

In order to determine if calcium-free seawater, 4-acetylpyridine, or ketanserin influenced the number of *C. teleta* larvae that metamorphosed in the presence of nicotinamide or pyrazinecarboxamide, a one-way analysis of variance was conducted on the numbers of *C. teleta* that had metamorphosed by 24 h after treatment. If the results of an ANOVA were significant, Bonferroni's post-hoc test was conducted to determine which treatments were significantly different from each other. For each ANOVA, each replicate was the arcsine transformed ratio of individuals that had metamorphosed at that timepoint. It is important to note that, although the figures represented in this paper reflect the percent of *C. teleta* larvae that had metamorphosed, all statistical analyses were conducted with arcsine transformed ratio of the larvae that had metamorphosed. All statistical analyses were carried out with GraphPad Prism version 5.0.

Acknowledgments

We thank Dr. Judith Grassle (Rutgers University) for sending us a population of *C. teleta* to begin our cultures here at Tufts University.

Author Contributions

Conceived and designed the experiments: RB JP WB. Performed the experiments: RB WB GS CL. Analyzed the data: RB JP WB. Contributed reagents/materials/analysis tools: RB JP WB. Contributed to the writing of the manuscript: RB JP WB.

References

1. Pechenik JA (1990) Delayed metamorphosis by larvae of benthic marine invertebrates: Does it occur? Is there a price to pay? Ophelia 32: 63–94.
2. Hadfield MG, Paul VJ (2001) Natural chemical cues for settlement and metamorphosis of marine invertebrate larvae. In: McClintock JB, Baker BJ, editors. Marine chemical ecology. Boca Raton: CRC Press. 431–461.
3. Pawlik JR (1992) Chemical ecology of the settlement of benthic marine invertebrates. Oceanogr Mar Biol Annu Rev 30: 273–335.
4. Rittschof D, Forward RB Jr, Cannon G, Welch JM, McClary M Jr, et al. (1998) Cues and context: Larval responses to physical and chemical cues. Biofouling 12: 31–44.
5. Swanson RL, Byrne M, Prowse TAA, Mos B, Dworjanyn SA, et al. (2012) Dissolved histamine: a potential habitat marker promoting settlement and metamorphosis in sea urchin larvae. Mar Biol 159: 915–925.
6. Swanson RL, De Nys R, Huggett MJ, Green JK, Steinberg PD (2006) In situ quantification of a natural settlement cue and recruitment of the Australian sea urchin *Holopneustes purpurascens*. Mar Ecol Prog Ser 314: 1–14.
7. Swanson RL, Williamson JE, de Nys R, Kumar N, Bucknall MP, et al. (2004) Induction of settlement of larvae of the sea urchin *Holopneustes purpurascens* by histamine from a host alga. Biol Bull 206: 161–172.
8. Pawlik JR, Faulkner DJ (1986) Specific free fatty acids induce larval settlement and metamorphosis of the reef-building tube worm *Phragmatopoma californica*(Fewkes). J Exp Mar Biol Ecol 102: 301–310.
9. Woodin SA (1991) Recruitment of infauna: positive or negative cues? Am Zool 31: 797–807.
10. Blake JA, Grassle JP, Eckelbarger KJ (2009) *Capitella teleta*, a new species designation for the opportunistic and experimental *Capitella* sp. I, with a review of the literature for confirmed records. Zoosymposia 2: 25–53.
11. Grassle J, Grassle JF (1976) Sibling species in the marine pollution indicator *Capitella* (polychaeta). Science 192: 567–569.
12. Butman CA, Grassle JP, Webb CM (1988) Substrate choices made by marine larvae settling in still water and in a flume flow. Nature 333: 771–773.
13. Dubilier N (1988) H2S-A Settlement Cue or a Toxic Substance For *Capitella* sp. I Larvae? Biol Bull 174: 30–38.
14. Pechenik JA, Cerulli TR (1991) Influence of delayed metamorphosis on survival, growth, and reproduction of the marine polychaete *Capitella* sp. I. J Exp Mar Biol Ecol 151: 17–27.
15. Cohen RA, Pechenik JA (1999) Relationship between sediment organic content, metamorphosis, and postlarval performance in the deposit-feeding polychaete *Capitella* sp. I. J Exp Mar Biol Ecol 240: 1–18.
16. Thiyagarajan V, Soo L, Qian P-Y (2005) The role of sediment organic matter composition in larval habitat selection by the polychaete *Capitella* sp. I. J Exp Mar Biol Ecol 323: 70–83.
17. Biggers WJ, Laufer H (1992) Chemical induction of settlement and metamorphosis of *Capitella capitata* sp-I (polychaeta) larvae by juvenile hormone-active compounds. Invert Reprod Devel 22: 39–46.
18. Biggers WJ, Laufer H (1999) Settlement and metamorphosis of *Capitella* larvae induced by juvenile hormone-active compounds is mediated by protein kinase C and ion channels. Biol Bull 196: 187–198.
19. Ricker T, Poesnecker R, Deats S, Biggers WJ (2011) Biochemical regulation of settlement and metamorphosis of larvae of *Capitella teleta* by serotonin. Integr Comp Biol 51: e243.
20. Biggers WJ, Pires A, Pechenik JA, Johns E, Patel P, et al. (2012) Inhibitors of nitric oxide synthase induce larval settlement and metamorphosis of the polychaete annelid *Capitella teleta*. Invert Reprod Devel 56: 1–13.
21. Bishop CD, Pires A, Norby S-W, Boudko D, Moroz LL, et al. (2008) Analysis of nitric oxide-cyclic guanosine monophosphate signaling during metamorphosis of the nudibranch *Phestilla sibogae* Bergh (Gastropoda: Opisthobranchia). Evol Dev 10: 288–299.
22. Leise EM, Thavaradhara K, Durham NR, Turner BE (2001) Serotonin and nitric oxide regulate metamorphosis in the marine snail *Ilyanassa obsoleta*. Am Zool 41: 258–267.
23. Pechenik JA, Cochrane DE, Li W, West ET, Pires A, et al. (2007) Nitric oxide inhibits metamorphosis in larvae of *Crepidula fornicata*, the slippershell snail. Biol Bull 213: 160–171.
24. Bishop CD, Brandhorst BP (2001) NO/cGMP signaling and HSP90 activity represses metamorphosis in the sea urchin *Lytechinus pictus*. Biol Bull 201: 394–404.
25. Bishop CD, Bates WR, Brandhorst BP (2001) Regulation of metamorphosis in ascidians involves NO/cGMP signaling and HSP90. J Exp Zool 289: 374–384.
26. Comes S, Locascio A, Silvestre F, d'Ischia M, Russo GL, et al. (2007) Regulatory roles of nitric oxide during larval development and metamorphosis in *Ciona intestinalis*. Dev Biol 306: 772–784.
27. Gremare A, Marsh AG, Tenore KR (1988) Short-term reproductive responses of *Capitella* sp. I (Annelida: Polychaeta) fed on different diets. J Exp Mar Biol Ecol 123: 147–162.
28. Marsh AG, Gremare A, Dawson R, Tenore KR (1990) Translocation of algal pigments to oocytes in *Capitella* sp. I.(Annelida: Polychaeta). Mar Ecol Prog Ser 67: 301–304.
29. Gotthold ML, Koch J (1974) Further development of an artificial medium for the axenically cultured annelid *Enchytraeus fragmentosus*. Comp Biochem Physiol 48: 307–314.
30. Hatt H, Franke C (1987) Taste receptors in crayfish: recording of single nicotinamide-activated channels. Neurosci Lett 73: 137–142.
31. Hatt H, Schmiedel-Jakob I (1984) Electrophysiological studies of pyridine-sensitive units on the crayfish walking leg. J Comp Physiol A 154: 855–863.
32. Hatt H, Schmiedel-Jakob I (1985) Specific antagonists at the pyridine receptor: evidence from electrophysiological studies with acetylpyridines. Chem Senses 10: 317–323.
33. Panzeca C, Beck AJ, Tovar-Sanchez A, Segovia-Zavala J, Taylor GT, et al. (2009) Distributions of dissolved vitamin B12 and Co in coastal and open-ocean environments. Estuar Coast Shelf S 85: 223–230.
34. Giovannoni SJ (2012) Vitamins in the sea. Proc Natl Acad Sci USA 109: 13888–13889.
35. Sañudo-Wilhelmy SA, Cutter LS, Durazo R, Smail EA, Gómez-Consarnau L, et al. (2012) Multiple B-vitamin depletion in large areas of the coastal ocean. Proc Natl Acad Sci USA 109: 14041–14045.
36. Tsukamoto S, Kato H, Hirota H, Fusetani N (1999) Lumichrome. Eur - J Biochem 264: 785–789.
37. Kato T, Kumanireng AS, Ichinose I, Kitahara Y, Kakinuma Y (1975) Active components of *Sargassum tortile* effecting the settlement of swimming larvae of *Coryne uchidai*. Experientia 31: 433–434.
38. Tsutsumi H, Fukunaga S, Fujita N, Sumida M (1990) Relationship between growth of *Capitella* sp. and organic enrichment of the sediment. Mar Ecol Prog Ser 63: 157–162.
39. Brown MR, Mular M, Miller I, Farmer C, Trenerry C (1999) The vitamin content of microalgae used in aquaculture. J Appl Phycol 11: 247–255.
40. Marsili E, Baron DB, Shikhare ID, Coursolle D, Gralnick JA, et al. (2008) *Shewanella* secretes flavins that mediate extracellular electron transfer. Proc Natl Acad Sci USA 105: 3968–3973.
41. Sims GK, O'Loughlin EJ (1992) Riboflavin production during growth of *Micrococcus luteus* on pyridine. Appl Environ Microbiol 58: 3423–3425.
42. Mitra S, Thawrani D, Banerjee P, Gachhui R, Mukherjee J (2012) Induced biofilm cultivation enhances riboflavin production by an intertidally derived *Candida famata*. Appl Biochem Biotechnol 166: 1991–2006.
43. Ohwada K, Taga N (1972) Distribution and seasonal variation of vitamin B 12, thiamine and biotin in the sea. Mar Chem 1: 61–73.
44. Pellegrino M, Nakagawa T (2009) Smelling the difference: controversial ideas in insect olfaction. J Exp Biol 212: 1973–1979.
45. Meunier N, Marion-Poll F, Lucas P (2009) Water taste transduction pathway is calcium dependent in *Drosophila*. Chem Senses 34: 441–449.
46. Eckelbarger KJ, Grassle JP (1987) Interspecific variation in genital spine, sperm, and larval morphology in six sibling species of *Capitella*. Bull Biol SocWash 7: 62–76.

47. Launay J-M, Birraux G, Bondoux D, Callebert J, Choi D-S, et al. (1996) Ras involvement in signal transduction by the serotonin 5-HT2B receptor. J Biol Chem 271: 3141–3147.
48. Lindsay SM (2009) Ecology and biology of chemoreception in polychaetes. Zoosymposia 2: 339–367.
49. Carr WE, Ache BW, Gleeson RA (1987) Chemoreceptors of crustaceans: similarities to receptors for neuroactive substances in internal tissues. Environ Health Perspect 71: 31–46.
50. Altner H, Hatt H, Altner I (1986) Structural and functional properties of the mechanoreceptors and chemoreceptors in the anterior oesophageal sensilla of the crayfish, *Astacus astacus*. Cell Tissue Res 244: 537–547.
51. Baloun AJ, Morse DE (1984) Ionic control of settlement and metamorphosis in larval *Haliotis rufescens* (Gastropoda). Biol Bull 167: 124–138.

Permissions

All chapters in this book were first published in PLOS ONE, by The Public Library of Science; hereby published with permission under the Creative Commons Attribution License or equivalent. Every chapter published in this book has been scrutinized by our experts. Their significance has been extensively debated. The topics covered herein carry significant findings which will fuel the growth of the discipline. They may even be implemented as practical applications or may be referred to as a beginning point for another development.

The contributors of this book come from diverse backgrounds, making this book a truly international effort. This book will bring forth new frontiers with its revolutionizing research information and detailed analysis of the nascent developments around the world.

We would like to thank all the contributing authors for lending their expertise to make the book truly unique. They have played a crucial role in the development of this book. Without their invaluable contributions this book wouldn't have been possible. They have made vital efforts to compile up to date information on the varied aspects of this subject to make this book a valuable addition to the collection of many professionals and students.

This book was conceptualized with the vision of imparting up-to-date information and advanced data in this field. To ensure the same, a matchless editorial board was set up. Every individual on the board went through rigorous rounds of assessment to prove their worth. After which they invested a large part of their time researching and compiling the most relevant data for our readers.

The editorial board has been involved in producing this book since its inception. They have spent rigorous hours researching and exploring the diverse topics which have resulted in the successful publishing of this book. They have passed on their knowledge of decades through this book. To expedite this challenging task, the publisher supported the team at every step. A small team of assistant editors was also appointed to further simplify the editing procedure and attain best results for the readers.

Apart from the editorial board, the designing team has also invested a significant amount of their time in understanding the subject and creating the most relevant covers. They scrutinized every image to scout for the most suitable representation of the subject and create an appropriate cover for the book.

The publishing team has been an ardent support to the editorial, designing and production team. Their endless efforts to recruit the best for this project, has resulted in the accomplishment of this book. They are a veteran in the field of academics and their pool of knowledge is as vast as their experience in printing. Their expertise and guidance has proved useful at every step. Their uncompromising quality standards have made this book an exceptional effort. Their encouragement from time to time has been an inspiration for everyone.

The publisher and the editorial board hope that this book will prove to be a valuable piece of knowledge for researchers, students, practitioners and scholars across the globe.

List of Contributors

Kingsley O. Odigie and A. Russell Flegal
WIGS Laboratory, Environmental Toxicology, University of California Santa Cruz, Santa Cruz, California, United States of America

Davi Pedroni Barreto
Instituto de Microbiologia Prof. Paulo de Góes, Universidade Federal do Rio de Janeiro, Rio de Janeiro, Brazil

Ralf Conrad, Melanie Klose and Peter Claus
Max-Planck Institute for Terrestrial Microbiology, Marburg, Hessen, Germany

Alex Enrich-Prast
Instituto de Biologia, Universidade Federal do Rio de Janeiro, Rio de Janeiro, Brazil
Department of Water and Environmental Studies, Linköping University, Linköping, Sweden

João Manuel Bernardo
Departamento de Paisagem Ambiente e Ordenamento, Escola de Ciências e Tecnologia, Universidade de Évora, Évora, Portugal

Maria Ilhéu and Paula Matono
Departamento de Paisagem Ambiente e Ordenamento, Escola de Ciências e Tecnologia, Universidade de Évora, Évora, Portugal
Instituto de Ciências Agrárias e Ambientais Mediterrânicas, Universidade de Évora, Évora, Portugal

Bing Wang, Fenxiang Wen, Jiangtao Wua and Xiaojun Wang
College of Environmental Science and Resources, Shanxi University, Taiyuan, China

Yani Hu
Library, Hebei University of Science and Technology, Shijiazhuang, China

Ulrike Braeckman, Carl Van Colen, Katja Guilini, Dirk Van Gansbeke, Magda Vincx and Jan Vanaverbeke
Ghent University, Department of Biology, Marine Biology Research Group, Ghent, Belgium

Karline Soetaert
Netherlands Institute for Sea Research, Department of Ecosystem Studies, Yerseke, The Netherlands

Desirée D. Tullos and Cara Walter
Water Resources Engineering, Oregon State University, Corvallis, Oregon, United States of America

Debra S. Finn
Department of Integrative Biology, Oregon State University, Corvallis, Oregon, United States of America

Björn Berglund
Department of Clinical and Experimental Medicine, Linköping University, Linköping, Sweden

Ghazanfar Ali Khan, Richard Lindberg and Jerker Fick
Department of Chemistry, Umeå University, Umeå, Sweden

Per-Eric Lindgren
Department of Clinical and Experimental Medicine, Linköping University, Linköping, Sweden
Department of Microbiology, Medical Services, County Hospital Ryhov, Jönköping, Sweden

Janet Macfall and Paul Robinette, David Welch
Center for Environmental Studies, Elon University, Elon, North Carolina, United States of America

Elizabeth C. MacDonald, Elisabeth H. Frost and Diana J. Hamilton
Department of Biology, Mount Allison University, Sackville, New Brunswick, Canada

Stephanie M. MacNeil and Myriam A. Barbeau
Department of Biology, University of New Brunswick, Fredericton, New Brunswick, Canada

Xiao-yong Zhang, Gui-ling Tang, Xin-ya Xu, Xu-hua Nong and Shu-Hua Qi
Key Laboratory of Tropical Marine Bio-resources and Ecology/RNAM Center for Marine Microbiology/ Guangdong Key Laboratory of Marine Material Medical, South China sea Institute of Oceanology, Chinese academy of sciences, Guangzhou, China

Martina Demuro
Institute for Photonics and Advanced Sensing, School of Chemistry and Physics, The University of Adelaide, Adelaide, Australia
Centro Nacional de Investigación sobre la Evolución Humana, Burgos, Spain

Lee J. Arnold
Institute for Photonics and Advanced Sensing, School of Chemistry and Physics, The University of Adelaide, Adelaide, Australia
Centro Nacional de Investigación sobre la Evolución Humana, Burgos, Spain
The Environment Institute and School of Earth and Environmental Sciences, The University of Adelaide, Adelaide, Australia

Josep M. Parés, Alfredo Pérez-González, Ana I. Ortega and José M. Bermúdez de Castro
Centro Nacional de Investigación sobre la Evolución Humana, Burgos, Spain

Juan L. Arsuaga
Centro Mixto Universidad Complutense-Instituto de Salud Carlos III de Evolución y Comportamiento Humanos, Madrid, Spain
Departamento de Paleontología, Facultad de Ciencias Geológicas, Universidad Complutense de Madrid, Madrid, Spain

Eudald Carbonell
Institut Catalá de Paleoecologia Humana i Evolució Social, Àrea de Prehistòria, Universitat Rovira i Virgili, Tarragona, Spain

Wenzhong Tang, Baoqing Shan, Wenqiang Zhang, Hong Zhang and Yuekui Ding
State Key Laboratory on Environmental Aquatic Chemistry, Research Center for Eco-Environmental Sciences, Chinese Academy of Sciences, Beijing, China

Lishuo Wang
State Key Laboratory on Environmental Aquatic Chemistry, Research Center for Eco-Environmental Sciences, Chinese Academy of Sciences, Beijing, China
China University of Mining & Technology (Beijing), Beijing, China

Christoph Zielhofer and Eva Leitholdt
Institute of Geography, Leipzig University, Leipzig, Germany

Lukas Werther and Peter Ettel
Chair of Prehistory and Early History, Friedrich-Schiller University, Jena, Germany

Andreas Stele and Jens Bussmann
Institute of Geography, Osnabruck University, Osnabruck, Germany

Sven Linzen and Michael Schneider
Leibniz Institute of Photonic Technology (IPHT), Jena, Germany

Cornelius Meyer
Eastern Atlas, Berlin, Germany

Stefanie Berg-Hobohm
Bavarian State Department of Cultural Heritage, Munich, Germany

Michaela Schratzberger
Centre for Environment, Fisheries and Aquaculture Science, Lowestoft Laboratory, Lowestoft, Suffolk, United Kingdom

Piers Larcombe
RPS MetOcean, Subiaco, Western Australia, Australia

Nathalie Pülmanns and Inga Nordhaus
Department of Ecology, Leibniz Center for Tropical Marine Ecology (ZMT), Bremen, Germany

Karen Diele
School of Life, Sport and Social Sciences, Edinburgh Napier University, Edinburgh, United Kingdom

Ulf Mehlig
Instituto de Estudos Costeiros, Universidade Federal do Pará - Campus de Bragança, Bragança, PA, Brazil

Daniel Elias and Melody J. Bernot
Ball State University, Department of Biology, Muncie, Indiana, United States of America

Kirill S. Golokhvast
Scientific Educational Center of Nanotechnology, Far Eastern Federal University, Vladivostok, Russian Federation

Anna A. Shvedova
Pathology and Physiology Research Branch/NIOSH/CDC, Morgantown, West Virginia, United States of America
Department Physiology and Pharmacology, School of Medicine, West Virginia University, Morgantown, West Virginia, United States of America

Robert T. Burns and Jan A. Pechenik
Department of Biology, Tufts University, Medford, Massachusetts, United States of America

William J. Biggers, Gia Scavo and Christopher Lehman
Department of Biology, Wilkes University, Wilkes-Barre, Pennsylvania, United States of America

Index

R

S

T

V

W

X